PLANT GENETIC RESOURCES OF LEGUMES IN THE MEDITERRANEAN

Current Plant Science and Biotechnology in Agriculture

VOLUME 39

Aims and Scope
The book series is intended for readers ranging from advanced students to senior research scientists and corporate directors interested in acquiring in-depth, state-of-the-art knowledge about research findings and techniques related to all aspects of agricultural biotechnology. Although the previous volumes in the series dealt with plant science and biotechnology, the aim is now to also include volumes dealing with animals science, food science and microbiology. While the subject matter will relate more particularly to agricultural applications, timely topics in basic science and biotechnology will also be explored. Some volumes will report progress in rapidly advancing disciplines through proceedings of symposia and workshops while others will detail fundamental information of an enduring nature that will be referenced repeatedly.

The titles published in this series are listed at the end of this volume.

Plant Genetic Resources of Legumes in the Mediterranean

Edited by

NIGEL MAXTED
University of Birmingham

and

SARITA JANE BENNETT
Centre for Legumes in Mediterranean Agriculture (CLIMA)

KLUWER ACADEMIC PUBLISHERS
DORDRECHT / BOSTON / LONDON

Library of Congress Cataloging-in-Publication Data

ISBN 0-7923-6707-3

Published by Kluwer Academic Publishers,
P.O. Box 17, 3300 AA Dordrecht, The Netherlands.

Sold and distributed in North, Central and South America
by Kluwer Academic Publishers,
101 Philip Drive, Norwell, MA 02061, U.S.A.

In all other countries, sold and distributed
by Kluwer Academic Publishers,
P.O. Box 322, 3300 AH Dordrecht, The Netherlands.

Printed on acid-free paper

Printed in the Netherlands

Contents

Preface

Genetic erosion, that is the loss of genetic diversity, of the Earth's biological resources is notoriously difficult to quantify, but few would disagree that the Twentieth Century has seen a massive and exponential loss of native plant diversity in the Mediterranean Basin. This careless eradication of species and loss of genetic diversity from a botanical and agricultural 'hot-spot' of diversity is a result of human intervention and continues apace today. It can be directly linked to the introduction of modern agricultural practices and lack of sustainable development, which in turn have negative economic, political and social consequences for the region. Plant genetic diversity forms the biological basis for sustainable agriculture and directly contributes to food security for the regions growing human population, but while recent conservation efforts in the Mediterranean have had beneficial effects, the still diminishing Mediterranean flora can accurately be described as a 'Flora in Crisis'.

However, since the early 1960s there has been a movement to halt the loss of plant diversity and enhance its utilisation. This initiative was under-scored by the Convention on Biological Diversity that not only expounds the need to conserve biological diversity but links conservation to exploitation and development for the benefit of all. Article 8 of the Convention clearly states the need to develop more effective and efficient guidelines to conserve biological diversity, while Article 9, along with the FAO International Undertaking on Plant Genetic Resources, promotes a complementary approach to conservation incorporating both *in situ* and *ex situ* techniques.

It has been argued that the most important component of biodiversity is the genetic diversity of plants, those crop and forage species used to feed humans and livestock. Plant genetic diversity in both cultivated and wild species provides the raw material for further selection and improvement, and has enabled breeders to develop cultivars to feed to the expanding human population within the Mediterranean region and throughout the temperate world. However, formal plant breeding is a relatively recent activity, and farmers have been nurturing the crop and forage species that originated in the Mediterranean Basin and North West Asia for millennia.

The careless eradication of species and loss of genetic diversity is threatening and clearly impeding crop improvement. The Convention on Biological Diversity recognises the link between conservation and sustainable utilisation, as well as the need to improve the efficiency of conservation activities. As a result the Consultative Group on International Agricultural Research (CGIAR), which has a major international responsibility for global food security, established the System-wide Genetic Resources Programme (SGRP) to enhance the efficiency and effectiveness of the CGIAR's contribution to Plant Genetic Resources for Food and Agriculture (PGRFA) conservation and sustainable utilisation. The sixteen CGIAR centres currently hold about half a million accessions in trust for humanity. To help increase transparency and ensure maximum use of conserved diversity, the CGIAR published 'Biodiversity in Trust: Conservation and Use of Plant Genetic Resources in CGIAR Centres' in 1997. This text reviewed the status of the CGIAR collections, providing details of botany, distribution and agronomy of the most important PGRFA species. Although the CGIAR system contains very large and significant collections of PGRFA species, their collections are not exhaustive. The World Information and Early Warning System on Plant Genetic Resources (WIEWS) database of the Food and Agriculture Organization of the United Nations, which contains information on national PGRFA programmes, indicates that there are 6.1 million PGRFA accessions held worldwide. Therefore, there is a need to further enhance transparency in a complementary manner, by focusing attention on particular regions and plant families. As with the CGIAR volume, we hope to stimulate more effective genetic conservation of legume diversity in the Mediterranean region, and ensure improved and sustainable utilisation.

Why legumes of the Mediterranean? Leguminosae species are of major economic importance (second only to the Gramineae), and the diverse herbaceous species of sub-family Papilionoideae provide a particularly rich source of protein for human and animal foods. Numerous Leguminosae species are exploited throughout temperate and tropical agro-ecosystems. Within the temperate regions of the world, the highest concentration of legume species is found in the Mediterranean (91 genera, 1,956 species and 495 subspecific taxa (Med-Checklist Database: Berendsohn, pers. comm.). The Mediterranean Basin with South-West Asia is undoubtedly a "hot-spot" for legume PGRFA diversity. It is the centre of diversity for the four major temperate legume crops: peas, chickpeas, lentils and faba beans, which all originated in this region. The region also contains two of N.I. Vavilov's centres of major crop plant diversity. The Asia Minor Centre is important for *Cicer arietinum* (secondary centre), *Lens culinaris, L. orientalis*, *Vicia ervilia*, *Pisum sativum*, various forage plants such as *Medicago sativa, Trifolium resupinatum, Trigonella foenum-graecum, Onobrychis* spp., *Lathyrus cicera*, and several *Vicia* spp.; and the Mediterranean Centre is important for *Vicia faba, V. sativa, Lathyrus ochrus,* large-seeded *Cicer arietinum, Hedysarum coronarium* and *Ornithopus sativus*. Therefore, legumes and the Mediterranean region are a vital resource for world agriculture.

We have divided the book into four sections and 19 chapters. Section 1 introduces the subjects of conservation and use of legume diversity, the Mediterranean ecosystem and taxonomy of legumes. Section 2 reviews the taxonomic and genetic diversity of legume species found in the Mediterranean, largely based on generic summaries. Each chapter summarises generic taxonomy, centre of diversity, ecogeographic distribution, genetic diversity distribution, conservation status, conservation gaps and research needs. There is also a discussion of the importance of rhizobia to the maintenance of legume diversity. Section 3 reviews *ex situ* and *in situ* conservation activities in the region, while Section 4 reviews how legume genetic diversity is currently being used both within the Mediterranean and other regions of world where Mediterranean genetic resources are exploited. Finally, in conclusion we attempt in an Afterword to summarise the future priorities for *ex situ* and *in situ* plant genetic conservation for Mediterranean legumes and highlight some important development in terms of utilisation.

It should be noted that the many chapters contain summaries of current *ex situ* gene bank holdings. When editing this volume it was noted that in certain cases different figures are quoted for the same taxa in a gene bank. As editors we could have standardised these figures, however, we preferred to leave the figures as stated by the contributing authors. It should be remembered that any figures included in a text of this kind should be seen as being indicative rather than absolute, therefore small variations in quoted figures have little consequence. By their dynamic nature, gene bank holdings change regularly as accessions are added or lost. Therefore figures on current holding are valid only for a short period, thus if accurate figures are required they should be obtained from the collection managers or their web sites.

We wish to gratefully acknowledge the following grain and forage legume gene bank curators for supplying information on their holdings: Australian Temperate Field Crop Collection, Horsham, Australia; Australian Trifolium Genetic Resource Centre, Perth, Australia; Centre for Legumes in Mediterranean Agriculture (CLIMA), Australia, University of Western Australia, Nedlands, WA, 6907, Australia; South Australian Research and Development Institute, Waite Campus. GPO Box 397. Adelaide SA 5001 Australia; Institute for Plant Genetic Resources "K. Malkov", Sadovo, Plovdiv, Bulgaria; INRA-SGAP Domaine de Melgueil, F34130 Mauguio, France; IBEAS, Unviersité de Pau et des Pays de l' Adour, Pau, France; BAZ Gene Bank, Bundesallee 50, 38116 Braunschweig, Germany; Institut für Pflanzengenetik und Kulturpflanzenforschung (IPK), Gatersleben, Germany; Institute of Agrobotany, Tápiószele, Hungary; Istituto del Germoplasma, Bari, Italy; Centre for Genetic Resources, Centre for Plant Breeding and Reproduction research, PO Box 16, Wageningen, Netherlands; N.I. Vavilov Research Institute of Plant Industry, St. Petersburg, Russia; Instituto Nacional de Investigaciones Agrarias, Madrid, Spain; Nordic Gene Bank, Box 41, Alnarp, Sweden; Aegean Agricultural Research Institute, Menemen, Turkey; John Innes Centre, Norwich Research Park, Colney, Norwich, NR4 7UH, UK; Institute of Grassland and Environmental Research, Plas Gogerddan, Aberystwyth, Ceredigion SY23 3EB, UK; USDA-SEA, Plant Genetic Resources Unit, Washington State University, Pullman, Washington State, 99163, USA; USDA-ARS, Plant Genetic Resources Unit, Cornell University, Geneva, New York 14456-0462, USA; International Centre for Agricultural Research in Dry Areas, Aleppo, Syria; and International Crops Research Institute for the Semi-Arid Tropics, Patancheru 502 324, Andhra Pradesh, INDIA; the ICARDA and ICRISAT data sets were obtained from the System-wide Information Network for Genetic Resources (SINGER). We also thank ILDIS for supplying data on the distribution of legume species.

We have attempted to encourage our contributors to look forward rather than simply reviewing past and current activities. As highlighted above, any text which details current germplasm holding or locates current conservation sites alone is in the long term doomed to lose relevance as these data are by their nature ephemeral; in this sense both *ex situ* and *in situ* conservation activities are dynamic. However, the discussion contained within will, it is hoped, identify sources of genetic erosion, locate taxonomic and genetic diversity that is not conserved, and promote more efficient utilisation of conserved material.

Nigel Maxted
Sarita Jane Bennett
July 2000

List of Contributors

Abd El Moneim, Ali	International Centre for Agricultural Research in Dry Areas, Aleppo, SYRIA.
Açikgöz, Nevin	Aegean Agricultural Research Institute, Menemen, Izmir, TURKEY.
Ambrose, Mike	John Innes Centre, Norwich Research Park, Colney, Norwich, NR4 7UH, UK
Amri, Ahmed	International Centre for Agricultural Research in Dry Areas, Aleppo, SYRIA.
Auricht, Geoff	South Australian Research and Development Institute, PRC, GPO Box 397, Adelaide, SA 5001, Australia.
Bennett, Sarita Jane	Co-operative Research Centre for Legumes in Mediterranean Agriculture, University of Western Australia, Nedlands, WA 6907, AUSTRALIA.
Christiansen, Scott	Facilitation Unit c/o International Centre for Agricultural Research in Dry Areas, P.O. Box 2416, Cairo, EGYPT.
Cowling, Wallace	Plant Sciences, The University of Western Australia, Nedlands, WA 6907 AUSTRALIA.
Enneking, Dirk	Weuert 15a, D-49439 Steinfeld, GERMANY.
Erskine, William	International Centre for Agricultural Research in Dry Areas, Aleppo, SYRIA.
Ferguson, Morag	International Crops Research Institute for the Semi-Arid Tropics, Patancheru 502 324, Andhra Pradesh, INDIA.
Francis, Clive	Co-operative Research Centre for Legumes in Mediterranean Agriculture, University of Western Australia, Nedlands, WA 6907, AUSTRALIA.
Génier, G.	INRA-SGAP Domaine de Melgueil, F34130 Mauguio, FRANCE.

Hannan, Richard	USDA-SEA, Plant Genetic Resources Unit, Washington State University, Pullman, Washington State, 99163, USA.
Howieson, John	Centre for *Rhizobium* Studies, School of Biological Sciences and Biotechnology, Murdoch University, Murdoch, WA 6150, AUSTRALIA.
Hughes, Steve	South Australia Research and Development Institute, PRC, GPO Box 397, Adelaide, SA 5001, Australia.
Johnson, Richard	USDA-SEA, Plant Genetic Resources Unit, Washington State University, Pullman, Washington State, 99163, USA.
Khan, T.N.	Co-operative Research Centre for Legumes in Mediterranean Agriculture, University of Western Australia, Nedlands, WA 6907, AUSTRALIA.
Lamont, Emma-Jane	Wye College, University of London, Ashford, Kent TN25 5AM, UK
Le Houérou, Henri	327 Rue de Jussieu, F34090 Montpellier, FRANCE
Maxted, Nigel	School of Biosciences, The University of Birmingham, Edgbaston, Birmingham B15 2TT, UK
Prosperi, Jean Marie	INRA-SGAP Domaine de Melgueil, F34130 Mauguio, FRANCE.
Reid, Bob	Department of Agriculture, Stanley, FALKLAND ISLANDS.
Robertson, Larry D.	USDA-ARS, Plant Genetic Resources Unit, Cornell University, Geneva, New York 14456-0462, USA.
Sackville Hamilton, Ruaraidh	Institute of Grassland and Environmental Research, Plas Gogerddan, Aberystwyth, Ceredigion SY23 3EB, UK
Sarker, Ashutosh	International Centre for Agricultural Research in Dry Areas, Aleppo, SYRIA.
Siddique, K.H.M.	Co-operative Research Centre for Legumes in Mediterranean Agriculture, University of Western Australia, Nedlands, WA 6907, AUSTRALIA.
Street, Kenneth	International Centre for Agricultural Research in Dry Areas, Aleppo,

SYRIA.

Tan, Ayfer — Aegean Agricultural Research Institute, Menemen, Izmir, TURKEY.

Valkoun, Jan — International Centre for Agricultural Research in Dry Areas, Aleppo, SYRIA.

van de Wouw, Mark — School of Biosciences, The University of Birmingham, Edgbaston, Birmingham B15 2TT, UK

Zoghlami, Aziza — Institut National de la Recherche Agronomique de Tunisie, Rue Hedi Karray, 2049 Ariana, TUNISIA.

Color Plates

Figure 4.1. Distribution of collecting sites of 14211 accessions of *Trifolium* species collected with latitude/longitude data from Mediterranean countries. The map shows the number of accessions collected per cell in a grid of 50km x 50km cells. Albers equal-area conic projection, with central meridian at 20°E and standard parallels at 30°N and 46°N. Map prepared with ArcView® and ArcInfo® GIS, and reproduced with permission of ESRI™ (UK). (see also p. 93)

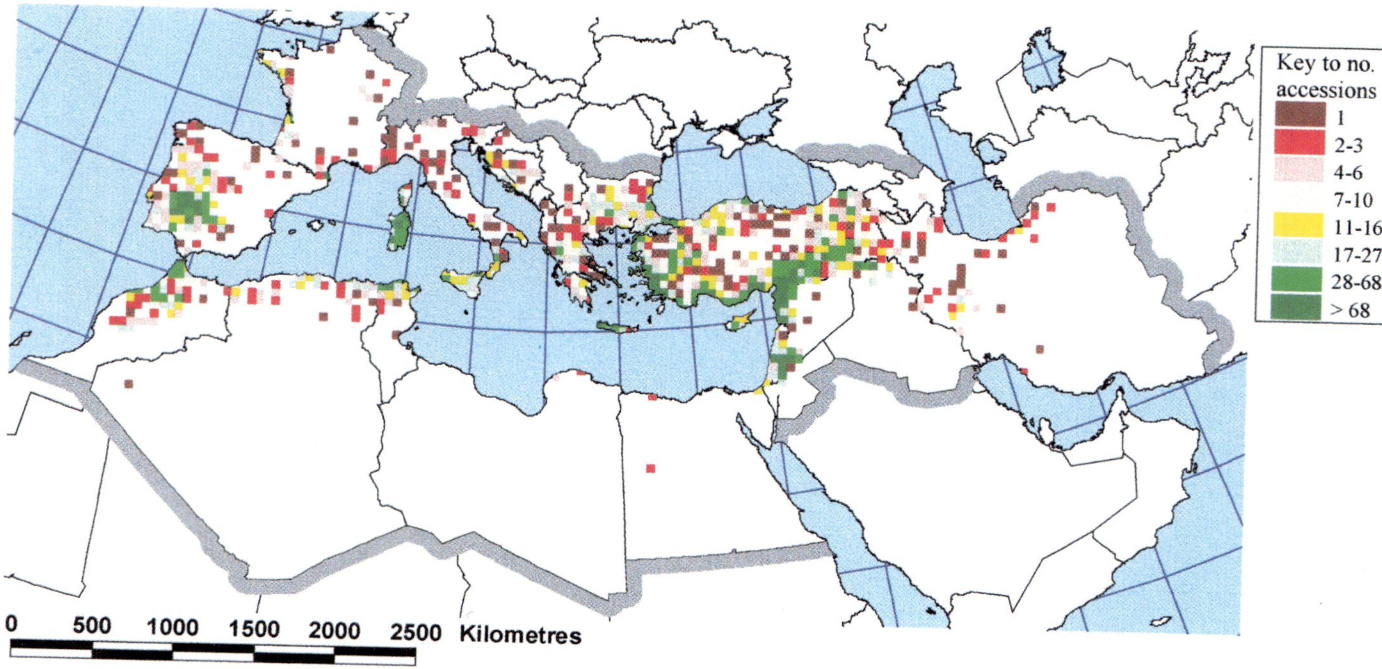

(a)

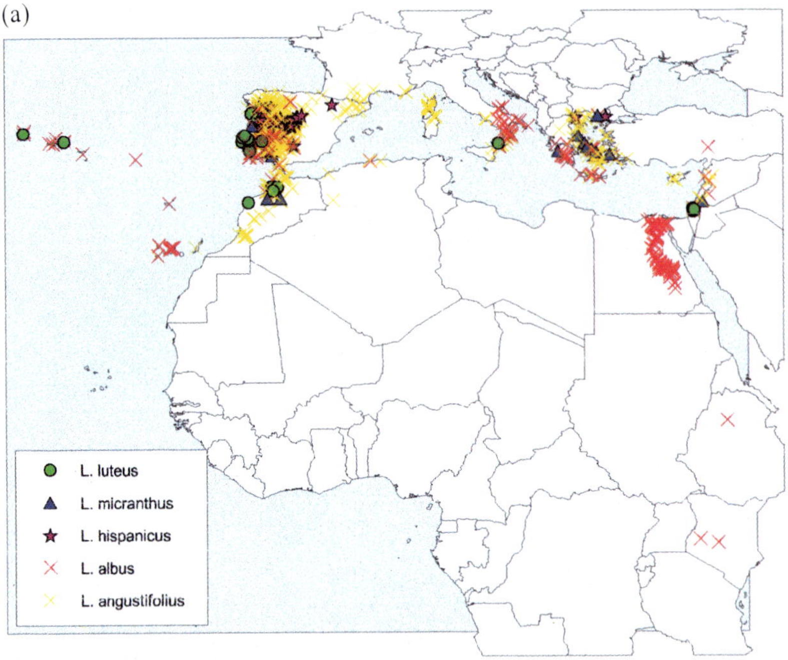

(b)

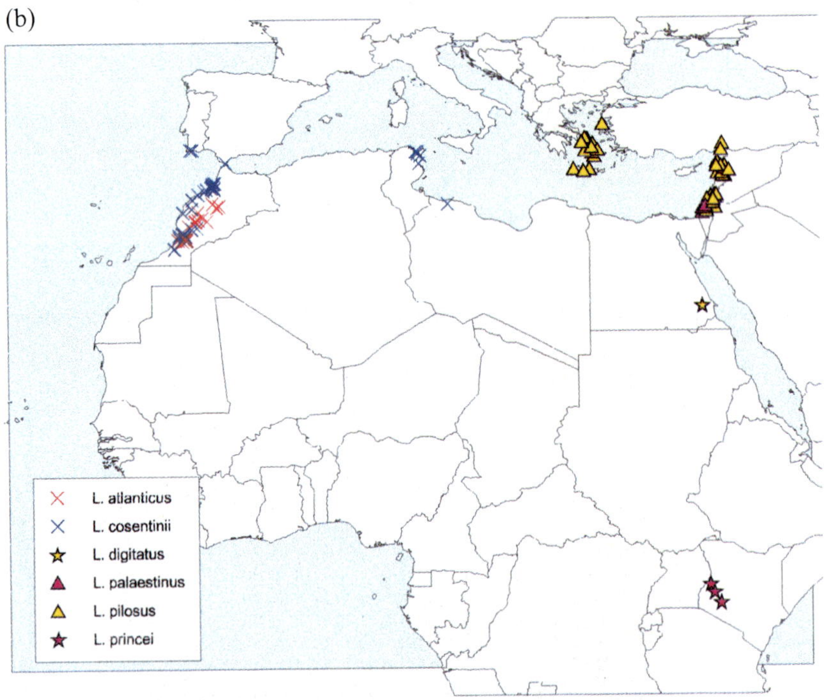

Figure 11.1. Map of precise locations in the Mediterranean region of collections of (a) the smooth-seeded and (b) the rough-seeded lupins held in the International Lupin Collection at Agriculture Western Australia, Perth. (see also pp. 196 and 197)

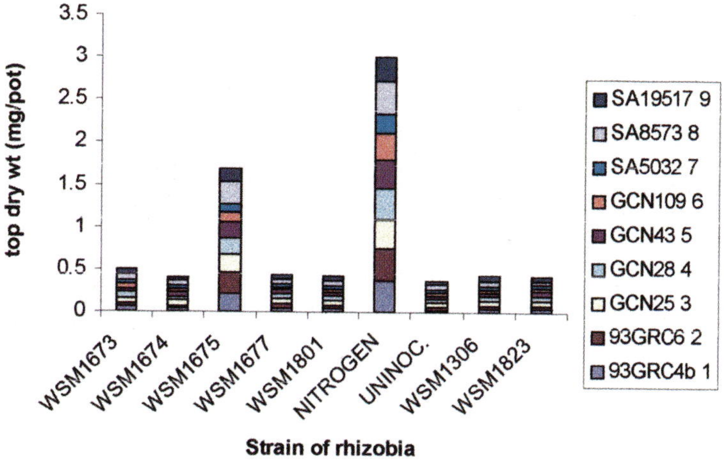

Figure 13.1. The relationship between rhizobial isolates from *H. Circinnatus* and plant genotypes of this legume, with all material collected from the Cyclades group of Greek islands. Only strain WSM1675 effectively fixed nitrogen, despite all strains nodulating all nine hosts (Howieson and Yates, unpublished data). (see also p. 234)

Plate 1. Collecting wild legume germplasm in Tadzhikistan (Photograph N. Maxted)

Plate 2. Farmers threshing cultivated legumes in Nepal (Photograph L.D. Robertson)

Figure 15.3. Distribution of collecting sites of 48082 accessions of forage legumes collected with latitude/longitude data from Mediterranean countries. The map shows the number of accessions collected per cell in a grid of 50km x 50km cells. Albers equal-area conic projection, with central meridian at 20°E and standard parallels at 30°N and 46°N. Map prepared with ArcView® and ArcInfo® GIS, and reproduced with permission of ESRI™ (UK). (see also p. 287)

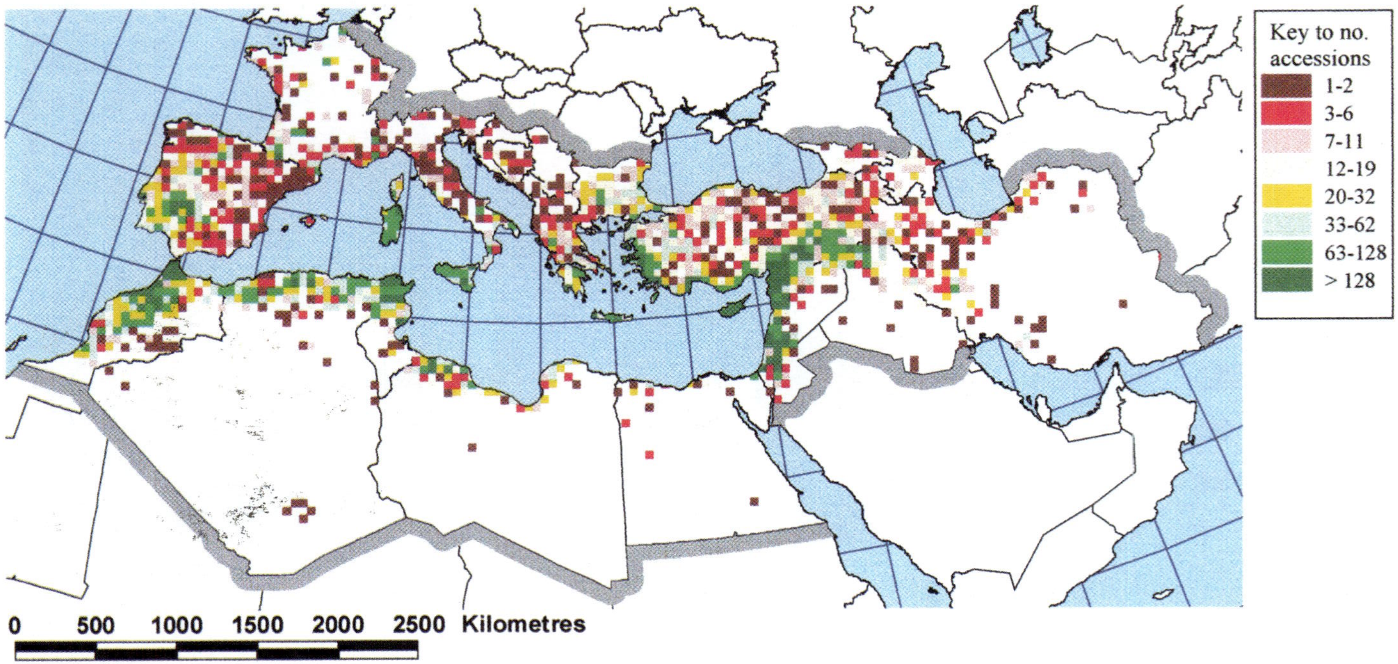

Figure 15.4. Distribution of collecting sites of 48082 accessions of forage legumes collected with latitude/longitude data from Mediterranean countries. The map shows the number of species that have been sampled at least once per cell in a grid of 50km x 50km cells. Albers equal-area conic projection, with central meridian at 20°E and standard parallels at 30°N and 46°N. Map prepared with ArcView® and ArcInfo® GIS, and reproduced with permission of ESRI™ (UK). (see also p. 288)

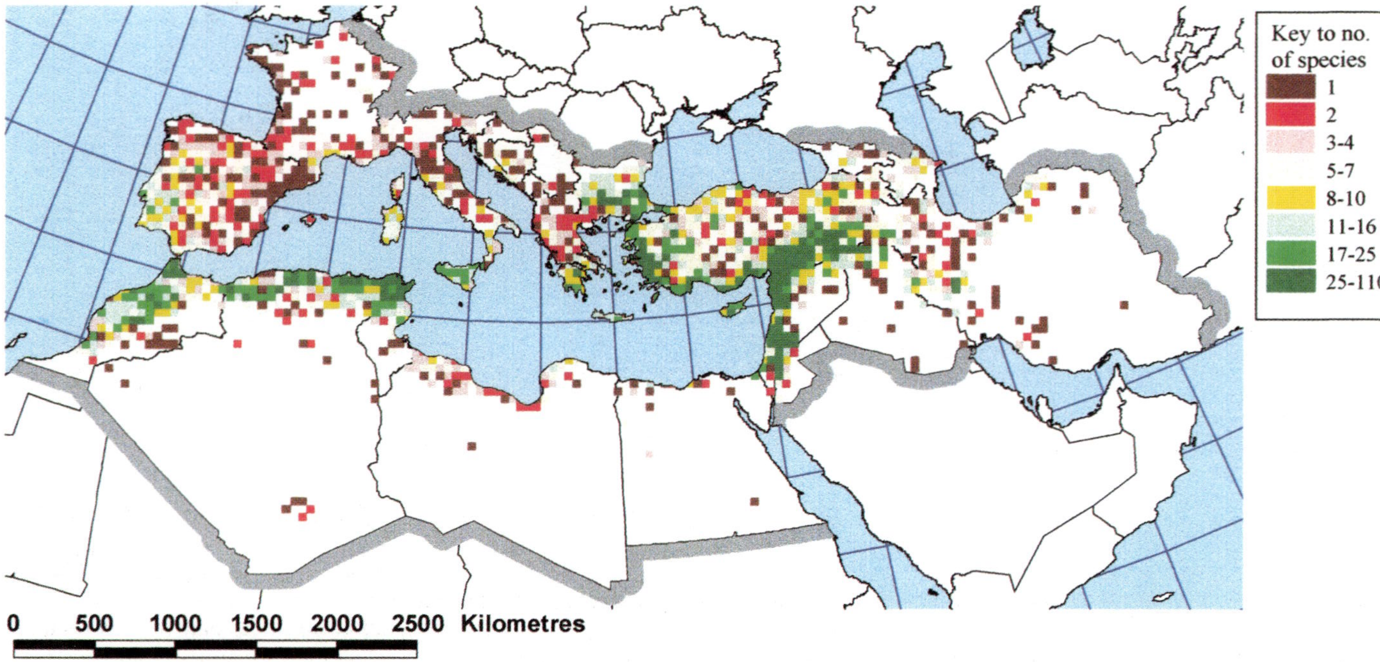

Section 1

Introduction

Chapter 1

Conservation, Diversity and Use of Mediterranean Legumes

Nigel Maxted and Sarita Jane Bennett

1.1 INTRODUCTION

Why conserve food and forage legumes? The conservation and use of legume species, and for that matter all biodiversity, is of increasing national and regional importance in the Mediterranean. We need the genetic diversity of peas, chickpeas, lentils and faba beans, as well as the various forage legume species to provide food security for humans and their domesticated animals, both in the Mediterranean region itself and in other regions of the world with Mediterranean climates. This is necessary because of the current loss of ecosystems, habitats, species and genetic diversity worldwide, and particularly the Mediterranean region, but also because of the ever present need to feed the increasing human population (Maxted *et al.*, 1997a). There is a catastrophic loss of plant genetic diversity occurring: species, gene combinations and alleles are being lost for perpetuity (Diversity, 1999). This degradation (i.e. genetic erosion) is increasing exponential and shows no signs of abatement. We need to conserve plant diversity because of the direct benefits to humankind that can arise from its exploitation in new agricultural and horticultural crops, the development of medicinal drugs but also because of the pivotal role played by plants in the functioning of all natural ecosystems.

The Mediterranean has a comparatively rich vascular plant flora, comprising 24,000 ± 600 species (Greuter, 1991), which is about 10% of the total world species. This is particularly impressive considering that this is only about 1.5% of the world's land surface and about half of this is classified as desert. The region is particularly rich in endemic taxa, with about half of the plant species endemic to the region (Quézel, 1985; Greuter, 1991; Heywood, 1995) and many of these narrowly restricted to ecological, altitudinal, geological or geographic 'islands'. The richest national flora of the Mediterranean basin is found in Turkey (Davis, 1965) with 9,000 of which approximately a third are endemic. The highest levels of plant diversity are found in the highly fragmented environments of the islands and mountain ranges of Greece and Turkey (Greuter, 1991). Turkey has a particularly high number of taxa because it houses the interface for the Euro-Siberian, Mediterranean and Irano-Turanian floras (Davis, 1965). The region contains two of Vavilov's (1997) centres of origin of major groups of crop plants: Asia Minor (wheats, rye, oats, grain and forage legumes, fruits, etc.), and Mediterranean Centre (wheats, barleys, forages, fruits and vegetables, spices and ethereal oil plants, etc.). It also contains numerous wild crop relatives and as such is one of the world's richest centres for wild progenitors of cultivated plants (Zohary, 1991; Zohary and Hopf, 1993; Heywood and Zohary, 1995). Harlan (1995) considers several factors to be responsible for the diversity of wild and crop species in the Mediterranean region:

- *Socio-political* - Ancient agriculture and conservation was promoted by diverse groups of rulers in the region, and extensive trading both within and with other regions, promoting the migration and diffusion of diverse crop species.
- *Agro-climatic* - In general, the harsh climate, moderately cool (>15°C) and wet Winters, warm (20-25 °C) Springs and hot Summers stresses the plants, leading to increased frequency of resistant alleles at individual loci, high total genetic diversity and the ability for diverse evolutionary development.

- *Ecological* - ecological diversity is often correlated with genetic diversity, and in a region with numerous distinct habitats and geographic barriers between habitats diverse evolutionary development is promoted.
- *Genetic* - the natural genetic diversity of native species resulting from the above factors was maintained and enhanced by conscious human selection.

The Fertile Crescent of the Near East is recognised as one of the four centres of agricultural development and it is thought that Neolithic hunter-gathers here first established the sedentary agricultural life around 10,000 - 9,500 BP (Ucko and Dimbleby, 1969; Willcox, 1998). The long history of plant and animal domestication in the region has had its effect on the environment and the majority of Mediterranean ecosystems have been modified by human actions (Cowling *et al.* 1996). Willcox (1995) states that the wild crop progenitors of pulses were exploited in the Epipalaeolithic era (12,000-10,000) several thousand years before their counterparts were finally domesticated between 9,600 - 9,200 BP.

Why link conservation of plant diversity to use? The Convention on Biological Diversity (CBD) recognises the fundamental importance of and link between these issues in its objectives:

> "the conservation of biological diversity, the sustainable use of its components and the fair and equitable sharing of the benefits arising out of the utilisation of genetic resources..."
>
> Article 1: Objectives (CBD, 1992)

Maxted *et al.* (1997) and Hawkes *et al.* (2000) argue vigorously that the prime justification for conservation must be human exploitation. Any programme of active plant conservation has a real and often substantial cost (Maxted *et al.*, 2000). Hawkes *et al.* (2000) estimate the cost of US $ 5.3 billion for collecting and conserving the world's germplasm currently held in gene banks. This cost is commonly met by international or national agricultural or developmental agencies, and thus indirectly by the public. Therefore, it is not unreasonable for the public to expect some reward for their investment, and specifically in the case of Mediterranean legumes, that legume conservation will be linked to some form of utilisation. However, the point should be stressed that use need not be just defined in terms of pure dollar income. I gain enjoyment from hiking in wilderness areas or walking my dog in the countryside on Sunday afternoons, these are just as much uses of conserved plant diversity as producing new crop varieties or plant-based medicines.

The CBD (1992) also specifically recognises this need to link conservation to use. Article 10 states:

> "Each Contracting Party shall, as far as possible and as appropriate: Integrate consideration of the conservation and sustainable use of biological resources into national decision making ...."

Thus when discussing the conservation of Mediterranean legumes the link is naturally made to human exploitation. This may appear a particularly 'natural' linkage, in the case of Mediterranean case because of the region's rich native flora, it being the site of the origin of many crop plants, as well as being one of the centres for the origin of agriculture.

The Leguminosae family is a major component of all habitats and sub-regions in the Mediterranean, and of all the temperate regions of the world, the highest concentration of legume species are found here. 91 genera, 1,956 species and 495 subspecific legume taxa are found in the Mediterranean region (Med-Checklist Database: Berendsohn, pers. comm.), with over 1,300 species being found in Turkey alone, of which about a quarter are endemic (Davis, 1970). This makes the Mediterranean region a major centre of diversity for legume species. Legumes are also an essential component of all Mediterranean agro-ecosystems, both within the Mediterranean region but also within the other regions of the world with a Mediterranean climate (California; Central Chile; Western Cape, South Africa; and Southern Australia). They provide an important superior source of human and animal nutrition, contribute to the nitrogen economy of pastures (50.4Mt of N_2 is fixed worldwide annually; Hauck, 1988) and act as a break crop in cereal production systems.

However, severe genetic erosion of the entire legume gene pool is currently occurring in the Mediterranean Basin (IBPGR, 1985; Bennett *et al.*, 1998; Maxted, 1999). Forage legumes and wild relatives of food legumes are threatened, primarily through over-grazing of natural grasslands by increasing numbers of livestock in many areas, and conversely through a decline in sheep production as a profitable economic activity (Stanners and Bordeau, 1995). Natural pastures are under greater threat now in the northern Mediterranean because European Union farming policies favour increased production and yields, and the denigration of natural pastures (Onorati, 1995; Porqueddu and Sulas, 1997). A general decline in land areas used for cultivation of legumes, agricultural improvements to natural and semi-natural grasslands through re-seeding and increasing use of herbicides are also factors contributing to an accelerating pace of genetic erosion (Munzer, 1989). This erosion presents a serious threat to the efficient and effective exploitation of legumes.

The causes of erosion are largely anthropogenic, and may be broadly grouped under the general headings of:

- destruction, degradation and fragmentation of natural habitats
- changes in agricultural practices and land use
- over-exploitation and incidental take
- exotic species, which compete with, prey on or hybridise with native species
- human socio-economic change and upheaval
- calamities, both natural and man-made
- genetic pollution

The current debate over the introduction of genetically modified (GM) crops has highlighted another threat to genetic diversity: that of genetic pollution of native plant genetic diversity from alien germplasm. Whatever the effects of GM crops, it is inevitable that there will be some cross-pollination between them and their wild relatives, thus polluting local gene pools. GM crops are, however, not the only cause of genetic pollution. Sackville-Hamilton (1999) has shown that there has already been severe genetic pollution of wild *Lolium perenne* populations in the UK from conventionally bred varieties. The few wild populations of this species investigated to date have completely lost the genetic distinction that would normally be associated with geographical isolation. The most likely explanation for this lack of genetic differentiation is gene flow between wild populations and widespread cultivars of the same species (Warren *et al.*, 1998). Although yet untested, it seems likely that this process of genetic pollution is also occurring in temperate legume populations.

There are no specific figures for loss of species or genetic diversity within temperate legumes, particularly within the Mediterranean Basin. However, Lugo (1988) estimated that 15-20% of all species could become extinct in the next 10 years and based on these figures Maxted, *et al.* (1997b) extrapolated that 25-35% of all plant genetic diversity could be lost over the same period, a level of genetic erosion unseen by previous generations. A rough estimate of the threat to legume species in the Mediterranean Basin may be gained from the 1997 IUCN Red List of Threatened Plants (Walters and Gillett, 1998). The threat figures included in Table 1.1 indicate total numbers of threatened legume species per genus in each Mediterranean Basin country using the IUCN Red List Categories. These figures indicate that 322 species from 15 legume genera are threatened in the Mediterranean basin, compared to 2,206 for all legume species worldwide (Walters and Gillett, 1998).

Table 1.1. 1997 IUCN red list of threatened legume plants of the Mediterranean region (Data extracted from Walters and Gillett, 1998).

Country	Threatened Legume Species Per Genus																All Species in Flora	
	Anthyllis	*Astragalus*	*Cicer*	*Lathyrus*	*Lens*	*Lotus*	*Lupinus*	*Medicago*	*Melilotus*	*Onobrychis*	*Ornithopus*	*Pisum*	*Trifolium*	*Trigonella*	*Vicia*	Total	Total Number	% Threat-ened
Portugal	2	0	0	0	0	3	0	0	1	0	0	0	0	0	3	9	5050	5.3
Spain	4	4	0	0	0	15	0	2	0	0	0	0	0	0	5	30	5050	19.5
France	0	0	0	0	0	0	0	0	0	0	0	0	1	0	0	1	4630	4.2
Italy	0	5	0	0	0	0	0	1	0	0	0	0	2	0	1	9	5599	5.6
Yugoslavia (former)	0	1	0	1	0	0	0	1	0	1	0	0	1	0	1	6	5351	2.9
Albania	0	2	0	0	0	0	0	0	0	0	0	0	0	0	0	2	3031	2.6
Greece	1	8	1	1	0	2	0	2	0	2	0	0	3	1	1	22	4992	11.4
Bulgaria	0	4	0	1	0	0	0	1	0	0	0	0	0	0	1	7	3572	3
Romania	0	4	0	0	0	0	0	0	0	0	0	0	0	0	0	4	3400	2.9
Georgia	0	1	0	0	0	0	0	1	0	0	0	0	0	0	1	3	NA	NA
Armenia	0	0	0	0	0	0	0	0	0	0	0	0	0	0	0	0	NA	NA
Azerbaijan	0	1	0	0	0	0	0	0	0	0	0	0	0	0	1	2	NA	NA
Turkey	1	134	4	12	0	1	0	1	0	23	0	0	7	11	4	198	8650	21.7
Cyprus	0	2	0	0	0	0	0	0	0	0	0	0	0	0	0	2	1682	3
Malta	0	0	0	0	0	0	0	0	0	0	0	0	0	0	0	0	914	1.6
Iraq	0	0	0	0	0	0	0	0	0	0	0	0	0	0	0	0	NA	NA

Iran	0	0	0	0	0	0	0	0	0	0	0	0	0	0	0	0	8000	0
Syria	0	0	0	0	0	0	0	0	0	0	0	0	0	0	0	0	3000	0.3
Lebanon	0	0	0	0	0	0	0	0	0	0	0	0	0	0	0	0	1591	1.3
Israel	0	2	0	1	0	0	0	0	0	0	0	0	0	0	0	3	2317	1.4
Jordan	0	2	0	0	0	0	0	0	0	0	0	0	0	1	0	3	2100	0.4
Egypt	0	3	0	0	0	0	0	0	0	0	0	0	0	0	1	4	2076	3.9
Libya	0	2	0	0	0	0	0	1	0	0	0	0	0	0	0	3	1825	3.1
Tunisia	0	0	0	0	0	1	0	0	0	0	0	0	0	0	0	1	2196	1.1
Algeria	0	1	0	0	0	1	0	0	0	0	0	0	0	1	1	4	3164	4.5
Morocco	0	5	1	0	0	0	0	0	0	0	1	0	1	0	1	9	3675	5.1
Total	8	181	6	16	0	23	0	10	1	26	1	0	15	14	21	322	-	4.94

As legumes are subject to increased genetic erosion, and bearing in mind that we need this diversity for agricultural exploitation, there is an obvious need to increase the efficiency of food and forage legume conservation and use in their centre of diversity, the Mediterranean. Therefore, the aim of this text is to review current conservation and use of legume diversity, as well as indicating how the efficiency of future conservation and use might be improved.

1.2 THE SCOPE OF THE PROJECT

The project to conserve and use legume diversity, which is the central focus of this text, has taxonomic and geographical scope. And as argued above, the conservation of legumes from the Mediterranean region should be in turn linked to their exploitation. But why focus on legumes of the Mediterranean?

The Leguminosae provide an important genetic resource throughout the world, second only to the Gramineae in world economic importance (Heywood, 1978). Leguminous crops, that is those species of the family Leguminosae exploited by humankind, may be divided into grain legumes, where the grain is eaten by humans, pasture legumes, where they are grown to feed domestic animals, and those grown in temperate or tropical climates. Some examples of tropical grain crop species are: groundnut (*Arachis hypogaea* L.), jack bean (*Canavalia ensiformis* (L.) DC.), hyacinth bean (*Lablab purpureus* (L.) Sweet), Kersting's groundnut (*Macrotyloma geocarpum* (Harms) Maréchal & Baudet), yam bean (*Pachyrrhizus erosus* (L.) Urban), tepary bean (*Phaseolus acutifolius* A. Gray), scarlet runner bean (*P. coccineus* L.), lima bean (*P. lunatus* L.), polyanthus bean (*P. polyanthus* Greenm.), common bean (*P. vulgaris* L.), winged bean (*Phsophocarpus tetragonolobus* (L.) DC.), moth bean (Vigna acontifolia (Jacq.) Maréchal), adzuki bean (*V. angularis* (Willd.) Ohwi & Ohashi), black gram (*V. mungo* (L.) Hepper), green gram (*V. radiata* (L.) Wilczek), bambara groundnut (*V. subterranea* (L.) Verdc.), rice bean (*V. umbellata*), cowpea (*V. unguiculata*), pigeon pea (*Cajanus cajan* (L.) Millsp.) and soya bean (*Glycine max* (L.) Merr.) (the latter is of tropical origin but is now grown in both tropical and temperate environments). Some examples of tropical pasture are: joint vetch (*Aeschynomene falcata* (Poir) DC.), calopo (*Calopogonium mucunoides* Desv.), jack bean (*Canavalia ensiformis* (L.) DC.), centro (*Centrosema pubescens* Benth.), butterfly pea (*Clitoria ternatea* L.), desmodium (*Desmodium* spp.), glycine (*Glycine wightii* (R. Grah. ex Wight and Arn.) Verdc.), puero (*Pueraria phaseoloides* (Roxb.) Benth.), and stylos (*Stylosanthes* spp.). Examples of browse species are: acacias (*Acacia* spp.), siris (*Albizia* spp.), desmodiums (*Desmodium* spp.), lead tree (*Leucaena leucocephala* (Lam.) de Wit.), algaroba (*Prosopis* spp.) and tamarind (*Tamarindus indica* L.). For detailed reviews of tropical grain and forage legume species see Skerman (1977), Duke (1980), Summerfield and Bunting (1980) and Smartt (1990). A companion text focussing on tropical legume conservation and exploitation is required, but this text focuses on the conservation and exploitation of temperate grain and pasture legumes.

As already stated, legumes are an essential component of Mediterranean agro-ecosystems, and this is the ecosystem from which one third of the foodstuffs of humankind have been domesticated (Harlan, 1995). Although many regions of the world are characterised by the Mediterranean climate and share a similar Mediterranean ecosystem, it is from the Mediterranean Basis itself that the species used as grain or forage species and where temperate legume agriculture itself originates, and these species remain today the majority of legume species cultivated today in temperate environments worldwide.

However, the geographical scope of this project cannot be restricted too narrowly to those political countries that border the Mediterranean Sea. Plants know nothing of international boundaries. The distribution of native legume genera found around the Mediterranean extends eastwards to include Iraq and Iran, as well as the three Caucasian republics of Georgia, Armenia and Azerbaijan. So, throughout this text the term Mediterranean is extended to include these five countries to the east of the Mediterranean *senso stricto*. Therefore, the species included will be those native legume taxa of socio-economic value to humans from this extended Mediterranean region, as shown in Figure 1.1.

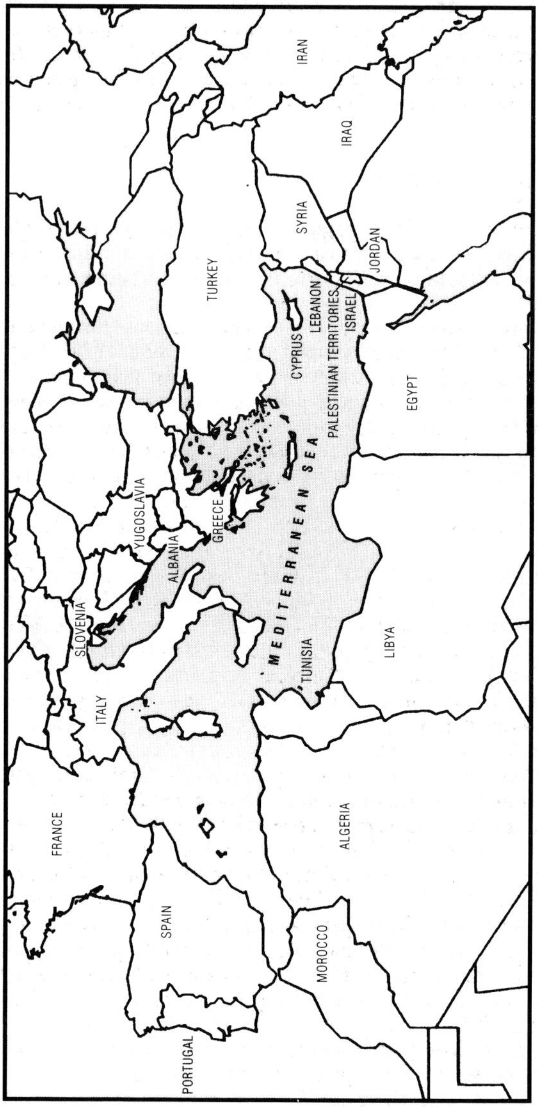

Figure 1.1. Delineation of the Mediterranean region (Diversity, 1995)

1.3 PLANT GENETIC RESOURCE CONSERVATION

How are we to conserve legume diversity? Hawkes *et al.* (2000) define botanical diversity conservation as the:

> "maintenance of the genetic, specific and ecosystem diversity of plants, and the interrelationships between plants, other organisms and their environment."

They also stress the necessary link between botanical conservation and use. To achieve their goals conservationists apply a range of skills derived from numerous disciplines, including genetics, ecology, geography, taxonomy, economics, and breeding, to conserve representative plant populations that constitute the genetic diversity of a species. Maxted *et al.* (1997a) proposed a model for plant genetic diversity conservation (see Figure 1.2) and this may be applied to the conservation and use of legumes. The raw materials of genetic conservation are genes within gene pools, the total diversity of genetic material of the particular taxon being conserved. The product of the gene pool is either preserved or utilised genetic diversity. The processes linking the raw matter and the utilised gene pool are referred to as conservation.

Figure 1.2. Model for plant genetic diversity conservation (After Maxted *et al.*, 1997a)

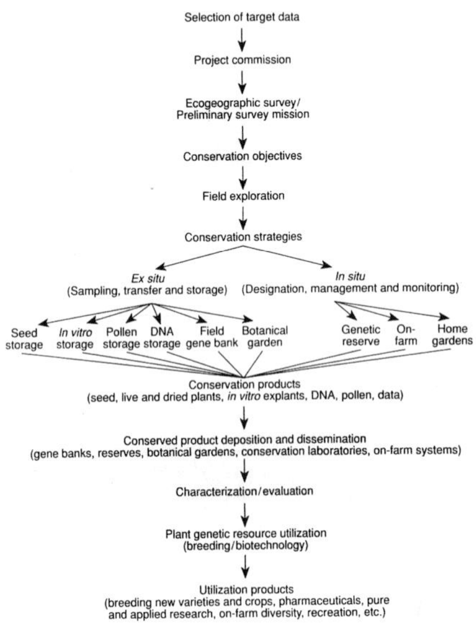

1.3.1 Preparing to Conserve

All conservation has a real cost to society (Maxted *et al.*, 2000) and with limited human, financial and technical resources available, the conservationist is forced to prioritise which taxa to conserve first. This choice should be objective, based on logical, scientific and economic principles related to the perceived value of the species (Maxted *et al.*, 1997b). The sort of factors that will be taken into account when attributing values and conservation priority to species include: current conservation status, threat of genetic erosion, taxonomic distinctiveness, ecogeographic distinctiveness, actual economic use, potential economic use, biological importance within ecosystems, cultural importance, aesthetic value, cost of conservation, and likely sustainability of the conservation effort. Rarely will one of the above factors alone provide sufficient weight to result in some form of conservation intervention. More commonly, all or a range of these factors will be assessed for a particular taxon, which will then be given a certain threshold level of national, regional or world conservation priority. Species with socio-economic use such as foods, medicines or industrial products will commonly be accorded highest value and conservation priority, especially in developing economies. Thus, both food and forage legume species and their wild relatives are given a high priority for conservation within national and international plant genetic resource conservation programmes.

Once the target taxon or group of taxa have been selected and possibly the target area delimited by the commissioning agency, the conservationist will begin to amass and synthesize the basic data necessary to begin the process of formulating an appropriate, comprehensive conservation strategy. This process is often referred to as an ecogeographic survey. It is only on the basis of at least a basic understanding of the taxonomy, genetic diversity, geographic distribution, ecological adaptation and ethnobotany of a plant group, and of the geography, ecology, climate and human communities of the target region, that sensible conservation decisions can be made. Ecogeographic studies involve the use of large and complex data sets obtained from experts, various published and unpublished media, herbarium specimens and germplasm collections. Analysis of these data results in the identification of priority areas in which to focus conservation efforts. These areas contain:

- high and complementary genetic and specific diversity
- crops and wild/weedy relatives growing together
- specific traits of interest for future exploitation
- taxa or phenotypes which are under-represented in collections
- germplasm not previously conserved, either *ex situ* or *in situ*, and
- germplasm at risk of (or already suffering from) genetic erosion

Several ecogeographic surveys have been published in recent years for important temperate grain and forage legume groups, both on a taxon or regional basis. These include: forage legumes in Syria (Ehrman and Cocks, 1990; Ehrman and Cocks, 1995); *Vicia* subgenus *Vicia* (Maxted, 1995); grain, forage and pasture legumes of Crete (Francis *et al.*, 1995); Mediterranean islands (Delanoë, 1996); the *Vicia narbonensis* complex (Bennett and Maxted, 1997); *Lathyrus sativus* (Campbell, 1997); *Vicia* in Turkey (Maxted and Kell, 1998); Lens (Ferguson *et al.*, 1998a, 1998b, 2000); south-west Turkey (Bennett *et al.*, 1998); Lupinus (Cowling *et al.*, 1998); north-west Turkey (Bennett *et al.*, 1999) and forage legumes (Woodgate *et al.*, 1999).

If the available ecogeographic data for the target taxon are unavailable or limited, the conservationist will not have sufficient background biological knowledge to formulate an effective conservation strategy and therefore will need to undertake a survey mission to gather fresh baseline biological data.

1.3.2 Application of Conservation Techniques

The products of the ecogeographic survey and preliminary exploration provide a basis for the conservationist to set out a clear, concise statement of the proposed conservation plan for the target gene pool. There are two conservation strategies, each with a number of diverse techniques, which can be combined to conserve genetic diversity in the most efficient, effective and complementary manner. The two strategies are *ex situ* and *in situ* conservation. Article 2 of the Convention on Biological Diversity (CBD, 1992) gives the following definitions:

> "*Ex situ* conservation means the conservation of components of biological diversity outside their natural habitats"

"*In situ* conservation means the conservation of ecosystems and natural habitats and the maintenance and recovery of viable populations of species in their natural surroundings and, in the case of domesticates or cultivated species, in the surroundings where they have developed their distinctive properties."

The fundamental difference between these two strategies is that *ex situ* conservation involves the location, sampling and transfer of genetic material away from the location where it was found to a remote location for conservation, while *in situ* conservation retains the material in the original location where it was found. Genetic diversity can be conserved *ex situ* as seed, pollen or even DNA in cold stores, explants *in vitro*, and whole plants in field gene banks and botanic gardens. The definition of *in situ* conservation used in the Convention on Biological Diversity encompasses two quite distinct techniques: conservation of primarily wild species in genetic reserves and of crops and weedy species 'on farm' in traditional farming systems. Each strategy has its advantages as well as disadvantages (for discussion see Maxted *et al.*, 1997a), no one strategy can be said to provide complete conservation, therefore it is expedient to apply multiple techniques in a complementary manner.

The most widely used technique for legume species conservation is seed storage in a gene bank because legume seed is generally known to be desiccation and freezing tolerant (Hong *et al.*, 1998), though some tropical species of *Inga* are recalcitrant. There are few practical examples of the application of other conservation techniques, but in recent years, genetic reserve and on farm conservation has begun to be used in the Middle East (see Chapter 16). Whatever conservation technique is applied, the analysis of ecogeographic data will enable the conservationist to know when, where and how to collect germplasm, where genetic reserves might best be placed and how they would need to be monitored and managed, and the relative contribution of *ex situ* and *in situ* approaches to an overall conservation strategy.

1.3.3 Conservation Products and Their Dissemination

The products of conservation activities are primarily conserved germplasm, living plants, associated reference specimens (e.g. herbarium) and data. Conserved material is ideally associated with a range of passport data, which details the taxonomic, geographical and ecological provenance of the material. Passport data associated with conserved material are normally entered into a data management system and made available for the management of the material, the formulation of future conservation priorities and strategies, and for use. Conservation products are either maintained in their original environment or deposited in a range of *ex situ* storage facilities. Whether the germplasm is conserved *in situ* or *ex situ*, to ensure its safety it should be duplicated in more than one location. The distribution of duplicate sets of material avoids inadvertent loss of the material due to fire, economic or political difficulties, warfare, cessation of project funding or other unforeseen circumstances. Duplication of passport data is relatively easy from the conservation database, and the commissioning agency, relevant host country institutes and other interested parties should hold copies.

Rhizobia (*Rhizobium* sp., *Bradyrhizobium* sp., *Sinorhizobium* and *Azorhizobium* sp.) are root-nodule bacteria that have a microsymbiotic relationship with their legume host (Date, 1995). Each legume species, and even ecotypes within a species, have a symbiotic relationship with a particular species or ecotype of rhizobia. Rhizobia persist naturally in the soil in the Mediterranean Basin, but the required types are not present in other areas of the world where legume crops or pasture species are to be sown. Without specific rhizobia being present in the soil, the legume crop will not persist, or will persist only poorly, as it is unable to fix nitrogen effectively.

Therefore, any legume collecting mission should include a collection of the rhizobia that occur in the nodules of the plants collected (see Date 1995). This can be difficult, as the soil is often very dry and hard during the time when seeds are collected, but an effort should be made to collect nodules where possible. The rhizobia are then stored *ex situ* in rhizobia gene banks.

For more details on rhizobia and their importance in legume conservation and utilisation see Chapter 13, which is dedicated the root-nodule bacteria from the Mediterranean basin.

1.3.4 Botanical diversity utilisation

As discussed above there should be an intimate linkage between conservation and utilisation. The products of conservation, whether 'living' or 'suspended' should be made available for utilisation by humankind. Conservation can be seen as the safe keeping of preserved material, so that the material is available for utilisation at a future date. There is an enormously wide range of uses to which plants have been put by humankind. Cook (1995) groups plant use into twelve broad categories: foods, clothing, medicines, ritual, fuel, pharmaceuticals, construction, cosmetics, recreational, transport, paper and paper products, and decoration. Although it is undoubtedly true that humankind has used plants in numerous diverse ways, in terms of *ex situ* plant conservation, the primary use remains to provide the raw materials for crop improvement programmes. Therefore, if the conserved plant diversity is not to become mere museum items of no more than historical interest they must be evaluated and used. Commonly, the first stage of utilisation will involve characterisation, the recording of genetically controlled characteristics, and evaluation, during which the material may be grown out under diverse environmental conditions to screen for example for drought or salt tolerance, or the deliberate infection of the material with diseases or pests to screen for particular biotic resistance. The biotechnologist screens for single genes that once located may be transferred into a host organism. The biochemist or bioprospector screens for particular chemical products that may be of use to the pharmaceutical industry. The results of use are therefore numerous, including new varieties, new crops and new pharmaceutical products.

1.4 MEDITERRANEAN LEGUME CONSERVATION

1.4.1 Seed Conservation

In terms of Mediterranean legume species, conservation most commonly takes the form of an ecogeographic survey, followed by a seed collection expedition and seed storage in a gene bank at -20°C. There have been hundreds of legume germplasm collection missions to the Mediterranean basin since N.I. Vavilov's first mission to the region in 1926 (Vavilov, 1997). Initially, seed collection expeditions in the region were largely international, with collectors travelling from the USA, Australia, Soviet Union and Germany. As an illustration of the extent of these missions, there have been 92 Australian funded seed collection missions to the Mediterranean since 1951, involving the collection of approximately 25,000 legume accessions and 1,000 rhizobia (Francis, pers. comm.). Latterly, however, there has been an increasing focus on national collecting teams collecting their native plant genetic resources, or the CGIAR institutes (notably ICARDA, the International Centre for Agricultural Research in the Dry Areas, based at Aleppo, Syria; IPGRI WANA, International Plant Genetic Resources Institute, West Asia and North Africa office based in Aleppo; and CIMMIT, Centro Internacional de Mejoramiento de Maiz y Trigo, based in Mexico) with regional responsibilities, systematically collecting their mandate species. Two other institutes have also played an important role in the conservation and use of legumes, ACSAD (Arab Centre for Studies of Arid Zones and Dry Lands) based in Damascus, Syria and CLIMA (Centre for Legumes in Mediterranean Agriculture) based in Perth, Australia.

ICARDA, IPGRI (then IBPGR) and FAO together with the various countries of the region established in 1992 the West Asia and North Africa Plant Genetic Resources Network (WANANET) with the objectives of:

- identifying common constraints to effective conservation and utilisation of

PGR in the region,

- assisting National Research Systems to develop genetic resource programmes,
- formulating and prioritising collaborative research and strategies for collecting, conservation, documentation, germplasm exchange, and training, and
- formulating recommendations for specific regional cooperative programmes.

WANANET established six working groups (Adham *et al.* 1995) to assist in meeting these objectives: cereals, horticultural crops, industrial crops, food legumes, pasture and forages, *in situ* and biodiversity. WANANET has passed through various development phases, working throughout to enhance effective conservation and utilisation of PGR in the region. A similar approach was taken by the European Cooperative Programme for Crop Genetic Resources Networks (ECP/GR) established in 1980 (Frison and Ager, 1995) following the recommendation of the European Association for Research on Plant Breeding (EUCARPIA), but ECP/GR only includes those Mediterranean countries that are also part of Europe.

The precise details of what food and forage temperate legume germplasm is held *ex situ* and in which gene banks will be reviewed in subsequent chapters, but a summary of current holding of the major *ex situ* collections for the Mediterranean is contained in Table 1.2. It should be remembered that figures included in a text of this kind should be seen as being indicative rather than absolute because by their very nature they are changing regularly. Accurate figures on actual current holding should always be obtained from the collection managers or their web sites.

1.4.2 Non-seed Based Conservation

Storing legume seed in gene banks is so relatively inexpensive and effective that few other conservation techniques have been routinely applied, except in the experimental sense. However, this is not to say that other techniques do not need to be used. Maxted *et al.* (1997a) stress the need to apply both *ex situ* and *in situ* strategies, as well as a range of techniques within each strategy, in a complementary fashion if the gene pool is to be considered safely conserved.

Botanic gardens offer an *ex situ* alternative to seed conservation. In the traditional sense, the first botanic gardens established in Italy in the mid-14th century collected living material from one location and transferred and maintained it at a second location as living plant collections of species in a garden or tree species in an arboretum. Botanic gardens and arboreta, in this strict traditional sense, are therefore living collections of plants held for public display, education, economic exploitation and scientific enquiry. Although botanic gardens often have a broad taxonomic coverage, they often hold only a few accessions of each species conserved, which limits their effectiveness in genetic conservation. However, botanic gardens are increasingly involved in additional conservation activities (Hawkes *et al.*, 2000):

- *Seed gene banks* - approximately 700 gardens have associated gene banks, but only about a third of these apply internationally accepted gene banking standards.
- *Micro-propagation and tissue culture* - for *in vitro* conservation of recalcitrant seeded or poor seed producing species.
- *Re-introduction programmes* - the garden may have specialised nurseries for propagation of live plants for re-introduction into the wild.
- *DNA storage* - e.g. the Royal Botanic Gardens at Kew has recently established a DNA bank (see below).

Table 1.2. Temperate food and forage in the six major *ex situ* collections.

	Crop Species	Crop Latin Name	Number of Accessions						
			ICARDA	USDA	Australian	IPK-I	AARI	VIR	Bari
Food Legumes	Pea	*Pisum sativum*	4893	4486	3683	3059	96	7983	4400
	Chickpea	*Cicer arietinum*	10158	4444	3114	296	1711	2631	336
	Faba bean	*Vicia faba*	5143	548	119	1580	339	1707	1950
	Lentil	*L. culinaris* subsp. *culinaris*	8267	2725	3954	329	564	3288	315
	Grass pea	*Lathyrus sativus*	1554	243	572	227	11	750	190
	Narbon bean	*Vicia narbonensis*	248	12	178	68	24	53	NA
Total			30263	12446	11620	5559	2745	16412	7191
Forage Legumes	Kidney vetches	*Anthyllis* sp.	47	75	4	21	0	NA	NA
	Milk vetches	*Astragalus* sp.	913	632	346	31	117	NA	NA
	Biserrula	*Biserrula pelecinus*	45	10	100	0	0	NA	NA
	Chickpeas	*Cicer* sp.	261	178	159	310	27	338	NA
	Crown vetches	*Coronilla* sp.	220	36	52	8	20	NA	NA
	Vetchlings	*Lathyrus* sp.	2877	674	139	496	730	906	NA
	Lentils	*Lens* sp.	467	158	138	352	39	26	NA
	Trefoils	*Lotus* sp.	138	983	172	83	13	NA	NA
	Lupins	*Lupinus* sp.	168	1273	29	861	3	2970	104
	Medics	*Medicago* sp.	7674	7567	23336	289	740	2909	1154
	Sweet clovers	*Melilotus* sp.	131	920	26	48	48	NA	NA
	Sainfoins	*Onobrychis* sp.	803	597	309	72	179	NA	NA
	Serradellas	*Ornithopus* sp.	51	37	786	62	0	NA	NA
	Peas	*Pisum* sp.	4988	34	52	3218	55	NA	NA
	Clovers	*Trifolium* sp.	496	5965	9089	463	671	NA	487
	Fenugreeks	*Trigonella* sp.	648	247	216	139	35	NA	86
	Vetches	*Vicia* sp.	5334	1414	634	3264	1646	3138	2400
Total			25261	20269	35587	9717	4323	10287	4231

Note: ICARDA - International Centre for Agricultural Research in the Dry Areas, Syria; USDA, W-6 / Regional Plant Introduction Station, Washington, USA; Australian - incorporates the Australian Temperate Field Crop Collection, Horsham, Australia, South Australian Research and Development Institute, Adelaide, Australia and Centre for Legumes in Mediterranean Agriculture University of Western Australia, Perth, Australia; IPK, Institut für Pflanzengenetik und Kulturpflanzenforschung (IPK), Gatersleben, Germany; AARI - Aegean Agricultural Research Institute, Menemen, Turkey; VIR - Vavilov Research Institute of Plant Industry, St. Petersburg, Russia; and Bari - Istituto del Germoplasma, Bari, Italy. NA - not available.

- *Reserve management expertise* - building on staff expertise in field collection in the wild and management of live plants in the gardens they may also be involved in the writing of management plans for wild populations in various kinds of reserves.
- *Public awareness and education* - one of the traditional roles of a botanic garden has been to provide a 'spectacle' of exotic plants, as well as a display of economic plants. This function can be expanded with the help of contemporary media techniques, informative demonstrations and presentations, which could perform a very useful role in increasing public awareness of botanical diversity and the importance of its conservation.

Du Puy and Wyse Jackson (1995) state that there are approximately 100 botanic gardens in the Mediterranean, though the majority of these are in European countries. These gardens are likely to contain important legume resources, but establishing which gardens have which species may be problematic, though Botanic Garden Conservation International (BGCI) may be able to assist with basic collection information. The application of the additional conservation activities outlined above is likely to provide an important opportunity for genetic conservation of legume diversity.

Although much progress has been made in collecting and conserving legume genetic resources *ex situ*, there has been no systematic attempt made to conserve temperate food or forage species *in situ* in either genetic reserves or on farm. Undoubtedly, existing protected areas in the Mediterranean Basin, such as in the Ammiad reserve in eastern Galilee, Israel (Anikster, 1995; Anikster *et al.* 1997), contain legume species, but their conservation is 'passive'. Such protected areas are likely to have been established to protect a landscape or conserve a rare animal, not to conserve legume plant genetic resources. The management regime is not likely to include monitoring of legume diversity, so any deleterious trend in legume populations is likely to pass unnoticed. This form of conservation can be said to be passive as it involves no active conservation effort, leaving legume populations susceptible to genetic erosion and taxonomic extinction.

Maxted (1995) proposed the first sites for the establishment of genetic reserves to conserve Vicieae species *in situ* in Syria and Turkey, but these proposals were not initially acted upon. The first positive initiative to conserve legumes in genetic reserves resulted from the Global Environment Facility (GEF) project in Turkey (Ertug Firat and Tan, 1997; Kaya *et al.*, 1998). Three reserves were established at Kaz Dag (Aegean Anatolia), Amanos, (southern Turkey) and Ceylanpinar (in south east Turkey). At each site, legume species populations were surveyed and are still being monitored. The recently established GEF project in the Fertile Crescent countries of Syria, Lebanon, Jordan and Palestine (Conservation and Sustainable Use of Dry Land Agro-biodiversity) is in the process of establishing genetic reserves, and one group of target taxa to be conserved within these reserves will be food and forage legumes. Interestingly, for Syria the sites selected will include those proposed by Maxted (1995). The newly established GEF project for the Fertile Crescent countries will also involve the establishment of *in situ* on farm conservation projects for food legumes within the region. There are pockets of traditional agriculture in all four countries included in the project but this is the first attempt to conserve legume genetic diversity on farm within the Mediterranean region. There are also several other current GEF projects (http://www.undp.org/gef/portf/bio.htm) in the region, see Table 1.3, which are less directly related to PGR conservation but within which general PGR and legume PGR are undoubtedly 'passively' conserved. Similar allied plant conservation with a more ecological or habitat, rather than genetic, conservation bias is being promulgate by the UNESCO Man and Biosphere Program. Figure 1.3 shows where existing biosphere reserves are placed in the Mediterranean region, and these will undoubtedly foster 'passive' legume diversity conservation.

Figure 1.3. UNESCO biosphere reserves found in the Mediterranean region (www.unesco.org/mab/brlist.htm).

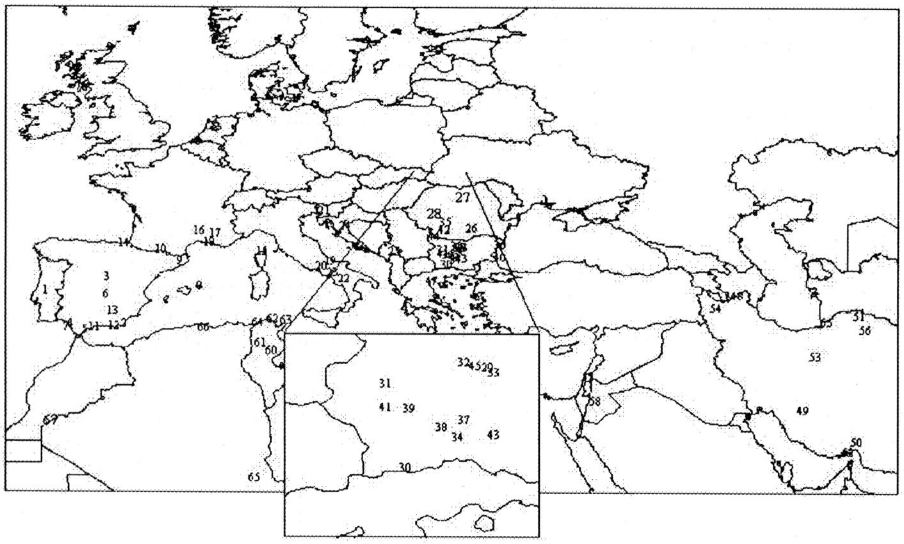

Note. Biosphere reserves are as follows: Algeria - El Kala (64), Tassili (65), Djurdjura (66); Bulgaria - Steneto (29), Alibotouche (30), Bistrichko Branichte (31), Boatione (32), Djendema (33), Doupkata (34), Doupki-Djindjiritza (35), Kamtchia (36), Koupena (37), Mantaritza (38), Maritchini ezera (39), Ouzounboudjak (40), Parangalitsa (41), Srebarna (42), Tchervenata stena (43), Tchouprene (44), Tsaritchina (45); Croatia - Velebit Mountain (24); Egypt - Omayed Experimental Research Area (59); France - Vallee du Fango (15), Cevennes (16), Luberon (17), Camargue (18); Greece - Gorge of Samaria (46), Olympos (47); Iran - Arasbaran (48), Arjan (49), Geno (50), Golestan (51), Hara (52), Kavir (53), Lake Oromeeh (54), Miankaleh (55), Touran (56); Israel - Mount Carmel (57); Italy - Collemeluccio-Montedimezzo (19), Circeo (20), Miramare (21), Cilento and Vallo di Diano (22), Somma-Vesuvio and Miglio d'Oro (23); Jordan - Dana (58); Morocco - Arganeraie (67); Portugal - Paul do Boquilobo (1); Rumania - Danube Delta (26), Pietrosul Mare (27), Retezat (28); Spain - Cabo de Gata-Nijar (2), Cuenca Alta del Rio Manzanares (3), Donana (4), Grazalema (5), Mancha Humeda (6), Marisma del Odiel (7), Menorca (8), Montseny (9), Ordesa-Vinamala (10), Sierra de las Nieves y su Entorno (11), Sierra Nevada (12), Sierras de Cazorla y Segura (13), Urdaibai (14), Tunisia - Bou-Hedma (60), Chambi (61), Ichkeul (62), Zembra and Zembretta (63); and Yugoslavia - Tara River Basin (25). Numbers in brackets indicates reserve numbers in Figure 1.3.

Table 1.3. Global Environmental Facility Plant Conservation Projects in the Mediterranean region.

Country	Title	Project Summary
Armenia	*In-situ* Conservation and Sustainable Use of Agrobiodiversity	The objective of the project is to develop an integrated approach to *in-situ* conservation of plant agrobiodiversity in Armenia. The project will combine *in situ* conservation of crop wild relatives in protected areas with the strengthening of conservation within traditional agricultural systems through farmer extension work on landrace management. The approach to be developed will remove barriers that are preventing the conservation of biodiversity importance to agriculture.
Egypt	Plantation of Green-belts Around 6th October City	Project to introduce greater diversity of trees and vegetation around the city.
	Surveying, Recording, Planting and Conserving Medicinal Plants	Project to survey and record medicinal plants in north Sinai, provide seeds and seedlings to local farmers, and assist farmers marketing medicinal plants.
	Cultivation of a Forest at Wady El Rayan	Project to include tree planting, animal protection and raising environmental awareness.
	The Conservation and Sustainable Use of Medicinal Plants in Arid and Semi-Arid Ecosystems of Egypt	Project's strategy is to protect medicinal plant diversity in three national priority areas focusing on habitats and species presenting global significance, and facilitating information exchange and to incorporate the conservation of medicinal plants into development plans and policies.
Georgia	Arid and Semi-arid Ecosystem Conservation in the Caucasus	Protection of biodiversity in the arid and semi-arid zone in Eastern Georgia is the objective of this project. The project is designed to ensure local land users participation in the design of alternative land uses, and their integration in its implementation. The project will complement and enhance proposed protection activities in the target area and coordinate these with neighbouring countries sharing sections of the ecosystem.
Jordan	Conservation of the Dana and Azraq Protected Areas	The project provides support for environmental management at two key protected areas and enables RSCN (NGO) to initiate similar initiatives elsewhere in the country and to provide environmental education to all segments of Jordanian society.
	Final Consolidation and Conservation of Azraq Wetlands and Dana Wild-lands	Conservation of Dana and Azraq Protected Areas to address the new threat (previously unforeseen) and pressures brought about by the Peace Accord in the Dana and Mujib Reserves (as strategic nodes in the Rift Valley ecosystem) and the Azraq Wetland Reserves.
	Wild Flowers of Jordan	Project to survey the wild flowers of Jordan and use the result to raise environmental awareness
	Conservation of Biodiversity in Petra National Park	Project to inventory the region, identify threats to biodiversity and assess socio-economic needs in the region.
Lebanon	Strengthening of National Capacity and Grassroots In-Situ Conservation for Sustainable Biodiversity Protection	The project will put into place an effectively managed system of protected areas to safeguard endemic and endangered species of flora and fauna, conserve their habitats and incorporate biodiversity conservation as an integral part of sustainable human development. The project will promote both the long term ecological and the short term economic objectives of wildlife conservation.
	National Biodiversity Strategy and Action Plan and Report to the CBD	The project will assist the Government of Lebanon, the Ministry of Environment, in developing a National Biodiversity Sustainable Action Plan within the framework of the CBD.

Morocco	Integrated Pastoral Range Management for Biodiversity Conservation and Sustainable Development	This project aims at formulating a project brief for biodiversity conservation and sustainable development through integrated pastoral range management in the drylands of the Atlas Mountains of Morocco
	Trans-humans for Biodiversity Conservation in the Southern High Atlas	The project will conserve globally significant biodiversity in the southern flank of the High Atlas through an adaptive management scheme integrating pastoral range management with biodiversity conservation in a grazing-dependent ecosystem. Threats to biodiversity are rooted in imbalanced incentives towards indiscriminate settlement, conversion of wetlands and common pastures for crops, reduced mobility of livestock, and lack of awareness.
Syria	National Biodiversity Strategy and Action Plan and Report to the CBD	The project aims at defining priority action to conserve and sustainably use Syria's biodiversity in the context of its economic development.
Tunisia	Management of the Touati Nature Reserve	Project to raise environmental awareness and mobilize community action to assist conservation management of the reserve.
	Environmental Program for Boukorine	Project to survey Boukorine National Park, develop a GIS, raise environmental awareness and provide community based conservation training.
	Contribution to the Creation of a Reserve at Djebel El Faoura at Hammamet	Project to inventory the Djebel El Faoura forest and establish a reserve, raise environmental awareness and provide community based conservation training.
Turkey	*In Situ* Conservation of Genetic Diversity	Project to inventory potential reserve sites and establish a reserves, raise environmental awareness and provide conservation training.
Regional	Germplasm for Arid Lands	Germplasm for Arid Lands This project will focus on the identification and conservation of endemic species of arid and semi-arid lands which are under threat with extinction, especially from excessive and unmanaged rangeland use and overgrazing.
	Participatory Management of Plant Genetic Resources in Oases of the Maghreb	The project focuses on *in situ* conservation of the indigenous (land races) species in the region that is important for maintaining oasis ecosystem stability and also contributes to soil stabilization, shade, and shelter from high winds. The project will provide *in situ* conservation of within-species diversity through partnerships with farmers, scientists, and government authorities.
	Conservation, Management and Sustainable Use of Dryland Biodiversity within the Near East	This project will promote the conservation, management and sustainable use of drylands agro-biodiversity in the Near East, through the development of productive and sustainable resource conservation and management strategies in priority agro-ecosystems.
	Biodiversity Strategy and Action Plan and Report to the CBD	The primary objective of the project is the elaboration of a Biodiversity Strategy and Action Plan for the conservation and sustainable use of biodiversity.
	Conservation and Sustainable Use of Dryland Agro-Biodiversity of the Fertile Crescent	The overall, long term global objective of the project is to ensure the continuous availability of agro-biodiversity in the Fertile Crescent that is essential, not only to the sustainable development of agriculture in that region, but also for global food security and production.

The Species Survival Commission of IUCN - The World Conservation Union has established a Specialist Group for the Mediterranean Islands and they have generated a Strategy for Action (Delanoë *et al.*, 1996). The group's objectives are:

- to evaluate and monitor changes in native plant diversity,
- to establish, co-ordinate and implement conservation action plans,
- to promote sustainable conservation of flora and habitats and increase public awareness of conservation issues.

Obviously, the work of the Mediterranean Islands Plant Specialist Group is not restricted to conserving legume species, but the group's specific recommendations and individual action plans will promote the conservation of the Mediterranean flora and thus all legume species.

1.4.3 Locating Conserved Legume Material

To exploit conserved legume diversity it is necessary to first know what is conserved, what legume accessions are currently conserved *ex situ* or *in situ* and where. This conserved material could be held in botanic gardens, conservation laboratories, genetic reserves and on farm systems, but for legume species is most likely to be found in gene banks. What material is currently held in which gene banks can be determined by consulting individual gene bank catalogues, directories and databases. Perry and Bettencourt (1995) review how to obtain information from these sources.

IBPGR (now IPGRI) published a series of international catalogues of *ex situ* germplasm holdings: these include volumes on Food Legumes (Bettencourt *et al.*, 1989), which includes information on *Arachis*, *Cajanus*, *Cicer*, *Lens*, *Lupinus*, *Phaseolus*, *Pisum*, *Psophocarpus*, *Vicia* and *Vigna* collections, and Forages (Bettencourt *et al.*, 1992), including information on *Adenocarpus*, *Anthyllis*, *Astragalus*, *Biserrula*, *Coronilla*, *Ebenus*, *Hedysarum*, *Hippocrepis*, *Hymenocarpos*, *Lathyrus*, *Lotononis*, *Lotus*, *Medicago*, *Melilotus*, *Onobrychis*, *Ononis*, *Ornithopus*, *Securigera*, *Scorpiurus*, *Trigonella*, *Trifolium* and other legume collections. These directories were compiled from data provided by curators of collections in national and international institutions and from data held in the international *ex situ* collections database administrated by the IPGRI. Searching each of these individually may prove time consuming. To help alleviate this problem national and international directories and databases of conserved material from multiple collections have been and are being established by a variety of national and international agencies, see below for details.

The other CG Centres continue to publish detailed breeder's catalogues, which contain more detailed passport and characterisation data, but these are usually focussed on a single crop. However, the unfortunate feature of published catalogues or even databases (if they are not updated regularly) is that they rapidly become out of date. Therefore, it is difficult to compile a complete and up-to-date table of the world's conserved gene bank holdings for legumes, because by their very nature, these collections are housed in numerous national, regional and international gene banks throughout the world, and if they are operated efficiently the collections will be dynamic; dynamic in the sense that new accessions are added frequently and other accessions are being sent out to users.

However, published catalogues and databases are at least useful pointers to which gene banks contain significant legume collections. Querying an international collection database, possibly via the World Wide Web, is a more efficient means of locating current holdings, providing it is regularly updated. Example of such databases are:

- Consultative Group In Agricultural Research (CGIAR) - www.cgiar.org/
- European Cooperative Programme on Plant Genetic Resources (ECP/GR) - http://www.cgiar.org/ecpgr/platform/index.htm)
- Genetic Resources Information Network (GRIN) - http://www.ars-

grin.gov/npgs/acc/acc_queries.html

- International Centre for Agricultural Research in Dry Areas (ICARDA) - www.cgiar.org/icarda/
- International Crops Research Institute for the Semi-Arid Tropics (ICRISAT)- - www.cgiar.org/icrisat/
- International Plant Genetic Resources Institute (IPGRI) - www.cgiar.org/ipgri/
- System-wide Information Network for Genetic Resources (SINGER) - http://www.cgiar.org/singer
- World Information and Early Warning System on Plant Genetic Resources (WIEWS) - www.fao.org/ag/agp/pgr/wiews/

The holdings of all of the CGIAR centres are recorded in and can be queried through the SINGER database, which is available on CD-Rom or via the Internet.

As these and other databases become more widely available, so the potential germplasm user will be able to more easily assess where a taxon is conserved, and the actual amount of genetic variation conserved. The potential germplasm user should also check the recent Plant Genetic Resources literature (e.g. Genetic Resources and Crop Evolution, Plant Genetic Resources Newsletter, "Beanbag") to ascertain where recent legume collection missions have been focussed, what was collected and where the collected material has been deposited.

However, care must be taken when interpreting information on current gene bank holdings or other forms of conserved legume diversity (Maxted *et al.*, 1995). The material held may be misidentified, though it should be possible to check the identification by consulting voucher material or asking for the material to be grown-out and identified from living material. The actual size and number of collections available could also be misleading: gene banks and botanical gardens are encouraged to duplicate their collection in other gene banks or botanical gardens, therefore duplicated accessions can give a false impression of the genetic diversity conserved. Some questions that need to be addressed when assessing conserved diversity are:

- what is the level of diversity contained in the collections?
- do different accession numbers indicate truly different accessions or are they just duplicate under a different name?
- is there adequate passport data is available? (older collections usually are associated with poor background data and these accessions are less likely to be utilised)
- are the collections 'good' genetic samples?
- were they sampled in an effective manner to maximise genetic variation or has variation been lost since sampling through
  a) genetic drift when sampling small samples,
  b) unsuitable regeneration conditions (selection in regeneration plots / out pollination),
  c) poor gene bank storage resulting in differential erosion,
  d) human errors, such as mislabeling of samples?

The potential germplasm user should consider each of these factors when choosing which accessions to utilise.

1.5 MEDITERRANEAN LEGUME CROP EXPLOITATION

A summary of the basic information for temperate grain legumes is given in Table 1.4, including their common name, wild relatives, centres of cultivation, uses and latest production values, and similar information for forage legume are listed in Table 1.5.

Taking the utilisation activities of ICARDA as an example of the utilisation of conserved temperate legume diversity, we can see that germplasm collections at ICARDA of faba bean, chickpea, lentil and forage legumes have been extensively exploited by national breeding programmes in South West Asia and North Africa and elsewhere, for the production of improved cultivars. The first lines distributed were selected directly from the germplasm collections after initial evaluation. This formed the first flush of releases by national programmes, for example, 56 cultivars of temperate legumes have been released by ICARDA from landraces or ecotypes from the germplasm collections (see Table 1.6). Many of these are selections that have been stabilised for such traits as seed size, disease resistance, frost tolerance, non-shattering, etc. More recently the majority of breeders lines distributed through the ICARDA International Testing Network are the product of hybridisation programmes based on crosses involving germplasm accessions found resistant or tolerant to one or more biotic and/or abiotic stress factors.

As discussed above, before conserved germplasm can be utilised it must first be characterised and evaluated. A systematic evaluation of food legumes for a wide range of morpho-agronomic characters based on the IBPGR / ICARDA descriptors, has led to publication and widespread distribution of catalogues which, in turn, have led to extensive use of the germplasm by national breeding programmes. The germplasm collections of faba bean, kabuli chickpea and lentil have largely been evaluated for their respective IBPGR / ICARDA descriptor lists and the results have been summarised in published catalogues (Erskine and Witcombe, 1984; Robertson and El-Sherbeeny, 1988; Robertson *et al.*, 1995; Singh *et al.*, 1983, 1991). Descriptors such as seed size, pods per plant, seeds per pod, seed shape, seed colour, testa colour, testa pattern, cotyledon colour, growth habit, and protein quantity have been evaluated and summarized in the catalogues. Germplasm has been scored for biotic stress resistances such as reaction to aphids, ascochyta blight, and *Orobanche crenata* and on seed and straw yield. This characterisation and evaluation has enhanced the value of the conserved germplasm and obviously resulted in increased germplasm use by national breeding programmes.

The most important use made of the germplasm collections has been their exploitation as a source for resistance and tolerance to biotic stresses. Selections for disease resistance sources in faba bean for chocolate spot (induced by *Botrytis fabae* Sard.), ascochyta blight (induced by *Ascochyta fabae* Speg.), rust (induced by *Uromyces fabae* (Pers.) de Bary) and stem nematodes (*Ditylenchus dipsaci* (Khhn) Filipjev) had led to the development of homogeneous resistance sources (see various annual ICARDA reports). This material was tested internationally and several sources were found to have durable resistances for chocolate spot, rust and ascochyta blight (Hanounik and Robertson 1987, 1988). Intensive efforts have been made to identify ascochyta blight resistant lines of chickpea and useful sources have been found (Singh and Reddy, 1993a), as with fusarium wilt (induced by *Fusarium oxysporum* Schlecht. emend. Synd. and Hans. f. sp. *ciceri* (Padwck) Synd. and Hans.) resistance (Jiménez-Díaz *et al.*, 1991, 1993). Rust (induced by *Uromyces fabae* (Pers.) de Bary) is the most important foliar disease of lentil and resistance has been found in Ethiopia, Morocco and Pakistan. While vascular wilt (induced by *Fusarium oxysporum* f.sp. *lentis* Vasd. and Srin.) is the most important soil-borne disease of lentil and useful sources of resistance have also been identified.

The Mediterranean Basin and South West Asia regions are characterised by extremes of temperature and moisture supply, and to deficiency or toxicity of mineral nutrients in the soil, and where winter planting could increase yield, cold tolerance was required and cold tolerant lines have been identified. Recently, effort has also been focused on drought tolerance in several crops and tolerance has been found.

Seed protein content varied between 18.6 to 30.2% in lentil (Erskine and Witcombe 1984), 18.0 to 31.0% for faba bean (Robertson and El-Sherbeeny, 1992) and in chickpea, 16.0 to 24.8% (Singh *et al.*, 1983). Protein content has not generally been a breeding objective for food legumes; the major activity has been monitoring of new lines developed to maintain the protein content found in existing cultivars.

Table 1.4. Major and minor temperate grain legume crops. (Production figures are for 1999 and taken from www.fao.org).

Name	Crops	Wild relatives	Centre of Cultivation	Uses	Production
Pea	*Pisum sativum* L.	*P. fulvum* Sibth. & Sm. *P. abyssinicum* A. Br.	Temperate world, but particularly in North America, Europe, Central and Western Asia, Mediterranean basin, Indian sub-continent and China and to a lesser extent some tropical regions.	Peas are eaten green, dried or preserved, green pods and foliage. It is also used for forage, ground cover, hay, silage and green manure.	Dry peas: 11.3Mt grown on 6.0Mha Green peas: 7.1Mt grown on 11.3Mha
Chickpea	*Cicer arietinum* L.	*C. echinospermum* P.H. Davis *C. reticulatum* Ladiz.	Temperate world, but particularly in Central and Western Asia, Ethiopia and Indian sub-continent and more recently the Americas and Australia.	Chickpeas are eaten fresh, dried, boiled or roasted while the green pods and young shoots can also be eaten as a vegetable	9.1Mt grown on 11.1Mha
Faba bean	*Vica faba* L.	*Vicia* sect. *Narbonensis*, *Vicia* sect. *Hypechusa*	Europe and Asia, North and South America and extending into Tropical East Africa	Beans are eaten green, dried or preserved, green pods and foliage, also as fodder, forage and as green manure	Dry peas: 3.0Mt grown on 2.1Mha Green peas: 1.0Mt grown on 0.2Mha
Lentil	*Lens culinaris* Medik. subsp. *culinaris*	*L. culinaris* subsp. *orientalis* (Boiss.) Ponert subsp. *tomentosus* (Ladiz.) M.E. Ferguson *et al.* subsp. *odemensis* (Ladiz.) M.E. Ferguson *et al.*	Indian subcontinent, Ethiopia, Mediterranean Basin and South West Asia and in recent times the Americas and Australia.	Lentils are eaten fresh, dried, boiled or roasted and are often consumed as dhal, soups or flour	3.0Mt grown on 3.4Mha
Lupin	*Lupinus* spp.	Other *Lupinus* species	Temperate world, but particularly in Central and Western Asia, Ethiopia and Indian subcontinent, the Americas and Australia.	Lupins are mainly grown for their high protein seed for animal and human consumption	1.5Mt grown on 1.3Mha
Grasspea	*Lathyrus sativus* L.	*Lathyrus* L. sect. *Lathyrus*	Indian subcontinent, Southern and south-central Europe, Near East and Ethiopia	Minor crop noted for its drought resistance, whose use has been limited by the presence of neuro-toxins	NA
Narbon bean	*Vica narbonensis* L.	*Vicia* L. subgen. *Vicia*	Southern Europe and South West Asia	Minor crop consumed as green or dried seeds	NA

Table 1.5. Temperate forage legume crops.

Common Name	Crops and Wild Relatives	Centre of Diversity	Ecology	Uses
Vetch Kidney	*Anthyllis* L. *A. vulneraria* L.	Europe and North Africa	Meadows and pastures, rocky ground, scrub and woodland.	Forage
Milkvetch Cicer Gum - -	*Astragalus* L. *A. cicer* L. *A. gummifer* Lab. A. hamosus L. *A. sinicus* Boiss.	Temperate Northern and Southern hemisphere, extending into the Arctic zone and the mountains of South America and Asia	Most species prefer open habitats with light, porous soils, steppe, desert, fields, roadsides, stony areas and woods.. Many are drought resistant.	Fodder, forage, land stabilisation, ornamental cultivation and gums
Biserrula	*Biserrula* L. *B. pelecinus* L.	Mediterranean basin	Sandy fields, waste ground	Forage, land stabilisation, ground cover
Chickpeas - -	*Cicer* L. *C. echinospermum* P.H. Davis *C. reticulatum* Ladiz.	Central and Western Asia. Endemics occur in Morocco, Ethiopia, Greece and Crete	Often drought-resistant and do not tolerate heavy rains, sandy or wet soils or hot climates, from sea level to 5600m, rocky ground, scrub, deciduous woodland, coniferous woodland, fallow, meadow or pasture.	Wild relative of crop
Vetch Crown	*Coronilla* L. *C. varia* L.	Europe and North Africa	Waste areas, roadsides, grasslands, rocky ground and scrub. They are best adapted to well-drained, fertile soils with a pH of 6 or more.	Forage, land stabilisation, ground cover
Vetchlings Annual - Hairy pea Broad everlasting pea - Sweetpea Narrow everlasting pea	*Lathyrus* L. *L. annuus* L. *L. cicera* L. *L. hirsutus* L. *L. latifolius* L. *L. ochrus* (L.) DC. *L. odoratus* L. *L. sylvestris* L.	Temperate northern and southern hemispheres, extends into tropical East Africa and South America	A wide range of habitats, including forests, macchie, rocky slopes, grassland and pastures, cultivated land (as a crop and weed), hedges, roadsides, water meadows, by streams or marshes. In general they are well adapted to rather dry areas but can tolerate water-logging, cool weather and soil with poor nutrients.	Fodder, forage, land stabilisation and ornamental cultivation

Lentils	*Lens* Miller	Mediterranean Basin and South Western Asia, while *L. ervoides* has outlying populations in Ethiopia and Uganda.	Species grow well on poor soils and loams. Typical habitats include rocky ground, steppe, scrub, coniferous woodland, fallow, meadow or pasture, and cultivated land.	Wild relative of crop
	L. culinaris Medik.			
-	subsp. *orientalis* (Boiss.) Ponert			
-	subsp. *tomentosus* M.E. FERGUSON *et al.*			
-	subsp. *odemensis* M.E. FERGUSON *et al.*			
Trefoils	*Lotus* L.	North temperate regions, extends to North and East Africa, Australia (2 endemics) and South America (1 species)	Diverse habitats, ranging from rocky ground, steppe, macchie, coniferous woodland, fallow, meadow or pasture, cultivated land to marsh or damp areas, and seashore.	Grown for hay, silage and pasture forage, it is winter-hardy, drought resistant, long-lived and is tolerant of less fertile soils
Birdsfoot	*L. corniculatus* L.			
Bigfoot	*L. ugliginosus* Schkuhr.			
Narrowleaf	*L. tenuis* Kit. ex Willd.			
Lupins	*Lupinus* L.	Two centres: the Andes and Rockies mountain chains, and secondly in the Mediterranean Basin and into tropical highlands of Africa.	Diverse habitats but most commonly found on dry sites, stony slopes, and calcareous and sandy-loam soils	Grown for ornament, fodder, forage, green manure, soil improvement
White	*L. albus* L.			
Blue	*L. angustifolius* L.			
European yellow	*L. luteus* L.			
Pearl	*L. mutabilis* Sweet			
Medics	*Medicago* L.	Europe, central and West Asia, and north Africa to Ethiopia, with a few species in tropical and South Africa	Species prefer deep, well-drained, light soils with a pH range of 6.5 - 7.5. They are well adapted to many climatic conditions and tend to be drought-resistant. Typical habitats include roadsides, cultivated fields and grasslands.	Important forage and fodder crops of warm temperate and cool sub-tropical regions with high yield and quality, wide soil and climatic adaptation and general flexibility
Spotted burclover	*M. arabica* (L.) Huds.			
Sickle	*M. falcata* L.			
Black	*M. lupulina* L.			
Disk	*M. orbicularis* (L.) Bart.			
Toothed burclover	*M. polymorpha* L.			
Alfalfa	*M. sativa* L.			
Snail	*M. scutellata* (L.) Miller			
Barrel	*M. truncatula* Gaertn.			
Sweet-clover	*Melilotus L.*	Europe, central and South West Asia, and North Africa	Species prefer temperate to warm climates, are moderately winter-hardy and drought-resistant. Typical habitats include rocky ground, cultivated land, and the seashore.	Cultivated as forage, fodder and silage, land stabilisation and green manure
White	*M. alba* Desr.			
Indian	*M. indica* (L.) All.			
Yellow	*M. officinalis* (L.) Desr.			

Sainfoin	*Onobrychis* Adans.	Europe, central and South West Asia, and North America	Banks and grassy places (robust perennials), stony dry places, steppe, woodland, cultivated land, seashore and roadsides (slender annuals). The plants prefer light, well-drained calcareous soils in areas with low precipitation.	Grown for fodder and forage in areas of low rainfall. They tend to be resistant to frost.
Common	*O. viciifolia* Scop.			
Serradella	*Ornithopus* L.	Europe, central and South West Asia, and North America	Species prefer sandy loam and slightly acidic soils. Typical habitats include fallow, meadow or pasture and seashore.	Grown for forage, green manure and as a cover crop
Common	*O. sativus* Brot.			
Peas	*Pisum* L.	Mediterranean region, Western Asia and Ethiopia	Species prefer well-drained soils, reasonable fertility, with a pH between 5.5 and 7. Habitats include rocky ground, scrub, fallow, meadow or pasture, cultivated land, and roadside.	Wild relative of crop
-	*P. fulvum* Sibth. & Sm.			
Abyssinian	*P. abyssinicum* A. Br.			
Clovers	*Trifolium* L.	Europe, North, West and Central Asia, North Africa, tropical and South Africa, North and South America	Within this large genus are species adapted to most habitats, ranging from sub-Arctic to tropical. Each species tends to be restricted to specific habitats. At least two-thirds of the species are limited to natural habitats. The genus is found from sea level to over 4000m.	Cultivated as forage, fodder, green manure and soil erosion control species, they are important nitrogen-fixers, improving the quality of pasture in both natural and cultivated grassland
Honey	*T. ambiguum* Bieb.			
Strawberry	*T. fragiferum* L.			
Rose	*T. hirtum* All.			
Alsike	*T. hybribum* L.			
Crimson	*T. incarnatum* L.			
Zigzag	*T. medium* L.			
-	*T. michelianum* Savi			
Ball	*T. nigrescens* Viv.			
Red	*T. pratense* L.			
White	*T. repens* L.			
Persian	*T. resupinatum* L.			
Subterraneum	*T. subterraneum* L.			
Arrowleaf	*T. vesiculosum* Savi			
Fenugreeks	*Trigonella* L.	Europe, Mediterranean Basin, Central and South West Asia	Species prefer loam soils which are well-drained, rocky ground, steppe, scrub, macchie, coniferous woodland, fallow, meadow or pasture, cultivated land, roadsides, and seashores.	Cultivated as fodder, forage and as green manure
Common	*T. foenum-graecum* L.			

Vetches	*Vicia* L.	Europe and Asia, North and South America and extending into Tropical East Africa. Production of common vetch is 1.0Mt grown on 1.1Mha	Species of *Vicia* thrive in many habitats and from sea level to over 3500m. They prefer well-drained, sandy soils and upland loams. Typical habitats include rocky ground, scrub, meadows or pasture, cultivated land, waste areas, roadsides, by rivers and in open woodland.	Cultivated as fodder, forage and as green manure
Purple	*V. benghalensis* L.			
Bitter	*V. ervilia* (L.) Willd.			
-	*V. noeana* Reuter ex Boiss.			
Hungarian	*V. pannonica* Crantz			
Common	*V. sativa* L.			
Winter	*V. villosa* Roth			

Table 1.6. Distribution of legume germplasm from ICARDA collection during 1990-1996. (Maxted *et al.*, 2000).

Crop	1990	1991	1992	1993	1994	1995	1996	Total
Food Legumes								
Faba bean	293	60	1278	2705	413	814	221	5784
Chickpea	1103	220	1578	1239	3019	4047	4392	15604
Wild *Cicer* sp.	273	181	613	385	748	568	514	3282
Lentil	1098	2006	1941	1889	2869	1807	2137	13747
Wild *Lens* sp.	103	340	464	33	50	228	166	1384
Sub-total	**2870**	**2807**	**5880**	**6251**	**7099**	**7464**	**7430**	**39801**
Forage Legumes								
Lathyrus sp.	228	89	769	854	877	265	569	3651
Annual *Medicago* sp.	313	500	145	987	1645	571	184	4345
Pisum sp.	21	56	74	15	93	107	143	509
Trifolium sp.	96	74	39	101	353	28	385	1076
Vicia sp.	398	778	701	3292	1761	3310	483	10723
Other forages sp.	-	10	-	255	432	715	569	1981
Sub-total	**1056**	**1507**	**1728**	**5504**	**5161**	**4996**	**2333**	**22285**
Total	**3926**	**4314**	**7608**	**11755**	**12260**	**12460**	**9763**	**62086**

Special attention has been given in *Lathyrus* spp. to evaluation for low neuro-toxin content in the germplasm collection (Aletor *et al.*, 1994). Screening results indicated that none of the *Lathyrus* spp. lines was ODAP-free, although in several the ODAP content was very low. This also seems to be a problem in related species, since samples of *L. cicera* ranged from 0.01 to 0.22% with a mean of 0.16% (Aletor *et al.*, 1994). *L. sativus* showed the biggest range from 0.16 to 0.74% with a mean of 0.48%, while *L. ochrus* lines were highest in ODAP, ranging from 0.46 to 0.67% with a mean of 0.57%. There appears to be, however, good potential for breeding *L. sativus* and *L. cicera* lines with low ODAP content.

In addition to maintaining collections of the cultigens, several gene banks have also assembled large collections of wild relatives and progenitors of lentil and chickpea, and wild and weedy forms of the forage legume species. In the past five years the wild *Lens* and *Cicer* collections at ICARDA have been evaluated for resistances to biotic and abiotic stresses. Sources of resistance have been found in wild *Cicer* species for ascochyta blight, fusarium wilt, leaf miner, seed beetle, cyst nematode, cold, and drought (Di Vito *et al.*, 1988; Kaiser *et al.*, 1994; Singh and Reddy, 1993b; Singh and Weigand, 1994; Singh *et al.*, 1989, 1990). Wild species have proven to be the only source of resistance so far found for seed beetle and cyst nematode; and they have a higher level of resistance than the cultivated species for fusarium wilt, leaf miner, and cold. Accessions of the wild progenitor of lentil *Lens culinaris* subsp. *orientalis* were found to have high levels of winter-hardiness than have been found within *L. culinaris* ssp. *orientalis* (Hamdi *et al.*, 1996); as well as being more drought resistant than the cultigen with their low relative reduction in yield with drought stress (Hamdi and Erskine, 1997). There are also sources of resistance to vascular wilt and Ascochyta blight in wild *Lens* species (Bayaa *et al.*, 1994; 1995).

1.6 CONCLUSIONS

There is a clear need to improve the link between those conserving and using legume diversity. Simmonds (1962) was the first to point out that mismanaged germplasm collections were largely worthless and were effectively no more than poorly displayed museum exhibits. How can we expect the public, who fund the major legume collection's to continue their support and provide sustainable conservation unless we ensure a better link between conservation and use? Given (1994) has estimated that 65% of conserved germplasm lacks basic passport data, 80% lacks characterisation and 95% lack evaluation data - which leaves only approximately 1% which are correctly catalogued and ready for use. Perhaps one of the best ways of ensuring this linkage is to ensure a seamless gradation of conservation into utilisation. Maxted *et al.* (2000) provide a list of recommendations that they believe will assist in meeting this goal. They are summarised below:

- **Conservation Planning**
  - *Choice of Target Taxa and Area* - The more efficient these choices, the more likely it is that the material collected will be utilised. It is often advisable to include potential germplasm users in the group that actually decide target taxa and areas, as well as ensuring current conservation gaps are filled.
- **Field Conservation**
  - *Efficient Population Sampling* - The better the quality of the sample transferred *ex situ* the more likely it is to be utilised.
  - *Collect Associated Materials* - Seed collecting is not simply restricted to collecting seed; an efficient collector will also collect passport data, herbarium voucher specimens and possibly vegetative plants and rhizobia cultures.
  - *Quality Collection of Passport Data* - The more complete the passport data collected, the more useful the germplasm accession will prove to users.
- **Gene Bank Based Conservation**
  - *Collected Material Processing* - The materials collected require appropriate and effective storage.
  - *Collecting Reports and Other Publications* - Following the expedition the commissioning agency will require an expedition report, but the existence of novel diversity can be signalled to potential users by publishing collecting reports.
  - *Efficient Storage* - Samples from mis-managed collections are less likely to be of use.
  - *Duplication* - Use of the collected material can be increased by duplication in national, regional and international collections.
  - *Collection Advertising* - Once the collection is safely held in the gene bank, the gene bank manager must be pro-active and advertise the presence of the new accessions in the collection. *Characterisation and Evaluation* - Any pre-breeding will enhance the utilisation of the accessions.
  - *Developing Core Collections* - With the increasing size of many collections and the limited funds available for characterisation and evaluation, it may be necessary to develop a core collection of representative samples.
  - *Germplasm Holdings Reports and Other Publications* - Any characterisation and evaluation results should be published to ensure the potential users are aware of the diversity of material available and attention should be drawn to any extreme forms.
  - *Quality Service to Users* - The quality of the service provided to germplasm users by germplasm curators will affect the potential for utilisation.
- **Linking Conservation and Utilisation Sites**
  - Unfortunately the people who conserve and utilise germplasm are seen as distinct professions, often located in two distinct locations; however, utilisation can be improved by bringing conservationists and germplasm users physically as well as professionally together.
- **Staff training**
  - All of the above actions to improve utilisation of conserved germplasm are only feasible if the gene bank is staffed by appropriately trained personnel.

The points discussed above are extensive, and in many cases, it is possible that the majority of 'good' plant collectors, gene bank managers and germplasm users are already applying them.

Sustainability in the sense of continuance is a fundamental concept for all conservation. Whether practising legume conservation *ex situ* in a gene bank or *in situ* in a reserve, each option has a financial cost and it would be wasteful of limited conservation funds not to ensure the conservation project is sustainable at least in the medium term. If the material is lost from a reserve the resources expended on establishing the reserve would have been wasted and the cost of rehabilitating populations using materials stored *ex situ* would have to be considered. The latter option is commonly expensive and may require extensive research to ensure the reintroduced materials do not likewise go extinct. Unfortunately, currently many conservation projects are funded on a short-term basis, and it is therefore essential that an effective conservation strategy be developed so the conservation programme itself is sustainable. This should ensure the public will continue their long term support for field conservation and germplasm collections, as well as underwriting conservation and sustainable utilisation for the long term benefit of all humankind.

REFERENCES

Adham, Y., Bari, A. and Jaradat, A.A., 1995. West Asian and North African countries establish WANANET - a regional genetic resources network. *Diversity*, 11(1 and 2): 29-30.

Aletor, V.A., A. M. Abd El Moneim and A. V. Goodchild, 1994. Evaluation of the seeds of selected lines of three *Lathyrus* spp. for BOAA, tannins, trypsin inhibitor activity and certain *in vitro* characteristics. *Journal of Science, Food and Agriculture*, 65: 143-151.

Anikster, Y., 1995. Ammiad: a model for in situ conservation ... and peace in the Middle East. *Diversity*, 11(1 and 2): 59.

Anikster, Y., Feldman, M., and Horovitz, A., 1997. The Ammiad experiment. In: *Plant genetic conservation: the* in situ *approach* (eds. Maxted, N., Ford-Lloyd, B.V. and Hawkes, J.G.), pp. 239-253. Chapman and Hall, London.

Bayaa, B. and W. Erskine, 1990. A screening technique for resistance to vascular wilt in lentil. *Arab Journal of Plant Pathology*, 8: 30-33.

Bayaa, B., W. Erskine and A. Hamdi, 1994. Response of wild lentils to *Ascochyta fabae* f. sp. *lentis* from Syria. *Genetic Resources and Crop Evolution*, 41: 61-65.

Bayaa, B., W. Erskine and A. Hamdi, 1995. Evaluation of a wild lentil collection for resistance to vascular wilt. *Genetic Resources and Crop Evolution*, 42: 231-235.

Bennett, S. and Maxted, N., 1997. An ecogeographic analysis of the *Vicia narbonensis* complex. Genetic resources and crop evolution. *Genetic Resources and Crop Evolution*, 44: 411-428.

Bennett, S.J., Maxted, N., Kitiki, A. and Francis, C., 1999. Collecting and ecogeography of pasture, forage and grain legumes in Northwest Turkey. *Plant Genetic Resources Newsletter*, 119: 33-36.

Bennett, S., Maxted, N. and Sabanci, C.O., 1998. The ecogeography and collection of grain, forage and pasture legumes in south-west Turkey. *Genetic Resources and Crop Evolution*, 45: 253-269.

Bettencourt, E., J. Konopka, and A.B. Damania, 1989. *Directory of Crop Gerrmplasm Collections. 1. I. Food Legumes. Arachis, Cajanus, Cicer, Lens, Lupinus, Phaseolus, Pisum, Psophocarpus, Vicia* and *Vigna*. International Board for Plant Genetic Resources, Rome.

Bettencourt, E., Th. Hazekamp and M.C. Perry, 1992. *Directory of Germplasm Collections. 7. Forages (Legumes, grasses, browse plants and others)*. International Board for Plant Genetic Resources, Rome.

Campbell, C.G., 1997. *Grasspea (*Lathyrus sativus *L.)*. Promoting the conservation and use of underutilised and neglected crops, 18. Institute of Plant Genetics and Crop Plant Research, Gatersleben / International Plant Genetic Resources Institute, Rome, Italy.

Convention on Biological Diversity, 1992. *Convention on Biological Diversity: Text and Annexes*. pp. 1-34. Secretariat of the Convention on Biological Diversity, Montreal.

Cook, F.E.M., 1995. *Economic Botany Data Collection Standards*. Royal Botanic Gardens, Kew.

Cowling, R.M., Runcel, P.W., Lamont, B.B., Arroyo, M.K. and Arianoutsou, M., 1996. Plant diversity in Mediterranean-climate regions. *Tree*, 11: 362-366.

Cowling, W.A., Buirchell, B.J. and Tapia, M.E., 1998. *Lupin (*Lupinus *L.)*. Promoting the conservation and use of underutilised and neglected crops, 23. Institute of Plant Genetics and Crop Plant Research, Gatersleben / International Plant Genetic Resources Institute, Rome, Italy.

Date, R.A., 1995. Collecting *Rhizobium, Frankia* and mycorrhizal fungi. In: *Collecting plant genetic diversity: technical guidelines* (eds. Guarino, L., Ramanatha Rao, V. and Reid, R.). pp. 551-560. CAB International, Wallingford.

Davis, P.H., 1965. *The Flora of Turkey and East Aegean Islands, Vol.1.* Edinburgh University Press: Edinburgh.

Davis, P.H., 1970. *The Flora of Turkey and East Aegean Islands, Vol.3.* Edinburgh University Press: Edinburgh.

Delanoë, O., de Montmollin, B. and Olivier, L., (compilers)., 1996.*Conservation of Mediterranean Island plants: 1. Strategy for action.* IUCN / SSC Mediterranean Islands Plants Specialist Group. IUCN, Gland, Switzerland and Cambridge, UK.

Di Vito, M., N. Greco, K.B. Singh and M.C. Saxena, 1988. Response of germplasm lines to *Heterodera ciceri* attack. *Nematol. Medit.*, 16: 17-18.

Diversity, 1999. International Botanical Congress calls for new UN agency to oversee biodiversity conservation. *Diversity*, 15(3): 7.

Duke, J.A., 1981. *Handbook of legumes of world economic importance.* Plenum Press, New York.

Du Puy, B. and Wyse Jackson, P., 1995. Botanic gardens offer key component to biodiversity conservation in the Mediterranean. *Diversity*, 11(1 and 2): 47-50.

Ehrman, T. and Cocks, P.S., 1990. Ecogeography of annual legumes in Syria: distribution patterns. *Journal of Applied Ecology*, 27: 578-591.

Erskine, W. and J.R. Witcombe, 1984. *Lentil Germplasm Catalogue.* ICARDA, Aleppo, Syria. 284 pp.

Ertug Firat, A. and Tan, A., 1997. *In situ* conservation of genetic diversity in Turkey. In: *Plant genetic conservation: the* in situ *approach* (eds. Maxted, N., Ford-Lloyd, B.V. and Hawkes, J.G.). pp. 254-262. Chapman and Hall, London.

Ferguson, M.E., Robertson, L.D., Ford-Lloyd, B.V., Newbury, H.J. and Maxted, N., 1998. Contrasting genetic variation amongst lentil landraces from different geographical origins. *Euphytica*, 102: 265-273.

Ferguson, M.E., Ford-Lloyd, B.V., Robertson, L.D., Maxted, N. and Newbury H.J., 1998. Mapping the geographical distribution of genetic variation in the genus *Lens* for the enhanced conservation of plant genetic diversity. *Molecular Ecology*, 7: 1743-1755.

Ferguson, M.E., Maxted, N., van Slageren, M. and Robertson, L.D., 2000. A re-assessment of the taxonomy of *Lens* Mill. (Leguminosae, Papilionoideae, Vicieae). *Botanical Journal of the Linnean Society*, 133: 41-59.

Francis, C., Maxted, N., Kyriakakis, S. and Bennett, S., 1995. Crete: observations on the ecogeography of grain, forage and pasture legumes. *CLIMA Occasional Papers,* No. 11. Centre for Legumes in Mediterranean Areas, Perth.

Frison, E. and Ager, H., 1995. European cooperative program: a focus on Mediterranean biodiversity. *Diversity*, 11(1 and 2): 30-32.

Given, D.R., 1994. *Principles and practice of plant conservation.* Chapman and Hall, London.

Greuter, W., 1991. Botanical divcersity, endemism, rarity, and extinction in the Mediterranean area: an analysis based on the published volumes of Med-Checklist. *Botanika Chronika,* 10: 63-79.

Hamdi, A., I. Kusmeasglu and W. Erskine, 1996. Sources of winter hardiness in wild lentil. *Genetic Resources and Crop Evolution*, 43: 63-67.

Hamdi, A. and W. Erskine, 1997. Reaction of wild species of the genus *Lens* to drought. *Euphytica* 91: 173-179.

Hanounik, S.B. and L.D. Robertson, 1987. New sources of resistance in *Vicia faba* to chocolate spot caused by *Botryis fabae. Plant Disease*, 72: 696-698.

Hanounik, S.B. and L.D. Robertson, 1988. Resistance in *Vicia faba* germplasm to blight caused by *Ascochyta fabae. Plant Disease*, 73: 202-205.

Harlan, J.R., 1995. Agricultural origins and crop domestication in the Mediterranean region. *Diversity*, 11: 14-16.

Hawkes, J.G., N. Maxted and B.V. Ford-Lloyd, 2000. *The* ex situ *conservation of plant genetic resources.* Kluwer, Dordrecht.

Hauck, R.D., 1988. A human ecosphere perspective of agriculture nitrogen cycling. In: *Advances in nitrogen cycling in agricultural ecosystems.* pp. 13-17. CAB International, Wallingford.

Heywood, V.H., 1978. *Flowering Plants of the World.* Oxford University Press: Oxford.

Heywood, V.H., 1995.The Mediterranean flora in the context of world diversity. *Ecologia Mediterranea,* 21: 11-18.

Heywood, V.H. and Zohary, D., 1995. A catalogue of the wild relatives of cultivated plants native to Europe. *Flora Mediterranea*, 5: 375-415.

Hong, T.D., Linington, S. and Ellis, R.H., 1998. *Compendium of Information on Seed Storage Behaviour*. Royal Botanic Gardens, Kew.

IBPGR, 1985. *Report from a Working Group on Forages for Mediterranean and Adjacent Semi-arid/Arid Type Areas.* International Board for Plant Genetic Resources, Rome.

Jiménez-Díaz, R.M., K.B. Singh, A. Trapero-Caspas, A. and J.L. Trapero-Casas, 1991. Resistance in kabuli chickpea to fusarium wilt. *Plant Disease*, 75: 914-918.

Jiménez-Díaz, R.M., P. Crino, M.H. Halila, C. Mosconi and A.T. Trapero-Casas, 1993. Screening for resistance to fusarium wilt and ascochyta blight in chickpea. In: *Breeding for Stress Tolerance in Cool Season Food Legumes.* (eds. Singh, K.B. and Saxena, M.C.). pp. 97-106. John Wiley and Sons, Chichester, UK.

Kaiser, W.J., A.R. Alcal-JimJnez, A. Hervs-Vargas, J.L. Trapero-Casas and R.M. JimJnez-DRaz, 1994. Screening of wild *Cicer* species for resistance to races 0 and 5 of *Fusarium oxysporum* f.sp. *ciceri*. *Plant Disease*, 78:962-967.

Kaya Z., Kün, E. and Güner, A. 1998. National plan for in situ conservation of plant genetic diversity in Turkey. In: *The proceedings of international symposium on* in situ *conservation of plant diversity* (eds. Zencirci, N., Kaya, Z., Anikster, Y. and Adams, W.T.), pp. 333-47. Central Research Institute for Field Crops, Ankara, Turkey.

Lugo, A.E., 1988. Estimating reductions in the diversity of tropical forest species. In: *Biodiversity* (ed. Wison, E.O.). pp. 58-70. National Academy Press, Washington DC.

Maxted, N., 1995. *An ecogeographic study of Vicia subgenus Vicia.* Systematic and Ecogeographic Studies in Crop Genepools 8. IBPGR, Rome. Pp. 184.

Maxted, N., van Slageren, M.W. and Rihan, J., 1995. Ecogeographic surveys. In: *Collecting plant genetic diversity: technical guidelines* (eds. Guarino, L., Ramanatha Rao, V. and Reid, R.). pp. 255-286. CAB International, Wallingford.

Maxted, N., Erskine, W., Robertson, L.D. and Asthana, A.N., 2000. *Are Our Germplasm Collections Museum Items?* Conference paper given at the International Food Legume Research Conference III, Adelaide, Australia, September, 1997.

Maxted, N., B.V., Ford-Lloyd, and J.G. Hawkes, 1997a. Complementary Conservation Strategies. In: *Plant genetic conservation: the* in situ *approach* (eds. N. Maxted, B.V. Ford-Lloyd and J.G. Hawkes), pp. 20-55. Chapman and Hall, London.

Maxted, N., J.G., Hawkes, L. Guarino and M.C. Sawkins, 1997b. The selection of taxa for plant genetic conservation. *Genetic Resources and Crop Evolution*, 44: 337-348.

Maxted, N. and Kell, S., 1998. Ecogeographic techniques and *in situ* conservation: a case study for the legume genus *Vicia* in Turkey. In: *The proceedings of international symposium on* in situ *conservation of plant diversity* (eds. Zencirci, N., Kaya, Z., Anikster, Y. and Adams, W.T.), pp. 323-344. Central Research Institute for Field Crops, Ankara, Turkey.

Munzer, M., 1989. *Fodder development and rangeland improvement.* Tarla Bitkilen Merkez, Arastioma Enstitusus.

Onorati, A., 1995. The bottom line on farmer's rights: A view from the NGO community. *Diversity*, 11: 35-36.

Perry, M.C. and E. Bettencourt, 1995. Sources of information existing germplasm collections. In: *Collecting plant genetic diversity: technical guidelines* (eds. Guarino, L., Ramanatha Rao, V. and Reid, R.). pp. 121-129. CAB International, Wallingford.

Porqueddu, C. and Sulas, L., 1997. Grasslands of the Mediterranean. In: *Ecological Aspects of Grassland Management* (eds. Nagy, G and Peto, K). 17th Mtg. Grassland Science in Europe Vol. 3. Grasslands. pp. 335-345.

Quézel, P. 1985. Definition of the Mediterranean region and the origin of its flora. In: *Plant conservation in the Mediterranean area* (ed. Gómez-Campo, C.) W. Junk, Dordrecht.

Robertson, L.D. and M. El-Sherbeeny, 1988. *Faba Bean Germplasm Catalogue: Pure Line Collection.* ICARDA, Aleppo, Syria. 140 pp.

Robertson and El-Sherbeeny, 1992. *Faba Bean Germplasm Catalogue: Pure Line Collection.* ICARDA, Aleppo, Syria. 140 pp.

Robertson, L.D., K.B. Singh and B. Ocampo, 1995. *A Catalogue of Annual Wild* Cicer *Species.* Aleppo, Syria, ICARDA. 171 p.

Robertson, A.M., Maxted, N. and Van Der Maesen, L.J.G., 2000. A phenetic and phylogenetic investigation of the genus *Cicer* L. (Leguminosae, Cicereae). *Botanical Journal Of The Linnean Society*, In Press.

Sackville-Hamilton, N.R., 1999. Genetic erosion issues in temperate grasslands. In: *Proceedings of the technical meeting on the methodology of the FAO World Information and Early Warning System on Plant Genetic Resources.* FAO, Rome.

Simmonds, N.W., 1962. Variability in crop plants, its use and conservation. *Biological Reviews*, 37: 422-465.

Singh, K.B. and M.V. Reddy, 1993a. Resistance to six races of *Ascochyta rabiei* in the world germplasm collection of chickpea. *Crop Science*, 33: 186-189.

Singh, K.B. and M.V. Reddy, 1993b. Sources of resistance to ascochyta blight in wild *Cicer* species. *Netherlands Journal of Plant Pathology*, 99: 163-167.

Singh, K.B. and S. Weigand, 1994. Identification of resistance sources in *Cicer* species to *Liriomyza cicerina. Genetic Resources and Crop Evolution*, 41: 75-79.

Singh, K.B., L. Holly and G. Bejiga, 1991. *A Catalogue of Kabuli Chickpea Germplasm.* ICARDA, Aleppo, Syria. 398 pp.

Singh, K.B., R.S. Malhotra and M.C. Saxena, 1990. Sources of tolerance to cold in *Cicer* species. *Crop Science*, 30: 1136-1138.

Singh, K.B., R.S. Malhotra and J.R. Witcombe, 1983. *Kabuli Chickpea Germplasm Catalogue.* ICARDA, Aleppo, Syria. 284 pp.

Singh, K.B., M. Di Vito, N. Greco and M.C. Saxena, 1989. Reaction of wild *Cicer* species lines to *Heterodera ciceri*. Nematol. Medit. 17: 113-114.

Skerman, P.J., 1977. *Tropical forage legumes.* FAO Plant Production and Protection Series 2. FAO, Rome

Smartt, J., 1990. *Grain Legumes. Evolution and Genetic Resources.* Cambridge University Press: Cambridge.

Stanners, D. and Bordeau, P., 1995. *Europes Environment: The Dobris Assessment.* European Environment Agency, The Stationary Office Agencies, Brussels.

Summerfield, R.J. and Bunting, A.H., 1980. *Advances in legume science.* pp. 1-667. Royal Botanic Gardens: Kew.

Ucko, P.J., and Dimbleby, G.W.,1969. *The domestication and exploitation of plants and animals.* Duckworth, London.

Vavilov, N.I., 1997. *Five continents.* International Plant Genetic Resources Institute, Rome, Italy.

Walters, K.S. and Gillett, H.J., 1998. *1997 IUCN Red List of Threatened Plants.* Compiled by the World Conservation Monitoring Centre. pp. 1-862. IUCN - The World Conservation Union, Gland, Switzerland and Cambridge, UK.

Warren, J.M, Raybould, A.F., Ball, T., Gray, A.J. and Hayward, M.D., 1998. Genetic structure in the perennial grasses *Lolium perenne* and *Agrostis curtisii. Heredity*, 81: 556-562.

Willcox, G. 1995. Archaeobotanists sleuth out origins of agriculture from early Neolithic sites in the Eastern Mediterranean. *Diversity*, 11(1 and 2): 141-142.

Willcox, G., 1998. Archaeobotanical evidence for the beginnings of agriculture in Southwest Asia. In: *The origins of agriculture and crop domestication* (eds. Damania, A.B., Valkoun, J., Willcox, G. and Qualset, C.O.). pp. 25-38. ICARDA, Aleppo, Syria.

Woodgate, K., Maxted, N. And Bennett, S, 1999. A generic conspectus for the forage legumes of the Mediterranean basin. In: *Genetic resources of Mediterranean Pasture and Forage Legumes* (eds. Bennett, S.J. and Cocks, P.S.) pp. 182-226. Kluwer Academic Publishers, Dordrecht.

Yakovlev, G.P., Sytin, A.K. and Roskov, Y.R., 1996. *Legumes of Northern Eurasia.* pp. 1-724. Royal Botanic Gardens, Kew.

Zohary, D., 1991.Conservation of wild progenitors of cultivated plants in the Mediterranean Basin. *Botanika Chronika*, 10: 465-474.

Zohary, D. and Hopf, M., 1993. *Domestication of plants in the Old World*, Second Edition. Clarendon Press, Oxford.

Chapter 2

Ecogeographic Environment of the Mediterranean

Sarita Jane Bennett and Nigel Maxted

2.1 INTRODUCTION

The different climatic and edaphic factors, relief, vegetation cover, soils and the influence of man, that combine to make up the ecogeography of the Mediterranean region will be briefly described. Within each, the diversity present will be highlighted and the major reasons for the variation explained. It is important to understand that when looking at the ecogeography of the Mediterranean as a whole, that nearly any combination of these edaphic factors can be present. Thus, when planning a plant collecting mission, for example, if a particular species has previously been collected from a maquis area, or on terra rossa soils, it can not be assumed that it will be found in another part of the Mediterranean with maquis vegetation, or terra rossa soils. It is the whole ecogeography of both areas that has to be taken into account. This chapter is therefore intended to provide an introduction, and in some senses, an explanation for the diversity of Leguminosae in the Mediterranean. When reading chapters within the following section, 'Genetic diversity of species in the Mediterranean', be aware that in most cases the distribution of any species is dependent on the ecogeography of the area(s) in which it inhabits.

The Mediterranean Basin is a relatively small area by land mass that encompasses the coastal areas of three continents: Europe, Asia and North Africa, which border the Mediterranean Sea. It is characterised by warm to hot summers and mild, wet winters, with a summer drought of varying severity. This climate is dominated by shrub-land vegetation known as maquis (Di Castri 1981). However this is a broad generalisation as the ecosystems of the Mediterranean are very diverse in terms of vegetation structure and soil types, as well as the relief of the land. The Mediterranean climate is not however restricted to the Mediterranean region *sensu stricto*; this climate is also found in western North America, Central Chile, the Cape region of South Africa and south-western and southern Australia. Each of these five regions support a similar evergreen sclerophyll vegetation, through which fires often burn and which contain a high proportion of species with lignotubers. Each region has, however evolved its own distinctive flora, and there has been little direct floristic exchange between them (Raven 1971). For the purposes of this publication, we restrict our discussion to the Mediterranean Basin and surrounding region, an area that is particularly rich in temperate legume diversity.

There are a number of reasons for the diversity of this region, the main points being listed below (adapted from Di Castri 1981):

1. It is an open transitional zone between temperate and tropical regions, thus both moist and arid ecosystems occur;
2. The region is fragmented into isolated valleys or areas, as a result of the accentuated relief;
3. The range of geological eras that the Mediterranean has been subjected to, has resulted in colonization by plant and animal species from widely different biogeographic origins,

which have adapted, and now coexist, with the numerous other species that have evolved in response to the Mediterranean climate itself;

4. The wide range of geological substrata, altitudinal gradients and long history of occupation by man, along with associated erosion processes, has resulted in a multitude of soil types being found in the region.

Despite this, in geological terms the Mediterranean is one of the more recent climatic zones to develop, as it was not present until the Pleistocene era (Axelrod 1973). It has been postulated (Raven 1973) that if climate change eliminates the cold ocean currents that the Mediterranean region depends on for its survival, then the Mediterranean will become a transient episode on the geological scale. The region is composed of a mosaic of different ecosystems generated by a combination of differences in climate, relief, vegetation cover, rocks and soils, as well as the localised influence of man. Each of these factors has contributed to the rich floristic diversity of the Mediterranean, and thus the importance of the region for plant genetic resources (Zohary, M., 1970; 1972). For many plant groups, including the temperate Leguminosae, it is the centre of diversity both for the family itself and for the majority of the genera within. It is also the centre of origin of numerous crop species (Zohary, D., 1970), containing two of Vavilov's 'centres of genetic diversity'(Vavilov, 1997). In order to understand this diversity and to utilise it more efficiently it is important to understand the ecogeographic environment of the Mediterranean. Chapter two is therefore devoted to the description of the various factors that contribute to the ecogeography of the Mediterranean. It is the combination of all factors described here that constitutes the diversity of the region; no single factor can be taken in isolation from the others when attempting to explain the distribution of a particular plant taxa.

2.2 THE CLIMATE OF THE MEDITERRANEAN

The Mediterranean climate is transitional between cold temperate and dry tropical and is unique in its bimodal nature (Blondel and Aronson, 1999). Accordingly, the short spring and autumn are critical periods for plant growth, but the plants must also be able to withstand unexpected cold spells out of season. Snow falls rarely in the Mediterranean, except high in the mountains, but relatively hard frosts are not rare and do restrict plant distribution.

Rainfall varies between less than 100mm / year at the edge of the Sahara and Syrian deserts to 4 m in the coastal massifs of southern Europe. In general, rainfall is higher in the northern than southern coastal areas, though particular levels of rainfall are strongly influenced by elevation. In the winter, moist winds blowing into the Basin from the Atlantic result in higher rainfall on the western sides of land masses.

Although the general climate is described as warm to hot summers and mild, wet winters, with a summer drought of varying severity, there is a large variation within this pattern. The climatic influences described in this section bear direct relevance to the plant communities that are affected by them. These are: summer and winter temperatures, rainfall patterns, days of frost and snow cover, and the resulting variation in the length of the growing season across the Mediterranean.

2.2.1 The True Mediterranean Climate

The Mediterranean Sea, the worlds largest inland (the Strait of Gibraltar is only 13 kms wide) water mass at 2,969,000 km^2, has a marked effect on the climate of the coastal zone. The coast of the Mediterranean Sea (Figure 2.1) is characterised by:

1. A winter dominant rainfall, but with some rain during the summer months. Rainfall is usually in the range of 500 to 850 mm per year.
2. Hot summers with the warmest month reaching over 21°C, and warm winters with the coldest month reaching below 6°C. Frost and snow, although not unknown, are relatively rare.
3. A high percentage of sunshine, both in winter and summer months.
4. Rainfall tends to fall as heavy storms of short duration.

Figure 2.1. The climatic sub-regions of the Mediterranean (adapted from Robinson, 1964). 1- The 'true' Mediterranean climate, 1a - The Oceanic Mediterranean climate, 1b - The Temperate Mediterranean climate, 1c - The Semi-arid Mediterranean climate and 2 - the Continental Semi-arid Mediterranean climate.

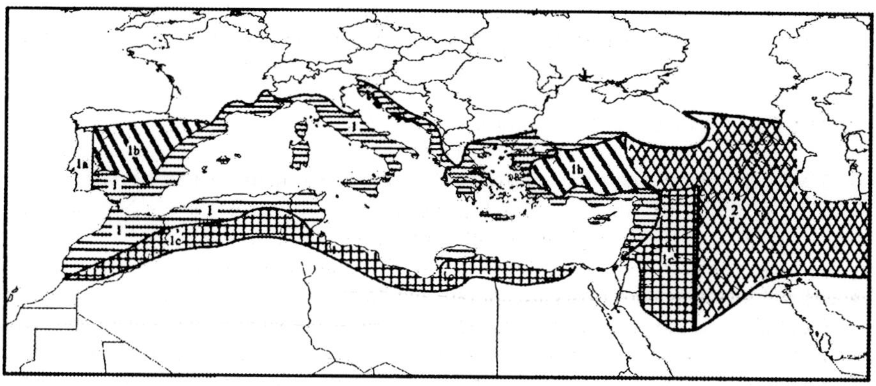

An example of a Mediterranean climate is shown in Figure 2.2a (Sassari, in Sardinia). This clearly shows the dominant winter rainfall and high summer temperatures. Although the eastern and south-eastern shores of the Mediterranean are still characterised as a 'true Mediterranean climate', they are further from the Atlantic influences and therefore typically have higher temperatures all year round, and lower annual rainfall. Annual rainfall is less than 500 mm, often below 250 mm, and is strictly confined to the short winter season (Robinson 1964).
There are three main deviations from the true Mediterranean climate, and although conforming to the majority of Mediterranean characteristics, they display some features of other climatic

zones. These are the plateaus of Iberia and Asia Minor, the semi-desert Mediterranean climate of North Africa and the Middle East, and the Atlantic coastal lands of Iberia (Figure 2.1).

2.2.2 The Temperate Mediterranean Climate

The plateaus of Iberia and Asia Minor (Figure 2.1, 1b) are influenced by the continent, as well as the Mediterranean, and so have lower mean winter temperatures with short spells of severe cold. The days of frost and snow cover are also greatly increased. However the summer temperatures still follow the true Mediterranean pattern and rainfall is still winter dominant. Annual rainfall is lower, and ranges between 250 and 400 mm. The differences in climate are explained by the mountain ranges that enclose these plateaus, thus increasing the temperate continental influence. An example of this type of climate is found at Konya, Turkey and is shown in Figure 2.2b. Here the winter temperatures drop below freezing during December, January and February. There is however still a winter dominant rainfall, albeit lower than that of the 'true Mediterranean climate', with high summer temperatures.

2.2.3 The Semi-arid Mediterranean Climate

This climate occurs in a narrow belt in the southern and south-eastern margins of the 'true Mediterranean' in North Africa and the Middle East (Figure 2.1, 1c), where the climatic conditions are transitional between the 'true Mediterranean' and the desert. The main differences in this climatic regime are the very high summer temperatures and the very low, erratic winter rainfall. Winters can be warm, as in Egypt and Libya, or cold as in Syria. Figure 2.2c shows the climate pattern found in Damascus, Syria, closely matching that described above. Rainfall is almost non-existent from May to August, and maximum summer temperatures are up to 40°C. The semi-desert conditions of Syria also result in low winter temperatures of less than 10°C.

2.2.4 The Oceanic Mediterranean Climate

This climate is typical of most of Portugal (Figure 2.1, 1a) where the weather has a strong maritime influence. This results in lower summer temperatures and milder winter temperatures. The greatest difference, however, is in the increase in rainfall, which falls as an autumn maximum followed by a secondary spring maximum. The climate of Coimbra, Portugal is shown in Figure 2.2d, where it can be seen that there are two maximum periods of rainfall, December and March, and a total average annual rainfall of 950 mm, much higher than is found in any of the other Mediterranean climate types. Both summer and winter temperatures are moderated by the ocean, with winter temperatures not dropping much below 15°C.

2.2.5 The Continental Semi-arid Mediterranean Climate

To the east of the true Mediterranean is an area that is very important for legume diversity, domestication and evolution. This includes Iran, Iraq, Georgia, Azerbaijan, Armenia and the Caucasus, and contains five of the diversity 'hotspots' described in Chapter 1.

Papadakis (1975) describes this area as having a continental semi-arid Mediterranean environment, in which the coldest month is less than 10°C, and the warmest month is up to 40°C. Rainfall is similar to the semi-arid Mediterranean climate occurring in western Syria, with no rain falling for over 7 months, and the wettest months occurring in winter (Dallman, 1998). In this area, cropping is almost impossible without irrigation, but there is good grazing in the short growing season in spring.

Figure 2.2. The climates of the Mediterranean. a. The 'true Mediterranean climate'; b. The temperate Mediterranean climate; c. The semi-arid desert; d. The oceanic Mediterranean climate.

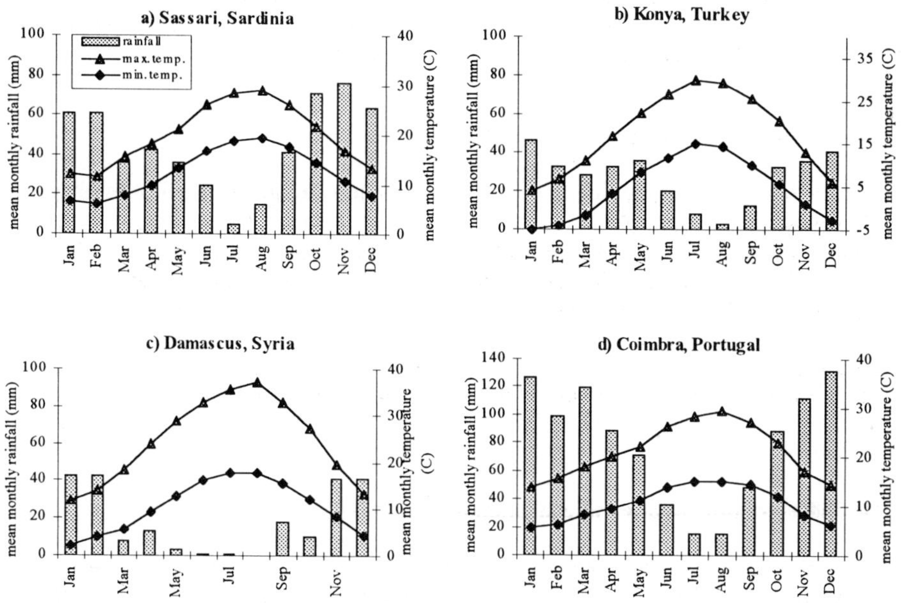

2.3 TOPOGRAPHY

One of the reasons for the rich floristic diversity of the Mediterranean is the topographic diversity. There are mountain chains all around the sea, except in the southeast and on the major islands. The Mediterranean region was subject to much folding during the late Tertiary and early Quaternary periods, followed by glaciation and the outwash stream effects during the Pleistocene period (Bradbury 1981). As a result, the Mediterranean can generally be described as consisting of relatively high fold mountains (up to 3000m with a Mediterranean climate, but to over 4500m in adjacent areas) that rise relatively close to the shore, banded by narrow coastal plains and small, discontinuous lowlands. Highland areas are found in the Alpine regions of France, Italy and Yugoslavia, the Lebanon, the Atlas mountains of Morocco, and the interior plateaus of Spain and Turkey. The mountain systems are the source of most rivers in the region, which drain into the Mediterranean Basin. Notable examples are the Guadalquivir, Ebro, Rhône, Pô, Evros, Nestos and Ceyhan, the Nile being 'non-native' in that the headwaters are outside the region (Blondel and Aronson, 1999). There are also numerous islands and archipelagos (almost 5,000), and most of the larger islands have been disconnected from the mainland since at least the Miocene era. As well as true islands there are 'biological' islands, which are effectively isolated from the land to which they are attached. For instance, the Maghreb (Morocco, Algeria and Tunisia) of North Africa, isolated from the rest of Africa by the Sahara, Cyrenaica in Libya, and further east, the Elburz mountains south of the Caspian Sea (Blondel and Aronson, 1999). It is these 'geological' or 'biological' islands that contain the bulk of the endemic species of the region (Heywood, 1995).

Altitude has a dramatic effect on the climate due to temperature decrease with increasing altitude, thus creating a zoning effect on mountain vegetation. The direction of the mountain range also influences rainfall, temperature, and sunlight hours. South-facing slopes are in sunlight for much longer periods than north-facing slopes, which receive only a short period of sunlight. In turn, this influences the type of vegetation cover on slopes: plants with sunny preference grow on south-facing slopes and vice versa. Rainfall is also affected by altitude, especially ranges that run in a north-south direction and therefore have their slopes facing west and east. Slopes facing the direction of prevailing winds (usually westerlies) receive higher rainfall than slopes lying in the rain shadows of mountains. Again, this determines the vegetation type of the two slopes.

2.4 THE MAJOR SOIL TYPES

As any plant collector will be able to attest, plant distribution in the Mediterranean is closely associated with rock and soil types. Therefore efficient collecting is often associated with knowledge of local geology. The major soil types of an area are dependent on climate, and in semi-arid and arid landscapes, the underlying parent rock is also a reliable indicator of soil properties (Bradbury 1981). The soils of the Mediterranean are inherently complex due to variations in climate, altitude and vegetation cover. Despite this, some broad generalisations can be made. Most Mediterranean soils are derived from limestone of marine origin, however there are pockets of volcanic and metamorphic rock throughout the region.

On steep mountain slopes, there is usually a thin layer of soil due to years of soil erosion. The coastal and river deltas, in contrast, contain soils that are usually deep and fertile as they contain high quantities of alluvium that has washed down from the highlands. The sparse vegetation cover in much of the Mediterranean results in soils that are poor in humus, although leaching is rare because of the low to moderate rainfall pattern. Most of the Mediterranean is covered by soils with an alkaline pH, many are ferruginous brown soils known as terra rossa but dolomitic clays, marls, rendzines, loess, regisols, lithosols, and gypsum outcrops also occur, scattered throughout the region (Blondel and Aronson, 1999).

2.4.1 Terra Rossa

This is the most prevalent soil throughout the Mediterranean. The red earth, as it's name implies, develops on calcareous rocks (Robinson, 1964), it's red colour developing from the mineral decomposition of iron sesquioxides (Fe_2O_3). It is a heavy, clay-rich soil (50-70%) containing minor amounts of humus, and is often very stony; pH values vary between 7.0 and 8.0 (Fisher 1978). As a result, it is not very fertile. An important characteristic of terra rossa soils is their high moisture-holding capacity, which allows them to store considerable amounts of rainfall. However, they are also very susceptible to erosion and therefore their distribution has been substantially reduced by habitat loss or conversion. They are most common in the European Mediterranean areas of southeast Spain, Corsica, Sardinia, southern Italy, Sicily, Greece and the mountainous coastal regions of Turkey (Robinson 1964).

2.4.2 Dark Brown and Chestnut Soils

Dark brown and chestnut soils occur on the plateaus of Spain and the Atlas region of Morocco. They develop on the steppe where only a light vegetation cover consisting of grasses and shrubs is supported. Although these soils are low in humus content, this is a result of low rainfall rather then lack of organic matter. They are rarely cultivated and are generally used as permanent pastoral areas.

2.4.3 Desert Soils

Desert soils are characteristic of much of Libya, Egypt, southern Israel and Jordan, and the western interior plateau of Turkey (Robinson 1964). They are found in areas where the rainfall is sporadic and very low. Soil moisture is therefore continually drawn to the surface by capillary action, bringing dissolved mineral salts to the surface. A saline crust is often formed at, or just below, the surface of the soil.

2.4.4 Granitic and Basaltic Soils

Soils that are formed on either granite or basalt tend to be either neutral or acidic. Acid soils are rare in the Mediterranean region, as both granite and basalt are uncommon. The main areas in which they are found are Sardinia, Corsica, and the Cyclades Islands of Greece (Howieson 1994), with small pockets of mildly acid soils also occurring in Morocco, Tunisia and Spain (Gladstones 1976).

Where granite outcrops protrude, the soils have a tendency to be acid (pH 5.0 to 7.0) (Francis and Gillespie 1981). They are grey-white soils, well-drained, coarse and sandy, containing quartz particles. The low clay and humus content of these soils means they are very infertile, and as such usually form pastures, often heavily grazed.

Soils forming on basalt are of more recent volcanic origin. They are typically brown, or chocolate-brown loams and sandy loams with a pH of 5.0 to 7.5. These soils are more fertile and so cropping is frequent in areas where the rocks or slopes are not excessive.

2.5 THE DIVERSITY OF VEGETATION

Broadly speaking, sclerophyll evergreen trees and shrubs dominate the vegetation of the Mediterranean (Raven, 1971), although the particular vegetation cover is very closely associated with climate. Plants growing in this region have to cope with two major physiological stresses: drought during the summer months, along with high daily temperature maxima and low minimum temperatures during the winter months limiting plant activity, which coincides with the period of highest rainfall. The main periods of plant growth are therefore autumn and in particular the spring months, when both requirements of warmer temperatures and water availability are met.

Consequently, evergreen trees, shrubs and scrub, which can withstand long hot summers without rain, dominate the vegetation of the Mediterranean. Most herbaceous species die back in the summer, with perennials having dormant buds in the soil, and annuals completing their life cycles by mid-summer (Polunin and Walters, 1985). Annual species, at about 40%, form a much larger proportion of the flora than elsewhere in the world, where annuals form approximately 10% of the flora (Zohary, 1962). The climatic conditions of mild, wet winters and dry summers suit annuals and sclerophyll evergreen trees and shrubs, both strategies evading the major problem of summer drought.

Inland in the hilly sub-Mediterranean areas, rainfall is higher and here the evergreen species are replaced by more deciduous species, such as deciduous oaks, hornbeam, maples, manna ash, beech, sweet chestnut, walnut and plane. The grassland of the Mediterranean is largely steppe-like with feathered grasses, which are maintained as grassland by grazing.

Despite this, the Mediterranean is an area of great species richness. This is partly because it largely escaped the last ice age 10,000 years ago, but also because the area encompasses such a wide range of climates that the number of different plant communities is extensive. For example, the Mediterranean region contains over 100 species of trees, compared to European temperate forests, which contain only 30 (Quézel 1976). In total, the region is estimated to contain over 15,000 plant species, about three times that of temperate Europe (Zohary, M., 1972). The Mediterranean is also home to many species that are endemic to very small areas, such as the Balkan Peninsula, which contains 700 species alone, and of these one in four is endemic to the region (Polunin and Huxley 1987).

Due to the influence of man in the Mediterranean, there are few areas of natural vegetation left in the region. Plant communities that persist are either scattered woodlands that exist where they have either avoided or survived human interference, or more commonly dense evergreen scrub known as maquis, or even more commonly dwarf, scattered evergreen scrub known as garrigue (Polunin and Walters, 1985). The final stage of this depauperisation of the natural vegetation is grass-like steppe, commonly with exposed areas of rock or stones. The dominant tree species here are evergreen oaks and pines, commonly associated with smaller evergreen shrubs such as juniper, heathers, cistus, spiny broom, strawberry trees, olive and carob. There are four classic natural vegetation types of the Mediterranean, each closely associated with particular climate and soil characteristics, along with the transitional vegetation types found at the limits of the Mediterranean region. These are briefly described below. The four vegetation types form a spectrum from natural climax community vegetation to degraded communities (Figure 2.3) and can occur in clines or mosaics. Those wishing a more comprehensive treatise on the vegetation of the Mediterranean are referred to Polunin and Walters (1985), and Polunin and Huxley (1987).

2.4.1 Evergreen Forests

The most extensive forests of the Mediterranean region are those dominated by the holm oak (*Quercus ilex*). These grow to 15-18 m and usually, because of selective felling, form open woodland with a dense shrub and herb flora of either a well developed maquis or open glades where many of the garrigue species are common. Other species that are commonly associated with Holm Oak are; Myrtle (*Myrtus communis*), Wild Olive (*Olea europaea* ssp. *oleaster*), Terebinth (*Pistacia terebinthus*) and the Strawberry Tree (*Arbutus unedo*), plus other species of evergreen oak.

In other, more specific regions, there may be other dominant trees. For example, the Cork oak (*Q. suber*) is common in the western and central Mediterranean on siliceous soils, where it is found in plantations as well as growing in the wild. It prefers a warmer, more humid climate with higher rainfall (1000mm / year) than the Holm oak, although they are often found growing together. In the eastern Mediterranean the Holm oak is replaced by the Kermes oak (*Q. coccifera*), but it only forms a forest on Crete and the Peloponnese peninsula. Inland, in the Apennine and Balkan peninsulas, away from the north Mediterranean shore, White oak (*Q. pubescens*) along with Turkey oak (*Q. cerris*) and Hungarian oak (*Q. frainetto*) dominate. Here there are less severe, more continental or higher altitudinal climatic conditions. Laurel (*Laurus nobilis*) woodlands are rare but individual specimens are often found in maquis throughout the Mediterranean. The hotter regions of the Mediterranean, particularly in coastal zones, are dominated by the Aleppo pine (*Pinus halepensis*), a very drought resistant pine that forms climax communities on limestone and littoral sands. The Maritime pine (*P. pinaster*) occurs along the coast from Spain to Italy where it forms woodlands on siliceous rocks (Polunin and Huxley 1987); it has also been extensively planted on coastal dunes in the eastern Mediterranean. Stone pine (*P. pinea*) is also distributed along the coast, primarily on the Iberian peninsular but with significant stands in Greece and Turkey-in-Europe.

Figure 2.3. Flow chart to show the change in vegetation across the Mediterranean.

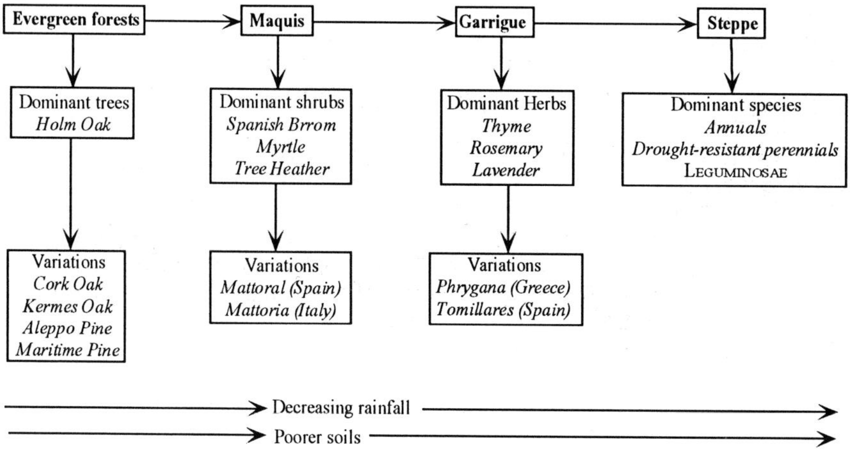

2.4.2 Maquis

Maquis is the most common and well-known type of vegetation in the Mediterranean today and corresponds to the chaparral in California and matorral in Chile. It typically consists of tall, dense, evergreen shrubs and low trees. Many areas covered by maquis would have been originally forested but due to clearance by man and subsequent grazing, it is now found where low rainfall and poor soils prevent the re-growth of woodlands. These are known as secondary maquis. Maquis is a scrub forest, which can form an impenetrable thicket of scrub and taller trees of 2m or more. The common taller trees are the evergreen oaks, such as Holm oak (*Quercus ilex*) and Kermes oak (*Q. coccifera*), Judas tree (*Cercis siliquastrum*), wild Olive (*Olea europaea* subsp. *oleaster*) and Aleppo pine (*Pinus halepensis*). These are typically well spaced with large shrub species forming the thicket between. Commonly occurring shrub species include Spanish Broom (*Spartium junceum*), Myrtle (*Myrtus communis*), Tree Heather (*Erica arborea*) and Wig Tree (*Phillyrea media*). The name "maquis" originates from France, although it is so common in the Mediterranean that many other countries in the region have their own names for this type of vegetation, such as "mattoral" in Spain and "macchia" in Italy (Robinson 1964).

2.4.3 Garrigue

Garrigue is associated with the thinner soils of calcareous outcrops or other more degraded soils. It forms a thinner, more open cover than maquis, with open ground, rocks or sand occurring in between the vegetation. Most commonly, garrigue occurs in the hotter and drier terrains of the Mediterranean Middle East, such as Cyprus, northwest Cyrenaica (Libya) and the western Levant (Spain), or areas degraded from evergreen forest or maquis. It is a dwarf shrub community, usually less than 0.5 m high, which is rich in aromatic and mostly evergreen scrubs (Polunin and Walters, 1985). However, despite the difficult growing conditions, garrigue is usually rich in species. Many shrubs of the garrigue are thorny, and many of our culinary herbs are from this vegetation type: thyme, lavender, rosemary, sage and garlic among others. It is also the home of many ornamental bulbs such as tulips, crocuses and irises, and after the spring rains, these areas form a carpet of flowers and herbs for some weeks. A number of plant families are particularly well adapted to surviving in garrigue: Leguminosae, Euphorbiaceae, Thymelaeaceae, Labiatae, Compositae, Liliaceae, Orchidaceae and Gramineae.

There are many variations on garrigue around the Mediterranean, each defined by the community composition and geographical location. "Phrygana", for example, is restricted to western Asia Minor and the Aegean (Fisher 1978). The name was used by the Ancient Greeks and refers to areas that consist principally of thorn bushes, which, although only of moderate growth and height, are very spiny and close set. "Tomillares" is another form of garrigue that occurs in Spain, and refers to thyme and other members of the Mint family (*Labiatae*), such as lavender, rosemary and sage, that are so common in this area (Polunin and Huxley 1987).

2.4.4 Steppe

Steppe vegetation occurs in areas of wide seasonal variation of temperature, and lower rainfall, such as the Anatolian Plateau in Turkey, the interior plateau of the Atlas Mountains in

Morocco and in many areas of Spain. In the majority of cases, steppe is semi-natural, containing some native species but maintained as steppe-grassland by intense grazing or other disturbances, causing an inability to regenerate to a more natural state. Although many of these communities require grazing to halt natural vegetation succession, these grasslands have existed for centuries and so remain reasonably stable (Polunin and Walters, 1985). Steppe vegetation differs from true grassland in that the grass species are adapted to semi-arid conditions. The feather grasses, *Stipa* species, are characteristic of the driest areas, where the annual rainfall is less than 500 mm/year. In spring, the area is covered with a wide range of flowers and grasses that present a brilliant splash of colour and appear almost luxuriant. However these soon set seed and dry leaving only a few remaining hardy bushes, with large areas of bare ground. Most of these species are annuals, but perennials that have either a deep root system or underground storage organs to allow them to survive the harsh summers, are also characteristic. This is where many of the Leguminosae are most commonly found, along with other areas of grassland that are presently under the influence of man.

2.4.5 Transitional Vegetation Types

There are three main types of transitional vegetation: desert, mountain and cold temperate. They do not strictly form part of the Mediterranean vegetation but will be described in brief here, as it is important to understand the transition from the Mediterranean vegetation into other types.

Desert vegetation occurs when the climate becomes too arid to support steppe or garrigue, such as in North Africa and the Middle East away from the coast. Trees, although still present, become sparse and in many cases deciduous, losing their leaves as a method of reducing water loss during the height of the dry season. Most other plants of this region also display forms of adaptation to drought and saline conditions, the most common one being the camel thorn or jujube (*Zizyphus lotus*), a member of the Leguminosae (Fisher 1978). Other plants complete their growing cycle within a few weeks after the winter rains.

The transition to mountain vegetation occurs with increasing altitude and correspondingly lower temperatures. It is characterised by a change from evergreen forests, to deciduous forests at altitudes of 800 to 1800m, and then to coniferous forests at altitudes above this (Polunin and Huxley 1987). The deciduous zone includes trees such as deciduous Oaks (*Quercus* sp.), ash (*Fraxinus ornus*), and Sweet Chestnut and Beech (*Fagaceae* sp.). The coniferous zone includes Silver Fir (*Abies cephalonica*), Pines (*Pinus* sp.) and in some areas Cedars (*Cedrus* sp.).

Cold temperate vegetation takes over as latitude increases in Europe, and the maximum winter and summer temperatures decrease. This vegetation type is similar to the deciduous forests of the mountain zone, although in many areas it has been cleared by man and is now used for agriculture.

2.5 THE MEDITERRANEAN FLORAS

The Mediterranean flora is a complex mixture of taxa with different biogeographic origins, respective ages and evolutionary histories (Blondel and Aronson, 1999). The floras are not all well known and certainly those on the northern shores are better understood than in the eastern or southern. The region is generally accepted, however, to have five basic floral elements: Afro-tropical, Holoarctic ('Euro-Siberian'), Irano-Turanian, Saharo-Arabian and 'true''Mediterranean (Figure 2. 4). The major elements of these floras will be summarised here, but those wishing a more detailed account are referred to Zohary, (1973), Quézel (1985) and Blondel and Aronson (1999).

2.5.1 Afro-Tropical

Historically, in the era of the Tethys Sea, before the Americas were separated from Eurasia, there was a much closer relationship between the Mediterranean flora and those further south, but now the Mediterranean Basin is effectively isolated from Afro-tropical regions by the Saharo-Arabian desert. The Mediterranean and Afro-tropical (Sudanian) floras have been effectively isolated since the Miocene-Pliocene period, 5-6 Myr BP. An effective isolation that the Mediterranean Basin does not have with the floras to its north. The Afro-tropical flora is predominantly a hot, dry desert flora. Even though the Mediterranean Basin has effectively been isolated from Afro-tropical regions for such a long period, there are still numerous Afro-tropical taxa present north of the Saharo-Arabian desert. These have been called the 'palaeo-tropical' relicts (Raven, 1971; Quézel, 1985) and include the evergreen *Asparagus*, *Capparis*, *Ceratonia*, *Chamaerops*, *Cotinus*, *Jasminum*, *Laurus*, *Myrtus*, *Nerium*, *Olea*, *Paliurus* and *Phillyrea*. There are numerous examples of tropical plant families that have a small number of endemic species found north of the Sahara. The so-called 'Rand-flora' also links the Mediterranean Basin more remotely with the flora of Southern Africa; Raven (1973) cites the *Olea europaea* complex and the endemic *Argania spinosa* (an endemic of south-west Morocco) as examples. More recently, this link would only be possible using the mountain ranges of Africa as 'stepping stones', but the fossil record provides evidence of such a historic migration route (Blondel and Aronson, 1999).

Figure 2.4. The Floras of the Mediterranean (After Zohary, 1973).

2.5.2 Holoarctic

The Holoarctic or Euro-Siberian flora is historically derived by the migration and mixing of taxa from the northern rim of the Mediterranean Basin. These species appear to have reached the Mediterranean by the upper Pliocene, before the first glaciation, and today persist primarily in the colder, wetter region of the 'true' Mediterranean. This flora is typified by Oriental Plane (*Platanus orientalis*), Walnut (*Juglans regia*), Hazelnut (*Corylus avellana*) and Beech (*Fagus sylvatica*). Many of the Holoarctic taxa are most diverse in the northeast region, particularly in the Pontic and Hyrcanian provinces of the Euro-Siberian flora. The Pontic area is rich in Fir (*Abies*), Pine (*Pinus*) and Spruce (*Picea*), and *Rhododendron* species (Blondel and Aronson, 1999), while the Hyrcanian lacks these species and characteristically contains Ironwood (*Parottia persica*) and the Witch-hazel family (Hamamelidaceae).

2.5.3 Irano-Turanian

The Irano-Turanian flora has possibly been the most invasive of the Mediterranean Basin. It is typified by *Artemisia*, *Ephedra*, *Haloxylon*, *Pistacia*, *Salsola* and *Suaeda*, all of which have centres of diversity and possible origins in Central Asia, along with the Judas tree (*Cercis siliquastrum*), Storax tree (*Styrax officinalis*) and Rosaceous trees (Blondel and Aronson, 1999). Zohary (1973) considers there to have been waves of interpluvial penetrations by the Irano-Turanian flora since the Upper Tertiary. The influence of the Irano-Turanian flora is, not surprisingly, greater in the eastern Mediterranean, closer to its origin, but Irano-Turanian taxa are found throughout the whole Mediterranean. Many taxa in this flora have undergone extensive evolutionary radiation with large numbers of taxa per genus, e.g. *Astragalus*, *Acantholimon*, *Cousinia*, *Centaurea* and many monotypic genera (Zohary, 1973). The hotspots of diversity are located in the Kurdo-Zagrosian, the Taurus and the Turco-Armenian mountains, along with the sub-alpine Iranian plateau. Most of the cultivated fruit and nut trees in the Mediterranean are of Irano-Turanian origin, and the deciduous oaks of the eastern Mediterranean show strong affinities with those further east.

2.5.4 Saharo-Arabian

The Saharo-Arabian flora is less diverse (less than 1,500 species) but contributes strongly to the overall xerophytic flora of the Mediterranean region, although few have extending their distribution significantly north of the Mediterranean Sea. Because of its origin in the Saharan and Arabian deserts the species characteristic of the Saharo-Arabian flora tend to be drought resistant and are found in frost-free arid pockets. Typically, this flora is exemplified by saltbushes (Chenopodiaceae), the Zygophyllaceae (*Balanites*, *Peganum*, *Tribulus*, *Zygophyllum*) and a few perennial grasses (Zohary, 1973; Quézel, 1985). Their distributional limit roughly coincides with the 150 mm contour that makes the edges of the desert (Blondel and Aronson, 1999).

2.5.5 'True' Mediterranean

The True Mediterranean flora is composed of several thousand taxa, with 50% of endemics native to the Mediterranean Basin (Zohary, 1973). Typically, this flora is exemplified by the Strawberry tree (*Arbutus*), the various evergreen sclerophyll oaks (e.g. *Quercus ilex*), Tree Heather (*Erica arborea*), along with *Helianthemum*, *Lavatera*, *Salvia*, *Cupressus*, *Pinus* and *Juniperus* (Quézel, 1985). In general, annual and ruderal species are particularly well represented in this flora, probably because by fruiting early in the season they can avoid the long dry summers. Zohary (1973) argues that the annual ruderal species form a distinct sub-flora of about 1,500 species, and it is to this flora that the majority of crop relatives and specifically forage legume species of the region belong. He comments that many have a surprisingly restricted, endemic distribution considering that they have been weeds of cultivation and disturbed land for such a long period. This group includes a large number of genera with a small number of species; e.g. Crucifers, *Bunias* and *Calepina*, the Composites, *Cardopatium* and *Ridolfia*, the Umbellifers, *Bifora*, *Exoacantha* and *Smyrniopsis*, along with the legume genera, *Lens*, *Pisum*, *Biserrula*, *Scorpiurus*, *Hammatolobium*, *Hymenocarpus* and *Cytisopsis*.

Blondel and Aronson (1999) summarise the history of the flora, saying that from the Tertiary until the Oligocene it was typically tropical with a mixture of forest and savanna ecosystems. At the beginning of the Oligocene and into the Pliocene the tropical climate changed to what we now know as Mediterranean (about 3.2 Myr BP), bounded by arid and semi-arid areas in the south and east, with the Atlantic to the west and more mesic areas to the north. It is within this enclosed region that the flora we currently recognise evolved. The climatic crisis at the end of the Pliocene and beginning of the Pleistocene, when the mouth of the Mediterranean was temporarily sealed and water levels fell, resulted in the disappearance of the bulk of the tropical species; though the Palaeogene (original Mediterranean) flora at least partially remains in the Macronesian archipelago (Canary Islands and Madeira), and to a lesser extent on some of the larger Mediterranean islands (Crete and Cyprus).

2.7 HUMAN INFLUENCE

Humans have played an important part in the evolution of the vegetation of the Mediterranean, through cutting, burning, cropping and grazing, and their activities have modified most of the region's ecosystem (Cowling *et al.*, 1996). Heywood (1995) lists numerous recent developments that have had a detrimental effect on the environment:

- changes in agriculture towards large-scale, intensive, high-input operations;
- merging of farms into larger units;
- loss of boundaries with a consequent loss of biodiversity;
- abandonment of terracing;
- movement of the population to urban centres;
- crop and cultivation system substitution;
- introduction of new and intensive cash crops (strawberries, tomatoes, peppers, ornamentals);
- rapid spread of exotics alien and invasive species (e.g. *Amaranthus*, *Conyza*, *Phytolacca*, *Oxalis*, *Agave*, *Opuntia*);
- genetic erosion and pollution of native species.

Intensification of agriculture and felling of native forest has resulted in soil erosion which in turn has led to the exposure of bedrock. This can only support sparse steppe-like vegetation, and consequently, little remains today of the natural Mediterranean vegetation (Polunin and Walters, 1985). It has been estimated that between 3000 and 1500 BC, 50 to 60% of the Mediterranean region was cultivated, while cropped areas occupy 48% of the region today (le Houérou 1981). The Mediterranean region, and in particular the eastern Mediterranean, is the centre of diversity and dispersal of many cultivated plants in use today (Zohary, M. 1970; Zohary, D., 1970), such as wheat, barley, sorghum, lentil and chickpea, plus many fruit trees, amongst others.

For the last 2500 to 4000 years, burning has played an essential role in the preparation of land for cropping, or to attain earlier and better pastures (le Houérou 1981). Subsequent over-grazing reduces the seed bank, and prevents or slows down regeneration. Also, a burning sequence of approximately 25 years, which is the estimated frequency in Mediterranean France (le Houérou 1981), restricts vegetation from ever returning to full climax, either evergreen forest or primary maquis (see Figure 2.5), and encourages pyrophytic plants to dominate. Other detrimental effects of fire are soil erosion, flooding and silting up of dams, which in turn reduces the amount of water available and the fertility of the soil, all of which have an effect on the vegetation type that will regenerate in that area.

Grazing affects the majority of the Mediterranean region. Those areas that have not been cropped are used as natural pastures, and those areas that have been cropped are under pasture for some part of the year, either as cereal stubble or fallow. Thus it has been estimated that 75% of the agricultural area of the Mediterranean is pastureland in any one year (Le Houérou 1981). In many areas, animals remain on the land for the whole year, which leads to over-grazing. Unfortunately, this encourages those species that are unpalatable or inedible to the livestock to increase, and those species that have potential forage value to decrease. However, grazing experiments have shown that a certain level of grazing is beneficial for species diversity. Naveh and Whittaker (1979) found a fourfold reduction in the number of species in northern Israel if grazing animals were completely excluded in garrigue, and succession to a closed canopy of maquis shrubs resulted. The diversity of forage and pasture legume species is covered in Chapter 3, and so will not be described here, other than to say that the Mediterranean, and in particular the eastern Mediterranean, is an area of high diversity of, in many cases, under-utilised legume species that are in grave danger of genetic erosion.

Deforestation of the Mediterranean Basin is an increasing problem. The regions biomes are now made up of a patchwork of natural and semi-natural vegetation in a mosaic of agricultural land, planted forests, wasteland, roads, industrial landscapes and urbanisation (Heywood, 1995). The growth of human population in the region has made enormous demands on wood for fuel and construction. The high demand in virtually treeless Egypt has led to the denuding of the once extensive forests of Crete, Cyprus and Lebanon (Dallman, 1998). The largest forests remain in Spain, France and Italy, but there are extensive planting schemes in the eastern and southern Mediterranean, some funded by the Global Environment Facility.

A final point on the influence of humans, is that, as well as dispersing many of the worlds major crops around the world from the Mediterranean, he or she has also introduced many other plants from other parts of the world as ornamentals, fruits and vegetables. With this has been the introduction of many weeds from other areas of the world with a Mediterranean climate, such as the Bermuda Buttercup (*Oxalis pes-caprae*) from South Africa and the Spiny Clotbur (*Xanthium spinosum*) from South America (Polunin and Huxley 1987). These are only two examples among many, and the number is increasing with time. As these 'weeds' often have few competitors, they will increasingly play a part in the description of vegetation types of the Mediterranean region.

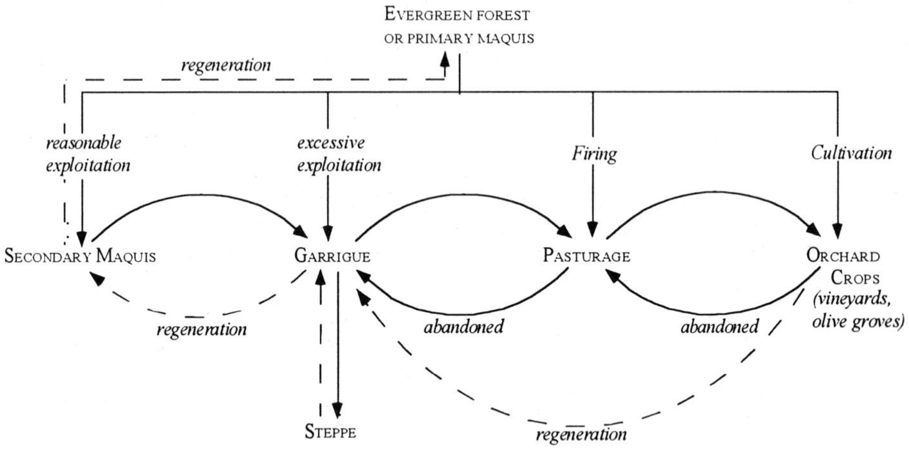

Figure 2.5. Stages in the degeneration and regeneration of plant communities, and the effect of man on Mediterranean vegetation (from Polunin and Huxley 1987).

REFERENCES

Axelrod, D.I., 1973. History of the Mediterranean ecosystem in California. In: *Mediterranean type ecosystems, origin and structure* (eds. Di Castri, F and Mooney, H.A.). pp. 225-277. Springer-Verlag, Berlin.

Blondel, J. and Aronson, J., 1999. *Biology and wildlife of the Mediterranean region*. pp. 327. Oxford University Press, Oxford.

Bradbury, D.E., 1981. The physical geography of the Mediterranean lands. In: *Ecosystems of the world 11: Mediterranean type shrublands* (eds. Di Castri, F., Goodall, W. and Specht, R.L.). pp.53-62. Elsevier Scientific Publishing Co., Amsterdam.

Cowling, R.M., Rundel, P.W., Lamont, B.B., Arroyo, M.K. and Arianoutsou, M., 1996. Plant diversity in Mediterranean-climate regions. *Tree*, 11: 362-366.

Dallman, P.R., 1998. *Plant life in the world's Mediterranean climates*. University of California Press, Berkeley.

Di Castri, F, 1981. Mediterranean-type shrublands of the world. In: *Ecosystems of the world 11: Mediterranean type shrublands* (eds. Di Castri, F., Goodall, W. and Specht, R.L.). pp.1-52. Elsevier Scientific Publishing Co., Amsterdam.

Fisher, W.B., 1950. *The Middle East. A physical, social and regional geography*. Methuen and Co. Ltd., Cambridge.

Francis, C.M. and Gillespie, D.J., 1981. Ecology and distribution of subterranean clover and *Medicago* species in Sardinia. *Australian Plant Introduction Review*, 13:15-25.

Gladstones, J.S., 1976. Observations on the distribution and ecology in Iberia and North Africa of some annual legumes adapted to neutral and acid soils. *Australian Plant Introduction Review*, 11: 9-23.

Heywood, V.H., 1995. The Mediterranean region a major centre of plant diversity. *Cahiers Options Méditerranéennes*, 38: 5-16.

Howieson, J.G. and Loi, A., 1994. The distribution and evaluation of alternative pasture legumes and associated root-nodule bacteria collected from acid soils of Greece (Seriferos), Morocco, Sardinia and Corsica. *Agriculture Mediterranean*, 124:170-186.

le Houérou, H.N., 1981. Impact of man and his animals on Mediterranean vegetation. In: *Ecosystems of the world 11: Mediterranean type shrublands* (eds. Di Castri, F., Goodall, W. and Specht, R.L.). pp. 479-522. Elsevier Scientific Publishing Co., Amsterdam.

Nahal, I., 1981. The Mediterranean climate from a biological viewpoint. In: *Ecosystems of the world 11: Mediterranean type shrublands* (eds. Di Castri, F., Goodall, W. and Specht, R.L.). pp. 63-72. Elsevier Scientific Publishing Co., Amsterdam.

Naveh, Z. and Whittaker, R.H., 1979. Structural and floristic diversity of shrublands and woodlands in Northern Israel and other Mediterranean areas. *Vegetatio*, 43: 171-190.

Papadakis, J. 1975. *Climates of the world and their potentialities*. Buenos Aires, Argentina.

Polunin, O. and Huxley, A., 1987. *Flowers of the Mediterranean*. Chatto and Windus Ltd, London.

Polunin, O. and Walters, M., 1985. *A guide to the vegetation of Britain and Europe*. Oxford University Press, Oxford.

Quézel, P, 1976. Les forêts du poutour Mediterraneén. *Technical Man and Biosphere*, 2: 9-34.

Quézel, P, 1985. Definition of the Mediterranean region and the origin of its flora. In: *Plant conservation in the Mediterranean area* (ed. Gomez-Campo, C). pp. 9-24. Dr. W. Junk, Dordrecht.Raven, P.H., 1971. The relationship between 'Mediterranean''floras. In: Plant life of South-west Asia (Davis, P.H., Harper, P.C. and Hedge, I.C.). pp. 119-134. Botanical Society of Edinburgh, Edinburgh.

Raven, P.H., 1973. The evolution of Mediterranean floras. In: *Ecosystems of the world 11: Mediterranean type shrublands* (eds. Di Castri, F., Goodall, W. and Specht, R.L.). pp. 213-224. Springer-Verlag, Berlin.

Robinson, H., 1964. *The Mediterranean lands. Book 2 (2nd Ed.).* University Tutorial Press Ltd, London.

Vavilov, N.I., 1997. *Five continents.* International Plant Genetic Resources Institute, Rome, Italy.

Zohary, D., 1970. Centres of diversity and centres of origin. In: *Genetic resources in plants - their exploration and conservation* (eds. Frankel, O.H. and Bennett, E.). IBP Handbook, No. 11, pp. 33-42. Blackwell Scientific Publications, Oxford.

Zohary, M., 1962. *Plant life of Palestine.* Ronald Press, New York.

Zohary, M., 1970. The phytogeographical foundations of the Middle East. In: *Plant life of South-west Asia* (eds. Davis, P.H., Harper, P.C. and Hedge, I.C.). pp. 43-52. Botanical Society of Edinburgh, Edinburgh.

Zohary, M., 1972. *Geobotanical foundations of the Middle East (2 Vols.).* Fischer-Verlag, Stuttgart.

Chapter 3

Legume Diversity in the Mediterranean Region

Nigel Maxted and Sarita Jane Bennett

3.1 Introduction

The Mediterranean has one of the richest floras of the world. It contains some 25,000 species, including a high percentage of endemic species, and it is the centre of diversity for many socio-economically important species (temperate cereals, legumes, olives, vines and salad vegetables fruit trees) and their secondary and tertiary gene pools. There are several hypotheses to explain the floral abundance of the region: it is set at the cusp of three continents and three distinct floras (Euro-Siberian, Mediterranean and Irano-Turanian); it is deeply dissected, with peninsulas, extensive mountain ranges and numerous islands (Heywood, 1995; Akeroyd, 1999), which has led to reproductive isolation of populations and evolutionary divergence. Stebbins (1967) points out that if areas with different environmental and floristic characters are juxtaposed, it may act as a catalyst to evolution. Many temperate legume genera would appear to be a good examples of taxa to have evolved in such an evolutionary "melting pot".

There is evidence, however, that taxa with centres of distribution in the Balkans, South-west and South-east Asia are generally centred in these areas due to climatic deterioration linked to floristic movement from more northerly distributed floras (Takhtadjan, 1969); it is believed that the Quaternary Ice Age drove the Boreal-tertiary Flora south towards these three floristic refugia in the south. It seems likely that many temperate legume taxa evolved in the early tertiary and the centre of origin is likely to have been much further north than is indicated by contemporary concentrations of taxa. Stebbins (1967) suggests that the most genotypically and phenotypically plastic species of a genus (e.g. *Medicago polymorpha*, *Lathyrus aphaca*, *Trifolium campestre*, *Vicia sativa*) have evolved rapidly away from their centre of diversity, while the more primitive species remain at the centre of origin (e.g. *Medicago carstiensis*, *Lathyrus mulkak*, *L. roseus*, *Trifolium eximium*, *Vicia biennis*, *V. sepium*). There are, however, also counter arguments. Willis (1921) argues that evolution and plant dispersal occur at the same rate, therefore the oldest and most primitive species are most widespread and are currently found on the distributional periphery, while the most recently evolved species occur at the centre of origin and are localised. It is difficult to reach a firm conclusion on the centre of origin of temperate legume species, but Stebbins suggestion seems to better correspond with the current distribution of species in the Mediterranean.

3.2 Centres of Diversity

Whatever the truth concerning the centre of origin for legume species found in the Mediterranean, the region is a centre of diversity for temperate legume species. It is well know that plant diversity is not evenly distributed over the surface of the Earth. Plant species diversity increases as one moves away from the poles towards the equator, but even at similar latitudes, some geographic areas are more diverse than others. The areas that have very high levels of plant species richness and endemism are referred to as plant "hot-spots". They are primarily tropical and in the Southern Hemisphere, but in terms of the Northern Temperate Hemisphere, the Mediterranean Basin is the most important "plant hot-spot". Plant species richness and endemism for the countries of the Mediterranean region is provided in Table 3.1.

Over 1,300 legume species are found in Turkey alone, about a third of these are endemic to Turkey, and the majority are endemic to the region (Davis, 1970). Many of the endemic species in Turkey and in the broader Mediterranean region have extremely narrow distributions, often restricted to a single hillside, mountain or mountain range (e.g. *Vicia eristalioides*, *V. kalakhensis*, *Lathyrus belinensis*, *Medicago heyniana*, *Trifolium dolopium*). Certain areas surrounding the Mediterranean emerge as "hot-spots" of species richness and these are usually limestone mountains (Akeroyd, 1999). The "hot-spots" of legume diversity for the Mediterranean region (see Figure 3.1) are based on the general floristic observation of Davis *et al.* (1994), Akeroyd (1999) and from personal observation. These hot-spots are:

- Andalucia (Spain)
- Balearic Is. (Spain)
- Massifs of Gudar and Javalambre (Spain)
- Pyrenees (Spain and France)
- Maritime Alps (France and Italy)
- Tyrrhenian Is. (France and Italy)
- Central and South Apennines (Italy)
- Dalmatian coast (Slovenia and Croatia)
- Balkan and Rhodope Massifs (Bulgaria, Greece and Serbia)
- Mt. Olympus and nearby mountains (Greece)
- South and Central Greece (Greece)
- Mountains and gorges of Crete (Greece)
- Troödos mountains (Cyprus)
- The Dodecanese islands, Aegean and Southern Turkey (Greece and Turkey)
- Southern Turkey and the Levantine mountains (Israel, Palestine, Syria and Turkey)
- North east Anatolia (Turkey and Georgia)
- Greater Caucasus (Georgia, Azerbaijan and Russian Federation)
- Eastern Turkey and Lesser Caucasus (Turkey, Armenia, Azerbaijan and Iran)
- North-east Iraq and North and West Iran (Iraq and Iran)
- Cyrenaica (Libya)
- Atlas Mountains (Morocco)

Table 3.1. Plant species richness and endemism for the Mediterranean region (Davis *et al.*, 1994, 1995).

Country	Native Vascular Plant Species	Endemic Species	% Species Endemism
Portugal	2600	3	0.1
Spain	5050	941	18.6
France	4650	133	2.9
Italy	5600	712	12.7
Yugoslavia (former)	5350	137	2.6
Albania	3000	24	0.8
Greece	5000	742	14.9
Bulgaria	3600	320	8.8
Romania	3400	41	1.4
Caucasus: Georgia Armenia Azerbaijan	6000	1200	20
Turkey	8650	2675	30.9
Cyprus	1650	88	5.3
Malta	914	5	0.5
Iraq	3000	190	6.3
Iran	8000	1400	17.5
Syria	3100	395	13
Lebanon	2600	311	12
Israel	2225	165	7.4
Jordan	2100	145	7.3
Egypt	2076	70	3.4
Libya	1825	134	7.3
Tunisia	2196	?	?
Algeria	3164	250	7.9
Morocco	3675	625	17

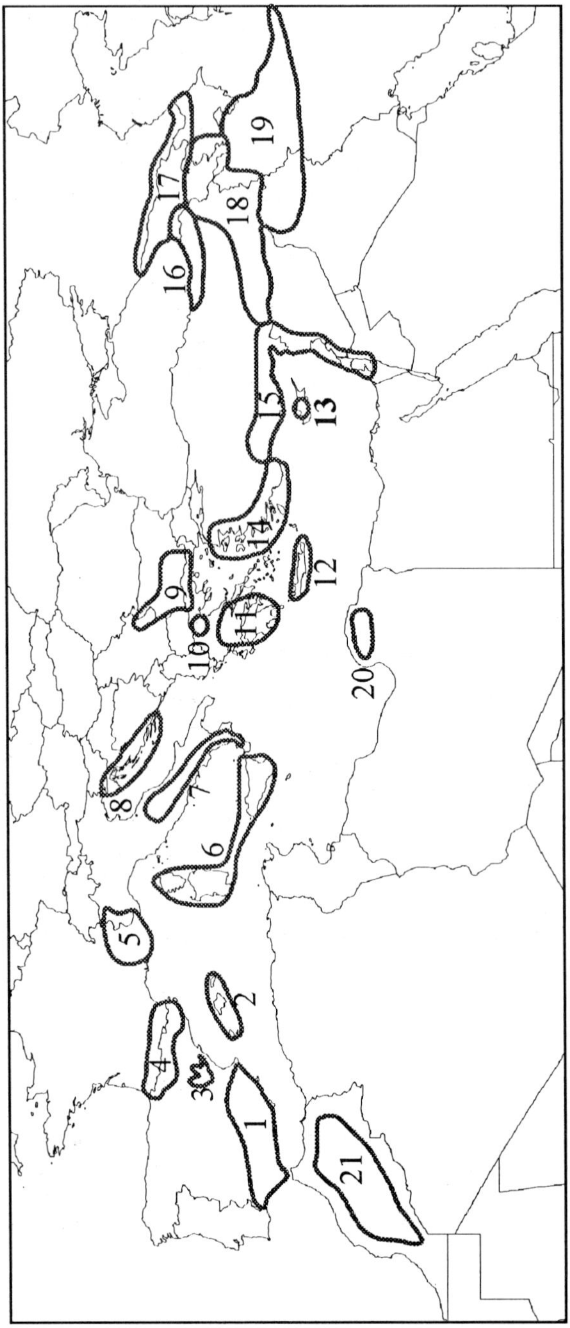

Figure 3.1. “Hot-spots” of Legume Diversity in the Mediterranean Region.

A similar picture of uneven geographic distribution worldwide emerges for the genetic diversity within crop gene pools. The Mediterranean region contains two of N.I. Vavilov's (1926, 1927, 1951, 1997) eight regions or centres of diversity of major groups of crop plants:

- Centre IV / Asia Minor Centre - This centre includes all of Transcaucasia, Iran and Turkmenistan, in addition to Asia Minor in the narrow sense. The most important legume crops included by Vavilov were: *Cicer arietinum* (secondary centre); *Lens culinaris, L. orientalis*; *Vicia ervilia*; *Pisum sativum*; various forage plants such as *Medicago sativa, Trifolium resupinatum, Trigonella foenum-graecum, Onobrychis* spp., *Lathyrus cicera*, and several *Vicia* spp.
- Centre V / Mediterranean Centre - Vavilov considered this centre of more limited importance than Asia Minor. In fact, many of its crops are closely related to the Asia Minor ones as subspecies or found there as secondary centre species. The legume species Vavilov included are: *Vicia faba, Lathyrus ochrus, Vicia sativa*, large-seeded *Cicer arietinum, Hedysarum coronarium, Ornithopus sativus.*

Therefore, the broader Mediterranean region can be seen to be the centre of diversity for most temperate legume genera.

3.3 Legume Taxonomy

The Leguminosae is a morphologically diverse family, ranging from trees to aquatics to xerophytes, and ranks second to the Gramineae in economic importance (Heywood, 1978). The family contains approximately 650 genera and 18,000 species, and is the largest family of plants after the Compositae and Orchidaceae (Polhill *et al.*, 1981). The family is normally divided into three sub-families, Mimosoideae, Caesalpinioideae and Papilionoideae, of which the latter is the largest. The Mediterranean grain and forage legumes are all members of the Papilionoideae, which is distributed through temperate, sub-tropical and tropical regions. Of the 400-500 genera and 10,000 Papilionoid species, most are herbs, but some are trees and shrubs (Polhill *et al.*, 1981). The leaves are usually pinnate; the flowers are irregular with lateral petals enclosed by the standard in the bud; there are usually 10 stamens, commonly diadelphous but sometimes monadelphous or free.

The Papilionoideae has been traditionally divided into 10 or 11 tribes, based on habit, vegetative and floral characters. More recently, legume taxonomists have tended to increase the number of tribes; Gillett *et al.* (1971) detail 17 tribes and Polhill (in Lackey 1977) suggests a complete break with prior tribal delimitations to form 31 tribes; while Polhill and Raven (1981) divide the sub-family into 32 tribes. Using this classification, the economically useful temperate forage and food legumes of the Mediterranean Basin are found in 9 tribes and 29 genera, listed in Table 3.2.

A sound taxonomic understanding of any taxon is prerequisite to efficient and effective conservation and utilisation (Maxted, 1995a). How could we be expected to collect and conserve *ex situ*, forage legume diversity, if we did not know how many species exist or how to distinguish them in the field? For many forage legume genera there is no recent taxonomic treatment. However, as Turkey has by far the largest number of forage legume taxa in the Mediterranean region, the Flora of Turkey (Davis, 1970) remains the standard taxonomic work for forage legume conservation and use. However, other useful taxonomic information can be obtained from the major Floras and reference books of the Mediterranean region, these include:

- Conspectus Florae Orientalis Vol. 5 (Heller and Heyn, 1990)
- Flora of the USSR Vol. XI (Schischkin, 1945)
- Flora Europaea, Vol. 2 (Tutin *et al.*, 1968)
- Flora Iranica. Papilionaceae I (Chrtkova-Zertova *et al.*, 1979)
- Flora Iranica. Papilionaceae II (Rechinger, 1984)
- Flora of Iraq, Vol.3 (Townsend, 1974)
- Nouvelle Flore du Liban et de la Syrie. Vol. 2 (Mouterde, 1966)
- Flora Palaestina, Vol. 2 (Zohary, 1972)
- Flora of Turkey, Vol. 3 (Davis, 1970)
- Flora of Cyprus, Vol. 1 (Meikle, 1977)
- Flora of the USSR Vol. XIII (Schischkin and Bobrov, 1948)
- Flowers of Greece and the Balkans: A Field Guide (Polounin, 1987)
- Flowers of the Mediterranean (Polounin and Huxley, 1970)
- Flowers of South-west Europe (Polounin and Smythies, 1988)
- Mediterranean Wild Flowers (Blamey and Grey-Wilson, 1993)
- Med-Checklist, Vol. 4 (Greuter *et al.*, 1989)
- Legumes of West Asia. A Check-list (Lock and Simpson, 1991)
- Legumes of Northern Eurasia. A Check-list (Yakovlev *et al.*, 1996)

Unfortunately, most traditional keys currently available, including some of those in the Flora of Turkey, are difficult to use because of the technical terms they employ, their reliance on obscure characters that are difficult to record and their general lack of flexibility for the user (Maxted and Crust, 1995). Traditional dichotomous keys have the additional disadvantage that they force the user to commence identification in a pre-ordained sequence, however, multi-access keys allow entry through any characters in the data set, with no preordained sequence of questions needed to achieve an accurate identification. To assist the legume conservationist or germplasm user identify their material, both a traditional dichotomous key and multi-access key are provided in Appendix 1 and 2 respectively.

The Mediterranean flora has been well studied, and numerous taxonomic studies of the legume flora found in this region has led to abundant synonymy. As Maxted (1995b) states there are 149 synonyms for the field or common vetch alone, which was described originally by Linnaeus 250 years ago as *Vicia sativa*. Therefore, to avoid confusion, the International Legume Database and Information Service (ILDIS) list of accepted scientific names for legume species is used (see http://biol.soton.ac.uk/ildis/LegumeWeb) along with the standard Brummitt and Powell (1992) list of authors abbreviations.

Table 3.2. Economically useful temperate forage and food legumes of the Mediterranean basin.

Tribe (Genera in Tribe)	Genus	Crops	Number of Species	Useful Taxonomic Treatments
Cicereae Alef. (1 genus)	*Cicer* L.	*C. arietinum* L.	43 spp. (Robertson *et al.*, 2000)	Coles *et al.* (1998) Robertson *et al.* (2000) Van der Maesen (1972) Van der Maesen (1984)
Coronilleae (Adans.) Boiss. (6 genera)	*Coronilla* L.	*C. varia* L. *C. emerus* L. *C. scorpoides* (L.) Koch	20 spp. (Polhill, 1981a)	Chamberlain (1970a) Lassen (1989)
	Hammatolobium Fenzl	*H. lotoides* Fenzl	2 spp. (Polhill, 1981a)	Chamberlain (1970b)
	Hippocrepis L.	*H. multisiliquosa* L. *H. unisiliquosa* L.	21 spp. (Polhill, 1981a)	Chamberlain (1970c)
	Ornithopus L.	*O. sativus* Brot.	6 spp. (Polhill, 1981a)	Chamberlain (1970d)
	Scorpiurus L.	*S. muricatus* L.	2-4 spp. (Polhill, 1981a)	Heyn (1970a)
Crotalarieae (Benth.) Hutch. (6 genera)	*Lotononis* (DC) Eckl. & Zeyh.	L. genistoides (Fenzl) Benth.	100 spp. (Polhill, 1981b)	Chamberlain (1970e)
Galegeae - Astragalinae (Adans.) Benth. (6 genera)	*Astragalus* L.	*A. cicer* L. *A. gummifer* Lab. *A. hamosus* L. *A. sinicus* Boiss.	2,000 spp. (Polhill, 1981c)	Chamberlain and Matthews (1970)
	Biserrula L.	*B. pelecinus* L.	1 sp. (Polhill, 1981c)	Matthews (1970)
Genisteae - Lupininae (Hutch.) Bisby (1 genus)	*Lupinus* L.	*L. albus* L. *L. angustifolius* L. *L. luteus* L. *L. mutabilis* Sweet.	200 spp. (Bisby, 1981)	Bisby (1981) Plitmann and Heyn (1984) Cowling *et al.* (1998)
Hedysareae DC. (7 genera)	*Ebenus* L.	-	20 spp. (Polhill, 1981d)	Huber-Morath (1970a)

	Hedysarum L.	*H. coronarium* L. *H. hedysaroides* (L.) Schinz & Thell. *H. splendens* L.	100 spp. (Polhill, 1981d)	Hedge (1970a) Rechinger (1984)
	Onobrychis Mill.	*O. viciifolia* Scop.	130 spp. (Polhill, 1981d)	Hedge (1970b)
Loteae DC. (6 genera)	*Anthyllis* L.	*A. vulneraria*	20-25 spp. (Polhill, 1981e)	Cullen (1970)
	Cytisopsis Jaub & Spach	*C. dorycniifolia* Jaub & Spach	1 sp. (Polhill, 1981e)	Chamberlain (1970f)
	Hymenocarpus Savi	*H. circinnatus* (L.) Savi	1 sp. (Polhill, 1981e)	Chamberlain (1970g)
	Lotus L.	*L. corniculatus* L. *L. pedunculatus* *L. tenuis* Wald. & Kit ex Wild.	100 spp. (Polhill, 1981e)	Heyn (1970b) Rechinger (1984)
	Tetragonolobus Scop.	*T. purpureus* Moench.	130 spp. (Polhill, 1981e)	Heyn (1970c)
Trifolieae (Bronn) Benth. (7 genera)	*Medicago* L.	*M. sativa* L. *M. arabica* (L.) Huds. *M. falcata* L. *M. lupulina* L. *M. orbicularis* (L.) Bart. *M. polymorpha* L. *M. scutellata* (L.) Mill. *M. truncatula* Gaertn.	50 spp. (Heyn, 1981)	Heyn and Davis (1970) Lesins and Lesins (1979)
	Melilotus Mill.	*M. albus* Medik. *M. indica* (L.) All. *M. officinalis* Lam.	20 spp. (Heyn, 1981)	Chamberlain (1970h)
	Ononis L.	*O. spinosa* L. *O. fruticosa* L.	75 spp. (Heyn, 1981)	Huber-Morath (1970b)

	Trifolium L.	*T. pratense* L. *T. repens* L. *T. ambiguum* Bieb. *T. incarnatum* L. *T. subterraneum* L. *T. hybribum* L. *T. nigrescens* Viv. *T. resupinatum* L. *T. medium* L. *T. michelianum* Savi *T. vesiculosum* Savi *T. fragiferum* L.	250 spp. (Zohary and Heller, 1984)	Zohary (1970) Zohary and Heller (1984)
	Trigonella L.	*T. foenum-graecum* L.	80 spp. (Heyn, 1981)	Huber-Morath (1970c) Rechinger (1984)
Vicieae (Adans.) DC. (5 genera)	*Lathyrus* L.	*L. sativus* L. *L. latifolius* L. *L. sylvestris* L. *L. annuus* L. *L. cicera* L. *L. ochrus* (L.) DC. *L. hirsutus* L.	130 spp. (Kupicha, 1983)	Davis (1970) Kupicha (1983) Campbell (1997)
	Lens Mill.	*L. culinaris* Medik. subsp. *culinaris*	4 spp. (Ferguson *et al.*, 2000)	Ferguson *et al.* (2000) Ladizinsky *et al.* (1984)
	Pisum L.	*P. sativum* L.	3 spp. (Maxted and Ambrose, 2000)	Davis (1970) Polhill and van der Maesen (1985) Maxted and Ambrose (2000)
	Vavilovia A. Fed.	*V. formosa* (Stev.) A. Fed.	1 sp. (Maxted and Ambrose, 2000)	Davis (1970)

Vicia L.	*V. faba* L. *V. narbonensis* L. *V. sativa* L. *V. noeana* Reuter ex Boiss. *V. pannonica* Crantz *V. villosa* Roth *V. ervilia,* (L.) Willd.	140 spp. (Kupicha, 1976)	Davis and Plitmann (1970) Kupicha (1976) Maxted (1993)

References

Akeroyd, J., 1999. Conserving the Mediterranean flora: a way forward. *Plant Talk*, 18: 23-28.
Bisby, F.A., 1981. Genisteae (Adans. Benth). In: *Advances in Legume Systematics* (eds. Polhill, R.M. and Raven, P.M.). pp. 409-425. Royal Botanic Gardens, Kew.
Blamey, M. and Grey-Wilson, C., 1993. *Mediterranean wild flowers*. pp. 1-560. Harper-Collins, London.
Brummitt, R.K. and Powell, C.E., 1992. *Authors of plant names*. Royal Botanic Gardens, Kew.
Campbell, C.G., 1997. *Grasspea (*Lathyrus sativus *L.)*. Promoting the conservation and use of underutilised and neglected crops, 18. Institute of Plant Genetics and Crop Plant Research, Gatersleben / International Plant Genetic Resources Institute, Rome, Italy.
Chamberlain, D.F., 1970a. *Coronilla* L. In: *Flora of Turkey and East Aegean Islands, Vol. 3* (ed. P.H. Davis). pp. 538-544. Edinburgh University Press, Edinburgh.
Chamberlain, D.F., 1970b. *Hammatolobium* Fenzl In: *Flora of Turkey and East Aegean Islands, Vol. 3* (ed. P.H. Davis). pp. 547. Edinburgh University Press, Edinburgh.
Chamberlain, D.F., 1970c. *Hippocrepis* L. In: *Flora of Turkey and East Aegean Islands, Vol. 3* (ed. P.H. Davis). pp. 545-547. Edinburgh University Press, Edinburgh.
Chamberlain, D.F., 1970d. *Ornithopus* L. In: *Flora of Turkey and East Aegean Islands, Vol. 3* (ed. P.H. Davis). pp. 544-545. Edinburgh University Press, Edinburgh.
Chamberlain, D.F., 1970e. *Lotononis* (DC) Eckl. & Zeyh. In: *Flora of Turkey and East Aegean Islands, Vol. 3* (ed. P.H. Davis). pp. 36-37. Edinburgh University Press, Edinburgh.
Chamberlain, D.F., 1970f. *Cytisopsis* Jaub & Spach. In: *Flora of Turkey and East Aegean Islands, Vol. 3* (ed. P.H. Davis). pp. 511-512. Edinburgh University Press, Edinburgh.
Chamberlain, D.F., 1970g. *Hymenocarpus* Savi. In: *Flora of Turkey and East Aegean Islands, Vol. 3* (ed. P.H. Davis). pp. 533. Edinburgh University Press, Edinburgh.
Chamberlain, D.F., 1970h. *Melilotus* L. In: *Flora of Turkey and East Aegean Islands, Vol. 3* (ed. P.H. Davis). pp. 448-452. Edinburgh University Press, Edinburgh.
Chamberlain, D.F. and Matthews, V.A., 1970. *Astragalus* L. In: *Flora of Turkey and East Aegean Islands, Vol. 3* (ed. P.H. Davis). pp. 49-254. Edinburgh University Press, Edinburgh.
Chrtkova-Zertova, A., Van der Maesen, L.J.G. and Rechinger, K. H., 1979. *Flora Iranica. Papilionaceae I - Vicieae.* (Akademische Druck-u. Verlagsanstalt: Graz).
Coles, S., Maxted, N. and van der Maesen, L.J.G., 1998. Identification aids for *Cicer* (Leguminosae, *Cicereae*) taxa. *Edinburgh Journal of Botany*, 55: 243-265.
Cowling, W.A., Buirchell, B.J. and Tapia, M.E., 1998. *Lupin (*Lupinus *L.)*. Promoting the conservation and use of underutilised and neglected crops, 23. Institute of Plant Genetics and Crop Plant Research, Gatersleben / International Plant Genetic Resources Institute, Rome, Italy.
Cullen, J. (1970). *Anthyllis* L. In: *Flora of Turkey and East Aegean Islands, Vol. 3* (ed. P.H. Davis). Edinburgh University Press, Edinburgh. pp. 3: 534.
Davis, P.H., 1970. *The Flora of Turkey and East Aegean Islands, Vol.3.* Edinburgh University Press: Edinburgh.
Davis, P.H. and Plitmann, U., 1970. *Vicia* L. In: *Flora of Turkey and East Aegean Islands, Vol. 3* (ed. P.H. Davis). pp. 274-325. Edinburgh University Press, Edinburgh.
Davis, S., Heywood, V.H. and Hamilton, A.C., 1994. *Centres of plant diversity: a guide and strategy for their conservation. Volume 1, Europe, Africa, South West Asia and the Middle East.* WWF and IUCN, Gland.
Davis, S., Heywood, V.H. and Hamilton, A.C., 1995. *Centres of plant diversity: a guide and*

strategy for their conservation. Volume 2, Asia, Australia and the Pacific. WWF and IUCN, Gland.

Ferguson, M.E., Maxted, N., van Slageren, M. and Robertson, L.D., (2000). A re-assessment of the taxonomy of *Lens* Mill. (Leguminosae, Papilionoideae, Vicieae). *Botanical Journal of the Linnean Society*, 133: 41-59.

Gillett, J.B., 1971. Introduction to the Flora of Tropical East Africa. In: *Flora of Tropical East Africa, Leguminosae* (eds. J.B. Gillett, R.M. Polhill and B. Verdcourt). Whitefriars Press, London and Tunbridge.

Greuter, W., Burdet, H.M. & Long, G., (1989). *Med-Checklist, Vol. 4 - Dicotyledons (Lauraceae - Rhamnaceae).* Med-Checklist Trust of OPTIMA, Geneva.

Groombridge, B., 1992. *Global biodiversity: status of the Earth's living resources.* Chapman & Hall, London.

Hawkes, J.G., N. Maxted & B.V. Ford-Lloyd, 2000. *The* ex situ *conservation of plant genetic resources.* Kluwer, Dordrecht.

Heller, D. and Heyn, C., 1990. Papilionaceae. In: *Conspectus Florae Orientalis: An annotated catalogue of the Flora of the Middle East, Fascicle 5.* pp. 6 - 75. Israel Academy of Science and Humanities, Jerusalem.

Heywood, V.H., 1978. *Flowering Plants of the World.* Oxford University Press: Oxford.

Huber-Morath, A., 1970a. *Ebenus* L. In: *Flora of Turkey and East Aegean Islands, Vol. 3* (ed. P.H. Davis). pp. 590-596. Edinburgh University Press, Edinburgh.

Huber-Morath, A.1970b. *Ononis* L. In: *Flora of Turkey and East Aegean Islands, Vol. 3* (ed. P.H. Davis). pp. 373-384. Edinburgh University Press, Edinburgh.

Huber-Morath, A., 1970c. *Trigonella* L. In: *Flora of Turkey and East Aegean Islands, Vol. 3* (ed. P.H. Davis). pp. 452-482. 560. Edinburgh University Press, Edinburgh.

Kupicha, F.K., 1976. The infrageneric structure of *Vicia. Notes from the Royal Botanic Garden Edinburgh.*, 34: 287-326.

Kupicha, F.K., 1983. The infrageneric structure of *Lathyrus. Notes From The Royal Botanic Garden, Edinburgh*, 41: 209-244.

Lackey, J.A., 1977. A revised classification of the tribe Phaseoleae (Leguminosae, Papilionoideae), and its relation to canavanine distribution. *Botanical Journal of the Linnean Society*, 74: 163-178.

Ladizinsky, G., Braun, D., Goshen, D. and Muehlbauer F. 1984. The biological species of the genus *Lens. Botanical Gazette*, 145: 253-261.

Lassen, P., 1989. A new delimitation of the genera *Coronilla*, *Hippocrepis* and *Securigera. Willdenowia*, 19:49-62.

Lesins, K. A. & I., Lesins, 1979. *Genus* Medicago *(Leguminosae). A Taxogenetic Study.* Dr. W. Junk., The Hague.

Lock, J.M. and Simpson, K. 1991. *Legumes of West Asia. A Check-list.* Royal Botanic Gardens: Kew.

Maxted, N., 1993. A phenetic investigation of *Vicia* L. subgenus *Vicia* (Leguminosae, Vicieae). *Botanical Journal Of The Linnean Society*, 111: 155-182.

Maxted, N., 1995a. *Taxonomy and systematics.* In: Conservation Biology (ed. Spellerberg, I.), pp. 91-103. Longman Group, Harlow, Essex.

Maxted, N., 1995b. *An ecogeographic study of* Vicia *subgenus* Vicia. Systematic and Ecogeographic Studies in Crop Genepools 8. IBPGR, Rome. Pp. 184.

Maxted, N. and Ambrose, M. (2000). *Pisum* L. In: *The Plant Genetic Resources of Legumes in the Mediterranean* (eds. Maxted, N. and Bennett, S.). Kluwer, Dordrecht.

Maxted, N. and Crust, R.,J., 1995. Aids to taxonomic identification. In: *Collecting plant genetic diversity: technical guidelines* (eds. Guarino, L., Ramanatha Rao, V. & Reid, R.), pp. 181-194. CAB International, Wallingford.

Meikle, R.D., 1977. *The Flora of Cyprus, Vol. 1.* Bentham-Moxon Trust, Royal Botanic Gardens: Kew.

Mouterde, P., 1966. *Nouvelle Flore du Liban et de la Syrie. Vol. 2.* Librairie Orientale: Beirut, Lebanon.

Plitmann, U. and C.C. Heyn, 1984. *Old world* Lupinus*: taxonomy, evolutionary relationships and links with the new world species.* Proc. 3rd. Int. Lupin Conf. La Rochelle, June 1984, pp. 40-54.

Polhill, R.M., 1981a. Coronilleae (Adans.) Boiss. In: *Advances in Legume Systematics* (eds. Polhill, R.M. and Raven, P.M.). pp. 375. Royal Botanic Gardens, Kew.

Polhill, R.M., 1981b. Crotalarieae (Benth.) Hutch. In: *Advances in Legume Systematics* (eds. Polhill, R.M. and Raven, P.M.). pp. 399-402. Royal Botanic Gardens, Kew.

Polhill, R.M., 1981c. Galegeae - Astragalinae (Adans.) Benth. In: *Advances in Legume Systematics* (eds. Polhill, R.M. and Raven, P.M.). pp. 357-363. Royal Botanic Gardens, Kew.

Polhill, R.M., 1981d. Hedysareae DC. In: *Advances in Legume Systematics* (eds. Polhill, R.M. and Raven, P.M.). pp. 367-370. Royal Botanic Gardens, Kew.

Polhill, R.M., 1981e. Loteae DC. In: *Advances in Legume Systematics* (eds. Polhill, R.M. and Raven, P.M.). pp. 371-374. Royal Botanic Gardens, Kew.

Polhill, R.M. and Raven, P.H., 1981. *Advances in Legume Systematics, Part One*. Royal Botanic Gardens: Kew.

Polhill, R.M., Raven, P.H. and Stirton, C.H., 1981. Evolution and Systematics of the Leguminosae. In: *Advances in Legume Systematics* (Eds. R.M. Polhill and P.H. Raven). Royal Botanic Gardens: Kew.

Polhill, R.M. and L.J.G., van der Maesen, 1985. Taxonomy of grain legumes. In: *Grain Legume Crops*. pp. 3-36 (eds. R.J. Summerfield and E.H. Roberts). Collins, London.

Polunin, O., 1987. *Flowers of Greece and the Balkans. A Field Guide*. Oxford University Press: Oxford.

Polounin, O. and Huxley, A.,1970. *Flowers of the Mediterranean*. Chatto and Windus, London.

Polounin, O. and Smythies, B.E., 1988. *Flowers of South-west Europe. A Field Guide*. Oxford University Press: Oxford.

Rechinger, K. H., 1984. *Flora Iranica. Papilionaceae II.* Akademische Druck-u. Verlagsanstalt, Graz.

Schischkin, B.K., 1945. *Flora of the USSR Vol. XI.* Izdatel'stvo Akademii Nauk SSSR: Moscow.

Schischkin, B.K. and Bobrov, E.G., 1948. *Flora of the USSR Vol. XIII.* Izdatel'stvo Akademii Nauk SSSR: Moscow.

Townsend, C.C., 1974. *Flora of Iraq. Vol.3.* Ministry of Agriculture and Agrarian Reform of the Republic of Iraq: Bagdad.

Tutin, T.G., Heywood, V.H., Burges, N.A., Moore, D.M., Valentine, D.H., Walters, S.M., and Webb, D.A., 1968. *Flora Europaea, Vol. 2.* Cambridge University Press: Cambridge.

van der Maesen, L.J.G., 1972. *Cicer* L., *A Monograph of the Genus with Special Reference to the Chickpea (*Cicer arietinum *L.), its Ecology and Cultivation.* H. Veenman and Zonen N.V: Wageningen.

van der Maesen, L.J.G., 1984. Taxonomy, distribution and evolution of the chickpea and its wild relatives. In: *Genetic Resources and their Exploitation - Chickpea, Faba Beans and Lentils* (eds. J.R. Witcombe and W. Erskine), pp. 95-104. Martinus Nijhoff / Dr. W. Junk Publishers: The Hague.

van der Maesen, L.J.G., 1987. Origin, history and taxonomy of chickpea. In: *The Chickpea* (eds. Saxena, M.C. and Singh, K.B.). pp. 11-34. CAB/ICARDA, Oxon, England.

Vavilov, N.I., 1997. *Five continents*. International Plant Genetic Resources Institute, Rome, Italy.

Woodgate, K., Maxted, N. And Bennett, S, (1999). A generic conspectus for the forage legumes of the Mediterranean basin. In: *Genetic resources of Mediterranean Pasture and Forage Legumes* (eds. Bennett, S.J. and Cocks, P.S.) pp. 182-226. Kluwer Academic Publishers, Dordrecht.

Yakovlev, G.P, Sytin, A.K. and Roskov, Y.R., 1996. Legumes of northern Eurasia. pp. 1-724. Royal Botanic Gardens, Kew.

Zohary, M., 1970. *Trifolium* L. In: *Flora of Turkey and East Aegean Islands, Vol. 3* (ed. P.H. Davis). pp. 384-391. Edinburgh University Press: Edinburgh.

Zohary, M., 1972. *Flora Palaestina, Vol. 2.* pp. 158-162. Israel Academy of Sciences and Humanities: Jerusalem, Israel.

Zohary, M. and D. Heller, 1984. *The Genus* Trifolium. The Israel Academy of Sciences and Humanities.

Appendix 1. Bracketed Key To Temperate Forage Legume Genera (Woodgate *et al.*, 1999).

1(0).	Leaves simple	2
	Leaves compound	9
2(1).	Stipules present	3
	Stipules absent	8
3(2).	Stipules conspicuous	4
	Stipules small and inconspicuous	5
4(3).	Fruit turgid or inflated	*Astragalus*
	Fruit laterally compressed	*Coronilla*
5(3).	Fruit curved or coiled; inflorescence axillary	6
	Fruit straight; inflorescence terminal	*Anthyllis*
6(5).	Calyx tubular; calyx teeth bilabiate; fruit terete	*Scorpiurus*
	Calyx campanulate; calyx teeth subequal; fruit laterally compressed	7
7(6).	Bracts scarious; ovules numerous; seed oblong	*Coronilla*
	Bracts foliaceous; ovules few; seed reniform	*Hymenocarpus*
8(2).	Inflorescence axillary; fruit dehiscent; fruit exserted from calyx	*Cytisopsis*
	Inflorescence terminal; fruit indehiscent; fruit enclosed in calyx	*Anthyllis*
9(1).	Leaves with terminal leaflet; style glabrous	10
	Leaves usually with terminal tendril or mucro; style pubescent	67
10(9).	Anthers monomorphic	11
	Anthers dimorphic	66
11(10).	Leaflet arrangement digitate	12
	Leaflet arrangement trifoliate	14
	Leaflet arrangement pinnate	37
12(11).	Stipules present; fruit indehiscent	13
	Stipules absent; fruit dehiscent	*Cytisopsis*
13(12).	Stamens monadelphous; stipules small and inconspicuous; stem or leaf surface pubescent; stipules scarious; fruit septa present	*Hammatolobium*
	Stamens diadelphous; stipules conspicuous; stem or leaf surface glabrous; stipules foliaceous; fruit septa absent	*Trifolium*
14(11).	Stigma terminal	15
	Stigma lateral	34
15(14).	Stamens monadelphous	16
	Stamens diadelphous	18
16(15).	Stem or leaf surface glabrous; margins of leaf or leaflet not entire	*Melilotus*
	Stem or leaf surface pubescent or pilose; margins of leaf or leaflet entire	17
17(16).	Inflorescence axillary; standard orbicular	*Hammatolobium*
	Inflorescence terminal; standard ovate	*Anthyllis*
18(15).	Ovules few	19
	Ovules numerous	26
19(18).	Stem or leaf surface glabrous	20
	Stem or leaf surface pubescent or pilose	22
20(19).	Fruit turgid or inflated	*Astragalus*
	Fruit laterally compressed	21
21(20).	Fruit exserted from calyx	*Melilotus*
	Fruit enclosed in calyx	*Trifolium*

22(19). Stipules conspicuous 23
Stipules small and inconspicuous 24
23(22). Wing obovate .... *Cicer*
Wing oblong .... *Astragalus*
24(22). Margins of leaf or leaflet entire 25
Margins of leaf or leaflet not entire .... *Cicer*
25(24). Inflorescence axillary .... *Ebenus*
Inflorescence terminal .... *Anthyllis*
26(18). Inflorescence axillary 27
Inflorescence terminal 33
27(26). Style filiform 28
Style linear 29
Style subulate 30
Style incurved 31
28(27). Stipules fused with petiole; wing oblong .... *Trigonella*
Stipules free from petiole; wing obovate .... *Cicer*
29(27). Fruit septa present .... *Astragalus*
Fruit septa absent .... *Trigonella*
30(27). Leaves trifoliate; seed reniform .... *Medicago*
Leaves imparipinnate or upper trifoliate and lower simple; seed oblong .... *Coronilla*
31(27). Margins of leaf or leaflet entire 32
Margins of leaf or leaflet not entire .... *Cicer*
32(31). Bracts scarious .... *Coronilla*
Bracts foliaceous .... *Lotus*
33(26). Stipules conspicuous .... *Astragalus*
Stipules small and inconspicuous .... *Anthyllis*
34(14). Fruit exserted from calyx 35
Fruit enclosed in calyx .... *Trifolium*
35(34). Bracts scarious; style subulate .... *Medicago*
Bracts foliaceous; style incurved 36
36(35). Margins of leaf or leaflet entire; stipules small and inconspicuous; calyx .... *Lotus*
Margins of leaf or leaflet not entire; stipules conspicuous; calyx subulate
.... *Tetragonolobus*
37(11). Stipules conspicuous 38
Stipules small and inconspicuous 44
38(37). Fruit turgid or inflated 39
Fruit laterally compressed 40
39(38). Wing obovate .... *Cicer*
Wing oblong .... *Astragalus*
40(38). Fruit septa present 41
Fruit septa absent 73
41(40). Bracteoles present .... *Hedysarum*
Bracteoles absent .... *Coronilla*
42(40). Stipules fused with petiole .... *Trifolium*
Stipules free from petiole 43
43(42). Stem winged .... *Lathyrus*
Stem without wings .... *Vicia*

44(37). Inflorescence axillary 45
Inflorescence terminal 64

45(44). Fruit turgid or inflated 46
Fruit terete 48
Fruit laterally compressed 52
46(45). Margins of leaf or leaflet entire 47
Margins of leaf or leaflet not entire .... *Cicer*
47(46). Keel blunt .... *Dorycnium*
Keel beaked or pointed .... *Lotus*
48(45). Bracts scarious 49
Bracts foliaceous 50
49(48). Bracteoles present .... *Ornithopus*
Bracteoles absent .... *Hippocrepis*
50(48). Stamens monadelphous; fruit indehiscent .... *Hammatolobium*
Stamens diadelphous; fruit dehiscent 51
51(50). Keel blunt .... *Dorycnium*
Keel beaked or pointed .... *Lotus*
52(45). Fruit septa present 53
Fruit septa absent 60
53(52). Bracts scarious 54
Bracts setaceous .... *Hedysarum*
Bracts foliaceous 57
54(53). Bracteoles present 55
Bracteoles absent 56
55(54). Style filiform .... *Hedysarum*
Style incurved .... *Ornithopus*
56(54). Seed oblong .... *Coronilla*
Seed arcuate .... *Hippocrepis*
57(53). Stamens monadelphous .... *Hammatolobium*
Stamens diadelphous 58
58(57). Bracteoles present; fruit enclosed in calyx; style filiform .... *Ebenus*
Bracteoles absent; fruit exserted from calyx; style incurved 9
59(58). Fruit dehiscent; ovules numerous; lower leaflets stipuloid .... *Lotus*
Fruit indehiscent; ovules few; lower leaflets not stipuloid .... *Hymenocarpus*
60(52). Fruit arched 61
Fruit circular .... *Onobrychis*
Fruit linear 62
61(60). Bracts scarious; fruit indehiscent; ovules few; calyx tubular .... *Onobrychis*
Bracts foliaceous; fruit dehiscent; ovules numerous; calyx campanulate .... *Lotus*
62(60). Keel blunt .... *Vicia*
Keel curved .... *Securigera*
Keel beaked 63
Keel pointed .... *Lotus*
63(62). Fruit dehiscent; wing obovate .... *Lotus*
Fruit indehiscent; wing oblong .... *Securigera*

64(44). Fruit dehiscent .... *Dorycnium*
Fruit indehiscent 65

65(64). Stem or leaf surface glabrous; standard orbicular .... *Securigera*
Stem or leaf surface pubescent or pilose; standard ovate .... *Anthyllis*

66(10). Stipules conspicuous; glands on leaf lamina present; ovary stipitate .... *Ononis*
Stipules small and inconspicuous; glands on leaf lamina absent; ovary sessile 67

67(66). Leaves digitate or trifoliolate; 1, 3 or rarely 5 leaflets per leaf. Stipules free petiole. Flowers solitary, raceme or umbels .... *Lotononis*
Leaves palmate; 5-15 leaflets per leaf. Stipules adnate to base of petiole. Flowers raceme or spike .... *Lupinus*

68(9). Leaves glandular; style filiform, incurved .... *Cicer*
Leaves never glandular; style compressed with dilated apex 69

69(68). Stipules considerably bigger than the leaves; stems terete; leaves with terminal tendrils; style folded longitudinally .... *Pisum*
Stipules smaller than the leaves; stems angled or winged; leaves with or without terminal tendrils; style not folded longitudinally 70

70(69). Style pubescent all over or only on abaxial face (rarely); stem not winged *Vicia*
Style pubescent only on adaxial surface; stem may be winged 71

71(70). Dwarf perennials; leaflets 1-paired; style with reflexed margins .... *Vavilovia*
Annuals or perennials; leaflets 1-many paired; style without reflexed margins 72

72(71). Legumes usually more than 2-seeded; more than 15mm long; seeds more or less globose; annuals or perennials .... *Lathyrus*
Legumes 1-2 seeded; usually less than 15mm long; seeds compressed; annuals .... *Lens*

73(40). Leaves trifoliate; no terminal tendril or mucro .... *Trifolium*
Leaves imparipinnate or paripinnate; terminal tendril or mucro 74

74(73). Few leaflets; winged stem; style pubescent on upper side only .... *Lathyrus*
Many leaflets; stem not winged; style pubescent all over or on lower side only ... *Vicia*

Appendix 2. Tabular Key To Temperate Forage Legume Genera (Woodgate *et al.*, 1999).

NB. Character numbers and character state codes are explained in Appendix 1.

Astragalus	11a 19a 60a 47a
Coronilla	11a 19a 60a 47c
Scorpiurus	11a 19a 60b 25a 47b 35b 34a
Coronilla	11a 19a 60b 25a 47c 35a 34b 37a 61b 53c
Hymenocarpus	11a 19a 60b 25a 47c 35a 34b 37c 61a 53a
Anthyllis	11a 19a 60b 25b
Cytisopsis	11a 19b 25a 50a 51a
Anthyllis	11a 19b 25b 50b 51b
Hammatolobium	11b 43a 40a 62a 19a 50b 39a 60b 9b 20a 49a
Trifolium	11b 43a 40a 62a 19a 50b 39b 60a 9a 20b 49b
Lupinus	11b 43a 40a 62a 19a 50a
Cytisopsis	11b 43a 40a 62a 19b 50a
Melilotus	11b 43a 40a 62b 41a 39a 9a 16b
Hammatolobium	11b 43a 40a 62b 41a 39a 9b 16a 25a 28c
Anthyllis	11b 43a 40a 62b 41a 39a 9b 16a 25b 28d
Astragalus	11b 43a 40a 62b 41a 39b 61a 9a 47a
Melilotus	11b 43a 40a 62b 41a 39b 61a 9a 47c 51a
Trifolium	11b 43a 40a 62b 41a 39b 61a 9a 47c 51b
Cicer	11b 43a 40a 62b 41a 39b 61a 9b 60a 30b
Astragalus	11b 43a 40a 62b 41a 39b 61a 9b 60a 30d
Ebenus	1b 43a 40a 62b 41a 39b 61a 9b 60b 16a 25a
Anthyllis	11b 43a 40a 62b 41a 39b 61a 9b 60b 16a 25b
Cicer	11b 43a 40a 62b 41a 39b 61a 9b 60b 16b
Trigonella	11b 43a 40a 62b 41a 39b 61b 25a 42a 21a 30d
Cicer	11b 43a 40a 62b 41a 39b 61b 25a 42a 21b 30b
Astragalus	11b 43a 40a 62b 41a 39b 61b 25a 42b 49a
Trigonella	11b 43a 40a 62b 41a 39b 61b 25a 42b 49b
Medicago	11b 43a 40a 62b 41a 39b 61b 25a 42D 53a
Coronilla	11b 43a 40a 62b 41a 39b 61b 25a 42D 53c
Coronilla	11b 43a 40a 62b 41a 39b 61b 25a 42E 16a 37a
Lotus	11b 43a 40a 62b 41a 39b 61b 25a 42E 16a 37c
Cicer	11b 43a 40a 62b 41a 39b 61b 25a 42E 16b
Astragalus	11b 43a 40a 62b 41a 39b 61b 25b 60a
Anthyllis	11b 43a 40a 62b 41a 39b 61b 25b 60b
Medicago	11b 43a 40a 62b 41b 51a 37a 42d
Lotus	11b 43a 40a 62b 41b 51a 37c 42e 16a 60b 34b
Tetragonolobus	11b 43a 40a 62b 41b 51a 37c 42e 16b 60a 34c
Trifolium	11b 43a 40a 62b 41b 51b
Cicer	11b 43a 40a 62c 60a 47a 30b
Astragalus	11b 43a 40a 62c 60a 47a 30d
Hedysarum	11b 43a 40a 62c 60a 47c 49a 38a
Coronilla	11b 43a 40a 62c 60a 47c 49a 38b
Vicia	11b 43a 40a 62c 60a 47c 49b
Lathyrus	11b 43a 40a 62c 60a 47c 49b

Trifolium	11b 43a 40a 62c 60a 47c 49b
Dorycnium	11b 43a 40a 62c 60b 25a 47a 16a 31b
Lotus	11b 43a 40a 62c 60b 25a 47a 16a 31d
Lotus	11b 43a 40a 62c 60b 25a 47a 16a 31e
Cicer	11b 43a 40a 62c 60b 25a 47a 16b
Ornithopus	11b 43a 40a 62c 60b 25a 47b 37a 38a
Hippocrepis	11b 43a 40a 62c 60b 25a 47b 37a 38b
Hammatolobium	11b 43a 40a 62c 60b 25a 47b 37c 39a 50b
Dorycnium	11b 43a 40a 62c 60b 25a 47b 37c 39b 50a 31b
Lotus	11b 43a 40a 62c 60b 25a 47b 37c 39b 50a 31d
Lotus	11b 43a 40a 62c 60b 25a 47b 37c 39b 50a 31e
Hedysarum	11b 43a 40a 62c 60b 25a 47c 49a 37a 38a 42a
Ornithopus	11b 43a 40a 62c 60b 25a 47c 49a 37a 38a 42e
Coronilla	11b 43a 40a 62c 60b 25a 47c 49a 37a 38b 53c
Hippocrepis	11b 43a 40a 62c 60b 25a 47c 49a 37a 38b 53f
Hedysarum	11b 43a 40a 62c 60b 25a 47c 49a 37b
Hammatolobium	11b 43a 40a 62c 60b 25a 47c 49a 37c 39a
Ebenus	11b 43a 40a 62c 60b 25a 47c 49a 37c 39b 38a 51b
Lotus	11b 43a 40a 62c 60b 25a 47c 49a 37c 39b 38b 51a
Hymenocarpus	11b 43a 40a 62c 60b 25a 47c 49a 37c 39b 38b 51a
Onobrychis	11b 43a 40a 62c 60b 25a 47c 49b 46b 37a 50b 61a
Lotus	11b 43a 40a 62c 60b 25a 47c 49b 46b 37c 50a 61b
Onobrychis	11b 43a 40a 62c 60b 25a 47c 49b 46c
Vicia	11b 43a 40a 62c 60b 25a 47c 49b 46e 31b
Securigera	11b 43a 40a 62c 60b 25a 47c 49b 46e 31c
Lotus	11b 43a 40a 62c 60b 25a 47c 49b 46e 31d 50a 30b
Securigera	11b 43a 40a 62c 60b 25a 47c 49b 46e 31d 50b 30d
Lotus	11b 43a 40a 62c 60b 25a 47c 49b 46e 31e
Dorycnium	11b 43a 40a 62c 60b 25b 50a
Securigera	11b 43a 40a 62c 60b 25b 50b 9a 28c
Anthyllis	11b 43a 40a 62c 60b 25b 50b 9b 28d
Ononis	11b 43a 40b 60a 18a 45b
Lotononis	11b 43a 40b 60b 18b 45a
Vicia	11b 43b 13a 44a
Lathyrus	11b 43b 13a 44b 58a
Lens	11b 43b 13a 44b 58b 35a 9b
Pisum	11b 43b 13a 44b 58b 35b 9a
Cicer	11b 43b 13a 44c
Vicia	11b 43b 13b 44a
Lathyrus	11b 43b 13b 44b 60a 58a
Lens	11b 43b 13b 44b 60a 58b
Vavilovia	11b 43b 13b 44b 60b
Cicer	11b 43b 13b 44c

Generic Character Set Used In Appendix 2.

1-4. Curatorial information.

5. life span: a, annual; b, biennial; c, perennial.

6. habit: a, plant a herb <without persistent stem and definite wood structure>; b, plant a shrub <low, woody>.

7. height range / cm.

8. growth habit: a, plant erect <without climbing>; b, plant climbing; c, plant ascending <indirectly upwards>; d, plant decumbent <lying on ground but with leaf ascending>; e, plant prostrate <lying flat>; f, plant sprawling or creeping.

9. stem or leaf surface: a, glabrous; b, pubescent or pilose.

10. spines <on stems or leaves>: a, present; b, absent.

11. leaves <type>: a, simple; b, compound.

12. leaf apex: a, paripinnate; b, imparipinnate.

13. leaves with terminal: a, tendril (more than 5mm); b, mucro (less than 5mm).

14. leaflet number per leaf (range).

15. leaflet shape: a, linear or filiform <long and narrow>; b, lanceolate <wide at base, tapering at apex>; c, ovate or cordate <egg or heart shaped>; d, elliptic <widest in middle>; e, oblong <broad with sides nearly parallel>; f, obovate or cuneate <with apex broader than base>.

16. margins of leaf or leaflet: a, entire; b, not entire.

17. leaflet lamina: a, glabrous; b, pubescent or pilose.

18. glands on leaf lamina: a, present; b, absent.

19. stipules: a, present; b, absent.

20. stipules <form>: a, scarious <membrane-like>; b, foliaceous; c, tubercule-like.

21. stipules <fusion to petiole>: a, fused with petiole; b, free from petiole.

22. stipels: a, present; b, absent.

23. flowers arrangement: a, solitary; b, inflorescence.

24. inflorescence arranged in: a, heads; b, racemes; c, spikes; d, umbels.

25. inflorescence position: a, axillary; b, terminal.

26. flower number per inflorescence (range).

27. corolla colour: a, white; b, pink; c, purple; d, blue; e, yellow; f, red.

28. standard shape: a, obovate <terminal half broader than basal>; b, obcordate <lobed and wider at apex>; c, orbicular <circular>; d, ovate <egg-shaped>; f, acuminate <sides concave, to point>; g, oblong <longer than broad, sides parallel>;

29. claw description.

30. wing shape: a, falcate <sickle-shaped>; b, obovate; c, ovate; d, oblong; e, obtuse <blunt>.

31. keel shape: a, oblique <slanting>; b, blunt; c, curved; d, beaked; e, pointed.

32. relative sizes of petals.

33. peduncle: a, present <inflorescence pedunculate>; b, absent <inflorescence sessile>.

34. calyx shape: a, tubular; b, campanulate; c, subulate.

35. calyx teeth: a, subequal; b, bilabiate.

36. bracts: a, present; b, absent.

37. bracts type: a, scarious <membrane-like>; b, setaceous <bristled>; c, foliaceous.

38. bracteoles: a, present; b, absent.

39. stamens: a, monadelphous; b, diadelphous.

40. anthers: a, monomorphic; b, dimorphic.

41. stigma: a, terminal; b, lateral.

42. style shape: a, filiform <thread-like>; b, linear; c, spatulate <spoon-shaped>; d, subulate

<taper from base to apex>; f, incurved.

43. style hairiness: a, glabrous; b, pubescent.
44. style pubescence: a, complete; b, on adaxial surface; c, on abaxial surface.
45. ovary: a, sessile; b, stipitate.
46. fruit shape: a, oval; b, arched; c, circular; d, coiled; e, linear.
47. fruit transverse: a, turgid or inflated; b, terete <cylindrical>; c, laterally compressed.
48. fruit surface: a, smooth and glabrous; b, not smooth.
49. fruit septa: a, present; b, absent.
50. fruit dehiscence: a, dehiscent; b, indehiscent.
51. fruit exertion: a, exserted from calyx; b, enclosed in calyx.
52. pod length range / mm.
53. seed shape: a, reniform; b, compressed; c, oblong; d, globose; e, ovoid; f, arcuate.
54. seeds number per fruit (range).
55. flowering time.
56. altitude range / m asl.
57. typical habitats include: a, rocky ground; b, steppes; c, scrub; d, macchie; e, deciduous woodland; f, coniferous woodland; g, fallow, meadow or pasture; h, cultivated land; i, roadside; j, marsh or damp areas; k, seashore.
58. stem: a, winged; b, without wings.
59. lower leaflets: a, stipuloid; b, not stipuloid.
60. stipules: a, conspicuous; b, small and inconspicuous.
61. ovules <number>: a, few <up to 3>; b, numerous <more than 3>.
62. leaflet arrangement: a, digitate; b, trifoliate; c, pinnate.

Appendix 3. Glossary (definitions from Hickey and King, 1988 and Toothill, 1984).

Abaxial The side of an organ facing away from the main axis ie. the lower surface; dorsal
Acuminate An acute apex with concave sides, tapering to a protracted point
Adaxial The side of an organ towards the axis ie. the upper surface; ventral
Annual A plant which lasts for one year from seed to maturity and death
Arched Curved or bowed (as arcuate)
Arcuate As arched
Ascending Rising up obliquely or indirectly
Axillary In an axil (the upper angle that a petiole or peduncle makes with the stem)
Beak A long, prominent point
Biennial A plant that takes two years to complete its life cycle
Bilabiate Two-lipped
Blunt Without a sharp edge or point
Bracteoles A leaf-like organ subtending an inflorescence that is itself subtended by a bract
Bracts A leaf-like organ subtending an inflorescence
Calyx The collective term for the sepals, constituting the outer whorl of the perianth
Campanulate Bell-shaped
Circular Having form of a circle
Claw The long narrow petiole-like base of petals or sepals
Climbing Growing upwards with support by tendrils or twining with a surface
Coiled Twisted into circular or spiral shape; arranged in concentric rings
Complete Entire, having all parts
Compound(leaf) A leaf of two or more leaflets
Compressed Flattened, especially laterally
Corolla Collective term for the petals, constituting the inner whorl of the perianth
Cordate Heart-shaped with the lobe at the base
Cuneate Wedge-shaped or triangular with the narrow end at the point of attachment
Curved Surface with no part straight
Decumbent Lying on the ground but with end ascending
Dehiscent The splitting open of a plant organs such as anthers or pods
Dentate Toothed; with sharp notches perpendicular to the margin
Diadelphous An androecium comprised of two sets of stamens
Digitate Hand-like. A compound leaf with all the leaves arising from one point
Dimorphic Occurring in two forms
Elliptic Oval in outline, widest in the middle
Enclosed Surrounded, contained
Entire margin With an edge completely continuous; not indented at all
Exserted Sticking out; not included
Falcate Sickle-shaped
Filament The stalk of a stamen with the anther at the apex
Filiform Thread-like
Foliaceous Leaf-like
Glabrous Not hairy
Gland A secreting part or appendage
Glaucous Surface with a bloom or whitish-grey substance that rubs-off
Globose Rounded
Head A short, dense spike

Herb A plant without persistent stem or definite wood structure
Keel The two lower fused petals of a papilionaceous flower which forms a boat-like structure
Imparipinnate A terminal leaflet is present and the total number of leaflets is an odd number
Incurve Bend into a curve, curve inwards
Inflorescence A flowering system consisting of more than one flower; a cluster
Indehiscent Not regularly opening, as seed pod or anther
Lamina The bladelike portion of a leaf, usually flattened; a blade or expanded portion
Lanceolate Longer than broad; widens above the base and tapers to the apex
Lateral On or at the side
Linear Long and narrow, the sides parallel or almost so
Lobed Divided into curved or rounded parts
Monadelphous Stamens united into one group by connation of their filaments
Monomorphic Occurring in a single form
Mucro A fine small point arising abruptly, usually at the tip
Obcordate Deeply lobed at the apex
Oblique Slanting; unequal-sided
Oblong Broad with the sides nearly parallel
Obovate The terminal half broader than the basal
Obtuse Blunt, rounded
Orbicular Circular or disc-shaped
Ovary The swollen basal part of the carpel, containing the ovules
Oval Egg-shaped or ellipsoidal
Ovate Egg-shaped
Ovoid Oval in flat outline
Ovules The egg-containing unit of the ovary which becomes the seed after fertilisation
Paripinnate No terminal leaflet present and the total number of leaflets is an even number
Peduncle Stalk of a flower cluster
Perennial A plant that lives for many years
Petiole The stalk that attaches the leaf lamina to the stem
Pilose Long soft hairs
Pinnate Compound leaf in which the leaflets are arranged in two rows, either side of the midrib
Pitted With little depressions or cavities
Pointed With a sharp tip or end
Prostrate Lying flat
Pubescent Covered in short, soft hairs
Raceme A elongated, indeterminate inflorescence in which the flowers are formed on individual pedicels on the main axis
Reniform Kidney-shaped
Rugose Wrinkled
Scarious Dry, thin and membranous, often translucent
Septa A partition or cross wall
Sessile Not stalked
Setaceous Bearing bristles
Setose Covered in bristles
Shrub A low, woody plant

Simple (leaves) Not compounded into leaflets
Spatulate Spoon-shape
Spike An inflorescence in which the flowers are sessile and borne on an elongated axis
Spine A modified leaf or part of a leaf forming a sharp pointed structure; a sharp woody body arising from the wood of the stem
Sprawling Spreading out or straggling
Stamens The unit of the androecium and typically composed of anther and filament
Standard The upper, broad, often erect petal of a papilionaceous flower
Stigma The receptive part of the carpel that receives the pollen
Stipels Stipules of a leaflet
Stipitate Born on a short stalk
Stipules A pair of structures at the base of a leaf or along a petiole. They may be leaf-like, spines, glands or scales
Stipuloid Resembling stipules
Style The part of the pistil between the ovary and the stigma
Subequal More or less the same
Subulate Shaped like an awl ie. narrow, tapering from base to apex, more or less flattened
Tendril A threadlike process or extension by which a plant grasps an object. It may be a modified stem or leaf
Terete Circular in transverse section; imperfectly cylindrical
Terminal At the tip, apical or distal end
Trifoliate A compound leaf with three leaflets arising from the same point
Tubercule A small tuber or rounded protruding body
Tubular Shaped as a long hollow cylinder
Turgid Swollen from fullness
Umbel An inflorescence in which the flowers are borne on undivided pedicels originating from a common node on the main axis. The whole inflorescence is flat topped, resembling an umbrella
Villous Long soft hairs, shaggy
Wing The lateral petals of a papilionaceous flower / The membranous outgrowth of certain fruits / A flange running down a stem or stalk

Section 2

Genetic Diversity of Legumes in the Mediterranean

Chapter 4

Clovers (*Trifolium* L.)

Emma-Jane Lamont, Aziza Zoghlami, Ruaraidh Sackville Hamilton and Sarita Jane Bennett

4.1 INTRODUCTION

Natural grasslands that contain a range of *Trifolium* species are an important source of forage in most countries of the Mediterranean basin (Russi *et al.*, 1992). *Trifolium* species are also exploited in ley-farming systems, for which *T. subterraneum* cultivars have proved particularly successful because they are adapted to a wide variety of climatic and edaphic conditions (Ewing, 1996). Several species of *Trifolium* are cultivated in intensive agricultural systems in association with companion grass species in simple or complex seeds mixtures. A survey of 99 research stations (Russell and Webb, 1976) revealed that *Trifolium* species are among the most valuable forage legumes in all parts of the world, except the tropics. In temperate areas, *Trifolium* species are important nitrogen fixers, improving the quality of pasture in both natural and cultivated grassland. Alfalfa (*Medicago sativa*) typically out-yields clovers on an annual basis when harvested for hay, but clovers have a wider natural and cultivated distribution (Keuren and Hoveland, 1985).

Approximately 240 *Trifolium* species are recognised worldwide, of which 217 species names are accepted by ILDIS (International Legume Database and Information System). They are native to all the humid, temperate regions of the world except the Himalayas and Australasia (Zohary and Heller, 1984). The natural distribution of several clover species is difficult to determine because of confusion with their cultivated distribution. Man has both deliberately and unconsciously increased the distribution of species used in agriculture, although deliberate cultivation of pastures has only occurred since the 3rd century AD (Evans 1976). Centres of diversity in *Trifolium* occur in the Eastern Mediterranean, East Africa and South America. (Zohary, 1972). The Mediterranean centre contains the most species, and includes many species that have become commercially important. Approximately 150 *Trifolium* species occur in the Mediterranean, representing 7 out of 8 sections of the genus, and of these 7 sections all but one have their predominant number of species here (Zohary and Heller, 1984).

It would be extremely difficult to quantify the contribution made by *Trifolium* species to agriculture in the Mediterranean basin. Katznelson (1974) was unable to calculate the value of naturally occurring populations of *T. subterraneum* to Mediterranean countries. In addition to the large number of *Trifolium* species, the legume component of grasslands varies widely. A study of Syrian grassland composition found that legumes (including *Trifolium*) made up less than three percent of plants (Cocks and Osman, 1996). The legume composition of high altitude grassland communities studied in Georgia ranged from 0.5 up to 51% (Patterson and Espie, 1997).

4.2 TAXONOMIC SUMMARY

The genus *Trifolium* L. is a member of the Fabaceae (Leguminosae) family, subfamily Papilionaceae. *Trifolium* is grouped with closely related genera, *Trigonella, Medicago* and *Melilotus*, in tribe Trifolieae. Many revisions of the genus have been published and several detailed reviews of the taxonomic history are available, for example Hossain (1961), Gillet (1970) and Zohary and Heller

(1984).

In the fifth edition of Genera Plantarum, Linnaeus (1754) circumscribed the genus *Trifolium* by including Tournefort's *Trifolium* and *Melilotus*, Micheli's *Trifoliastrum* and Buxbaum's *Lupinaster*. Forty species of *Trifolium* divided into five groups were described in the first edition of Species Plantarum (Linnaeus, 1753). Four *Trifolium* species groups in the 'Species Plantarum' (Linnaeus, 1753) are essentially retained in more recent classifications. However, Linnaeus' groups cannot easily be recognised in current classifications because they have been subdivided many times.

In *Trifolium* species the claws of the wings and keel are fused to the stamens and the inflorescences are umbellate or capitate (or rarely a single flower). Species of the genus *Melilotus* are distinguished by their petal claws, which are free from the staminal tube, and their elongate inflorescences. *Trifolium* species have ovate or oblong pods that are no longer than the keel and are included in the calyx. In contrast *Trigonella* species have long pods that are straight, falcate or arcuate (Hossain, 1961). Other features characteristic of *Trifolium*, include the usually persistent corolla and often indehiscent pod that usually contains one seed. Species of the genus *Medicago* generally have non-persistent corollas and their pods contain more than one seed (Heyn 1981).

Presl (1832) identified nine groups of species, which he ranked as independent genera. The proposal that the genus *Trifolium* should be split into several genera was not subsequently accepted. However, the groupings recognised by Presl formed the basis of later infra-generic groupings of *Trifolium* species. Bobrov (1967) also proposed that *Trifolium sensu stricto* should be split, but into eleven genera. The revision was founded on Bobrov's comprehensive treatise on clovers of the USSR. Bobrov included two '*Trifolium*' genera in the tribe Trifolieae Bronn., the other nine genera were placed with the genus *Lupinus* L. in a newly proposed tribe; Lupineae Bobr.

Revisions of Mediterranean *Trifolium* species and of the entire genus have recognised the integrity of *Trifolium sensu stricto*. For example in the Flora Orientalis (Boissier, 1872) the genus *Trifolium* is divided into eight sections very similar in content to the species proposed by Presl (1832). Hossain's (1961) revision of *Trifolium* in the eastern Mediterranean counties grouped species in a similar way to Boissier (and thus Presl, 1832). However, Hossain ranked these groups mainly as subgenera rather than sections. The Flora Europaea (Tutin *et al.* 1968) divided *Trifolium* species in two subgenera, of which one contained a single species (subgenus Falcatula (Brot.) D.E. Coombe). Species in the other subgenus, *Lotoidea* Pers., are arranged in sections that correspond almost exactly to Presl's genera.

The most comprehensive study of *Trifolium* taxonomy is by Zohary and Heller (1984) in their world monograph of the genus. They divide the species into eight sections some of which are divided into sub-sections and series as is shown Table 4.1. Zohary and Heller provide more details of morphology, distributions and relationships between *Trifolium* species than any other authors. The groups of species recognised by Zohary and Heller at the level of sections within the genus, and of subsections within section *Lotoidea* Crantz, largely agree in content with Presl (1832). Presl identified nine groupings of clover species, but gave them independent generic rank, a recommendation supported by some authors, notably by Roskov (1990a and b), but the proposals that *Trifolium* should be split has not been internationally accepted (ILDIS, 1999).

All the taxonomic studies discussed above were conducted using traditional morphological methods. Molecular methods have also been used more recently to investigate the taxonomy of the genus. Sayed-Ahmed *et al.* (1996) used nuclear ribosomal DNA to investigate the taxonomic relationships between sixty-three European and Asian *Trifolium* species from seven of the eight sections described by Zohary and Heller (1984). The results were analysed using phylogenetic methods. Sayed-Ahmed *et al.* (1996) concluded that their results did not support the classification of Zohary and Heller because the *Trifolium* species studied were shown to form a monophyletic group, but none of the previously recognised sections appeared monophyletic. The taxonomic system proposed by Zohary and Heller (1984) is followed here because it is based on a study of more species and considers more characters than any other revision of the genus *Trifolium*. Sayed-Ahmed *et al.*

(1996) included only a subset of the genus in their analysis and did not propose an alternative taxonomy.

Table 4.1. A summary of the classification of *Trifolium* species (Zohary and Heller, 1984).

Section	**Subsection / Series**
Lotoidea 99[1]	Subsect: *Falcatula* 1; *Loxospermum* 5; *Lupinaster* 7; *Ochreata* 10; *Lotoidea* 47; *Oxalioidea* 1; *Platystylium* 22; *Calycospatha* 2; *Neolagopus* 4
Parmesus 2	
Mistyllus 9	
Vesicaria 7	
Chronosemium 17	Ser: *Stipitata* 1; *Comosa* 5; *Badia* 4; *Agraria* 4; *Filiformia* 3
Trifolium 72	Subsect: *Trifolium* 5; *Intermedia* 6; *Ochroleuca* 8; *Alpestria* 2; *Stellata* 3; *Stenosemium* 1; *Trichoptera* 2; *Scabroidea* 3; *Phleiodea* 3; *Lappacea* 5; *Arvensia* 4; *Angustifolia* 10; *Alexandrina* 6; *Squamosa* 2; *Urceolata* 7; *Echinata* 2; *Clypeata* 3.
Trichocephalum 9	
Involucrarium 22	Subsect: *Involucrarium* 16; *Physosemium* 6

[1] All numbers beside taxon names indicate number of species in that taxon.

4.3 GENETIC DIVERSITY

More than 80% of the *Trifolium* species for which chromosomes have been counted are diploid. The vast majority of these have a chromosome count of 8, although species with a basic chromosome number of 7, 6 or 5 also occur (Cleveland, 1985). Sixteen percent of species are polyploid, occurring in populations with pure or mixed numbers of chromosomes. Generally, annual species have low chromosome numbers of $2n = 10 - 32$, while perennial species possess chromosome numbers in the range $2n = 12 - 130$ (Taylor *et al.*, 1979). Approximately one-third of the *Trifolium* species are perennial, most of which are self-incompatible (Taylor *et al.*, 1979; Townsend and Taylor, 1985), however exceptions to this rule include *T. fragiferum* which occurs in self-incompatible and self-fertile forms (Evans, 1976).

Bulinska-Radomska (1994) assayed Polish populations of three *Trifolium* species at 15 enzyme systems to elucidate their pattern of genetic variability. It was shown that the annual, self pollinating species *T. campestre* had fewer polymorphic loci and a lower level of diversity than the two cross pollinated species, *T. fragiferum* and *T. montanum*. The isozyme variation present in the cross-pollinated species was mainly allocated within populations, unlike *T. campestre*. Similar results were obtained with the inbreeding species *T. arvense* compared to the cross pollinated species *T. hybridum* and *T. incarnatum* (Bulinska-Radomska, 1998).

In a study of the variation within and between populations of *T. hirtum* (Molina-Freaner and Jain 1992b) twelve isozymes were assayed for populations collected in Turkey, where it is native and California, where it has been introduced. Similar levels of total gene diversity were found in both sets of populations. Most of the variation present in the Californian populations of *T. hirtum* occurred within populations, while in the Turkish material most of the variation was allocated between populations. The Californian populations have been more intensively studied and are known to exhibit gynodioecious and hermaphroditic breeding systems (Molina-Freaner and Jain 1992a). If the *T. hirtum* populations from Turkey were autogamous this could explain their comparatively high level of diversity between populations.

Collins *et al.* (1984) investigated variation within and between 22 registered cultivars of *T. subterraneum* for 15 enzyme systems. They found that the Mediterranean derived cultivars were all homogeneous for the isozymes assayed, while three of the five artificially created cultivars were polymorphic for at least one locus. In other words very little within population variation exists and the variation mainly occurs between populations.

T. pratense is a naturally cross-pollinated species with a gametophytic self-incompatibility system and a requirement for insect pollination (Smith *et al.*, 1985). Studies using Random Fragment Length Polymorphic DNA (RFLPs)(Milligan, 1991) Random Amplified Polymorphic DNA (RAPDs) and isozymes (Kongkiatngam *et al.*, 1995) revealed high within population and high within cultivar variation.

4.4 ECOGEOGRAPHIC DISTRIBUTION

Trifolium species are native to all the humid, temperate regions of the world except the Himalayas and Australasia (Zohary and Heller, 1984). Approximately sixty percent of species occur in Eurasia. The largest number of species occurs in the northern Mediterranean basin (Table 4.2), particularly Turkey where just less than 100 species are known (Zohary 1970, Zohary and Heller, 1984). In north Africa 33 species have been recorded in Tunisia, and 41 species in Algeria (ILDIS, 1999), of which 40 species in Algeria are native.

Trifolium species often occur in semi-natural grasslands that have arisen from woodland or forest that has been heavily degraded by centuries of grazing (Cocks and Osman 1996). In Mediterranean areas, annuals, including species such as *T. scabrum* and *T. cherleri*, replace the trees and most other perennial plants (Cocks and Osman, 1996). Perennial *Trifolium* species such as *T. hybridum*, *T. repens* and *T. fragiferum* mainly occur in mountain meadows and areas with high rainfall (Osman *et al.*, 1990). Annual clover species are promoted by activities that create open grasslands, such as burning, forest clearance and grazing. Overgrazing can cause an increase in the proportion of small seeded annual species such as *T. campestre*, *T. stellatum* and *T. tomentosum*. This is because the seeds of these species can avoid damage during mastication, and their hard seed coat protects the embryo during rumination (Russi *et al.*, 1992).

Publications such as Acikgoz *et al.* (1998) describe the habitat type, soil type and altitude ranges within which Mediterranean *Trifolium* species are usually found. A great deal of ecogeographic data has been gained during germplasm collecting missions, for example in south-west and north-west Turkey (Bennett *et al.*, 1998; 1999) and in Greece and Crete (Francis *et al.*, 1995; 1977). Due to lack of space, discussion in this chapter is limited to selected species of economic importance.

Table 4.2. The distribution of *Trifolium* species in the Mediterranean. Key: N=native, I=introduced, U=uncertain. Countries are ranked in order of number of native *Trifolium* species. Species are ranked in order of their tendency towards Mediterranean endemism, as the ratio of number of Mediterranean to non-Mediterranean countries in which they have been recorded as native. (Source: ILDIS).

Agriculturally important species	**TR**	**GR**	**IT**	**YU**	**BG**	**FR**	**AL**	**ES**	**LB**	**MA**	**IN**	**PT**	**DZ**	**IL**	**IQ**	**TN**	**CY**	**LY**	**MA**	**EG**	**AN**	**Σ N**
T. hirtum	N	N	N	N	N	N	N	N	N	N	N	N	N	N	N	U	N	U	U	U	U	16
T. hybridum	N	N	N	N	N	U	N	N	U	N	N	U	U	U	U	U	U	U	U	U	N	10
T. michelianum	N	N	N	N	N	U	N	N	U	N	U	N	N	U	U	N	U	U	U	U	U	11
T. cherleri	N	N	N	N	N	N	N	N	N	N	N	N	N	N	N	N	N	N	N	U	U	19
T. nigrescens	N	N	N	N	N	N	N	N	N	N	N	N	N	N	N	N	N	U	N	N	U	19
T. resupinatum	N	N	U	U	N	U	U	U	N	N	U	I	N	N	N	U	N	N	N	N	U	12
T. lappaceum	N	N	N	N	N	N	N	N	N	N	N	N	N	N	N	N	N	N	N	U	U	19
T. glomeratum	N	N	N	N	N	N	N	N	N	N	N	N	N	N	N	N	N	U	U	U	N	18
T. subterraneum	N	N	N	N	N	N	N	N	N	N	N	N	N	N	N	N	N	N	N	U	U	19
T. striatum	N	N	N	N	N	N	N	N	U	N	N	N	N	U	N	N	N	U	U	U	U	15
T. vesiculosum	N	N	N	N	N	N	N	I	U	U	U	U	U	U	U	U	U	U	U	U	U	7
T. dubium	N	N	N	N	N	N	N	N	U	N	U	N	U	N	N	N	N	U	U	U	N	15
T. fragiferum	N	N	N	N	N	N	N	N	N	N	N	N	N	N	N	N	N	N	N	N	N	21
T. campestre	N	N	N	N	N	N	N	N	N	N	N	N	N	N	N	N	N	N	N	N	N	21
T. ambiguum	N	U	U	U	U	U	U	U	U	U	N	U	U	U	U	U	U	U	U	U	U	2

T. incarnatum	N	N	N	N	N	N	N	N	U	I	U	I	U	U	U	N	U	U	U	I	N	10
T. repens	N	N	N	N	N	N	N	N	N	N	N	N	N	N	N	N	N	U	N	N	U	19
T. arvense	N	N	N	N	N	N	N	N	N	N	N	N	N	N	N	N	N	N	N	N	N	21
T. pratense	N	N	N	N	N	N	N	N	I	N	N	N	N	N	N	N	U	U	U	U	U	15
T. aureum	N	N	N	N	N	N	N	N	N	U	N	N	U	U	U	U	U	U	U	U	N	12
T. medium	N	N	N	N	N	N	N	N	U	U	N	N	U	U	U	U	U	U	U	U	U	10
T. alexandrinum	I	U	U	I	U	U	I	U	I	U	I	I	U	I	U	I	U	U	U	I	U	0
Number of agricultural native species	21	20	19	19	20	17	19	18	12	16	16	16	14	13	14	14	13	7	9	6	8	
Number of other native species	60	55	47	46	43	42	37	31	37	30	26	25	26	27	19	18	13	13	11	9	3	
Total number of native *Trifolium*	81	75	66	65	63	59	56	49	49	46	42	41	40	40	33	32	26	20	20	15	11	

Table 4.3. Cultivated *Trifolium* species (Evans, 1976, Herman, 1953, Russell and Webb, 1976, and Speer and Allinson, 1985, Evans and Snowball, 1993).

Species	Common name	Specific use	Lifespan
T. alexandrinum	Egyptian/ Berseem	Green-chopped forage, pasture	Annual
T. ambiguum	Kura/ Caucasian	Of growing importance in New Zealand and Australia. Used for re-vegetating in the Australian Alps.	Perennial
T. arvense	Rabbit Foot	Adapted to infertile, dry, sandy soils	Annual
T. campestre	Low- / Large-hop	With grass for early spring grazing, hay	Annual
T. cherleri	Cupped	Grazing, cultivars released in Australia	Annual
T. dubium	Yellow suckling / Small hop / Drooping	With grass for early spring grazing, hay	Annual
T. fragiferum	Strawberry	Grazing, particularly close, continuous grazing, and phase farming.	Perennial
T. hirtum	Rose	Low intensity grazing, esp. in unproductive areas, soil stabilisation, cultivars released in Australia	Annual
T. hybridum	Alsike/ Swedish	Hay, pasture, particularly in cool moist areas	Perennial
T. incarnatum	Crimson	Winter pasture, in rotations, roadside stabilisation	Annual
T. lappaceum	Lappa	With grass for spring grazing on calcareous soils	Annual
T. medium	Zigzag	Not widely used, partly because of very poor seed set	Perennial
T. michelianum	Balansa Clover	Grazing, silage, green manure. Some water-logging tolerance. Cultivars released in Australia	Annual
T. nigrescens	Ball	Winter grazing	Annual
T. pratense	Red	In grass mixtures for hay, pasture and silage	Perennial
T. repens	White	With grass for grazing, perennial pastures	Perennial
T. resupinatum	Persian	For grazing, hay, silage. Some water-logging tolerance	Annual
T. semipilosum	Kenya White	commercial varieties available	Perennial
T. subterraneum	Subterranean	For grazing, particularly in dryer areas. Widely sown in Australia	Annual
T. vesiculosum	Arrowleaf	For grazing, hay, silage, cultivars released in Australia	Annual

4.5 SPECIES OF POTENTIAL VALUE

Approximately 30 species of *Trifolium* have been cultivated on a commercial scale (Herman, 1953; Russell and Web, 1976; Evans, 1976; Speer and Allinson, 1985; van Keuren and Hoveland, 1985). A world-wide survey carried out by Russell and Web (1976), showed that the most commonly cultivated *Trifolium* species are *T. repens*, *T. subterraneum*, *T. pratense*, *T. fragiferum* and *T. hybridum*. Evans (1976) included the same species in her list of the ten most commonly cultivated *Trifolium* species, plus *T. resupinatum*, *T. incarnatum*, *T. alexandrinum*, *T. dubium* and *T. ambiguum*. Specific uses of these, and other cultivated *Trifolium* species, are given in Table 4.3. All the species listed in Table 4.3, except *T. semipilosum*, are native to the Mediterranean (Zohary and Heller, 1984).

4.5.1 *Trifolium repens* L.

Trifolium repens is a tetraploid (2n=4x=32) that is usually described as a creeping perennial (e.g. Burdon, 1983). However, Hollowell (1966) postulated that in some environments this species behaves as an annual, or as a mixture of annual and perennial types, that persist through a combination of asexual propagation by stolons and re-seeding. *T. repens* occurs spontaneously throughout Europe to approximately 71°N (Tutin *et al.* 1968), and also in north-west Asia and north Africa. The distribution of *T. repens* in the Mediterranean is determined by its intolerance to drought conditions (Burdon, 1983). At low altitudes, large leafed varieties often grow along watercourses or under irrigation. Under drier conditions, hard grazing and at high altitudes only small leafed types can persist (Davies and Young, 1967). Caradus and Forde (1996) collected samples of *T. repens* from areas of eastern Turkey at altitudes ranging to more than 2000m that receive less than 600mm annual rainfall. This is lower than is normally considered adequate for persistence (Caradus and Forde, 1996).

T. repens is widely sown and naturalised throughout temperate regions of the world, and is commonly found in grasslands outside its natural range (Mathison, 1983). For example, it is the most important forage legume in perennial pastures in south-eastern USA (Pederson *et al.*, 1989). The species is also often cultivated within its natural range, for example in the UK approximately 10 million hectares of agricultural land is under pasture (frequently consisting of *L. perenne* and *T. repens*) (MAFF 1998). Agricultural use of *T. repens* in the Mediterranean normally utilises natural pastures rather than cultivated varieties, e.g. in eastern Turkey (Caradus and Forde, 1996). *T. repens* is a highly variable species and considerable differences have been observed within and between populations in a wide range of morphological characters (Burdon 1983). Agronomists recognise three types of white clover. The wild type, *T. repens* var. *repens*, occurs throughout the entire distribution of the species. Giant forms, var. *giganteum*, occur in several Mediterranean countries (France, Italy, Israel, Lebanon and Morocco). The best known giant ecotype, known as Ladino clover, originated in northern Italy and is now cultivated in many parts of the world (Williams, 1987). Dutch white clover, var. *hollandicum*, was the term used by early British agriculturalists to describe all non-wild types of white clover. Today the name is used occasionally to describe forms of *T. repens* that are morphologically intermediate between the British wild type and giant white clover (Williams, 1987).

4.5.2 *Trifolium pratense* L.

T. pratense has been valued as a pasture plant since the ancient Greek and Roman civilisations (Julén, 1959). It is a short lived perennial with 14 chromosomes (Mousset, 1990). Zohary and Heller (1984) describe this species as being very polymorphic, and state that a taxonomic revision of the species is required. The natural range of *T. pratense* includes north Africa, much of Asia and all of Europe except the extreme north and parts of the extreme south (Tutin *et al.*, 1968). In warm and dry regions *T. pratense* is restricted to high altitudes and humid areas. Modern *T. pratense* cultivars are thought to have a Spanish or Arabic-Spanish origin (Julén,

1959).

The upright habit of *T. pratense* means that this species is more suitable for conservation than grazing (Frame 1990). There are two agronomic types: the late, single cut type and the early, double cut type. The late type is cultivated in northern and continental Europe, while the early type is cultivated in southern Europe. Red clover is one of the most important species for forage production in Sweden (Andersson and Kristiansson, 1989), and in the UK it is the second most important pasture legume (Frame 1990). However, exploitation of this species is very limited in the Mediterranean basin even though it grows well there under irrigation (Villax 1963).

The isoflavones produced naturally in the leaves of *T. pratense* have been found to benefit menopausal women in need of hormone replacement therapy (Nestal *et al.*, 1999). The Australian based company Novogen Ltd. has recently released a product 'Promensil' that is the first standardised dietary isoflavone product, sourced from *T. pratense*. This product is designed specifically to help to manage the symptoms of the menopause naturally (Novogen, 1998a and b).

4.5.4 *Trifolium subterraneum* L.

The diploid species (2n=16) *T. subterraneum* is one of the most commonly exploited annual *Trifolium* species. *T. subterraneum* often dominates grazed, unseeded vegetation in the Mediterranean basin (Katznelson, 1974) and is also cultivated for example in Spain and Portugal (McGuire, 1985). In addition, it is widely naturalised and cultivated outside the original distribution of the species. It is the most important forage legume in Australia, where a large number of varieties have been released (Hopkins *et al*, 1994). *T. subterraneum*, and also *T. dubium*, are important winter annuals in the USA, where they are grown in temperate to subtropical climates (Speer and Allinson, 1985)

According to Katznelson (1974), human activity has caused *T. subterraneum* to become extremely common in the Mediterranean basin. Forest clearance and heavy grazing by livestock provide ideal conditions for the species. *T. subterraneum* has a branched prostrate habit and a seed burial mechanism, both of which give good resistance to grazing. *T. subterraneum* was found to be rare in regions of the Iberian peninsula and north Africa studied by Gladstones (1976). Stands (limited in extent) were only encountered in areas under permanent grazing such as along roadsides and in rocky outcrops. Francis *et al.* (1975) studied *T. subterraneum* in Mediterranean Turkey and found it to be a relatively uncommon component of the grasslands. However, differences in frequency of occurrence of *T. subterranean* in the Mediterranean are likely to be due to changes in soil acidity rather than grazing pressure. This has been found in other species of *Trifolium* around the Mediterranean (Bennett 2000).

Although *T. subterraneum* can provide highly nutritive feed for ruminants, some cultivars have the potential to cause reproductive disorders in sheep due to high phytoestrogenic activity. In an attempt to identify lines with low isoflavone traits for use in breeding programs, Gildersleeve *et al.* (1991) measured the isoflavone content of a large number of *T. subterraneum* accessions. The isoflavone content was found to vary from very low to levels that would be unacceptably high in a commercial cultivar. There are a few Mediterranean grasslands where the natural stands of *T. subterraneum* are sufficiently dominant and are exploited heavily enough to cause acute reproductive disorders. In this case, other sources of feed in addition to *T. subterraneum* rich grasslands, such as stubble grazing, must be provided to reduce the risk of isoflavone induced infertility (Katznelson, 1974). However, breeding for low isoflavones has been successful in Australia and a number of cultivars with low levels have been released (Nichols *et al.*, 1996; Croker *et al.*, 1994).

Katznelson (1974) proposed that *T. subterraneum* should be split into three species based mainly on differences in plant size and pH of the soil on which populations are found. Zohary and Heller (1984) rejected this, but retained the taxa as subspecies; subsp. *subterraneum*, subsp. *yanninicum* and subsp. *brachycalycinum*. Cultivars of all three subspecies have been released (Collins *et al.*, 1984).

4.5.5 *Trifolium incarnatum* L.

T. incarnatum is a diploid (2n=14) annual or, rarely, biennial (Zohary and Heller, 1984). Steiner *et al.* (1998) studied nearly 40 samples of this species, all of which produced seeds autogamously, but Picard (1959) found some self-incompatible individuals. Zohary and Heller (1984) described two varieties of *T. incarnatum*. The cultivated variety, *T. incarnatum* var. *incarnatum*, usually has a red corolla that is the same length as the calyx, while the yellowish-white or pink corolla of *T. incarnatum* var. *molinerii* is longer than the calyx. An analysis of variation in internal transcribed region sequences (Steiner *et al.*, 1998) concluded that the two varieties are sufficiently different to be classified as separate species.

T. incarnatum is cultivated in much of Europe, except the extreme north. The natural range of this species is southern and western Europe, but it has also become widely naturalised (Tutin *et al.* 1968). Like *T. pratense*, *T. incarnatum* is an erect-growing species that is often cultivated for hay (Steiner *et al.* 1998). The species is also cultivated for grazing in rain-fed annual meadows, as pure stands or in association with grasses (Mousset 1995). In Morocco, *T. incarnatum* is cultivated in coastal areas and in the high lands of Meknès and Fès. In France it is widely cultivated over an area of 200,000ha (Villax 1963).

4.5.6 *Trifolium alexandrinum* L.

Wild ancestors of the cultivated species *Trifolium alexandrinum* are not known (Zohary and Heller, 1984). It is thought to have originated in Asia Minor, later migrating southward to Syria, Palestine and Egypt (Singh 1993). The species is frequently cultivated in south-western Asia, for example in northern and southern Iran, particularly in the provinces of Mazandaran, Gilan and Khozistan. *T. alexandrinum* is grown in rotation with rice on more than 40 000 hectares (Attaran, 1989).

Most populations of *T. alexandrinum* are diploid (2n=16). However, tetraploid varieties have been found to give high yields in India (Singh, 1993). Villax (1963) described differences between landraces originating in Egypt and Palestine. Landraces, such as 'Mescawi' and 'Khadarwi', are grown under irrigation, while others, like 'Baali' and 'Fahl', do not require irrigation. 'Fahl' is cut only once for forage, but landraces grown under irrigation can be cut up to six times in year. *T. alexandrinum* is cultivated in regions with mild winters as it is not adapted to hard frosts or excessive heat. It can be used as winter or summer forage according to local conditions.

4.5.7 *Trifolium resupinatum* L.

Three botanical varieties of *T. resupinatum* (2n=14, 16 and 32) were recognised by Zohary and Heller (1984). The cultivated variety, *T. resupinatum* var. *majus*, is characterised by large fistulose stems and large flowers. Tutin *et al.* (1968) believed that the species was introduced to southern Europe through cultivation for fodder. *T. resupinatum* now occurs in all Mediterranean countries (Zohary and Heller, 1984). This highly nutritious species (Noble and Papineau, 1972) is cultivated in many temperate countries, including the USA. According to Villax (1963), *T. resupinatum* is widely cultivated in India and in Mediterranean countries such as Portugal. In Morocco, it is very common in low lands, and it is often cultivated in association with a forage grass, particularly on heavy wet soils (Knight, 1985). Depending on the local climate, this species is used as a summer or winter forage (Mousset 1995). It has been shown to have moderate water-logging tolerance, and a preference for alkaline soils (Evans and Snowball, 1993).

4.5.8 Other *Trifolium* species of economic importance

Several other *Trifolium* species of Mediterranean origin are cultivated on a commercial scale, mainly outside the Mediterranean. For example *T. hybridum* occurs naturally in countries such as Turkey and France but is cultivated mainly in northern Europe and Canada (Evans, 1976). Some species of *Trifolium* with recognised value are under-utilised. For example, *T. ambiguum* is a perennial forage adapted to cool climates with a deep root system that aids soil stabilisation. Its use in New Zealand in mountainous areas is under evaluation (Patterson and Espie (1997). However, widespread use of the species has been slow to establish due to the need for *Rhizobium* inoculation (Speer and Allinson, 1984). Investigation is continuing into *Trifolium* species with the potential to be brought into agriculture (Francis, 1999; Russell and Webb, 1976). An example of a species that has been brought into cultivation recently is *T. vesiculosum* (Speer and Allinson, 1984).

Other *Trifolium* species are common in Mediterranean permanent pastures. *T. cherleri* sometimes dominates heavily over-grazed grasslands in Greece and Crete, while *T. hirtum* occurs there where the grazing pressure is lower (Francis and Katznelson, 1977). In the Mediterranean area of Turkey, *T. subterraneum* was found to be a relatively uncommon component of the grasslands compared to species such as *T. lappaceum* by Francis *et al.* (1995). The legume species most commonly grazed by sheep in Syria are small-seeded *Trifolium* species, such as *T. stellatum, T. tomentosum* and *T. campestre* (Russi *et al.*, 1992). These have become widely naturalised across the world (Bennett, 1999). Other naturally occurring *Trifolium* species in the Mediterranean that are agriculturally useful include *T. striatum, T. angustifolium* (Mathison, 1983) and *T. echinatum* (Lorenzetti *et al.*, 1987).

4.6 CONSERVATION PRIORITIES

Genetic erosion of native floras has occurred throughout the Mediterranean, due to factors such as overgrazing, degradation of the native flora and inappropriate farming practices (Reid *et al.*, 1989). According to Cocks and Osman (1996), the most important factor causing continued deterioration of Mediterranean grasslands is overgrazing during seed production in the summer. Changes in the management of Mediterranean rangelands to conserve populations of *Trifolium in situ* are urgently required. Grazing pressure has been intense in north Africa and south-west Asia for centuries, but the recent increase in livestock populations has accelerated the degradation enormously (Reid, 1989). Changes in grazing practice will cause short-term reductions in returns, but is essential to avoid terminal degradation of grazing lands (Cocks and Osman, 1996).

Twenty one species of *Trifolium* have been identified by international agreement as priorities for *ex situ* conservation survey due to their economic importance and the threat of genetic erosion (Crawford, 1984; Anon, 1985). Twenty of these species (listed in Table 4.4) occur in the Mediterranean basin. The 1997 IUCN Red List of Threatened Plants names 39 species of *Trifolium*, 18 of which occur in the Mediterranean , which are priority species for *ex situ* conservation (Walters and Gillett, 1998). The species listed by the IUCN are different to those previously identified for priority treatment because of their negligible economic importance. Some of the species categorised as threatened by IUCN are known only from a type specimen, for example *T. euxinum*, and some are endemics with very narrow distributions, such as *T. pichisermollii* (Zohary and Heller, 1984).

The conservation needs identified by Crawford (1984), Anon (1985) and IBPGR (1985) have been partially addressed. As shown in Table 4.5, large collections of *Trifolium* germplasm are maintained *ex situ* and a number of ecogeographic surveys have been published. For example, legumes of south-west Turkey were surveyed by Bennett *et al.* (1998) and annual *Trifolium* species were surveyed in Morocco (Beale *et al.*, 1993). A series of papers (Ehrman and Cocks, 1990, Russi *et al.*, 1992, Cocks and Osman, 1996) investigated the botanical composition of communal grasslands in Syria and the contribution of legumes (including *Trifolium*) to herbage production. In west Asia and north Africa, more than 40 missions have taken place collecting more than three thousand accessions of *Trifolium* species (Bounejemate *et al.*, 1999).

Few publications describe *in situ* conservation projects relevant to *Trifolium* genetic resources. In Turkey a project to conserve wild crop relatives *in situ* has recently begun (Tan, 1998). Surveys of the proposed reserve areas have recorded the presence of a variety of *Trifolium* species. At Ceylanpinar (in south-east Turkey) the target species for conservation are wild wheat, barley and edible legumes, but *T. bullatum*, *T. campestre*, *T. leucanthum*, *T. pilulare*, *T. phleoides*, *T. purpureum* var. *purpureum*, *T. resupinatum*, *T. stellatum* and *T. sylvaticum* have been observed (Karagöz, 1998). The reserve will be managed to conserve crop relatives, which will also benefit populations of *Trifolium* species because they occur in similar habitats to the target species i.e. oak scrub, open pastures, disturbed steppe, fallow fields and so on (Davis and Plitmann, 1970, Davis, 1985, Tan, 1998). Surveys to monitor the reserves will continue to monitor populations of *Trifolium* species (Karagöz, pers. Comm.).

In Europe, the Institute for Agrobotany in Hungary has assembled a European Central Crop Database (ECCDB) documenting collections of *T. pratense* held in European Institutes (Horváth 1998). The database is served on the World-Wide Web by the Nordic gene bank (http://www.ngb.se), and documents 2140 accessions held at 20 European institutes. The Institute of Grassland and Environmental Research in the UK has assembled a ECCDB documenting collections of *T. repens* held in European Institutes. The database documents 1415 accessions held at 14 European institutes (http://www.igergru.bbsrc.ac.uk/ecpgr.html). The Servicio de Investigación y Desarrollo Tecnológico in Spain has assembled a ECCDB documenting collections of *T. subterraneum* held in European Institutes (López, 1998). The database documents 3077 accessions held at 18 European institutes.

4.7 RESEARCH NEEDS

Harlan (1984) stated that no collection of an annual forage species is complete, and that most are very inadequate. Large numbers of *Trifolium* accessions have been collected for *ex situ* conservation or breeding, but to date, collections have been dominated by the major cultivated species; *T. repens*, *T. pratense* and *T. subterraneum*, and many potentially valuable species have been neglected. In addition, many of the older accessions have insufficient passport data (such as latitude and longitude of the collecting site) which limits the usefulness of the conserved material (Taylor *et al.*, 1977).

Figure 4.1 shows the distribution of wild *Trifolium* accessions obtained from AARI, Turkey, AMGRC Australia, BARI Italy, *Trifolium* GRC, Australia, IPK Germany, IGER UK, Hebrew University of Jerusalem, Israel, SINGER and the European Central Crop Databases for *T. repens* and *T. pratense*. Approximately 80% of the accessions in these databases had associated latitudes and longitudes and therefore could be included on the figure. Figure 4.1 demonstrates that relatively intense collecting has occurred in Sardinia, Corsica, the Mediterranean coasts of Morocco, Tunisia and Algeria, and from the Syrian/ Turkish border, south into Jordan and north and east into Turkey. Gaps in the collecting activity are apparent in Italy, the Yugoslavian republics and Albania.

Changing agricultural practises or environmental problems influences the direction of research into the utilisation of *Trifolium* genetic resources. The current emphasis in Australian forage research is a search for perennial and deep-rooted annual legume species that will help to halt the rising water tables and salinity in the agricultural areas. Species that can tolerate saline or water-logged conditions are also required (Francis, 1999). There are a number of potential candidates such as *T. michelianum*, *T. resupinatum*, *T. fragiferum* and *T. tomentosum*, but more research is urgently needed to characterise the range of conditions that they can tolerate. Other areas requiring investigation are; evaluation for disease and pest resistance; species for phase-farming rather than the traditional ley-farming systems, and alternative uses for current species such as the phyto-oestrogens in *T. pratense*.

Table 4.4. *Trifolium* species identified as conservation priorities (Anon, 1985).

Species	High priority for *ex situ* conservation (Crawford, 1984; Anon, 1985)	High priority for ecogeographic survey and *in situ* conservation (IBPGR, 1985)
T. ambiguum	y	
T. fragiferum	y	y
T. hybridum	y	
T. repens	y	y
T. alexandrinum	y	
T. michelianum	y	
T. campestre	y	
T. canescens	y	
T. cherleri	y	y
T. glomeratum	y	
T. hirtum	y	y
T. israeliticum		
T. isthmocarpum	y	
T. nigrescens	y	
T. purpureum	y	
T. resupinatum	y	
T. scutatum	y	
T. stellatum	y	
T. subterraneum	y	y
T. vesiculosum	y	
T. pratense		y

Table 4.5. The numbers of accessions of economically important *Trifolium* species maintained *ex situ* in various genebanks.

Ex Situ Collection	*T. alexandrinum*	*T. incarnatum*	*T. pratense*	*T. repens*	*T. resupinatum*	*T. subterraneum*	Other Species	Total
AMGRC	31	18	68	113	246	823	2339	3638
Bari,	25	9	40	89	16	68	178	425
ECCDB			2140	862		3077		6079
Hokkaido,			368					368
IGER	10	12	235	553	39	13	687	1549
IPK	18	15	51	16	20	4	326	450
HUJ	36	1	7	10	3	212	39	308
Margot Forde			700	13000				13700
SINGER	36	2	9	23	233	63	3136	3502
GRC	186	81	13	34	695	7900	5426	14335
NPGS	118	34	1106	579	180	300	1983	4300
VIR			2913					2913
Total	462	179	7740	15436	1565	12547	16620	54549

AARI = Aegean Agricultural Research Institute, Izmir, Turkey
AMGRC = Australia *Medicago* Genetic Resource Centre, South Australian Research and Development Institute, Adelaide, Australia
Bari = Istituto del Germoplasma, Bari, Italy
ECCDB = European Central Crop Database. See below for more information.
Hokkaido = Hokkaido National Agricultural Experiment Station, Eniwa, Hokkaido, Japan
IGER = Institute of Grassland and Environmental Research, Aberystwyth, UK.
IPK = Institut für Pflanzengenetik und Kulturpflanzenforschung, Gatesleben, Germany
HUJ = Department of Botany, Institute of Life Sciences Hebrew University of Jerusalem, Israel
Margot Forde = Margot Forde Forage Germplasm Centre, Palmerston North, New Zealand
NPGS = National Plant Germplasm System, Regional Plant Introduction Station, Washington, USA.
SINGER = System-wide Information Network for Genetic Resources, (http://www.cgiar.org/singer)
GRC = *Trifolium* Genetic Resource Centre, Agriculture Western Australia, Perth, Australia.
VIR = Vavilov Research Institute of Plant Industry, St. Petersburg, Russia

Figure 4.1. Distribution of collecting sites of 14211 accessions of *Trifolium* species collected with latitude/longitude data from Mediterranean countries. The map shows the number of accessions collected per cell in a grid of 50km x 50km cells. Albers equal-area conic projection, with central meridian at 20°E and standard parallels at 30°N and 46°N. (For Color Plate version, see p. xvii)

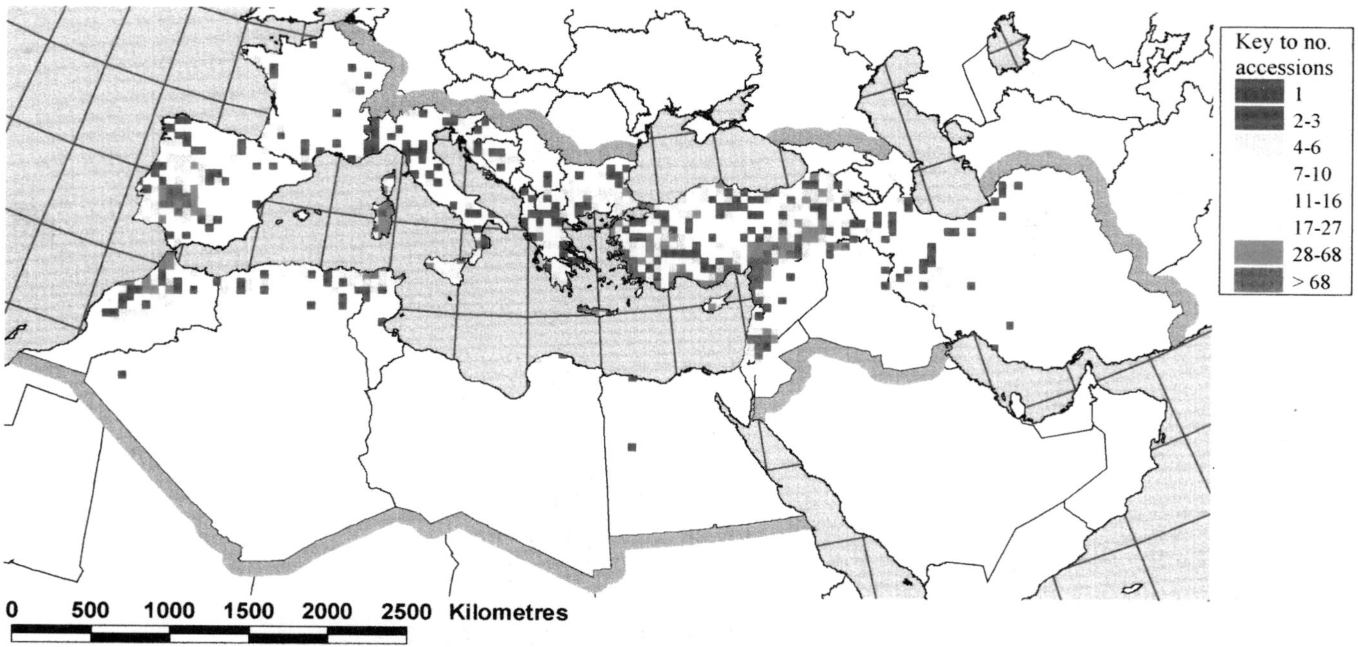

REFERENCES

Acikgoz, N., Sabanci, C.O. and Cinsoy, A.S., 1998. Ecogeography and distribution of wild legumes in Turkey. In: *Proceedings of International Symposium on* In situ *Conservation of Plant Genetic Diversity* (eds. Zenciri *et al.*). pp 87-91. CRIFC, Turkey.

Alapetite. G..P., 1979. *Flore de la Tunisie: Angiospermes-Dicotyledones. Apétales-Dialypétales. 1ère partie*. pp. 1-651. Imp Officielle de la Tunisie, Tunis

Andersson. B. and Kristiansson. V., 1989. *Breeding for resistance to root rot in red clover*. Proc. XVI International Grassland Congress, Nice, October, 4-11, pp.363-364.

Attaran, M..H., 1989. *Determination of the most suitable sowing date for Berseem clover*. In: Karaj. Proc. XVI. International Grassland Congress, Nice, October, 4-11, pp.381-382.

Beale, P.E., Bounejemate, M., Lahlou, A., Marx, D.B. and Christiansen, S., 1993. Distribution of annual *Trifolium* species in Morocco. *Australian Journal of Agricultural Research* 44: 1303-1310.

Bennett, S.J., 2000. Genetic variation of five species of *Trifolium* L. from south-west Turkey. *Genetic Resources and Crop Evolution*, 47: 81-91.

Bennett, S.J., 1999. Ecotypic variation between and within two populations of *Trifolium tomentosum* (woolly clover) from Syria and Western Australia: its success as a colonising species. *Australian Journal of Agricultural Research*, 50: 1443-1450.

Bennett, S.J., Maxted, N., Kitiki, A. and Francis, C., 1999. Collecting and ecogeography of pasture, forage and grain legumes in Northwest Turkey. *Plant Genetic Resources Newsletter*, 119: 33-36.

Bennett, S.J., Maxted, N. and Sabanci, C.O., 1998. The ecogeography and collection of grain, forage and pasture legumes in south-west Turkey. *Genetic Resources and Crop Evolution* 45: 253-262.

Bobrov, E.G., 1967. On the span of the genus *Trifolium* L. *Botanicheskii Zhurnal*, 52(11): 1593-1599.

Boissier, E., 1872. *Flora Orientalis* Volume II, Calyciflorae Polypetale, Geneva and Basel.

Bounejemate, M., Gintzburger, G., Christiansan, S., Robertson, L., Abd El Moneim, A. and Kanopka, J., 1999. Forage and Pasture legume genetic resources at ICARDA. In: Genetic resources of Mediterranean pasture and forage legumes (eds. Bennett, S.J. and Cocks, P.S.). pp. 163-170. Kluwer, Dordrecht.

Bulinska-Radomska, Z., 1994. Patterns of genetic variation in some *Trifolium* species. In: *Evaluation and exploitation of genetic resources pre-breeding* (ed. Balfourier, F.) Eucarpia Meeting Papers.

Bulinska-Radomska, Z., 1998. Genetic diversity and population structure of six *Trifolium* species. In: *The proceedings of International Symposium on* In situ *Conservation of Plant Genetic Diversity* (eds. Zenciri, N., Kaya, Z., Anikster, Y. and Adams, W.T.). pp. 129-134. CRIFC, Turkey.

Burdon, J.J., 1983. Biological Flora of the British Isles No 154 *Trifolium repens* L. *Journal of Ecology* 71: 307-330.

Caradus, J.R. and Forde, M.B., 1996. Characterisation of white clover populations collected from the Caucasus and high altitude regions of eastern Turkey. *Genetic Resources and Crop Evolution* 43: 143-155.

Cleveland, R.W., 1985. Reproductive cycle and cytogenetics. In: *Clover Science and Technology, Agronomy monograph, 25* (ed. Taylor, N.L.). pp 71-110. ASA-CSSA-SSSA, Madison, WI, USA.

Cocks, P.S. and Osman, A..E., 1996. Productivity and botanical composition of communally-owned Mediterranean grasslands in the marginal farming areas of north Syria. *Journal of Arid Environments* 33: 389-398.

Collins, W.J., Rossiter, R.C., Haynes, Y., Brown, A.H.D. and Marshall, D.R., 1984. Identification of Subterranean Clover Cultivars and their Genetic Relationships by Isozyme Analysis. *Australian Journal of Agricultural Research* 35: 399-411.

Crawford, E.J., 1984. *The potential for further acquisitions of alterative pasture legumes with desirable characteristics.* Workshop on Alternative legumes. Department of Agriculture South Australia, Struan, Oct 1984.

Croker, K., Nichols, P., Barbetti, M. and Adams, N., 1994. Sheep infertility from pasture legumes. *Farmnote* No. 79/99. Agriculture Western Australia, Perth.

Davis, P.H., 1985. Aegilops L. In: *Flora of Turkey and the East Aegean Islands, Vol. 9* (ed. Davis, P.H.). pp 233-245. Edinburgh University Press, Edinburgh.

Davis, P.H. and Plitmann, U., 1970. *Lens* Miller. In *Flora of Turkey and the East Aegean Islands, Vol. 3.* pp. 325-328. Edinburgh University Press, Edinburgh.

Davies, W.E.and Young, N.R., 1967. The Characteristics of European, Mediterranean, and other populations of white clover (*Trifolium repens* L.). *Euphytica* 16: 330-340.

Ehrman, T. and Cocks, P.S., 1990. Ecogeography of annual legumes in Syria: distribution patterns. *Journal of Applied Ecology* 27: 578-591.

Evans, A.M., 1976. Clovers. In: *Evolution of Crop Plants.* (ed. Simmonds, N.W.). pp. 175-179. Longman, London.

Evans, P.M. and Snowball, R., 1993. *Balansa and Persian clover lines outproduce registered cultivars, and subterranean clover and medics, in a 400mm annual rainfall zone in Western Australia.* pp. 53-56. Proceedings of the 7th Australian Agronomy Conference, Perth.

Ewing, M.A., 1996. Successful pasture legumes in ley farming systems. In: *Genetic Resources of Mediterranean Pasture and Forage Legumes.* Proceedings of International Workshop, Perth, Australia, 14-18 October1996.

Frame, J., 1990. The Role of Red Clover in United Kingdom Pastures. *Outlook on Agriculture* 19(1): 49-55.

Francis, C.M., 1999. The need to collect new pasture and forage species. In: *Genetic Resources of Mediterranean pasture and forage legume plants.* (eds. Bennett, S.J. and Cocks, P.). pp. 90-95. Kluwer Academic Publishers, Dordrecht.

Francis, C.M. and Katznelson, J.S., 1977. Observations on the distribution and ecology of subterranean clovers, some other clovers and medics in Greece and Crete. *Australian Plant Introduction Review* 12(1): 17-25.

Francis, C., Maxted, N., Kyriakakis, S. and Bennett, S., (1995). Crete: observations on the ecogeography of grain, forage and pasture legumes. *CLIMA Occasional Papers* No. 11. Centre for Legumes in Mediterranean Areas, Perth.

Francis, C.M., Katznelson, J.S. and Collins, W.J., 1975. Report on legume seed collection tour of Turkey. *Australian Plant Introduction Review* 10(3): 1-10.

Gildersleeve, R.R., Smith, G.R., Pemberton, I.J. and Gilbert, C.L., 1991. Screening Rose Clover and subterranean clover germplasm for isoflavones. *Crop Science* 31: 1374-1376.

Gillett, J.M., 1970. On the Taxonomy of the Genus *Trifolium* L. In: Report of the *Trifolium* research work conference. pp. 26-46 Clemson California.

Gladstones, J.S., 1976. Observations on the distribution and ecology in Iberia and North Africa of some annual legumes adapted to neutral and acid soils. *Australian Plant Introduction Review* 11(1 and 2): 9-23.

Herman, F.J., 1953. *A botanical synopsis of the cultivated clovers* (Trifolium) Agriculture Monograph No. 25. pp. 1-45. United States Department of Agriculture, Washington D.C.

Heyn, C.C., 1981. Tribe 23. Trifolieae (Bronn) Benth. In: *Advances in Legume Systematics* (eds. Polhill, RM and Raven, P.H.). pp. 383-385. Royal Botanic Gardens, Kew

Hollowell, E.A., 1966. White Clover *Trifolium repens* L. annual or perennial? In: *Proceedings of the 10th International Grassland Conference 1966, Helsinki, Finland.* pp. 184-187 Valtioneuvoston Kirjdpaino, Helsinki.

Hopkins, A., Davies, A. and Doyle, C., 1994. *Clovers and other grazed legumes in UK pasture land.* A literature review commissioned by MAFF Livestock Group, U.K.

Horváth, L., 1998. The European *Bromus*, *Trifolium pratense*, and other perennial forages databases. In: *Report of a Working Group on Forages. 6th Meeting, 6-8 March 1997, Beitostølen, Norway* (eds. Maggioni, L., Marum, P., Sackville Hamilton, N.R., Thomas,

I., Gass, T. and Lipman, E.). pp. 37-40. International Plant Genetic Resources Institute, Rome, Italy.

Hossain, M., 1961. A revision of *Trifolium* in the Nearer East, *Notes from the Royal Botanic Garden Edinburgh*, 23: 387-481

IBPGR, 1985. *Ecogeographic surveying and in situ conservation of crop relatives*. pp. 1-27. International Board for Plant Genetic Resources, Rome.

ILDIS, 1999. *International Legume Database and Information Service*. http://www.ildis.org

Julén. G., 1959. Rotklee-*Trifolium pratense* L., In: *Hanabuch der Pflan ZenzBchtung, Band, IV* (ed. Parey, P.) pp. 242-305 (traduction française réalisée par l'INRA).

Karagöz, A., 1998. *In situ*, conservation of plant genetic resources in the Ceylanpínar State Farm. In: *The proceedings of International Symposium on* In situ *Conservation of Plant Genetic Diversity* (eds. Zenciri, N., Kaya, Z., Anikster, Y. and Adams, W.T.). pp. 87-91. CRIFC, Turkey.

Katznelson, J., 1974. Biological Flora of Israel, 5: The subterranean clovers of *Trifolium* subsect. *Calycomorphum* Katzn. *Trifolium subterraneum* L. (*sensu lato*). *Israel Journal of Botany* 23: 69-108.

Keuren, Van R.W. and Hoveland, C.S., 1985. Clover Management and Utilization. In: *Clover Science and Technology* (ed. Taylor, N.L.) pp. 326-354. American Society of Agronomy, Wisconsin.

Knight, W.E., 1985. Miscellaneous annual clovers. In: *Clover Science and Technology, Agronomy Monograph 25*. (ed. Taylor, N.L.) pp. 547-562. ASA-CSSA, Madison, WI, USA.

Kongkiatngam, P., Waterway, M.J., Fortin, M.G. and Coulman, B.E., 1995. Genetic variation within and between two cultivars of red clover (*Trifolium pratense* L.): Comparisons of morphological, isozyme and RAPD markers. *Euphytica* 84: 237-246

Linnaeus, C., 1753. *Species Plantarum Volume II,* in a facsimile of the first edition printed in 1959 for the Ray Society.

Linnaeus, C., 1754. *Genera Plantarum*, 5th Edition. Stockholm

López, F.G., 1998. The European database of *Medicago* spp. (annual species) and *Trifolium subterraneum*. In: *Report of a Working Group on Forages. 6th Meeting, 6-8 March 1997, Beitostølen, Norway* (eds. Maggioni, L., Marum, P., Sackville Hamilton, N.R., Thomas, I., Gass, T. and Lipman, E.). pp. 60-63. International Plant Genetic Resources Institute, Rome, Italy.

Lorenzetti, S., Piccirilli, M., Veronesi, F. and Falcinelli, M., 1987. Collection and evaluation of annual self-reseeding legumes. In: *Proceedings of the AGRIMED Working Group on Annual Self-reseeding Legumes Meeting* (ed. Paino, E.). Cagliari (Italy) 20-30 April,

MAFF, 1998. *Agriculture in the United Kingdom 1997*. The Stationery Office, London.

Mathison, M.J., 1983. Mediterranean and temperate forage legumes, in *Genetic Resources of Forage Plants* (eds. McIvor, J.G. and Bray, R.A.) pp. 63-81. CSIRO, Melbourne, Australia.

McGuire, W.S., 1985. Subterranean clover. In: *Clover Science and Technology* (ed. Taylor, N.L.) American Society of Agronomy, Wisconsin.

Milligan, B.G., 1991. Chloroplast DNA diversity within and among populations of *Trifolium pratense*. *Current Genetics* 19: 411-416

Molina-Freaner, F. and Jain, S.K., 1992a. Breeding systems of hermaphroditic and gynodioecious populations of the colonizing species *Trifolium hirtum* All. in California. *Theoretical and Applied Genetics* 84: 155-160.

Molina-Freaner, F. and Jain, S.K., 1992b. Isozyme variation in Californian and Turkish populations of the colonizing species *Trifolium hirtum*. *Theoretical and Applied Genetics* 84: 160-165.

Mousset, D.C., 1990. Revue bibliographique sur l'hybridation interspécifique chez les trèfles. *Fourrages*, 121: 79-87.

Mousset, D.C., 1995. Les trèfles or le genre *Trifolium*. In: *Ressources Génétiques des Plantes Fourragères et à Gazon*. pp. 176-211. INGRA-BGR, France.

Nestel P.J., Pomeroy S.K.S., Komesaroff P., Behrsing J., Cameron J.D. and West L., 1999. Isoflavones from red clover improve systemic arterial compliance but not plasma lipids in menopausal women. *Journal of Clinical Endocrinology and Metabolism*. 84: 895-898

Nichols, P.G.H., Collins, W.J. and Barbetti, M.J., 1996. Registered cultivars of subterranean clover, their characteristics, origin and identification. *Bulletin No.* 4327. Agriculture Western Australia, Perth.

Noble, M. Le, and Papineau, J., 1972, Le trefle de perse, son intér^t comme fourrage annuel d'été. *Fourrages*, 52: 89-97.

Novogen, 1998a. *Annual Report 1998*. p.15. North Ryde, Australia.

Novogen, 1998b. *Promensil*. North Ryde, Australia.

Osman, A.E., Pagnotta, M., Russi, L., Cocks, P.S. and Falcinelli, M., 1990. The Role of Legumes in Improving Marginal Lands. In: *The Role of Legumes in the Farming Systems of the Mediterranean Areas* (eds. Osman, A.E. *et al.*) pp. 205-216, ICARDA, Aleppo, Syria.

Patterson, R.G. and Espie, P.R., 1997. Productivity and genotypic diversity of *Trifolium ambiguum* in the Central Caucasian Mountains, Georgia, *Proceedings of the XVIII International Grassland Congress*, Canada 26: 5-6

Pederson, G.A, Windham, M.M., Elesbury, M.R., Mclaughlin, Pratt R.G. and Brink G.E., 1989, White clover yield, quality and persistence: insecticide, fungicide, nematode and voirus affects. Proc. XVI Inter. pp. 723-724. *Grasland Congress*, Nice, October, 4-11.

Picard. J., 1959. Quelques résultats concernant l'amélioration du trèfle incarnat. *Annales d'Amélioration des Plantes*, 2: 371-328.

Presl, C.B., 1832. *Symbolicae Botanicae Volume 1*, Pragae.

Quezel. P and Santa. S., 1962, Nouvelle flore de l'Algérie et des régions désertiques méridionales. *Centre National de la Recherche scientifique (CNRS)*, 1: 505-516.

Reid, R., Kanopka, J and Rihan, J.R., 1989. Needs and priorities for further collection of Annula Medic germplasm. Proceedings of International Workshop on *Introducing Ley Farming to the Mediterranean Basin*. pp. 252-270. Perugia, Italy. 26-30 June 1989.

Roskov, Y.R., 1990a. *Revision of the genus* Trifolium *L. in the flora of the U.S.S.R.* Karmarov Botanical Institute, Leningrad.

Roskov, Y.R., 1990b. The new species and nomenclatural combinations in the genera *Lupinaster*, *Chrysaspis*, *Trifolium* and *Amoria* (Fabaceae). *Botaniceskij Zhurnal, (Leningrad)*, 75(5): 715-720.

Russell, J.S. and Webb, H.R., 1976. Climatic Range of Grasses and Legumes Used in Pastures: Results of a Survey Conducted at the 11th International Grassland Congress. *The Journal of the Australian Institute of Agricultural Science* 3: 156-166.

Russi, L., Cocks, P.S. and Roberts, E.H., 1992. The fate of legume seeds eaten by sheep from a Mediterranean grassland. *Jounal of Applied Ecology* 29: 772-778

Sayed-Ahmed, H, Watson LE, and Badr, A.F., 1996. Phylogeny of Old World *Trifolium* species, based on ITS sequences, *American Journal of Botany* 83(6): 208

Singh, V., 1993. Berseem (*Trifolium alexandrinum* L.) B a Potential Forage Crop. *Outlook on Agriculture* 22(1): 49-51.

Smith, R.R., Taylor, N.L. and Bowley, S.R., 1985. Red Clover. In: *Clover Science and Technology* (ed. Taylor, N.L.). American Society of Agronomy, Wisconsin

Speer, G.S. and Allinson, D.W., 1985. Kura clover (*Trifolium ambiguum*): Legume for forage and soil conservation. *Economic Botany* 39(2): 165-175.

Steiner, J.J., Piccioni, E., Falcinelli, M and Liston, A., 1998. Germplasm Diversity among cultivars and the NPGS Crimson Clover Collection. *Crop Science*: 38:263-271

Tan, A., 1998. Current status of plant genetic resources conservation in Turkey. In *The Proceedings of International Symposium on* In situ *Conservation of Plant Genetic Diversity* (eds. Zenciri *et al.*). pp. 5-16. CRIFC, Turkey.

Taylor, N.L. Gibson, P.B. and Knight, W.E., 1977. Genetic Vulnerability and Germplasm Resources of the True Clovers. *Crop Science*, 17: 632-634

Taylor, N.L., Quesenberry, K.H. and Anderson, M.K., 1979. Genetic systems relationships in *Trifolium*. *Economic Botany*, 33(4):431-441

Townsend C.E, and Taylor. N.L., 1985. Incompatibility and plant breeding. In *Clover Science and Technology* (ed. Taylor, N.L.). *Agronomy monograph* 25, 365-381. ASA-CSSA. Madison.WI, USA,

Tutin, T.G., Heywood, V.H., Burges, N.A., Moore, D.M., Valentine, D.H., Walters, S.M. and Webb D.A., 1968. *Flora Europaea, Vol. 2 Rosaceae to Umbelliferae.* Cambridge University Press, Cambridge.

Villax. J., 1963. La culture des plantes fourragères dans la région méditéranéenne occidentale: Maroc-Portugal-Tunisie-Algérie-Espagne-France. In: *Cahiers de la recherche agronomique, No17*, Imprimés aux Editions Marocianes Internationales. 11, Av de Rabat, 641p.

Walter, K.S. and Gillett, H.J., 1998. *1997 IUCN Red List of Threatened Plants*. Compile by the World Conservation Monitoring Centre. pp. 1-862. IUCN The World Conservation Union, Gland, Switzerland and Cambridge, UK

Williams, W.M., 1987. White clover taxonomy and biosystematics. In: *White Clover* (eds. Baker, M.J. and Williams, W.M.). CAB International, Wallingford, U.K.

Zaatout, M., Berrakia, R. and Abdelguerfi, A., 1989. Contribution à l'étude des espèces spontanées du genre *Trifolium* en Algérie: répartition en fonction de quelques facteurs du milieu. Proceedings of XVIth International Grassland Congress, Nice, France,4-11 October 1989.

Zohary, M. and Heller, D., 1984. *The genus* Trifolium, The Israel Academy of Sciences and Humanities, Jerusalem.

Zohary. M., 1972. Origin and evolution in the genus *Trifolium*. *Botaniska Notiser* 125: 501-511.

Zohary, M., 1970. Chapter 46. *Trifolium* L., in *Flora of Turkey and the East Aegean Islands, Vol. 3* (ed. Davis, P.H.). pp384-448. Edinburgh University Press, Edinburgh.

Chapter 5

Medics (*Medicago* L.)

Jean-Marie Prosperi, Geoff Auricht, G. Génier and Richard Johnson

5.1 INTRODUCTION

The genus *Medicago* is arguably the most important of the temperate pasture legume genera containing as it does the widely cultivated perennial plant *Medicago sativa*, (Lucerne or Alfalfa) along with a range of annual species important in specific regions. These include species such *as M. truncatula* (Barrel medic), *M. littoralis* (Strand medic), *M. polymorpha* (Burr medic) the others which are widely grown in cereal crop rotations in Australia. This chapter looks firstly at the evolving taxonomy of the Genus which in the past 10 years has grown significantly from 65 to 85 species with the transfer of 23 species from *Trigonella.* The genetic resources of the genus are then considered in terms of both their origin in the Mediterranean basin, their spread to Secondary Centres of diversity as far away as South America and Australia and their current ex-situ conservation status. *Medicago* germplasm is conserved in several major genebanks holding between 7,500 and 25,000 accessions in long term storage. Finally the chapter discusses the value of a range of perennial and annual *Medicago* species, for both agricultural and environmental purposes as well as raising a number of research needs.

5.2 TAXONOMY

The genus *Medicago* belongs to the Family *Fabaceae*, subfamily *Papilionoideae*. The first description of *Medicago* was by Linnaeus in *Species Planetarium* (Linnaeus, 1753, cited in Lesins and Lesins 1979). He described nine species, including several botanical varieties. The first synthesis of *Medicago*, although still relatively incomplete, was by Urban (1872, cited in Lesins and Lesins 1979). Despite this the genus has only been studied in detail since the 1950's. The taxonomy of *Medicago* is complex and has changed several times in the last century. The four most important descriptions are given by Nègre (1956), Heyn (1963), Lesins and Lesins (1979) and Small and Jomphe (1989).

The genus *Medicago* is close to both *Melilotus* and *Trigonella*, and some species have been placed at different times under both *Trigonella* and *Medicago*. Small (1987) transferred 23 species from *Trigonella* to *Medicago*. Recent data from molecular phylogenetic studies have confirmed the accuracy of this modification (Bena *et al.* 1998a and b). Following Small and Jomphe (1989), *Medicago* species can easily be distinguished from *Trigonella* species by their floral characters. These are characterised by a typical Papilionaceous corolla; nine stamens are fused to form a staminal column, the tenth stamen is free. The corolla and the staminal column form the tripping mechanism, which characterises the entire genus, both selfing and outbreeding species (Small *et al.* 1987).

The *Medicago* genus includes diploid and tetraploid species and unusually hexaploid species (*M. cancellata*, *M. saxatilis* and a few populations of *M. arborea*). The basic chromosome number is 8, although it is 7 for some annual species. The taxonomy, distribution and conservation status of each species of *Medicago* are detailed in Tables 5.1 and 5.2.

Table 5.1. Different species of annual *Medicago*. (Synthesis from Heyn 1963, Lesins and Lesins 1979, Small and Jomphe 1989 and contributions of authors).

Species	Section – Subsection	Synonymy	2x	Breeding System	Area of Distribution	Conservation status
M. arabica Huds.	*Spir. – Leptospireae*		16	Selfing	NM, E, Austr.	C
M. blancheana Boiss.	*Spir. – Rotatae*	*M. bonarotiana* Arc.	16	Selfing	EM	B
M. ciliaris Krock.	*Spir. – Intertextae*		16	Selfing	Med, Austr.	C
M. constricta Dur.	*Spir. – Pachyspireae*		14	Selfing	EM	B
M. coronata Bart.	*Spir. – Leptospireae*		16	Selfing	NM	B
M. disciformis DC.	*Spir. – Leptospireae*		16	Selfing	NM	B
M. doliata Carm.	*Spir. – Pachyspireae*	*M aculeata* Willd.	16	Selfing	Med	C
M. granadensis Willd.	*Spir. – Intertextae*		16	Selfing	EM	B
M. heyniana Greut.	*Heynianae*		16	Selfing	Greece	A
M. intertexta Mill	*Spir. – Intertextae*		16	Selfing	WM	B
M. italica Fior.	*Spir. – Pachyspireae*	*M. tornata* Mill.	16	Selfing	WM	B – Cult.
M. laciniata Mill.	*Spir. – Leptospireae*		16	Selfing	SM, Asia, Austr.	B
M. lanigera W. and F.	*Spir. – Leptospireae*		16	Selfing	Russia, Asia	A
M. laxispira Heyn.	*Spir. – Leptospireae*		?	?	EM	A
M. littoralis Rhode	*Spir. – Pachyspireae*		16	Selfing	Med	B – Cult
M. lupulina L.	*Lupularia*		16,32	Selfing	World	C – Cult
M. minima Bart.	*Spir. – Leptospireae*		16	Selfing	World	C
M. murex Willd.	*Spir. – Pachyspireae*	*M. lesinsii Small*	16	Selfing	Med	B – Cult
M. muricoleptis Tin.	*Spir. – Intertextae*		16	Selfing	Italy	A
M. noëana Boiss.	*Spir. – Rotatae*		16	Selfing	EM	A
M. orbicularis Bart.	*Orbiculares*		16	Selfing	Med, E. Asia, Austr	C
M. polymorpha L.	*Spir. – Leptospireae*		14	Selfing	World	C – Cult
M. praecox DC.	*Spir. – Leptospireae*		14	Selfing	NWM, Austr.	B
M. radiata L.	*Hymenocarpos*		16	Selfing	Med	C

M. rigidula All.	*Spir. – Pachyspireae*		14	Selfing	WM, E, Africa	C – Cult
M. rigiduloides Small.	*Spir. – Pachyspireae*		14	Selfing	EM, Middle East	C – Cult
M. rotata Boiss.	*Spir. – Rotatae*		16	Selfing	EM	B
M. rugosa Desr.	*Spir. – Rotatae*		30	Selfing	EM, WM	B
M. sauvagei Nègre	*Spir. – Leptospireae*		16	Selfing	Morocco	A
M. scutellata Mill.	*Spir. – Rotatae*		30	Selfing	Med	B – Cult
M. secondiflora Dur.	*Lupularia*		16	Selfing	WM	B
M. shepardii Post.	*Spir. – Rotatae*		16	Selfing	EM	A
M. soleirolii Duby	*Spir. – Pachyspireae*		16	Selfing	SWM	A
M. sphaerocarpos Bertol.	*Spir. – Pachyspireae*	*M . murex*	14	Selfing	Med	B – Cult
M. tenoreana Ser.	*Spir. – Leptospireae*		16	Selfing	NWM, rare	A
M. truncatula Gaertn.	*Spir. – Pachyspireae*		16	Selfing	Med, Austr.	C – Cult
M. turbinata All.	*Spir. – Pachyspireae*		16	Selfing	NEM	B
M. hypogaea Small.	*Geocarpa*	*T.aschersoniana* Urb.	?	?	SEM	A
M. ovalis Sirj.	*Platycarpae ?*	*T. ovalis* Boiss.	?	?	WM	?
M. plicata Sirj.	*Platycarpae*	*T. plicata* Boiss.	?	?	Turkey	?
4 Species	*Lunatae*	Moved from genus to *Trigonella* by Small and Jomphe (1989)	?	?	?	*Variable*
19 Species	*Buceras*	Moved from genus to *Trigonella* by Small and Jomphe (1989)	?	?	?	*Variable*

Abbreviations: **Section Spir.** = Spirocarpos ; **2x** : Number of chromosomes ; **Area of distribution**. Med. = Mediterranean basin, NM = north Mediterranean, EM = east Mediterranean, WM = west Mediterranean, SM = south Mediterranean, E = Europe, Austr. = Australia (unintentionally introduced in Australia, Cocks *et al.* 1980), World = Mediterranean climate areas all over the world ; **Conservation status**. A = Endangered species, endemic or rare, few accessions in gene banks, B = Species with threatened diversity due to habitat erosion or occur in regions threatened by climate change, few available accessions, C = Species widespread, large number of available accessions, Cult = Genetically improved for agriculture. ? = unknown or undetermined.

Table 5.2. The different species of perennial *Medicago* (Synthesis from Lesins and Lesins 1979, Small and Jomphe 1989 with contributions of authors).

Species – subspecies	Section	2x	Breeding system	Colour of flowers	Area of distribution	Conservation Status
M. arborea L.	*Dentrotelis*	32 (48?)	Out., Comp	Yellow	Mediterranean	B – Cult
M. archudis-nocolai Sirj.	*Platycarpae*	?	?	?	China	?
M. cancellata M. Bieb.	*Medicago*	48	?	Yellow	Caucasus	A
M. carstiensis Wulf	*Medicago*	16	Out., Comp	Yellow	Adriatic, endemic	A
M. cretacea M. Bieb.	*Platycarpae*	16	?	Yellow	Caucasus, Crimea, endemic	A
M. daghestanica Rupr.	*Medicago*	16	?	Violet	Caucasus, endemic	A
M. dzhawakhetica Bord.	*Medicago*	32, (16?)	?	Yellow	Caucasus, endemic	A
M. edgeworthii Sirj. [1]	*Platycarpae*	?	?	Yellow	Himalaya, Pakistan, India	?
M. hybrida Trautv.	*Medicago*	16	Self.	Yellow	South France, endemic	A
M. marina L.	*Medicago*	16	Self.	Yellow	Med., sea shores	B
M. papillosa Boiss.	*Medicago*	16, 32	?	Yellow	Anatolia, rare	A
M. pironae Vis.	*Medicago*	16	?	Yellow	North Italy, endemic	A
M. platycarpa Trauv.	*Platycarpae*	16, (32)	?	Yellow	Easter Europe, Asia	A
M. popovii Sirj. [1]	*Platycarpae*	?	?	?	Central asia, endemic	A
M. prostrata Jacq.	*Medicago*	16, 32	Out, ?	Yellow	Adriatic, rare	A
M. rhodopea Velen	*Medicago*	16	?	Yellow	Bulgaria, endemic	A
M. rupestris M. Bieb.	*Medicago*	16	?	Yellow	Crimea, endemic	A
M. ruthenica Led.	*Platycarpae*	16	?	Yellow	Siberia, Mongolia	A
M. sativa ssp. *coerulea* [2]	*Medicago*	16	Out.,?	Violet-blue	Turkey, Iran,	B
M. sativa ssp. *falcata* [2]	*Medicago*	16, 32	Out, Comp	Yellow	Europe ,Asia	B – Cult
M. sativa ssp. *glomerata* [2]	*Medicago*	16	Out, Comp	Yellow	Europe, Caucasus	A
M. sativa ssp. *glutinosa* [2]	*Medicago*	32	Out., ?	Yellow	Caucasus, endemic	A
M. sativa L. ssp. *sativa*	*Medicago*	16, 32	Out, Comp	Violet	World – cultivated	C – Cult
M. saxatilis M. Bieb.	*Medicago*	48	?	Yellow	Crimea, endemic	A
M. strasseri Greu.	*Dentrotelis*	?	?	Yellow	Crete, endemic	A
M. suffruticosa Ram.	*Medicago*	16	Self.	Yellow	West Mediterranean	B

Abbreviations: **2x** = Number of chromosomes, **Breeding system.** Self = Selfing, Out = Outcrossing, Comp = Self-compatible, ? unknown. **Distribution.** Med = Mediterranean basin, World = Mediterranean climates all over the world; **Conservation status.** A = Endangered species, endemic or rare, few accessions in gene banks, B = Species with threatened diversity due to habitat erosion or occur in regions threatened by climate change, few available accessions, C = Species widespread, large number of available accessions, Cult = Genetically improved for agriculture, cultivars registered mainly in Australia. ? = unknown or undetermined. 1: Species moved from genus Trigonella by Small and Jomphe 1989. 2: These species (Lesins and Lesins 1979) were moved at sub-species level by Small and Jomphe 1989. 3: *M. sativa ssp glutinosa* is cultivated in some regions of China.

Currently (Small 1987, Small *et al.* 1987 and Small and Jomphe 1989) *Medicago* contains 85 species of which two are shrubby (*M. arborea* and *M. strasseri*), 20 are herbaceous perennials, and 63 are herbaceous annuals. The pod remains the main characters to distinguish the species, particularly in the annuals. In general, the annual species are strongly self-pollinating and present strong barriers to inter-specific hybridisation; whereas most perennial species are allogamous, with variable levels of self-compatibility. Some are inter-compatible. Knowledge of crossability is essential for regeneration of accessions. A substantial portion of perennial *Medicago* stored in gene-banks are not identical to the original population due to insufficient pollen isolation during multiplication.

About ten species of the genus *Medicago* are cultivated and are grown in grazing or rangelands, mainly in the Mediterranean or the cereal zones of Australia. The best known species is certainly alfalfa (*Medicago sativa*), also known as lucerne. Some other species are associated with specific farming systems or particular climatic conditions. This is the case with the annual medics, cultivated for over a century in Australia. The most important annual medic species include *Medicago truncatula* (barrel medic), *M. polymorpha* (burr medic), *M. littoralis* (strand medic) and *M. scutellata* (snail medic). A short description of these and other important species is included in appendix 1.

5.3 CENTRE OF DIVERSITY AND DISTRIBUTION

According to Lesins and Lesins (1979), the ancestral *Medicago* species are thought to be perennial, probably woody, and preferentially allogamous. The annual species would have separated 6 to 7 million years ago during the Miocene, at the time of the marine transgressions of the Mediterranean basin. The Centre of origin of the genus is the 'fertile' crescent, and includes Turkey, Iran, Iraq, the area south of the Caucasus and the eastern Mediterranean basin. These annual species spread throughout the Mediterranean region and then, to the neighbouring steppes (Figure 5.1). From the 17th century, they were introduced to the 'New World', in particular the American and Australian continents in association with human migration.

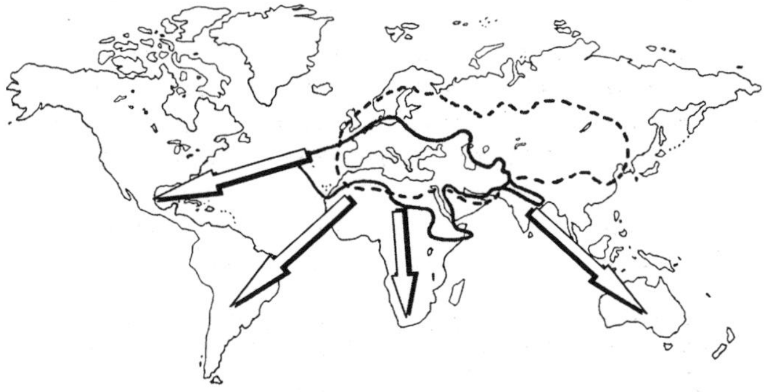

Figure 5.1 Distribution area of the annual (—) and the perennial (----) species of the genus *Medicago* and recent migrations (Partly after Heyn, 1963).

5.3.1 Annual Medics

The annual medics have highly variable distributions which are continually altering as a result of human interference (see table 5.1). Floras and herbarium collections provide information on the maximum range of a species, but do not give an indication of its frequency across the range, and therefore its conservation status. Rihan (1988) established the distribution of some annual species or varieties in the Mediterranean basin following analysis of herbarium specimens. A more precise picture of a species distribution can be obtained from records of germplasm collections. Prosperi *et al* (1991) gave an account of the frequency of some annual species in the western Mediterranean basin based on these data, and several categories of distribution are detailed below (Olivieri *et al.* 1991; Prosperi *et al.* 1995);

Species with a wide distribution and frequently occurring, for example *M. polymorpha, M. truncatula, M. orbicularis, M. minima* and *M. lupulina.* These species can be qualified as colonisers. They have all been introduced unintentionally to Australia (Cocks *et al.* 1980), are widely adapted and represent an important source of variability and genetic diversity.

Species frequent in particular regions. e.g. *M. arabica* is commonly found in Corsica, Sardinia and Greece, but is rare in Spain and north Africa. *M. intertexta* is localised in hot and rainy areas of southern Spain, northern Algeria and Sicily, but is only common on the heavy clay, sometimes salty soils, in these regions. *M. doliata* is abundant in north Africa and southern Spain but is rarer in the northern Mediterranean countries where, in similar habitats it is replaced by *M. rigidula. M. rugosa* is only found on red to brown clay loam soils and *M. littoralis* and *M. tornata* on sandy, mostly coastal soils.

Species rare or endemic. These include species such as *M. coronata*, plus others listed in Tables 5.1 and 5.2.

5.3.2 Alfalfa (*Medicago sativa sensu lato*)

It is difficult to qualify the first traces of alfalfa domestication. Ivanov (cited in Quiros and Bauchan 1988) suggests that alfalfa was cultivated 8,000 to 9,000 years ago. Hendry (cited in Bolton 1962) proposes that since 4,000 BC, maybe earlier, the Sumerian merchants transported alfalfa to the western Mediterranean. Sinskaya (1950) suggests that its distribution extended across to the Middle East by 1,000 BC, and from there, extended to China and India. In China, alfalfa was first introduced into the province of Shaanxi in 116 BC by an envoy of the Han Emperor sent to purchase horses from Central Asia (Chinese Herbage Cultivar Registration Board 1992). Alfalfa was introduced to Greece by the Xerxès and Medes armies (500 BC), then extended into Italy (200 BC) and spread all over the Roman Empire. Interest in alfalfa then regressed considerably in Europe after the collapse of the Roman Empire (around 300 AD), but remained in the wild in favourable climatic and edaphic zones. It was reintroduced into Spain through north Africa (around 700 AD) with the Moores and crossed the Pyrenees around the 8th century. These different paths of extension are summarised in Figure 5.2. It was introduced to South America (Mexico, Peru, Chile) by the Spanish in the 16th century, and then to the United States through California at the beginning of the 19th century as 'Chilean clover' (Michaud *et al.* 1988).

Escaped plants from agriculture are frequent in a variety of open habitats (roadsides, field borders), but *M. sativa* is rare in the wild, with the exception of the Iberian Peninsula and its centre of diversity (Caucasus, Turkey). In other west Mediterranean countries (southern France, Italy, Greece, and north Africa), plants of *M. x varia* are very common. Unlike *M. sativa, M. falcata* appears more frequently as a wild species, from north of the Mediterranean Sea (Bulgaria, Greece) to the northern limits of Russia. It is particularly well adapted to hard winters, as well as to the hot and dry summers typical of continental climates.

5.4. POTENTIAL VALUE

5.4.1 Annual medics.

The Australian farming system The annual medics are important on the alkaline soils of southern Australia, and constitute the basis of sheep production systems on these soils in Australian regions with a Mediterranean climate. Introduced unintentionally in the 1830's, the medics invaded fallow areas, following phosphate applications on cereal crops. A farming system based on the cereal-medic rotation (ley-farming) was developed as a result of hardseededness in the medic cultivars (Crawford *et al.* 1989). Progressively, the self seeding medic replaced the less productive classic fallow in the cereal rotation. The system is flexible and efficient; the seeding of a cereal crop for two or three consecutive years doesn't prevent the medic regenerating from the seed bank in the soil (Puckridge and French 1983). Viable seeds remain in the soil for over seven years in species such as *M. scutellata* and *M. italica* (syn. *M. tornata*) (Crawford and Nankivell 1989). Medics for the ley-farming system have been selected since the 1950's, and currently more than twenty registered cultivars are available in Australia (Oram 1990). Cultivar development in Australia has occurred through, the selection and commercialisation of locally adapted ecotypes from accidental introductions spread by farming, the deliberate introduction and selection of wild accessions collected in the Mediterranean or Chile (*M. polymorpha*), and the breeding programme associated with the Australian *Medicago* Genetic Resource Centre, centred in Adelaide, South Australia, since the late 1970's (Lake 1998). The breeding program has focused on developing high yielding, aphid resistant cultivars of *M. truncatula* and *M. littoralis* by back-crossing locally adapted 'ecotype' cultivars to aphid resistant material.

Mediterranean Europe and North Africa The conditions of the Mediterranean region (summer drought, stony soils, overgrazing, reduced agricultural activity and shrub fires) led to the establishment of a marginally productive woody vegetation (*Quercus* Sp., *Cistus* Sp., *Juniperus* Sp., *Pinus* Sp.) in Mediterranean Europe and to a poorer steppe in north Africa. To increase the quality and quantity of forage available, a management strategy has been developed in which the medics show great potential given their natural reseeding ability, their prostrate habit, and their tolerance to grazing and drought. Annual medic species, such as *M. rigidula*, *M. truncatula* and *M. polymorpha* could also be used as constituents of hardy meadows, and in pastures under grapevines and orchards to control the erosion of the soil. *M. scutellata* can also replace the pea and vetch in the annual forage rotations commonly used in the Mediterranean countries. For all these uses, annual medics should have better grazing and cold weather tolerance than currently available cultivars (Prosperi 1993). Medic genetic resources originating from northern Mediterranean regions are of great importance in helping to meet these objectives.

Medics in North America Although the annual medics currently have only a limited role in North American agriculture they have the potential for several important applications. As interest in sustainable agriculture has grown, interest in the annual medics has also developed. The potential benefits of annual medic utilisation include nitrogen fixation, green manure or forage production, reduced soil erosion, and weed control. They have a possible niche as a high quality forage during summer and fall, and may improve soil fertility in the north-central USA (Yanping-Zhu and Sheaffer 1997). Annual medics may also be applicable for inter-cropping systems in wide-spaced row crops such as maize (*Zea mays*) and sunflower (*Helianthus annuus*) (Kandel *et al.* 1997). Medics have been planted in vineyard and orchard alleyways to provide a low maintenance ground cover. In Montana, the black medic (*M. lupulina*) cultivar 'George' is used in rotation systems with spring wheat as a 'green fallow' or cover crop (Sims *et al.* 1985). A burr medic (*M. polymorpha*) cultivar has recently been released for use as a winter pasture in south and central Texas (Ocumpaugh *et al.* 1997). Both of these cultivars were developed from selections of local naturalised populations.

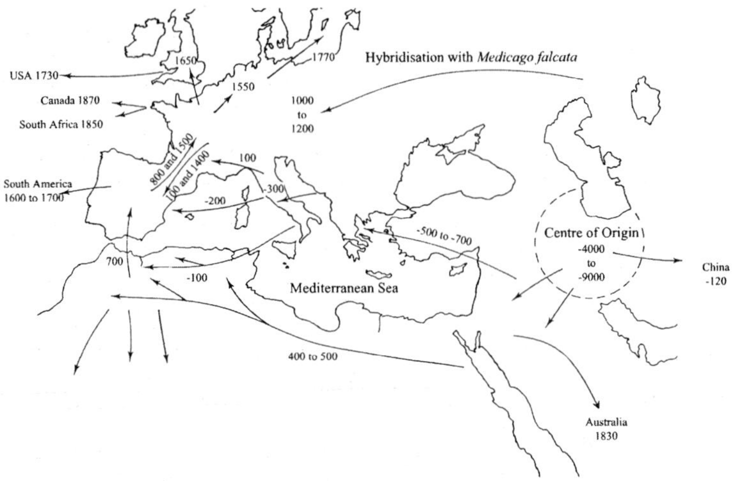

Figure 5.2 The different paths and approximate years BC (-) and AC of migration of alfalfa from its Centre of origin.

5.4.2 Alfalfa

Alfalfa is one of the world's most important forage legumes; in the 1980's, more than 33 million hectares were cultivated world-wide (Michaud *et al.* 1988). Alfalfa is grown in irrigated or dryland conditions across a wide range of environments from the Saudi-Arabian deserts to the sub-tropics of Australia, and through the cold temperate regions of Europe, west and central Asia, and South and North America.

Grown primarily for the production of high quality fodder or for direct grazing, it also provides a number of other benefits. These include nitrogen fixation, soil stabilisation (ground cover in row crops), soil enhancement (root channels, recycling nutrients from depth), hydrological benefits (lowering water tables, lessening dryland salinity, reduced nitrate leaching), rotational benefits (weed and disease break in crops) and for novel uses such as enzyme production.

Europe Alfalfa is commonly grown in Europe, but from the 1960's to 1980's, it decreased dramatically in western Europe due to intensive management of animal production mainly based on maize silage. Today, cultivation has stabilised. Alfalfa is also dehydrated to produced pellets which are used as a nitrogen supplement for feeding dairy animals (ewes or cows) or non-ruminant animals (rabbits, pigs, poultry). Since 1965 alfalfa breeding programs in Europe have been aimed primarily at increasing tolerance to diseases or pests (Raynal *et al. 1977*). In northern Europe, damage caused by Verticillium wilt (*Verticillium albo atrum*) brought about a 30% decrease in dry matter production during the 1970's. This problem was solved during the 1980's with the release of resistant varieties. Breeding for tolerance to leaf pests led to remarkable results against anthracnose (*Colletotrichum trifolii*) and to significant resistance to spring black stem (*Phoma medicagenis*) and common leaf spot (*Pseudopeziza medicagenis*). New varieties also have good tolerance to stem nematode (*Ditylenchus dispaci*), and in the near future, to pea aphid (*Acyrthosiphon pisum*) (Bournouville *et al.* 1999). A large number of these sources of resistance were found in old landraces of alfalfa, some of which came directly from wild populations.

Alfalfa in North America Alfalfa is the most widely grown forage legume in North America. More than 10 million ha are harvested in the USA annually, with an estimated 2 million ha in Canada and approximately 0.25 million ha in Mexico (Michaud *et al.* 1988). Barnes *et al.*, (1988) traced the origins of cultivated alfalfa in North America to nine basic germplasm sources. This includes the Spanish type alfalfa introduced to the south-western USA through South America, winter hardy types from Europe and Russia, intermediate winter-hardy types from south-west Asia and France, and non winter-hardy material from Peru and India. Multiple pest resistance has become a key feature of alfalfa plant breeding (Barnes *et al.* 1988). In the 1970's incorporation of phytophthora root rot (*Phytophthora megasperma* f. sp. *Medicaginis*), anthracnose and bacterial wilt (*Corynebacterium insidiosum*) resistance became routine in North American cultivars. In the 1980's, resistance to Fusarium wilt (*Fusarium oxysporium* f. sp. *medicaginis*) and Verticillium wilt became important breeding objectives. During the 1970's and 1980's breeding for insect resistance was mostly for spotted alfalfa aphid (*Therioaphis maculata*) and pea aphid resistance, with the potato leaf hopper (*Empoasca fabae*) continuing as the most problematic insect pest in the north-central USA (Barnes *et al.* 1988). In the 1980's it was found that exotic alfalfa germplasm with glandular-hairs provided resistance to potato leaf hopper and this has led to the development of potato leaf hopper resistant cultivars (Manglitz and Sorensen 1998).

Australia Alfalfa breeding in Australia commenced in the late 1970's following the devastation caused by the introduction of spotted alfalfa aphids and blue-green aphids (Auricht 1999). Since then breeding efforts have focused on persistence and productivity with a broad adaptation to soils, climate and management. This has resulted in the development of locally adapted cultivars with aphid, Phytophthora, Anthracnose and most recently, stem nematode resistance (*Ditylenchus dispaci*). The research focus is now shifting to the development of alfalfa cultivars for use in cereal rotations in Australia. The role of such cultivars is to lower watertables which are threatening crop production through increasing salinity and water-logging (Auricht 1999).

5.5 CONSERVATION RESOURCE STATUS

A number of gene banks exist that concentrate on *Medicago*. The three main ones for Mediterranean *Medicago* genetic resources are by order of importance : AMGRC Australia, USDA-ARS USA and ICARDA Syria. A large number of accessions of the genus are also available from the Vavilov Institute, Russia.

For annual *Medicago* species (medics), large '*ex-situ*' collections exist. Despite this, an effort must be made to preserve endangered species (see Table 5.1) and populations of common species originating from the regions surrounding of the 'centre of diversity' (Asia, Tropical Africa, Northern

Europe). In addition, preservation of materials from secondary centres of diversification (USA, South America, South Africa) is important. Given their adaptations to local climates, they may have some untapped potential for use in these regions without the greater effort that usually accompanies the development of exotic material. For example, the *M. lupulina* cultivar 'George' was based on parent seed collected from adapted ecotypes in the state of Montana (Sims *et al.* 1985). Local ecotypes were also used to develop a new *M. polymorpha* cultivar for Texas (Ocumpaugh *et al.* 1997). In France, three new medic cultivars tolerant to cold environment have been selected from local ecotypes (*M. rigidula* 'Ampus', *M. polymorpha* 'Mauguio' and *M. truncatula* 'Salernes'). Similarly in Australia, collections of naturalised *M. polymorpha* and *M. laciniata* are being assessed for their potential as cultivars in non-mediterranean environments.

For perennial *Medicago* species, collections are generally under-represented for species other than the *M. sativa* complex. For example, the *M. sativa* complex is represented in USDA collection by 3,816 accessions but the remaining perennials, representing 17 species, total only 265 accessions. Recent exploration in Inner Mongolia resulted in the addition of 105 *M. ruthenica* accessions to the USDA collection (Campbell *et al.* 1997), but several perennial species are either poorly represented or not represented, especially those from the centre of origin of the genus and from Asia. The diploid forms of the *M. sativa* complex species are also under-represented (i.e. subsp. *glomerata* or subsp. *coerulea*).

For the 23 new *Medicago* species moved from the genus *Trigonella*, only a few accessions are available. The status of these species is not well known, nor are their taxonomic characteristics, ecology and geographic distribution, or life-cycle and breeding system. An effort must be made to increase our knowledge of these species and to assess their potential for agronomic uses.

5.6 MAJOR *EX SITU* COLLECTIONS

Australian Medicago *Genetic Resources Centre (AMGRC), Adelaide, Australia.*

This Centre was established in the early 1960's when it was recognised that the production of annual medics could be improved through the deliberate introduction and evaluation of a diverse array of germplasm from the mediterranean region. The AMGRC now maintains the world's largest annual medic collection with 25,000 accessions along with a further 10,000 accessions of related genera. The Centre introduces an average of 1,000 new accessions each year and distributes seed of some 2,000 accessions globally to pasture research scientists (Hughes and McLachlan 1999). All data associated with the collection is maintained in one database. This is accessible to collection users on the World Wide Web (www.sardi.sa.gov.au) or directly from the curator. The AMGRC maintains samples of seed of each accession in a base collection at -20°C and in an active collection at 2°C.

United States Department of Agriculture (USDA) Agricultural Research Service (ARS)

The USDA-ARS collection of *Medicago* located in Pullman, WA is part of the USA National Plant Germplasm System (NPGS). The *Medicago* collection currently contains over 7,500 accessions representing 74 different taxa. About 4,000 of the accessions are perennial alfalfa. Overall the collection holds accessions from 91 countries, with 48% of the accessions originating directly or indirectly from countries bordering the Mediterranean Sea. Seed of the active collection is stored at 4 C and 30% relative humidity. Long-term storage is at the National Seed Storage Lab (NSSL) and is maintained at –20 C or in liquid nitrogen. Passport data and evaluation information is accessible on the Germplasm Resources Information Network (GRIN) (www.ars-grin.gov/npgs/). Evaluation data has been used to develop core subsets of the perennial (Basigalup *et al.* 1995) and annual (Diwan *et al.* 1994, Skinner *et al.* 1998) germplasm collections. During the five years period from 1993 to

1997, a total of 12,537 *Medicago* seed distributions were made to scientists both in and outside the USA. The collection remains an important source of disease and insect resistant genes.

International Centre for Agricultural Research in the Dry Areas (ICARDA), Aleppo, Syria

This collection includes more than 15,000 accessions of forage plants of which more than half are legumes, mainly medics, collected primarily in the Middle East. Of the 8,450 *Medicago* accessions, 7,800 are annual species and 650 are perennial species.

An example from EC Countries: The French Forage Genetic Resources Centre (FFGRC)

In Europe, there is no central gene bank; each country maintains its own plant genetic resources. However, data from the European Community is centralised through the European Cooperative Program for the Conservation and Exchange of Crop Genetic Resource (EC-IPGRI). Thus, GEVES[1] (France) is responsible for alfalfa and other perennial *Medicago* and SIA[2] Badajoz (Spain) for annual *Medicago* and subterranean clovers. The FFGRC operates as part of a network which includes the French INRA public forage plant breeding stations (Institut National de la Recherche Agronomique), private firms involved in plant breeding, and GEVES (the institution responsible for characterisation and registration of cultivars). Environmental conditions for mid-term seed storage are 3°C and less than 20% relative humidity. GEVES maintains a security back-up or long-term storage of accessions at –20°C. Collection activities concentrate on the western part of the Mediterranean basin (north and south). The FFGRC gene bank contains more than 2,800 accessions from 9 countries representing 32 species of annual medics, some 200 wild populations of perennial *Medicago* as well as 500 samples of cultivated material (mainly *M. sativa*).

5.7 RESEARCH NEEDS

Wild relatives of cultivated plants, have contributed to the breeding of most of our crop plants (Harlan 1976, Frankel *et al.* 1995). This is particularly true for forage plants where the difference between cultivated and wild forms is often minimal. In main cases, gene flows occurs between natural and cultivated forms(Jenczewski *et al.* 1999a,b). Within the genus *Medicago*, numerous cultivars have been obtained directly from natural populations with relatively little selection.

Changes in agriculture will continue to be limited by the availability of germplasm and the resources available for its conservation and evaluation. With the need for more sustainable systems, the prospect of transgenic crops and the erosion of natural habitats (climatic changes, industrialisation), the value of documented *ex-situ* collections will continue to increase.

Further research on the improvement, management, preservation and use of natural diversity will also be a significant challenge for the next century. *Medicago* is a key genus for these studies, with 85 species, all growing in symbiosis with *Rhizobium*. Wild and cultivated populations within the *M. sativa* complex and among the annual medics are of great agronomic interest, not only as a forage, but potentially for industrial uses as diverse as fibre crops for power generation, to the rehabilitation of degraded ecosystems (numerous authors, NAAIC 1998).

---

[1] Groupement d'Etudes et de contrôle des Variétés Et des Semences. Le Magneraud. Saint-Pierre d'Amilly, BP 52, 17700 Surgères, France.

[2] Servicio de Investigación Agraria. Junta de Extramadura. Apartado 22, 06080 Badajoz. Spain.

REFERENCES

Auricht, G.C. 1999. Lucerne: An Australian Breeding Success Story. In: *Australian Plant Breeding* Conference Proceedings (eds. Langridge, P. *et al.*). 1: 71-76.

Barnes, D.K., Goplen, B.P. and Baylor, J.E., 1988. Highlights in the USA and Canada. In: *Alfalfa and alfalfa improvement. Agronomy Monography 29* (eds. Hanson, A.A., Barnes, D.K. and Hill, R.R). pp. 1-24. American Society of Agronomy, Madison, USA.

Basigalup, D.H., Barnes, D.K. and Stucker, R.E., 1995. Development of a core collection for perennial *Medicago* plant introductions. *Crop Science*, 35: 1163-1168.

Bena, G., Prosperi, J.M., Lejeune, B. and Olivieri, I., 1998a. Evolution of annual species of the genus *Medicago*: A molecular phylogenetic approach. *Molecular Phylogenetics and Evolution*, 9: 552-559.

Bena, G., Prosperi, J.M., Lejeune, B. and Olivieri, I.,1998b. Molecular phylogenetic approach for studying life-history evolution: The ambiguous example of the genus *Medicago* L. *Proceeding of the Royal Society of London B.,* 265: 1141-1151.

Bolton, J.L., 1962. *Alfalfa botany, cultivation and utilisation.* Leonard Hill, London.

Bournouville, R., Landré, B., Aupinel, P., Girousse, C. and Badenhausser, I., 1999. Test en conditions contrôlées de la résistance variétale de la luzerne au puceron du pois. *Fourrages,* 158: 157-168.

Campbell, T.A., Bao, G., and Xia, Z.L., 1997. Agronomic evaluation of *Medicago ruthenica* collected in inner Mongolia. *Crop Science*, 37: 599-604.

Chinese Herbage Cultivar Registration Board, 1992. *Licensed cultivars of herbage crops in China.* Beijing Agricultural University press, China.

Cocks, P.S., Mathison, M.J. and Crawford, E.J., 1980. From wild plants to pasture cultivars, annual medics and subterranean clover in Southern Australia. In: *Advances in Legume Science* (eds. Summerfield, R.J. and Bunting, A.H.). pp. 569-596. Royal Botanic Garden, Kew.

Crawford, E.J. and Nankivell, B.J., 1989. Effect of rotation and cultivation systems on the establishment and persistence of annual medics. *Australian Journal of Experimental Agriculture*, 29:183-188.

Crawford, E.J., Lake, A.W.H. and Boyce, K.G., 1989. Breeding Annual Medicago. Species for semiarid Conditions in Southern Australia. *Advances in Agronomy*, 42: 399-437.

Diwan, N., Bauchan, G.R. and McIntosh, M.S., 1994. A core collection for the United States annual *Medicago* germplasm collection. *Crop Science*, 34: 279-285.

Frankel, O.H., Brown, A.H.D. and Burdon, J.J., 1995. *The conservation of plant diversity.* Cambridge University Press, Cambridge.

Harlan, J.R., 1976. Genetic resources in wild relatives of crops. *Crop Science*, 16: 329-333.

Heyn, C.C., 1963. The annual species of *Medicago. Scripta Hierosolymitana,* Hebrew University, Jerusalem.

Hughes, S.J. and McLachlan, D.R., (1999). *Australian Medicago Genetic Resources Centre Annual Report 1997.* Department of Primary Industries and Resources South Australia, SARDI Research Report No. 34. SARDI, Adelaide.

Jenczewski, E., Prosperi, J.M. and Ronfort. J., 1999a. Evidence for gene flow between wild and cultivated *Medicago sativa* (*Leguminosae*) based on allozyme markers and quantitative traits. *American Journal of Botany*, 86(5): 677-687.

Jenczewski, E., Prosperi, J.M. and Ronfort. J., 1999b. Differentiation between natural and cultivated populations of *Medicago sativa* (*Leguminosae*) from Spain: analysis with random amplified polymorphic DNA (RAPD) markers and comparison to allozymes. *Molecular Ecology*, 8: 1317-1330.

Kandel, H.J., A.A. Schneiter, and Johnson, B.L., 1997. Inter-cropping legumes into sunflower at different growth stages. *Crop Science*, 37: 1532-1537.

Lake, A.W.H., 1998. Breeding and Selection of Annual Medics for Resistance to Pests, Particularly Alfalfa Aphids. Proc. 36th *North American Alfalfa Improvement Conference (NAAIC)*, Montana, USA.

Lesins, K. and Lesins, I.,1979. *Genus* Medicago *(Leguminosae). A Taxogenic study*. Dr. W. Junk, Dordrecht.

Manglitz, G.R., and Sorensen, E.L., 1998. Alfalfa germplasm resources and insect resistance. In: *Global plant genetic resources for insect-resistant crops* (eds. Clement, S.L. and Queensberry, S.S.). CRC Press. Boca Raton, USA (in press).

Michaud, R., Lehman, W.F. and Rumbaugh, M.D.,1988. World distribution and historical development. In: *Alfalfa and alfalfa improvement* (eds. Hanson, A.A. Barnes, D.K. and Hill R.R.). Agronomy Monography 29. pp. 25-91. American Society of Agronomy, Madison, USA.

NAAIC, 1998. *Report of the thirty-sixth North American Alfalfa Improvement Conference*. pp. 1-172. August 2-6, 1998. Bozeman. Montana USA.

Nègre, R., 1956. Les luzernes du Maroc. *Travaux de l'institut chérifien. Série botanique n°5. Revised 1959 in le bulletin de la société d'histoire naturelle de l'Afrique du nord. Tome*, 50: 267-314.

Ocumpaugh, W.R., Hussey, M.A., Rahmes, J.N., Grichar, Jr., W.J., Sestak, D.C. and Smith, R.,1997. Burr Medic - A persistent cool-season legume for Texas. *American Forage and Grassland Council Proceedings*, 6: 59-63.

Olivieri, I., Gouyon, P.H. and Prosperi, J.M.,1991. Life cycles of some Mediterranean invasive plants. In: *Biogeography of Mediterranean Invasions* (eds. Groves, R.H. and Di Castri, F.). pp. 145-157. Cambridge University Press, Cambridge.

Oram, R.N. 1990. *Register of Australian Herbage Plant Cultivars 3rd Ed.* CSIRO, East Melbourne, Australia.

Prosperi, J.M., 1993. Selection of annual medics for French Mediterranean regions. In: *Introducing the ley farming system to the Mediterranean basin* (eds. Christiansen, S., Materon, L., Falcinelli M. and Cocks, P.). pp. 173-191. June 26/30, 1989. Perugia (Italy). ICARDA, Aleppo, Syria.

Prosperi, J.M., Guy, P., Génier, G. and Angevain, M.,1995. Les luzernes ou le genre *Medicago* In: *Ressources Génétiques des plantes fourragères et à gazon* (eds. Prosperi, J.M., Guy, P. and Balfourier, F.). pp. 131-168. INRA – BRG, Paris.

Prosperi, J.M., Boumard, P., Angevain, M. and Mansat, P.,1991. *Répartition et adaptation écotypique de Medicago annuelles en Méditerranée occidentale.* 4th Congrès International des Terres de Parcours. Montpellier, France.

Prosperi, J.M. and Ronfort, J.,1998. Constraints to the introduction of Medics in French Mediterranean farming systems. Invited conference. *36th North American Alfalfa Improvement Conference.* Bozeman, Montana August 2-6, 1998.

Puckridge, D.W. and French, R.J., 1983. The annual legume pasture in cereal-ley farming systems of Southern Australia, a review. *Agriculture, Ecosystems and Environment*, 9: 229-267.

Quiros, C.F., and Bauchan, G.R., 1988. The genus *Medicago* and the origin of *Medicago sativa* complex. In: *Alfalfa and alfalfa improvement* (eds. Hanson, A.A., Barnes, D.K. and Hill, R.R). pp. 93-124. Agronomy Monography. 29. American Society of Agronomy, Madison, USA.

Raynal, G., Bournouville, R., Gondran, J., Guy, P. and Jouannic, J.M., 1977. Maladies et ravageurs de la luzerne: Répartition en France et en Europe, Dégats et Valeurs des cultivars. *Fourrages*, 71: 3-26.

Rihan, J.R, 1988. A herbarium based on ecogeographic survey of forage *Medicago* species in the Mediterranean and adjacent arid/semi arid areas. Technical report. *International Legume Database and Information Service*, University of Southampton.

Saunder,s R., Howie, J.H. and Auricht, G.C., 1993. *Medicago orbicularis* (L) Bart. – A potential new medic for the cereal belt of southern Australia. Proceeding of 6th *Australian Agronomy Conference*. Australian Society of Agronomy, Parkeville.

Sims, J.R. Koala, S. Ditterline, R.L. and Wiesner, L.E., 1985. Registration of 'George' black medic. *Crop Science*, 25: 709-710.

Sinskaya, E. H., 1950. *Flora of cultivated plants of U.S.R.R. XIII: Perennial Leguminous Plants.* Israel Program for Scientific Translation (1961), Jerusalem.

Skinner, D.Z., Bauchan, G.R., Auricht, G.C. and Hughes, S.J., (1998). Development of a Core Collection for the Australian Annual Medic Collection. Proc. 36th *North American Alfalfa Improvement Conference (NAAIC)*, Montana, USA.

Small, E., 1987. A taxonomic study of the 'Medicagoid' Trigonella. *Canadian Journal of Botany*, 65: 1199-1211.

Small, E., and Jomphe, M.,1989. A synopsis of the genus *Medicago* (Leguminosae). *Canadian Journal of Botany*, 67: 3260-3294.

Small, E., Lassen, P. and Brookes, B., 1987. An expanded circumscription of *Medicago* (Leguminosae, *Trifolieae*) based on explosive flower tripping. *Willdenowia*, 16: 415-437.

Yanping, Z., and Sheaffer, C.C., 1997. Growth analysis of spring and summer seeded annual *Medicago* sp. *Crop Science*, 37: 1514-1519.

Appendix 1. Description of the most important species of the genus *Medicago*.

Annual *Medicago*

Medicago littoralis (Strand Medic). Several commercial cultivars released in southern Australia. Suited to low rainfall, light sandy soils and widely used in crop rotations in such areas. Early varieties may be found as dense pastures even 20 years after initial sowing and after up to ten cycles of cropping.

Medicago murex. Nodulates effectively on acidic soils well below the pH range of most of the *Medicago* species. The first commercial cultivar of this species was released in Western Australia in 1988.

Medicago orbicularis (Button medic). Frequent in deep and rich soils, in field borders or on fallow lands. No cultivars have been developed, although there has been evaluation in dryland South Australia (Saunders *et al.*, 1993).

Medicago polymorpha (Burr medic). As its name implies, it is the most polymorphic of the medic species. Spineless forms occur. The spiny pods represent a significant problem to wool producers by way of vegetable faults in wool clips. It is widely distributed. This species is frequent on acidic soils, or on the low and salty meadows of the Mediterranean seaside, and in the most arid steppes. Several cultivars of this species are commercially available. They are early maturing and relatively susceptible to cold conditions. New varieties with improved winter-hardiness are under selection (France, USA).

Medicago rigidula and *Medicago rigiduloïdes*. These species have recently been divided from the former *Medicago rigidula*, based mainly on geographic distribution They are adapted to the colder Mediterranean areas, which explains the interest in genetic selection of these species.

Medicago rugosa (Gama medic). It is an east Mediterranean species adapted to heavy clay soils. This species is of considerable importance in the heavier red brown soils of dryland South Australia (375 to 500 mm rainfall).

Medicago scutellata (Snail medic).This species has became relatively rare today and is found mainly in field borders, on fallow lands, and in rich and deep soils. It is susceptible to grazing and cold weather. It has a very irregular pattern of hard seed breakdown allowing seedlings to emerge at any stage throughout the growing season. As a result, snail medic is generally not favoured for use in cereal rotations in Australia as it becomes a weed in the cereal year. Recent cultivars however have been released for sub-tropical southern Queensland in Australia, extending the normal range of medics beyond the temperate zone.

Medicago sphaerocarpos (Sphere medic). One cultivar of this species was commercialised in Australia in 1993.

Medicago tornata (Disc medic). A large number of sub-species and forms are described. Well suited and productive on light sandy soils in medium to low rainfall districts. (325 to 475mm rainfall). However, lack of aphid resistance has limited the development and use of this species in southern Australia since the introduction of aphids in the late 1970's.

Medicago truncatula (Barrel medic). It is frequent on the medium to heavy, marly or clayey soils. This species has been extensively selected for agricultural uses in Australia: eleven cultivars exist; these mainly differ in their earliness, susceptibility to drought, soil adaptation and tolerance to aphids. These cultivars are quite susceptible to cold weather and to leaf diseases, but a large variability exists between the natural populations (Prosperi *et al.* 1989).

Perennial *Medicago*

Medicago sativa (Alfalfa or Lucerne). Instead of *Medicago sativa*, it is better to speak about the complex of species (or sub-species) *sativa / falcata*. Numerous features found in cultivars (winter-hardiness, prostrate habit, and tolerance to drought, grazing, pests and diseases) were introduced from wild accessions of *M. sativa* subsp. *sativa* or *M. sativa* subsp. *falcata*. Table 5.3 shows the ploidy and compatibility levels for the different subspecies of the sub-section *falcatae*. *Medicago sativa* is the most frequently used species of the genus. Several authors have noted its very large morphological variability. Wild accessions of *Medicago sativa* (mainly ssp *sativa* and *falcata*) represent an important reserve of genetic diversity, allowing the geographic extension of its cultivation and sources of disease and insect resistance.

Medicago lupulina. This species has a variable life-cycle and can be annual, biennial or perennial according to its origins.

Medicago arborea. Its life cycle follows a typically Mediterranean pattern with vegetative growth starting in autumn, after the first rains. In case of severe drought, the plant loses all its leaves. Its distribution is limited in the north by frosts (below -8°C). It is drought resistant and occurs in areas with annual rainfall below 250 mm, but due to its intolerance to grazing, its agronomic applications are limited to ornamental uses or prevention of soil erosion.

Table 5.3. Comparative inter-fertility between members of the sub-section *Falcatae* of the genus *Medicago* [● good fertility, ○ weak compatibility, □ sterility (barriers can be overcame by doubling the chromosome number)]. Adapted from Lesins and Lesins 1979, Quiros and Bauchan 1988, Prosperi *et al.*, 1995.

			M. sativa						*M. prostrata*	
			ssp. *falcata*		ssp. *sativa*		ssp. *glomerata*			
			2x	4x	2x	4x	2x	4x	2x	4x
M. sativa	ssp. *falcata*	2x	-	○	●		●		●	
		4x		-	□	○		●		●
	ssp. *sativa*	2x	●	□	-	○	●		●	
		4x		●	○	-	●	●		●
	ssp. *glomerata*	2x	●		●	●	-			
		4x		●		●		-		
M. prostrata		2x	●		○				-	
		4x		●		○				-

Chapter 6

Chickpeas (*Cicer* L.)

Richard Hannan, Nevin Açikgöz and Larry D. Robertson

6.1 INTRODUCTION

Cultivated chickpea (*Cicer arietinum* L.) is noted to be one of the first grain legumes to be domesticated in the old world, and currently ranks as the third most important grain legume in the world after dry bean (*Phaseolus vulgaris* L.) and dry peas (*Pisum sativum* L.) (Saxena 1990). According to the finds of palaeo-ethnobotanical material obtained from excavation of earliest Neolithic Village sites, Turkey is possibly one of the first domestication areas of chickpea, and probably from Hacilar near Burdur around 5450 BC (Helbaek, 1970). The crop centres of origin or diversity are correlated with the ancient human societies and their agricultural needs. Turkey is the centre of origin for *Cicer*, and chickpea is thought to have originated in southeastern Anatolia. Chickpea, the English name for *Cicer arietinum* L. is also known by the common names Bengal gram (India), Garbanzo (Latin America), Hommes and Hamaz (Arab world), Nohud and Lablabi (Turkey), Pois chiche (French) and Shimbra (Ethiopia). Commonly described as either 'kabuli' for the large (29-60 g/ 100 seed) cream coloured seed types, or 'desi' for the smaller (15-29 g/ 100 seed), pigmented, angular-shaped seed types, chickpeas are grown in semi-arid areas of tropical, sub-tropical and temperate regions of the world. The kabuli type is most often grown in temperate regions and the desi type is produced in the sem-arid tropics (Muehlbauer and Singh, 1987).

6.2 TAXONOMIC SUMMARY

The genus *Cicer* L. (Leguminosae, Cicereae) was classified by Popov (1929) who recognised four sections (*Monocicer* M.Pop., *Chamaecicer* M.Pop., *Polycicer* M.Pop., and *Achantocicer* M.Pop.) which were grouped under two subgenera (*Pseudononis* M.Pop. and *Viciastrum* M.Pop.) based on plant morphology, life cycle and geographic distribution. The genus was later revised by van der Maesen (1972) who added taxonomic information, but closely followed Popov's classification. Although the genus was first classified in the tribe Vicieae Alef., it was subsequently delimited by Kupicha (1977) from Vicieae to the monotypic tribe Cicereae Alef., subfamily Papilionoideae, family Leguminosae. Most recently Coles *et al.* (1998), in an identification aid for *Cicer* taxa, added new species clarification and identifications. It has been recently shown that the genus *Cicer* includes 42 species, nine annuals and thirty-three perennials (Coles *et al.*, 1998; Robertson *et al.*, 2000).

Due to the interest in the utilisation of wild germplasm in chickpea cultivar development, the stimulated interest in collecting new material has led to the recent description of several new species. Santos Guerra and Lewis (1985) described the new and very different species, *C. canariense* A. Santos & G.P.Lewis, collected from the remote Canary Islands. More recently, Rassulova and Sharipova (1992) added *C. luteum* Rassulova & B. A. Sharip. In a later discussion on gaps in conservation efforts, it is noted that there are species of *Cicer* that remain to be discovered and described.

After Ladizinsky (1975) described *C. reticulatum*, he proposed that the nine annual species be grouped into three classified groups, but four described groups (Ladizinsky and Adler 1976). The first group includes the cultivated chickpea (*C. arietinum* L.) as well as its closest relatives, *C.*

reticulatum Ladiz. and *C. echinospermum* P. H. Davis. At that same time, they offered supporting evidence that *C. reticulatum* is the putative progenitor of the cultivated chickpea, and could be taxonomically classified as a subspecies of *C. arietinum* L. The second group includes *C. bijugum* Rech., *C. judaicum* Boiss., *C. pinnatifidum*, Jaub. & Spach and *C. yamashitae* Kitam. *C. judaicum* and *C. pinnatifidum* are morphologically similar enough that some taxonomists group them under the same species (Post, 1932; Zohary, 1972), while others have kept them as two different species (Coles *et al.*, 1998; Maesen, 1972). *C. chorassanicum* (Bunge) Popov is in a group of its own, as is *C. cuneatum* Hochst.

Recent molecular genetic studies on the taxonomic relationships among the annual species of *Cicer* using isozyme variation support the conclusions based on previous cytogenetic studies. Using groupings based on allozyme variation Kazan and Muehlbauer (1991) concluded that the species groupings of Ladizinsky and Adler (1976) based on crossability and morphological similarities were correct. Labdi *et al.* (1996) found similar groupings, except the second group did not contain *C. yamashitae*, which grouped with *C. chorassanicum*. Furthermore, Kazan and Muehlbauer (1991) found correlations between genetic distances and the geographic distribution of the species studied. In a more comprehensive experiment using isozyme variation to study the genetic relationships among annual and two perennial *Cicer* species, Tayyar and Waines (1996) further clarified the previous groupings. The majority of their groupings were similar to the crossability studies, although there are several significant key points worth mentioning. *C. reticulatum* was genetically closest to *C. arietinum*, but the number of alleles shared between them varied significantly depending on the population of *C. reticulatum*. There was a relatively small genetic distance between *C. arietinum* and *C. echinospermum*, suggesting that *C. echinospermum* may in fact have played a role in the evolution of the cultivated species. The second grouping or cluster of the phenogram, contains all the species of the second crossability group of Ladizinsky and Adler (1976), except for *C. yamashitae*. In the study of Tayyar and Waines (1996) *C. judaicum* and *C. pinnatifidum* exhibited low genetic distance values suggesting conspecific populations. The two perennial species, *C. anatolicum* and *C. songaricum* clustered with *C. yamashitae* and *C. chorassanicum*, suggesting that the annual habit possibly arose at least twice from the perennial in the genus. *C. cuneatum* was found to be quite distinct and genetically distant from all the other species.

6.3 CENTRE OF DIVERSITY

The cultivated chickpea, *Cicer arietinum*, is known to be one of the first grain legumes to be domesticated in central Asia, and was first thought to have originated in the southern Caucasus and northern Persia (Maesen 1972). Although Candolle (1883) proposed the origin of chickpea to be an area south of the Caucasus Mountains down to northern Iran, Vavilov (1951, 1952) later identified two distinct centres of diversity based on his collections. Harlan (1992) classified chickpea as having an oligocentric centre of origin, where the crop does have a definable centre of diversity, but has been widely dispersed with more secondary centres. One centre of diversity was the Mediterranean region extending into the south-western reaches of central Asia. The other, a secondary centre, was in Ethiopia. According to Duke (1981) both botanical and archaeological evidence indicate that chickpea was first domesticated in the Middle East region (Turkey and Syria), and early on was cultivated in India, the Mediterranean civilisations, as well as in Ethiopia. Vavilov (1951, 1952) noted that in the Mediterranean region the large cream-coloured 'Kabuli' type of seed was most predominately used, and that as he journeyed eastward, utilisation changed to that of the smaller, dark-coloured 'Desi' type seeds. Ladizinsky (1975) narrowed the centre of origin to be in south-eastern Turkey, and later Maesen (1984) proposed that the south-eastern part of Turkey, near Syria, was the centre of origin of cultivated chickpea since the two closely related species, *C. reticulatum* and *C. echinospermum*, occur there.

6.4 ECOGEOGRAPHIC DISTRIBUTION

In general the most common species, *Cicer arietinum*, is only found in cultivated environments. It may exist as an escape or volunteer, but is essentially unable to colonize wild habitats successfully. Cultivation of this species occurs in semi-arid tropical, semi-tropical, and temperate regions around the world. Close relatives of the cultivated chickpea (e.g. *C. reticulatum, C. bijugum* and *C. echinospermum*) are found in both pristine ecosystems as well as disturbed habitats, such as roadsides, cultivated fields, mining sites, etc. Montane wild annual and perennial species are found both in very rocky areas (*C. pungens* and *C. yamashitae*) to heavily forested areas (*C. montbretii*). Species of *Cicer* occur from sea level to over 5000m elevation.

Figure 6.1. Origin of eight of the nine annual species of *Cicer*

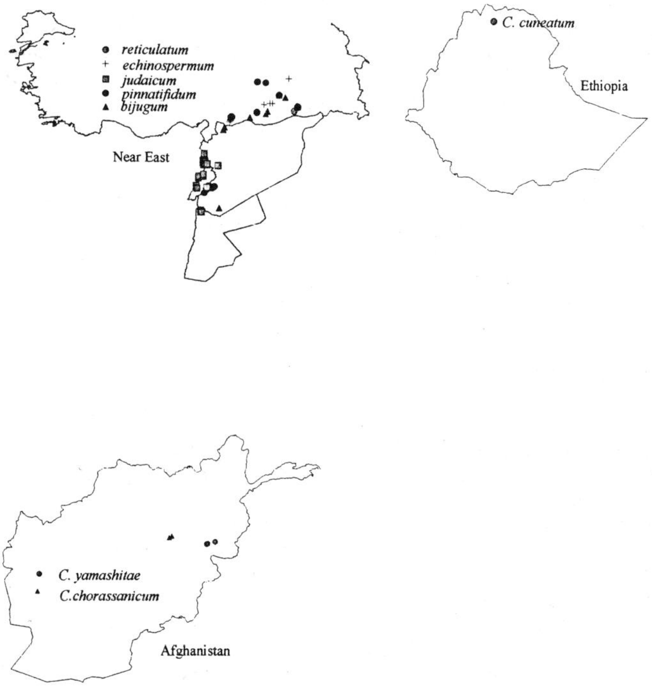

C. arietinum, the cultivated chickpea, has a wide distribution not only in the Mediterranean region but also in other parts of the world. It has been divided into 4 races by Popova (1937):

1. Race *Orientale* has 5 subraces, and is found in Ethiopia, Sudan, Egypt, India, Pamir,

Tadzhikistan, and Iran. The seeds are very small; 6-7 mm long, 5-6 mm wide, 100 seed weight 10-12 g, dark coloured; black, brown or brown-red, rarely lighter or cream. Seed shape is angular (ram's head). The flowers are small (up to 7 mm), pinkish, and the leaflets are small, 6-9 mm long. In general the plant is small with thin stems, except for the mountainous forms, and colouration induced by anthocyanins.

2. Race *Asiaticum* has 4 sub-races, and occurs in central Asia, Afghanistan, western China, and east Turkey. The seeds are larger than in race *orientale* with a 100 seed weight of 14-20 g. The seed shape is rounded (owl's head), and if coloured, angular (ram's head). The flowers are white or purple and the plant is of medium height.

3. Race *Mediterraneum* has only one sub-race and its general distribution is in Spain, Italy, Morocco, Algeria, Tunisia, west Turkey, central. America. Seeds are very large with a 100 seed weight of more than 35 g. They are cream coloured and seed shape is rounded, not angular or owl's head. The pods are large and inflated to a larger degree than in the other races. Flowers are white, and leaflets large. Plants are of medium height or very tall. The stems are green with no anthocyanin present.

4. Race *Euasticum* has four sub-races and is found in Syria, Palestine, Turkey, Armenia, Azerbaijan, and Ukraine. The seeds are medium large, 100 seed weight is 20-30 g and cream coloured. Seed shape is rounded, not angular or owl's head. Rarely are the coloured seeds of angular shape. The plant is large and partly ascending.

All four races of *C. arietinum* exist in the Mediterranean countries. A wide diversity exists in the cultivated chickpea in Turkey. Small seeded chickpeas are common in southeastern parts of the country, whereas the large seed-type exists in the west. In Turkey chickpeas have been grouped as; Spanish type (*C. arietinum* L ssp. *arieticeps*) which has very large, rams-head shaped, seeds; Leblebi type (*C. arietinum* L ssp *intermedium*) which has medium size, owl's shape, and; Ordinary type (*C. arietinum* L ssp. *pisiiforme*) which are pea-shaped or rounded.

Gençkan (1958) determined 34 groups of chickpeas in Anatolia according to both morphological and biological characteristics, and taxonomic studies continue to be conducted by Açikgöz and Cinsoy to further elucidate grouping within the cultivated species. Although the morphological characteristics of the wild annual species were found modest to poor when compared to modern cultivars (Robertson *et al.*, 1997), there probably exists unknown, unexploited useful alleles in the wild secondary germplasm pool. The wild annual species of *Cicer* provide a genetic resource for the future improvement of cultivated chickpea (Singh, 1998). Brief taxonomic and ecogeographic distribution for the diverse array of species, including both annual and perennial species, found in Turkey was best described by Davis (1970), and is as follows.

Table 6.1. Distribution of *Cicer* species (extant in collections).

Species	AF	BG	EG	ET	GR	IN	IL	IT	JO	LB	MA	PK	ES	SY	TJ	TR	UZ
anatolicum																✓	
arietinum		✓	✓		✓		✓	✓		✓	✓		✓	✓		✓	
atlanticum											✓						
bijugum														✓		✓	
canariense													✓				
chorassanicum	✓																
cuneatum			✓	✓													
echinospermum																✓	
flexuosum																	✓
floribundum																✓	
graecum					✓												
heterophyllum																✓	
incisum	✓			✓	✓					✓						✓	
isauricum																✓	
judaicum							✓		✓	✓	✓			✓		✓	
microphyllum						✓						✓					
monbretii		✓														✓	
oxyodon																✓	
pinnatifidum				✓						✓				✓		✓	
pungens	✓														✓		
rechingeri	✓																
reticulatum																✓	
songaricum															✓		
yamashitae	✓																

C. echinospermum (annual) is close to the cultivated chickpea but has echinate seeds. Pods do not shatter early. It is found in southeast Anatolia and has been collected from the provinces of Diyarbakir, Urfa and Mardin. The species is endemic to Turkey and into Iran and the habitat preference in Turkey is on rocky slopes, vineyards, fallow and cultivated fields, grass vegetation, and

Quercus scrub, at altitudes of 700-1100 m. *C. reticulatum* (annual) exists in the south-east of Turkey. Habitat preference is oak and scrub vegetation, vineyard borders, weedy habitats, and limestone hills. It is found at altitudes of 700-1000 m. *C. bijugum* (annual) is a close relative of the cultivated chickpea. It is found in the south east of Turkey, specifically from the provinces of Tunceli, Mardin and Urfa. Habitat preference is as weeds in very arid conditions of orchards, abandoned fields, and grazed areas, at altitudes of 500-1300 m. *C. pinnatifidum* (annual) has been widely collected from Ankara, southern and inner Anatolia. Its habitat preference is in *Pinus brutia* forests, on igneous or limestone slopes, at altitudes of 250-1400 m. With regard to the perennial species, *C. anatolicum* is one of the most frequently collected wild perennial chickpeas, and is found mainly in Inner Anatolia in the mountainous regions. The flowers are large, violet or pale violet. It is very variable in its leaves and length of its calix, particularly in Bitlis province. The preferred habitat is in *Pinus nigra* forests, scrub, rocky slopes and screes, at altitudes of 1200-3300 m. It is now threatened due to overgrazing by sheep and goats. *C. floribundum* (perennial) is an endemic wild species of the Toros mountains of southern Turkey and has rarely been collected. It is found in southern Turkey, from Karaisali-Adana, Köycegiz-Mugla, and Dörtyol-Hatay provinces. Its habitat preferences are on mountain slopes, in forests of *Fagus orientalis* and *Quercus crucifera*, at altitudes of 800-1400 m. *C. heterophyllum* (perennial) occurs in the Toros mountains of Turkey. Its habitat preference is *Pinus brutia* and *Quercus* scrub, at altitudes near 1100 m. *C. insicum* (perennial) exists in south and east Anatolia. Samples have been collected from Ovacik-Tunceli, Dogansehir-Malatya, Yahyali-Kayseri, Bozburun dagi-Antalya, Erzincan, Burdur, Içel and Adana provinces. *C. isauricum* (perennial) is endemic to a small area in the west of the Toros mountains. Accessions have been collected from Beysehir, Egridir-Isparta, Ermenek- Konya, Akseki-Antalya, and Kaz Dag. *C. montbretii* (perennial) has white flowers and is known to have seed dormancy. In Turkey, it has been found in Edremit-Balikesir, Tekirdag, Baltalimani and Göksu-Istanbul. In both Turkey and Bulgaria its habitat preference is on hilly slopes, in *Pinus nigra* or *Quercus* forests, rocky places and near water, at altitudes of 0-1200 m.

6.5 CONSERVATION RESOURCE STATUS (*EX SITU* AND *IN SITU*)

The primary world resources for *Cicer* germplasm are maintained by the International Crops Research Institute for the Semi-Arid Tropics (ICRISAT), Patancheru, India (17,244 accessions); the International Centre for Agricultural Research in the Dry Areas (ICARDA), Aleppo, Syria (9,682 accessions, 292 wild), (CGIAR 1998); the United States Department of Agriculture, Agricultural Research Service (USDA, ARS), Western Regional Plant Introduction Station (WRPIS), Pullman, WA (4,602 accessions), (GRIN, 1999).

The initial germplasm used to establish all three of these major collections was collected and donated in the 1960s by the former USDA Regional Pulse Improvement Project (RPIP) that was centred in Iran and India. The RPIP assembled 4,177 cultivated chickpea accessions, which were then contributed to the ICARDA and ICRISAT collections (Muehlbauer and Kaiser 1994). Much of that material was also added to the WRPIS collection, which now numbers 4,439 accessions of cultivated *C. arietinum* and 163 accessions of wild forms from over 20 countries. Many of the accessions of the wild *Cicer* species in the WRPIS collection were obtained from ICRISAT, and were largely the result of collections by L. J. G. van der Maesen, G. Ladizinsky, W. J. Kaiser, F. J. Muehlbauer, C. R. Sperling and N. Maxted. Of the 43 recognised species, less than half (17) of these are represented in the Pullman (WRPIS) collection as of fall 1999, and exist as seeds, plants or both. It is recognized by chickpea germplasm curators that the USDA collection in Pullman has the largest extant collection of viable seeds and/or plants of the wild species (Robertson, personal communication 1999). A long term perennial *Cicer* nursery is maintained at the Pullman location (Kaiser, *et al.*, 1997). Over 97% (3,860 accessions) of the WRPIS chickpea collection is also backed up in long term storage (-18°C) at the USDA, ARS, National Seed Storage Laboratory, Fort Collins, CO, the site of long term storage for the US National Plant Germplasm System. There are other collections of significant size maintained in Portugal (1,854) Australia (2,400 cultivated, 105 wild), Ethiopia (717), Turkey, Aegean Agricultural Research Institute (AARI) (1582), India (data not available), Mexico (1,600), Pakistan (626), Russia (2,631), Germany (369 cultivate, 9 wild), Peoples Republic of China (23), Bulgaria (285), Croatia (156), Slovak Republic (143), Cyprus (28), Israel (455 cultivated, 13 wild), Italy (data not available), and Spain (450), but these collections consist primarily of advanced materials of the cultivated species (Gass *et. al.*, 1996; Muehlbauer and Kaiser, 1994).

Collections of wild *Cicer* germplasm are much smaller than those of the cultivated chickpea. Currently, the largest collection of annual wild species is maintained at the International Centre for Agricultural Research in the Dry Areas (ICARDA) with 292 accessions (Robertson, ICARDA, email communication 1998). Two other significant collections are maintained at the International Crops Research Institute for the Semi-Arid Tropics (ICRISAT), and at the USDA, ARS, Western Regional Plant Introduction Station with 163 accessions (GRIN, 1999). Over the last two decades there has been continual collaboration and exchange between these three primary collections, and therefore, when reviewing the passport data and comparing the original plant collection identification numbers, it can be seen that the bulk of each of these collections have accessions in common. Multiplication of seeds of the eight wild annual species is generally possible under field, greenhouse, and/or growth chamber conditions at numerous locations worldwide.

Wild perennial *Cicer* species, however, are much more difficult to propagate, and until recently, there has been little information available on the proper regeneration protocols (Kaiser *et al.*, 1997). The largest extant collection is maintained at WRPIS with 32 accessions across twelve species. Collections are housed at ICRISAT and ICARDA, and other small collections exist in national programs throughout the world. Many of these locations do not, however, have adequate storage facilities.

There is little information available in the literature regarding *in situ* conservation of any wild *Cicer* species. There is one report of *in situ* conservation activities being undertake for the one rare species in Bulgaria (Gass *et al.*, 1996). However, there are repeated references to the actual loss of wild species habitat in the areas considered to be the centre of diversity for *Cicer*.

In situ conservation of plant genetic diversity is an important part of the National Plant Genetic Resources Research Project, which started in Turkey in 1993. In the Anatolian Diagonal and Ceylanpinar State Farm pilot conservation areas, *Cicer* species have been nominated as target species along with other legume cereal species.

6.6 CONSERVATION GAPS AND PRIORITIES

Over ten years ago Maesen (1984) suggested that with the large germplasm collections of chickpea at ICRISAT and ICARDA, in conjunction with the USDA, ARS collection at WRPIS, there was a firm genetic base with a few gaps, but that the wild species were insufficiently represented in the collections. Although the number of accessions of wild species have been increased to the present levels, significant gaps in the extant and available collections of wild species still exist, especially for the perennial species plus *C. yamashitae*, *C. chorassanicum* and *C. cuneatum* (which is currently only represented by one accession).

Central Asia has been identified as a priority area for collection activities to obtain both quality herbarium specimens for taxonomic purposes and more importantly viable seed for future *ex situ* conservation efforts (Coles *et al.*, 1998). They note that a number of species remain known only from their type collections, and that for others very few specimens exist.

Collection efforts by Maxted (Maxted, 1990; Maxted and Sperling, 1991) and others have indicated that our knowledge of the geographical distribution of the recognised species is too narrow, and that there are likely to be further *Cicer* species to be discovered and described.

6.7. RESEARCH NEEDS

One future research area is the continued development of molecular techniques and studies to augment traditional taxonomic studies of the genus *Cicer* (Tayyar and Waines, 1996). The traditional method for taxonomic classification has always been based primarily on morphological characteristics in conjunction with cytological data. In recent years it has been clearly shown that electrophoretic data are of significant value when applied to systematic studies (Buth, 1984; Crawford, 1983). Whether the banding patterns are the result of different alleles at a gene locus (allozymes) or of alleles at different loci (isozymes), the molecular characters elucidated provide an abundant source of interpretable data (Whitkus *et al.*, 1994). Allelic differences at a number of gene loci have been shown in *C. arietinum* (Twafe *et* al., 1988) and other *Cicer* species (Kusmenoglu *et al.*, 1992; Kazan and Muehlbauer, 1991). More recently, randomly amplified polymorphic DNA (RAPD) data has been used in the classification and phylogeny in the genus. From the molecular data of this last study, it was suggested that *C. reticulatum* be proposed as the wild progenitor of the cultivated species.

Including recent molecular data as a tool in marker assisted selection with traditional or new breeding methodologies, it may be possible to transfer exotic genes for tolerance or resistance to biotic and abiotic stresses into chickpea. It has already been shown that it is possible to cross *C. echinospermum* with the cultivated chickpea (Porto-Pugtia *et al.*, 1993) for introgression of genes for insect and pathogen resistance.

There remains today a considerable disparity between the potential yield (3-5 t ha^{-1}) and the current world average yield of 0.7 t ha^{-1} for commercial chickpea production (Robertson *et al.*, 1997). It is thought that chickpea plant breeders have reached the genetic yield potential of the crop (Singh *et al.*, 1998). However, there is a significant need for research to obtain stability of yield. In most cases, germplasm diversity will supply the solution as new crop niches and emerging cropping systems evolve.

REFERENCES

Buth, D.G., 1984. The applications of electrophoretic data in systematic studies. *Annual Review of Ecology and Systematics*, 15: 501.

CGIAR, 1997. *CGIAR Germplasm Collections*. http://noc1.cgiar.org/cgintrus.html.

Coles, S., Maxted, N. and van der Maesen, L.J.D., 1998. Identification aids for *Cicer* (Leguminosae, *Cicereae*) taxa. *Edinburgh Journal of Botany*, 55: 243-265.

Crawford, D.J., 1983. Phylogenetic and systematic inference from electrophoretic studies. In: *Isozymes in plant genetics and breeding. Part A.* (eds. Tanksley, S.D. and Orton, T.T.). Elsevier Science Publishers, The Netherlands.

Candolle, de A. 1883. *Origin of cultivated plants*. pp. 1-323. Reprinted in 1967. Hafner Publishing Company, New York.

Davis, P.H., 1970. *Cicer* L. In: *Flora of Turkey and the East Aegean Islands. Vol. 3.* (ed. Davis, P.H.). pp. 267-274. Edinburgh University Press, Edinburgh.

Duke, J.A., 1981. *Handbook of legumes of world economic importance*. Plenem Press, New York.

Gass, T., Ambrose, M. and LeGuen, J., Hadjichristodoulou, A. and Blixt, S., 1996. Review of national grain legume collections. In: *Report of First Working Group on Grain Legumes* (eds. Gass, T., Ambrose, M., Le Guen, J., Hadjichristoulou, A. and Blixt, S.). pp. 3-11. IPGRI, Rome.

Gençkan, S., 1958. Türkiye'nin önemli nohut çesitlerinin baslica vasiflari üzerinde arastirmalar. *Ege Üniversitesi Ziraat Fakültesi yayinlari*: 1. 105 s. Ege Üniversitesi Matbaasi.

GRIN, 1999. *Genetic resources information network*. www.ars-grin.gov/cgi-bin/npgs/

Helbaek, H., 1970. The plant husbandry at Hacillar. In: *Excavation at Hacillar* (ed. Mellaart, J.). pp. 189-244. Edinburgh University Press, Edinburgh.

Harlan, J.R., 1992. *Crops and Man*. ASA-CSA Press. Madison, USA.

Kaiser, W.J., Hellier, B.C., Hannan, R.M. and Muehlbauer, F.J., 1997. Growing techniques and conservation of wild perennial *Cicer* species in the U. S. Pacific Northwest. International *Chickpea and Pigeonpea Newsletter*, 4: 7-8.

Kazan, K. and Muehlbauer, F.J., 1991. Allozyme variations and phylogeny in annual species of *cicer* (*Leguminosae*). *Plant Systematics and Evolution*, 175: 11-21.

Kupicha, F.K., 1977. The delimitation of the tribe *Vicieae* and the relationship of *Cicer* L. Botanical *Journal of the Linnean Society*, 74: 131-162.

Kusmenoglu, I., Muehlbauer, F.J. and Kazan, K., 1992. Inheritance of isozyme variation in Ascochyta blight-resistant chickpea lines. *Crop Science*, 32: 121-127.

Labdi, M., Robertson, L.D. Singh, K.B. and Charrier, A., 1996. Genetic diversity and phylogenetic relationships among the annual Cicer species as revealed by isozyme polymorphism. *Euphytica*, 88: 181-188.

Ladizinsky, G., 1975. A new *Cicer* from Turkey. *Notes from the Royal Botanic Garden Edinburgh*, 34: 201-202.

Ladizinsky, G. and Adler, A., 1976. Genetic relationships among the annual species of *Cicer* L. *Theoretical and Applied Genetics*, 48: 197-203.

Maesen, van der L.J.G., 1972. Cicer *L., a monograph of the genus, with special reference to the chickpea (*Cicer arietinum *L.), its ecology and cultivation.* Mededelingen Landbouwhogeschool (Communications Agricultural University), Wageningen.

Maesen, van der L.J.G., 1984. Taxonomy, Distribution and Evolution of the Chickpea and Its Wild Relatives, In: *Genetic Resources and Their Exploitation--Chickpeas, Faba beans and Lentils* (eds. Witcombe, J.R. and Erskine, W.). pp. 95-101. Martinus Nijhoff / Dr. W. Junk Pub, Dordrecht.

Maxted, N., 1990. *Wild forage legume collection in the Central Asian region of the Soviet Union* - Part 1. Unpublished final report. IPGRI, Rome.

Maxted, N. and Sperling, C., 1991. *Wild forage legume collection in the Central Asian region of the Soviet Union - Part 2.* Unpublished final report. IPGRI, Rome.
Muehlbauer, F.J. and Singh, K.B., 1987. Genetics of chickpea. In: *The Chickpea* (eds. Saxena, M.C. and Singh, K.B.). pp. 99-125. CAB. International, Wallingford, UK.
Muehlbauer, F.J. and Kaiser, W.J., 1994. Using host plant resistance to manage biotic stresses in cool season food legumes, In: *Expanding the Production and Use of Cool Season Food Legumes* (eds. Muehlbauer, F.J. and Kaiser, W.J.). Kluwer Academic Publications, Dordrecht.
Popov, M.G., 1929. The genus *Cicer* and its species. Byull. Prikladnoi Botanike Genetike I Rastenievodstva. *Bulletin of Applied Botany, Genetics and Plant Breeding*, 21: 1-254.
Popova, G. M., 1937. Grain Leguminosae. In: *Flora of cultivated plants of the USSR* (eds. Vavilov, N.I. and Wulff, E.V.). pp. 25-71. Moscow-Leningrad.
Porta-Puglia, Singh, K.B. and Infantino, A., 1993. Strategies for multiple stress resistance breeding in cool season food legumes. In: *Breeding for stress tolerance in cool season food legumes* (eds. Singh, K.B. and Saxena, M.C.). ICARDA, Wiley Sayce Co. Pub.
Post, G.E., 1932. *Flora of Syria, Palestine and Sinai.* American Press. Beirut.
Rassulova, M.A. and Sharipova, B., 1992. New species of the genus *Cicer* from Tajikistan. *Izvestiya Akademii Nauk Respublic Tadzhikskoi SSR Otdelenie Biologicheskikh Nauk*, 1: 51-52.
Robertson, A.M., Maxted, N. and Maesen, L.G.D. van der, 2000. A phenetic and phylogenetic investigation of the genus *Cicer* L. (Leguminosae, Cicereae). *Botanical Journal of the Linnean Society* (in Press).
Robertson, L.D., Singh, K.B. and Ocampo, B., 1995. *A catalogue of annual wild* Cicer *species.* ICARDA, Aleppo, Syria.
Robertson, L.D., Singh, K.B. and Ocampo, B.,1997. Morphological variation in wild annual *Cicer* species in comparison to the cultigen. *Euphytica*, 95: 309-319.
Santos Guerra, A. and Lewis, G.P., 1985. A new species of *Cicer* (Leguminosae Papilionoideae) from the Canary Islands. *Kew Bulletin*, 41: 459-462.
Singh, K.B., Ocampo, B. and Robertson, L.D., 1998. Diversity for abiotic and biotic stress resistance in the wild annual Cicer species. *Genetic Resources and Crop Evolution*, 45: 9-17.
Saxena, M.C., 1990. Problems and potential of chickpea production in the nineties. In: Chickpea in the nineties (eds. van Rheenban, H.H. and Saxena, M.C.). *Proc. 2nd Intl. Workshop on Chickpea Improvement*. pp. 13-25. International Crops Research Institute for the Semi-Arid Tropics (ICRISAT). Patancheru, Andhra Pradesh, India.
Tayyor, R.I. and Waines, J.G., 1996. Genetic relationships among annual species of *Cicer* (Fabaceae) using isozyme variation. *Theoretical and Applied Genetics*, 92: 245-254.
Twafe, S., Khaler, A.L. Boe, A. and Ferguson, A., 1988. Inheritance of allozyme polymorphisms in chickpea (*C. arietinum* L.) *Journal of Heredity*, 79: 170-174.
Vavilov, N.I., 1951. The origin, variation, immunity an breeding of cultivated plants. *Chronica Botanica*, 13 (1/6): 14-54.
Vavilov, N.I. 1992. *On the origin of cultivated plants. Origin and geography of cultivated plants.* Translated by Doris Löve. Cambridge University Press, Cambridge.
Whitkus, R., Doebley, J. and Wendel, J.F., 1994. Nuclear DNA markers in systematics and evolution. In: *DNA based markers in plants* (eds. Phillips, R.L. and Vasil, I.K.). pp. 116-141. Kluwer Academic Publishers, Dordrecht.
Zohary, M. 1972. *Flora Palestine.* Israel Acad. Sci. Hum. Jerusalem, 2: 193-194.

Chapter 7

Lentils (*Lens* L.)

Morag Ferguson and William Erskine

7.1 INTRODUCTION

Lens is a small Mediterranean genus which contains the cultivated lentil (*L. culinaris* Medikus subsp. *culinaris*) and six related taxa (Ferguson *et al.*, 2000). Lentil is an ancient cool-season food legume that has been traditionally grown in the Middle East, South Asia, Ethiopia, North Africa and to a lesser extent, in southern Europe. More recently it has been introduced to the Americas and Australasia. Its ability to grow on poor soils under adverse environmental conditions yet produce a high quality protein has made it one of the principal pulse crops in the drier regions, particularly in the Middle East, South Asia and North Africa where it is a dietary mainstay. It is usually grown in rotation with cereals and in West Asia its straw is of great value as feed for small ruminants.

7.2 TAXONOMIC SUMMARY

The Latin word *Lens* describes the seed shape of the cultivated lentil. Tournefort (1719) was the first to use the word *Lens* to designate a specific genus. Subsequently *Lens* has been incorporated in a number of other genera. The use of the name *Lens* Miller and its generic delimitation, only excepted by the position of *L. montbretii*, became undisputed after Barulina's (1930) monograph of the genus.

The taxonomy of *Lens* has undergone numerous changes in recent years. The genus now consists of seven taxa split into four species. *L. culinaris* subsp. *culinaris* is grouped with its three wild, putative ancestor as a subspecies, as follows (Ferguson *et al.*, 2000):

culinaris Medikus
- subsp. *culinaris*
- subsp. *orientalis* (Boiss.) Ponert
- subsp. *tomentosus* (Ladizinsky) Ferguson, Maxted, van Slageren & Robertson
- subsp. *odemensis* (Ladizinsky) Ferguson, Maxted, van Slageren & Robertson

ervoides (Brign.) Grande
nigricans (M.Bieb.) Godron
lamottei Czefr.

The domestication of lentil occurred, together with that of emmer and einkorn wheat, barley, pea, chickpea, bitter vetch and flax, during the Neolithic Agricultural Revolution, which is thought to have taken place in the Eastern Mediterranean around the 8th and 7th millennia BC (Zohary and Hopf, 1973). Lentil cultivation spread rapidly with that of Neolithic agriculture to the Nile Valley, Europe and Central Asia. It was part of the Harappan crop assemblage in Pakistan and Northwest India between 2 250 and 1750BC (Zohary and Hopf, 1993). After 1500AD, the Spanish introduced lentil to South America via Chile (Sohl and Erkine, 1984). More recently it was introduced to

Mexico, Canada, the USA and Australia. The major morphological changes associated with domestication were a loss of pod shattering, increased seed size and biomass production. Breeding experiments have revealed three crossability groups among the six species of *Lens*: (1) *L. culinaris* subsp. *culinaris* and *L. culinaris* subsp. *odemensis;* (2) *L. ervoides*, *L. nigricans* and *L. lamottei;* and, (3) *L. culinaris* subsp. *tomentosus*, although the extent of crossability varies according to the accessions used (Ladizinsky *et al.* 1984, Ladizinsky 1997).

7.3 CENTRE OF DIVERSITY

Centres of diversity have been identified for each of the wild taxa by mapping the distribution of genetic variation as measured using random amplified polymorphic DNA (RAPD) (Ferguson *et al.* 1998a). For *L. culinaris* subsp. *orientalis* two centres exist, one in south-eastern Turkey and north-western Syria, the other in southern Syria and northern Jordan. The southern centre overlaps with the centre of diversity for *L. culinaris* subsp. *odemensis*. A region of diversity exists for *L. ervoides* along the eastern Mediterranean coast and for *L. nigricans* in south-western Turkey. Centres of diversity in *Lens* are also characterised by high population densities.

Areas of low diversity have also been located (Ferguson *et al.* 1998a). *L. culinaris* subsp. *orientalis* accessions from Iran, central Asia and northern Turkey are all very similar to each other, and correspond to the common cytotype identified by Ladizinsky *et al.* (1984). *L. ervoides* populations from the coastal region of the former Yugoslavia have a narrow genetic base, as do *L. nigricans* accessions from the same region as well as those from France and Spain. Maps illustrating the distribution of variation can be found in Ferguson *et al.* (1998a).

7.4 ECO-GEOGRAPHIC DISTRIBUTION

Wild lentils are distributed primarily in the Mediterranean basin (Figures. 7.1-7.3). *L. culinaris* subsp. *orientalis*, as the name suggests, has an eastern distribution from Turkey, Cyprus and Palestine across to Uzbekistan. *L. culinaris* subsp. *odemensis* has a more restricted distribution in the east, extending from Turkey southwards to Syria and Palestine. A single population has been found in Libya. Still more restricted is the distribution of *L. culinaris* subsp. *tomentosus*, in south-eastern Anatolia only. *L. ervoides* has a broad distribution from Spain to Ukraine and south to Jordan. It is relatively common along the eastern Mediterranean coast, the coast of Turkey and the Adriatic coast of the former Yugoslavia. Its distribution is more restricted in Spain and Algeria. Outlier populations have also been found in Ethiopia and Uganda. *L. nigricans* is mainly distributed in southern Europe from the Canary Islands to Turkey. It also occurs occasionally in Morocco and Algeria. *L. lamottei* is found in France and Spain, and has recently been reported from Morocco (Oss *et al.* 1997).

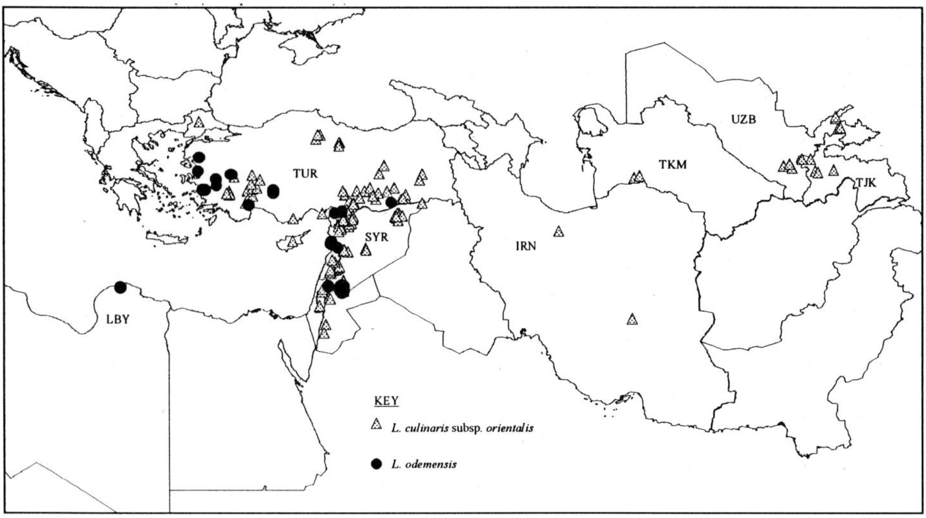

Figure 7.1. Geographical distribution of *L. culinaris* subsp. *orientalis* and *L. culinaris* subsp. *odemensis* (form ICARDA accessions).

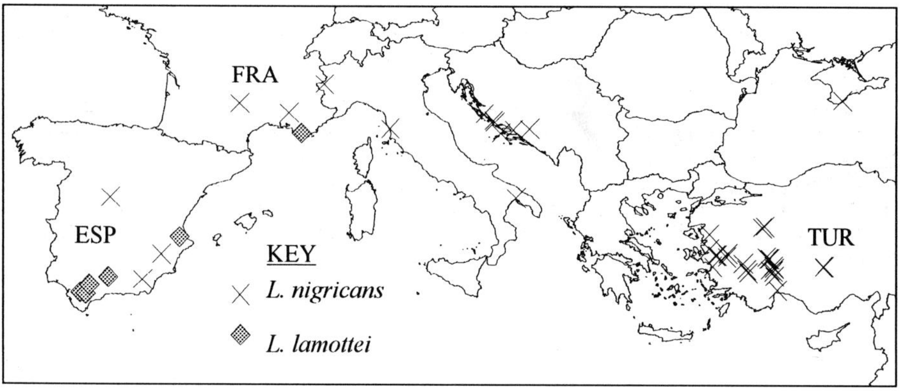

Figure 7.2. The geographical distribution of *L. nigricans* and *L. lamottei* (from ICARDA accessions)

Wild lentils are poor competitors and are highly palatable to grazing animals. They are found predominantly in primary, ungrazed habitats where they are not subject to competition by aggressive coloniser plants. They usually form small disjunct populations. The density of plants may vary dramatically between years apparently because of climatic conditions. *L. culinaris* subsp. *orientalis*, subsp. *tomentosus*, subsp. *odemensis* and *L. nigricans* are generally found in open or partially shaded habitats, on shallow stony soils originating from calcareous, metamorphic or basalt rocks with other short-stature annual legumes such as annual vetch, clover, medic and chickling species. *L. nigricans* and *L. lamottei* have also been found in habitats showing signs of earlier human disturbance such as abandoned terraces or plantations and ruins. Both *L. nigricans* and *L. ervoides* have been found from sea level to over 1000m. *L. culinaris* subsp. *orientalis* and subsp. *odemensis* have been found at even higher altitudes from 500m up to 1700m and from 700m up to 1400m respectively. Populations of *L. lamottei* have been located between 100 and 800m. Although all wild species (except *L. ervoides*) have similar habitat preferences, they are rarely found together. *L. ervoides* usually inhabits shady places and is frequently associated with pine forests often growing on calcareous bedrock. The ecological preferences of the wild taxa are reviewed by Ladizinsky (1993).

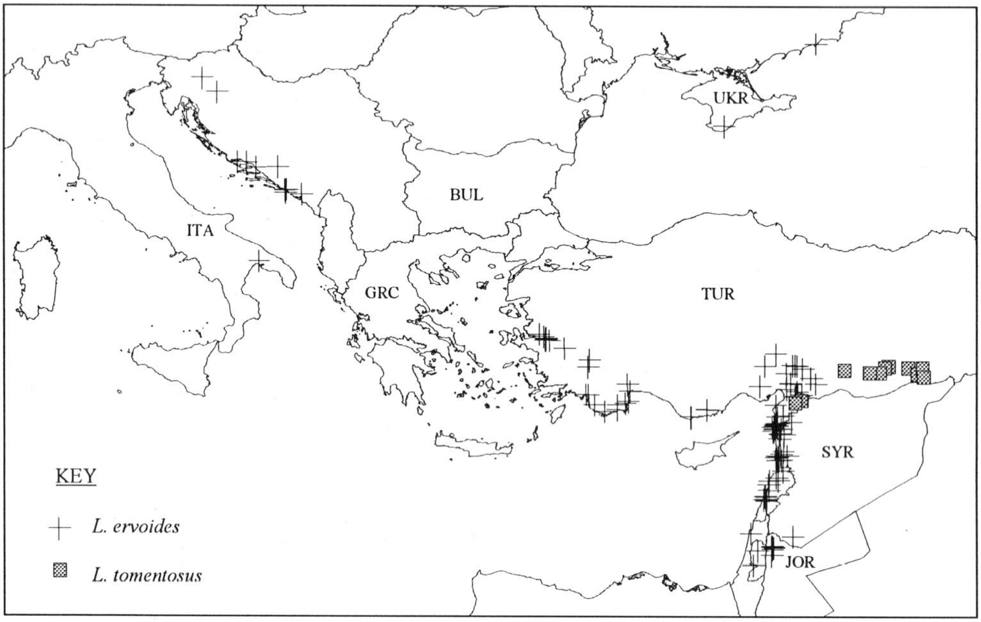

Figure 7.3. Geographical distributions of *L. ervoides* (from ICARDA accessions) and *L. culinaris* subsp. *tomentosus* (according to Ladizinsky 1997).

7.5 GENETIC DIVERSITY

High genetic diversity has been reported in *L. nigricans*, *L. culinaris* subsp. *orientalis* and subsp. *odemensis* compared with the cultivated lentil. *L. lamottei* and *L. ervoides* are the only species which have been reported as having a similar or more restricted genetic base than the cultivated lentil. This is evident from an isozyme study based on a large number of accessions (Ferguson and Robertson 1996) and studies based on randomly amplified polymorphic DNA (RAPD) (Abo-elwafa *et al.* 1995; Sharma *et al.* 1995; Ferguson *et al.* 1998a). Other studies, based on a few accessions, have produced conflicting results (isozyme: Pinkas *et al.* 1985, Hoffman *et al.* 1986, Rosa and Jouve 1992; RFLP: Havey and Muehlbauer 1989; cpDNA: Mayer and Soltis 1994).

Lentil landraces exhibit specific adaptation to their ecological environment and have evolved into distinct ecotypes in different geographical regions (Barulina 1930; Erskine *et al.* 1989). Adaptation for an appropriate phenology appears to be the main evolutionary force behind this differentiation (Erskine *et al.* 1989) and clear differences between ecotypes in response to temperature and photoperiod have been found (Erskine *et al.* 1990). Other factors conditioning the adaptation of the crop are climatic, for example cold tolerance (Erskine *et al.* 1981), and edaphic, for example iron efficiency (Erskine *et al.* 1993).

Lentils from South Asia (exclusively of the 'pilosae' ecotype) are quite distinct from lentils from other countries in terms of qualitative morphological characters, and group together with those from Ethiopia in terms of quantitative characters (Erskine *et al.* 1998). According to isozyme and RAPD studies, South Asian germplasm clusters tightly with that from Afghanistan (Ferguson *et al.* 1998b). This is not, however, the case for agro-morphological characters (Erskine *et al.* 1998). Associated with the uniqueness of South Asian germplasm is a very restricted genetic base, evident from agro-morphological and phenological characters (Erskine *et al.* 1989, 1990), as well as isozyme and RAPD studies (Ferguson *et al.* 1998b). This may be due to a bottleneck when lentils were first introduced into India around 2000 BC (Erskine *et al.* 1998). Barulina (1930) reports a high number of endemic types from the Hindu Kush-Himalaya border. However, an expedition to Afghanistan in 1974 failed to find the same range of variability (Solh and Erskine 1984) and recent studies have also shown low levels of variability from this region (Ferguson *et al.* 1998b).

Phenological and agro-morphological characters have been evaluated in both the (ICARDA) wild and cultivated lentil collections (Erskine and Witcombe 1984; ICARDA 1993; Ferguson and Robertson 1999). The greatest diversity in quantitative descriptors in lentil is for seed yield, biomass, straw yield and seed size (ICARDA 1993). The process of artificial selection during domestication has caused an increase primarily in plant height, rachis length, number of leaflets per leaf, leaf area, number of flowers, and pod and seed size (Cubero 1984). Certain *L. culinaris* subsp. *orientalis* accessions have a substantially greater number of leaves per plant, leaf area, number of peduncles per plant, pods per plant and seeds per plant than two cultivated lentil checks (Ferguson and Robertson 1999). In the same study, the total biomass obtained from both checks was comparable with the best *L. culinaris* subsp. *orientalis* and the harvest index of one check was comparable with the two best *L. culinaris* subsp. *orientalis* accessions. Accessions of *L. culinaris* subsp. *orientalis* from Uzbekistan, Turkmenistan and Tajikistan had amongst the largest biomass, the greatest number of peduncles per plant and many pods and seeds per plant. The latter was not offset by a decline in seed size. Of the wild taxa *L. lamottei* had the largest seeds.

Sources of resistance to lentil rust, the major foliar disease (induced by *Uromyces fabae* (Pers.) de Barry), vascular wilt, the most important soil-borne disease (induced by *Fusarium oxysporum* Schlecht. em. Snyd. a Hons.f.sp. *lentis* (Vasud. and Srin. Garden) and ascochyta blight (induced by *Ascochyta Fabae spog. F. sp. lentis* (Bond & Vassil.) Gossen et al.) have been identified in cultivated lentil (Bayaa *et al.* 1997; Erskine *et al.* 1994). Resistance to vascular wilt and ascochyta blight have also been found in *L. culinaris* subsp. *orientalis* (Bayaa *et al.* 1994, 1995). No useful source of resistance to the parasitic weed Broomrape (*Orobanche* sp.) has yet been found in either the wild or cultivated lentil. Greater cold tolerance has been found in *L. culinaris* subsp. *orientalis* than in the cultivated lentil (Hamdi *et al.* 1996). The wild lentils, particularly *L. culinaris* subsp. *odemensis* and *L. ervoides*, show a greater resistance to drought in terms of low relative reduction in yield with drought stress (Hamdi and Erskine 1997).

7.6 CONSERVATION RESOURCE STATUS (*EX SITU* AND *IN SITU*)

ICARDA holds the largest *ex-situ* germplasm collection of both lentil landraces and wild *Lens* species in the world. It houses 7758 landrace accessions from 64 countries, although more than half of these are from just three countries, namely India (1848), Syria (1246) and Iran (912). The cultivated-lentil collection is now safety duplicated in the National Bureau of Plant Genetic Resources (NBPGR) in New Delhi, India. The wild *Lens* collection contains 475 accessions, comprising 222 accessions of *L. culinaris* subsp. *orientalis*, 57 subsp. *odemensis*, 134 *L. ervoides*, 55 *L. nigricans* and 7 *L. lamottei*. The repository at USDA-ARS Western Regional Plant Introduction Station, Pullman, Washington also maintains 2720 accessions of cultivated lentil and 119 accessions of wild species. Most of the wild accessions are duplicates of seed held at Washington State University, which maintains 211 accessions, comprising 112 *L. culinaris* subsp. *orientalis*, 25 subsp *odemensis*, 51 *L. ervoides* and 23 *L. nigricans*. Most of these accessions are also duplicated at ICARDA.

The genetic structure of wild populations is typical of highly in-breeding species: They have low levels of heterozygosity (Hoffman *et al.* 1986; Ferguson and Robertson 1996) and one or a few discrete genotypes which tend to retain their integrity. Outcrossing is usually less than 1% in the cultivated lentil (Wilson and Law 1972), but has been reported to reach 6% (Erskine and Muehlbauer 1991). A study on the genetic variation within populations of the wild species using isozymes and RAPDs revealed limited variation within population, with between 78% and 99% of the variation attributable to between-population differences (except from isozyme evidence of *L. lamottei*) (Ferguson *et al.* 1998c). The fact that far greater genetic diversity exists between populations as opposed to within populations implies that priority should be given to *ex-situ* as opposed to *in-situ* conservation if the objective is to conserve genetic variation. Nevertheless, wild lentil populations are included in *in-situ* Gene Management Zones currently being established in Turkey. These areas are Ceylanpinar State Farm in south-east Turkey and the Anatolian Diagonal (Amanos Mountains) of south and central Turkey. Studies have, however, also revealed that substantial variation exists in the levels of diversity found within populations and implies that levels of diversity within populations must be measured and considered prior to targeting of specific populations for *in-situ* conservation (Ferguson *et al.* 1998c).

7.7 CONSERVATION GAPS AND PRIORITIES

Problems of incomplete passport data, especially from Iran and India, exist in the lentil collection at ICARDA. Emphasis is currently placed in tracing accessions to fill these gaps in documentation. Based purely on the number of wild *Lens* accessions available, germplasm from North African countries such as Algeria, Libya, Sudan and Tunisia appear to be under-represented in the collection, as does germplasm from the new central and west Asian republics of the former Soviet Union.

Collection priorities for the wild species have been determined on the basis of the distribution of genetic variation, measured using RAPDs (Ferguson *et al.* 1998a). Priority collection for all species was in south-west Turkey. A collection mission in that region in 1995 added 37 accessions. However, because of the importance of this region for wild lentil, further missions should be conducted, particularly in the region in and around the provinces of Burdur, Isparta and Afyon. In addition, for *L. culinaris* subsp. *orientalis*, collection priorities lie in two broad regions, the first being southern Syria, north and west Jordan and the West Bank of the Palestinian Authority, and secondly from south Turkey across to south-east Turkey. For *L. culinaris* subsp. *odemensis* additional priorities lie in the Levant and parts of North Africa. The region surrounding the coastal junction of Turkey and Syria is a priority for *L. ervoides*, and, although of a lower priority, the Ukraine may yield valuable *L. nigricans* germplasm.

7.8 RESEARCH NEEDS

The geographical distribution range of the two recently recognised species, *L. lamottei* and *L. culinaris* subsp. *tomentosus*, needs to be established. In addition, the extent of the distribution of other wild *Lens* taxa in North Africa needs clarifying. Crossability relations between all *Lens* taxa need to be established, particularly relating to *L. culinaris* subsp. *odemensis*, subsp. *tomentosus* and *L. lamottei* and *L. culinaris* subsp. *culinaris*. These studies should take into consideration within-taxon variation and perhaps follow on from the studies using the GA_3 hormone by Ahmad *et al.* (1995). From the perspective of the breeder, a major gap in the collection (in both cultivated and wild species) is for germplasm with a large biomass. There is also a lack of resistance to pea leaf weevil (*Sitona* sp.) and the parasitic weed broomrape (*Orobanche* spp.).

REFERENCES

Abo-elwafa, A., Murai, K. and Shimada, T., 1995. Intra- and inter-specific variations in *Lens* revealed by RAPD markers. *Theoretical and Applied Genetics*, 90: 335-340.

Ahmad, M., Fautrier, A.G., McNeil, D.L., Burritt, D.J. and Hill, G.D., 1995. Attempts to overcome post-fertilization barrier in interspecific crosses of the genus *Lens*. *Plant Breeding*, 114: 558—560.

Barulina, H., 1930. Lentils of the USSR and other countries. *Bulletin of Applied Botany, Genetics and Plant Breeding (Suppl) (Leningrad, English summary)*, 40: 225-238.

Bayaa, B., Erskine, W. and Hamdi, A., 1994. Geographic distribution of resistance to ascochyta blight in wild lentil. *Genetic Resources and Crop Evolution*, 41: 61-65.

Bayaa, B., Erskine, W. and Hamdi, A., 1995. Evaluation of a wild lentil collection for resistance to vascular wilt. *Genetic Resources and Crop Evolution*, 42: 231-235.

Bayaa, B., Erskine, W. and Singh, A., 1997. Screening lentil for resistance to fusarium wilt: methodology and sources of resistance. *Euphytica*, 98: 69-74.

Cubero, J. I. 1984. Taxonomy, distribution and evolution of the lentil and its wild relatives. In: *Genetic Resources and their Exploitation - Chickpeas, Faba beans and Lentils* (Eds. Witcombe, J.R. and Erskine, W.). Martinus Nijhoff / Dr W. Junk Publishers for ICARDA and IBPGR. The Hague.

Erskine, W. and Muehlbauer, F.J., 1991. Allozyme and morphological variability, out crossing rate and core collection formation in lentil germplasm. *Theoretical and Applied Genetics*, 83: 119-125.

Erskine, W. and Witcombe, J.R., 1984. *Lentil Germplasm Catalogue*. ICARDA, Aleppo, Syria.

Erskine, W., Myveci, K.T. and Izgin, N., 1981. Screening a world lentil collection for cold tolerance. *LENS Newsletter*, 8: 5-8.

Erskine, W., Adham, Y. and Holly, L., 1989. Geographical distribution of quantitative traits in a world lentil collection. *Euphytica*, 43: 97-103.

Erskine, W., Saxena, N.P. and Saxena, M.C., 1993. Iron deficiency in lentil: Yield loss and geographic distribution in a germplasm collection. *Plant and Soil*, 151: 249-254.

Erskine, W., Ellis, R.H., Summerfield, R.J., Roberts, E.H. and Hussain, A., 1990. Characterization of responses to temperature and photoperiod for time to flowering in a world lentil collection. *Theoretical and Applied Genetics*, 80: 193-199.

Erskine, W., Tufail, M., Russell, A., Tyagi, M.C., Rahman, M.M. and Saxena, M.C., 1994. Current and future strategies in breeding lentil for resistance to biotic and abiotic stresses. *Euphytica*, 73: 127-135.

Erskine, W., Chandra, S., Chaudhry, M., Malik, I.A., Sarker, A., Sharma, B., Tufail, M. and Tyagi, M.C., 1998. A bottleneck in lentil: widening the genetic base in South Asia. *Euphytica*, 101: 207-211.

Ferguson, M. 1998. *Studies of genetic variation within the genus* Lens. PhD Thesis. School of Biological Sciences, University of Birmingham, Birmingham, UK.

Ferguson, M. and Robertson, L.D., 1996. Genetic diversity and taxonomic relationships within the genus *Lens* as revealed by allozyme polymorphism. *Euphytica*, 91: 163-172.

Ferguson, M. and Robertson, L.D., 1999. Morphological and phenological variation in the wild relatives of lentil. *Genetic Resources and Crop Evolution*, 46: 3-12.

Ferguson, M., Ford-Lloyd, B.V., Robertson, L.D., Maxted, N. and Newbury, H.J., 1998a. Mapping the geographical distribution of genetic variation in the genus *Lens* for the enhanced conservation of plant genetic diversity. Molecular Ecology, 7:1743-1755.

Ferguson M.E., N. Maxted, van Slageren, M. and L. D. Robertson, (2000). A re-assessment of the taxonomy of *Lens* Mill. (Leguminosae, Papilionoideae, Vicieae). *Botanical Journal of the Linnean Society*. 133: 41-59.

Ferguson, M., Robertson, L.D., Ford-Lloyd, B.V., Newbury, H.J. and Maxted, N., 1998b. Contrasting genetic variation amongst lentil landraces from different geographical origins. *Euphytica*, 102:265-273.

Ferguson, M.E., Newbury, H.J., Maxted, N., Ford-Lloyd, B.V. and Robertson, L.D., 1998c. Population genetic structure of *Lens* taxa revealed by isozyme and RAPD analysis. *Genetic Resources and Crop Evolution*, 45: 549-559.

Hamdi, A. and Erskine, W., 1997. Reaction of wild species of the genus *Lens* to drought. Euphytica, 91: 173-179.

Hamdi, A., Küsmenoglu, I. and Erskine, W., 1996. Sources of winter hardiness in wild lentil. *Genetic Resources and Crop Evolution*, 43: 63-67.

Havey, M. and Muehlbauer, F.J., 1989. Variability for restriction fragment lengths and phylogenies in lentil. *Theoretical and Applied Genetics*, 77: 839-843.

Hoffman, D., Soltis, D., Muehlbauer, F.J. and Ladizinsky, G., 1986. Isozyme polymorphism in *Lens* (Leguminosae). *Systematic Botany*, 11: 392-402.

ICARDA, 1993. Legume Program: Annual Report for 1992. pp. 155-160. ICARDA, Aleppo, Syria.

Ladizinsky, G. 1993. Wild lentils. *Critical Reviews in Plant Science*, 12: 169-184.

Ladizinsky, G. 1997. A new species of *Lens* form south-east Turkey. *Botanical Journal of the Linnean Society*, 123: 257-260

Ladizinsky, G., Braun, D., Goshen, D. and Muehlbauer, F.J., 1984. The biological species of the genus *Lens*. *Botanical Gazette*, 145: 253-261.

Mayer, M.S. and Soltis, P.S., 1994. Chloroplast DNA phylogeny of *Lens* (Leguminosae): origin and diversity of the cultivated lentil. *Theoretical and Applied Genetics*, 87: 773-781.

Oss, Van H., Aron, Y. and Ladizinsky, G., 1997. Chloroplast DNA variation and evolution in the genus *Lens* Mill. *Theoretical and Applied Genetics*, 94: 452-457.

Pinkas, R., Zamir, D. and Ladizinsky, G., 1985. Allozyme divergence and evolution in the genus *Lens*. *Plant Systematics and. Evolution*, 151: 131-150.

Rosa, de la L. and Jouve, N., 1992. Genetic variation for isozyme genes and proteins in Spanish primitive cultivars and wild subspecies of *Lens*. *Euphytica*, 59: 181-189.

Sharma, S.K., Dawson, I.K. and Waugh, R., 1995. Relationships among cultivated and wild lentils revealed by RAPD analysis. *Theoretical and Applied Genetics*, 91: 647-654.

Solh, M. and Erskine, W., 1984. Genetic resources of lentil. In: *Genetic Resources and their Exploitation - Chickpeas, Faba Beans and Lentils* (Witcombe, J.R. and Erskine, W.). pp. 205-224. Martinus Nijhoff/Dr W Junk Publishers for ICARDA and IBPGR, The Hague.

Tournefort, De J.P., 1719. *Institutiones rei herbariae*. Editio tertia. Paris.

Wilson, V.E. and Law, A.G., 1972. Natural crossing in *Lens esculenta* Moench. Journal of the. *American Society of Horticultural Science*, 97: 142-143.

Zohary, D. and Hopf, M., 1973. Domestication of pulses in the Old World. *Science*, 182: 887-894.

Zohary, D. and Hopf, M., 1993. *Domestication of Plants in the Old World*. Clarendon Press, Oxford, UK.

Chapter 8

Vetches (*Vicia* L.)

Mark van de Wouw, Dirk Enneking, Larry D. Robertson and Nigel Maxted

8.1 INTRODUCTION

The genus *Vicia* comprises approximately 190 species (ILDIS, 1999), chiefly located in Europe, Asia and North America, extending to temperate South America and tropical East Africa. The genus is primarily located in the Mediterranean and Irano-Turanian regions (Kupicha, 1981). It is adapted to temperate regions but can also be found at high altitudes in tropical Africa. Endemic species are present on all continents, except Australia and Antarctica.

Perhaps the best known species of the genus is the faba bean or broad bean, *V. faba* L., an important pulse, fodder crop and vegetable. Throughout North Africa and West Asia faba beans are an important component of the diet and has been referred to as the "poor man's meat". Many fields in West Asian and North African (WANA) regions are harvested early for green pods and beans and later harvested for dry seed. In 1998 2.4 million ha was planted with this crop, of which the large majority was planted for pulse production (2.2 million ha, Figure 8.1). Approximately 600,000 ha of this crop was planted in the countries surrounding the Mediterranean sea (FAO, 1999). Compared to the other grain legumes world trade in faba bean is inconsequential, with most produce locally consumed. However, there is an expectation that faba bean may become a major export crop in Australia. World production of vetches in the same year was 1.3 million ha, of which 684,000 ha in the Mediterranean. Making up the major part of the reported production area is the common vetch (*V. sativa* L.). The extent of cultivation of vetches is therefore of equal importance in the Mediterranean to that of faba bean.

Many other species of *Vicia* are used as minor seed (Bellido, 1994), forage or fodder crops, these include *V. ervilia* Willd. (Esteban 1996), *V. narbonensis* L., members of the *V. sativa* L. aggregate, *V. benghalensis* L and *V. articulata* Hornem. The seed is traditionally used as an additive to voluminous feeds for ruminants (Bellido, 1994; Enneking, 1995), while the forage is often grown for hay in mixtures with cereal tutor crops, such as barley and oats.

The area of cultivation of traditional agricultural grain crops such as *V. ervilia* (Esteban, 1996) and *V. articulata* (Hammer and Perrino, 1995) has generally been declining. For example, *V. ervilia* cultivation in Spain declined from 105,000 ha in 1930 to as low as 12,000 ha in 1993 with a slight recent increase to 26,000 ha in 1995 (Esteban, 1996). The production area of recent domesticates such as *V. villosa*, *V. benghalensis* and *V. pannonica* appears to be increasing. Their widespread adaptation and excellent capacities to produce biomass make them very attractive to farmers. Hairy vetch (*V. villosa* Roth) is grown as a green manure crop and has been found to be an excellent mulch for tomato production (Baki *et al.*, 1997). *V. sativa,* the common vetch, is the most commonly grown forage crop, it is grown throughout the Mediterranean region. In Spain the area of cultivation for vetches has actually increased 4 fold (FAO, 1999) due to the support for minor crops under the EEC set-aside scheme. Expansion of this species as a pulse crop has been hindered by concerns about the toxicity of the seeds for human consumption (Tate and Enneking, 1992; Enneking, 1995). A covert trade of this species for human consumption however continues (Tate *et al.*, 1999), while at the same time significant progress has been made towards the selection and breeding of non-toxic cultivars (Tate and Rathjen, pers. comm).

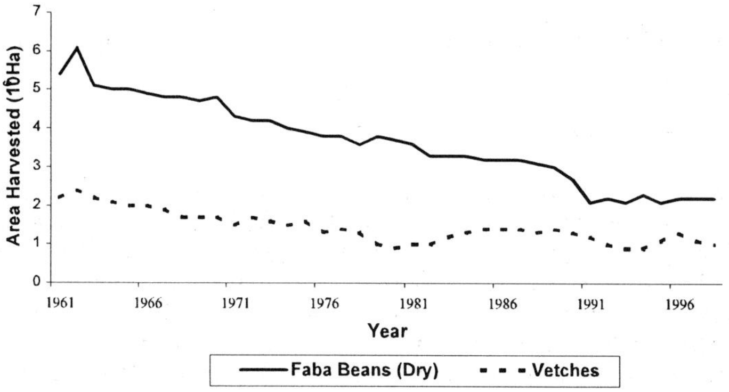

Figure 8.1. Global harvested area for dry faba beans and vetches (FAO, 1999).

Many farmers experience members of the genus as noxious weeds (Gams, 1924; Aarssen *et al.*, 1986). *V. cracca*, a perennial, is very persistent and particularly difficult to control. *V. sativa* subsp. *nigra* and *V. villosa* can cause large problems in cereals: their twining habit can make mechanical harvesting very difficult and may cause lodging of the crop, increasing the incidence of fungal diseases (Aarssen *et al.*, 1986). Among the other species which occur as weeds in agricultural field are *V. benghalensis, V. tetrasperma* (L.) Schreber, *V. tenuissima* (Bieb.) Schinz and Thell., *V. hirsuta* (L.) S.F.Gray, *V. disperma* DC., *V. ervilia, V. articulata, V. sepium* L., *V. bithynica* L., *V. narbonensis, V. peregrina* L., *V. lutea* L., *V. grandiflora* Scop. and *V. pannonica* Crantz (Hanf 1990). Some genotypes of *V. sativa* are known to mimic the shape of lentil seeds, and are therefore often found mixed with lentils (Barulina,1930; Erskine *et al.*, 1994).

The genus is well placed to help meet the increased global demand for animal feed and to provide crops for a diversity of farming systems (Francis *et al.*, 1998). It is therefore important to ensure that its genetic resources are conserved and available for further crop development. *Vicia* conservation has been given priority by the International Board for Plant Genetic Resources (IBPGR 1985) in recent years.

8.2 TAXONOMIC SUMMARY

The genus *Vicia* L. is a member of the legume tribe *Vicieae* of the Papilionoideae along with *Lathyrus* L., *Lens* Mill., *Pisum* L. and *Vavilovia* A. Fedorov. The precise generic boundaries between these genera have been much debated, which has led to an abundant and complex synonymy. The genus itself has proved a popular group to study, there being 20 major classifications of the genus since Linnaeus (Maxted, 1993a). However, certain species groupings recur throughout the history of the genus, notably "*Ervum*", "*Cracca*", "*Vicia*" and "*Faba*".

The first distinct entity recognised as *Vicia* was proposed by de Tournefort (1694) in his 'Elements de Botanique'. He recognized 6 species in *Vicia*, excluding the faba bean which he considered a distinct genus. Kupicha (1976) undertook the most comprehensive revision of the genus. She divided the four natural groups of species into two subgenera, "*Ervum*" and "*Cracca*" in her subgenus *Vicilla* and "*Vicia*" and "*Faba*" in her subgenus *Vicia*. The distinction between the two subgenera being primarily based on relative length of the inflorescence and the presence of nectarifous spots on the stipules (Table 8.1). Her subgenus *Vicilla*, which includes forage species such as *V. villosa, V. ervilia, V. benghalensis* and *V. hirsuta*, is further divided into 17 sections (Table 8.2). Its sectional classification presents some difficulties as the variation in the subgenus is very wide and there is only a small correlation between the variation in the different characters (Kupicha, 1976). Several of the annual species have been placed in monotypic or ditypic sections. Subgenus *Vicilla* is considered more primitive and diverse than subgenus *Vicia* (Maxted, 1993a). Kupicha's conception of subgenus *Vicia* is more narrow and coherent, containing 38 species divided into 5 sections. This subgenus contains the more agriculturally important species as *V. faba, V. sativa* and *V. narbonensis*. Maxted (1993a) further revised this subgenus, and enacted many of the suggestions made by Kupicha (1976), he split the species into nine sections (Table 8.2), giving *V. faba* monospecific sectional status. No such detailed review has yet been undertaken for subgenus *Vicilla*, to which *V. ervilia* Willd., *V. villosa* Roth, *V. monantha* Retz, *V. pannonica* Crantz and *V. benghalensis* L. belong.

Table 8.1. Characters used to distinguish the two *Vicia* subgenera.

Character	Subgenus *Vicilla*	Subgenus *Vicia*
Stipule nectary	absent	present on abaxial stipule face
Peduncle length	= or > leaf	< leaf or absent
Style type	terete, dorsally compressed pubescent or tufted, laterally compressed	dorsally compressed, tufted
Keel shape	edge curved round and above style	apical part encircled style and anthers
Legume	stipitate, sub-torulose	never stipitate
Canavanine	present	absent

Table 8.2. Summary of the sectional classification of subgenera *Vicilla* (*sensu* Kupicha 1976) and *Vicia* (*sensu* Maxted 1993a).

Taxon	Number of Species	Economically Useful Species	Geographical Distribution
Subgen. *Vicilla* (Schur) Rouy in Rouy & Fouc.			
Sect. *Vicilla* (Schur) Asch. & Graebner	15		Europe, West, East & North Asia
Sect. *Cassubicae* Radzhi	9		Europe, West, East & North Asia, North America and Chile
Sect. *Perditae* Kupicha	1		Azores
Sect. *Cracca* Dumort.	40	*V. villosa*, *V. benghalensis*	Europe, West, East & North Asia, North Africa, North America
Sect. *Variegatae* Radzhi	3		Europe, West and Central Asia
Sect. *Pedunculatae* Rouy in Rouy & Fouc.	3		Europe, North West Africa
Sect. *Americanae* Kupicha	1		North America, China
Sect. *Subvillosae* Kupicha	1		Central Asia
Sect. *Volutae* Kupicha	1		Southern Europe, Southern Russia, Caucasia
Sect. *Panduratae* Kupicha	3		South East Europe, West Asia
Sect. *Ervum* (L.) Taub. in Engl. & Prantl	3		Europe, West, East & North Asia, North Africa
Sect. *Ervoides* (Godron) Kupicha	1	*V. articulata*	South East Europe, West Turkey
Sect. *Ervilia* (Link) W. Kock	1	*V. ervilia*	Southern Europe, West Asia, North Africa
Sect. *Lentopsis* Kupicha	1		Turkey
Sect. *Trigonellopsis* Rech. fil.	3		South East Europe, West Asia
Sect. *Australes* Kupicha	13		Mexico, Central and South America
Sect. *Mediocinctae* Kupicha	1		New Mexico
Subgen. *Vicia* L.			
Sect. *Atossa* (Alef.) Asch. & Graebner	4		Europe, West & North Asia
Sect. *Microcarinae* Maxted	1		Syria
Sect. *Hypechusa* (Alef.) Asch. & Graebner	14	*V. noeana*, *V. hybrida*	Europe, West & Central Asia, North Africa
Sect. *Peregrinae* Kupicha	3		Southern Europe, West & North Asia
Sect. *Wiggersia* (Alef.) Maxted	2		Europe, West & North Asia, & North Africa
Sect. *Vicia*	5	*V. sativa*	Europe, Asia & North Africa
Sect. *Bithynicae* (B. Fedtsch. ex Radzhi) Maxted	1		Europe, West & North Asia, & North Africa
Sect. *Narbonensis* (Radzhi) Maxted	7	*V. narbonensis*	Europe, West & North Asia, & North Africa
Sect. *Faba* (Miller) Ledeb.	1	*V. faba*	Not known out of cultivation
Total =	138		

The current taxonomy of the genus has been generally accepted with only a few areas of continuing debate. Much of the current debate has focused around three species or species complexes, *V. sativa*, *V. narbonensis* and *V. faba.* Taxonomists disagree about the rank and number of taxa belonging to the *V. sativa* aggregate (Stankevich, 1978; Zohary and Plitmann, 1979; Maxted, 1995; Potokina, 1997) and the rank and number of taxa belonging to the *V. narbonensis* complex (Plitmann, 1967; Davis and Plitmann, 1970; Schäfer, 1973; Maxted, 1993a; Bennet and Maxted, 1997). The problems over the taxonomic distinction within both species complexes have in part been the result of authors adopting a regional approach to their taxonomy rather than studying each group throughout its natural range. The large variation in morphology and karyotypes and the conflicting results of morphological and other studies have added to the confusion (Hanelt and Mettin, 1989: Maxted *et al.*, 1991b).

The appropriate position and rank of *V. faba* within the genus and the search of the wild progenitor of this important species is another issue which attracts attention of taxonomists (Stankevich, 1982; Bueno Perez, 1979; Chrtkova-Zertova, 1979; Radzhi, 1971; Plitmann, 1967; Kupicha ,1976; Maxted *et al.*, 1991a; Maxted, 1993a). One reason why so much taxonomic study has been devoted to *Vicia* is the lack of an obvious progenitor for the faba bean. Even though there has been systematic collecting throughout the Eastern Mediterranean in recent years no close ally of the faba bean has been discovered (Maxted *et al.*, 1989). Schäfer (1973) hypothesised that *V. faba* originated from an extinct ancestor, though this would be impossible to prove. There is always a possibility that a progenitor may appear in areas not yet thoroughly surveyed (e.g. South East Turkey, Iran or Afghanistan) and therefore conservationists should continue to search for this agricultural 'holy grail'. *V. narbonensis* L. (2n=14) has received the greatest attention as the putative ancestor of faba bean. However, with a different karyotype and different chromosome number, it has never successfully been crossed with *V. faba*, and therefore cannot be regarded as a direct ancestor of faba bean. Based on the taxa currently known, there remains no close ally of *V. faba* and there is little agreement over the most appropriate position and rank for the faba bean within *Vicia*, although all would now agree that the species is quite distinct and isolated from all other *Vicia* taxa.

8.3 Centre of Diversity

The Mediterranean is the most important centre of current diversity for *Vicia*. Secondary centres of diversity exist in South America, North America and Southern Siberia (Kupicha, 1981; Nikiforova, 1988; Hanelt and Mettin, 1989; Maxted, 1995). Elucidation of the centre of origin of the genus is however more problematic. It can be argued that relative species concentrations can be used to indicate the centre of origin, which would therefore suggest a South Eastern Europe and North Western Asia origin. Over half of the *Vicia* taxa are endemic to this area. This region is very important floristically as it is at the cusp of three major temperate phyto-geographical regions, Euro-Siberian, Mediterranean and Irano-Turanian (Takhtadjan, 1969). Stebbins (1967) points out that if areas with different environmental and floristic characters are juxtaposed, they may act as a catalyst to evolution. There is evidence, however, that taxa with centres of distribution in the Balkans, South-west and South-east Asia are generally centred in these areas due to climatic deterioration and floristic movement from more northerly distributed Floras (Takhtadjan, 1969). The Boreal-tertiary Flora was driven South by the Quaternary Ice Age towards these three floristic refugia in the South. Kupicha (1974) concludes that *Vicia* evolved in the early tertiary and the centre of origin is likely to have been much further North than is indicated by contemporary concentrations of taxa.

Stebbins (1967) suggests that the most genotypically and phenotypically plastic species evolve rapidly away from the centre of diversity, while the more primitive species will remain at the centre. This would account for the current distribution and various forms of *V. sativa* and the localised nature of some very restricted primitive species, such as *V. dionysiensis*. Contrary to this hypothesis, however, Willis (1921) argues that evolution and plant dispersal occur at the same rate. Therefore the oldest and most primitive species are likely to be the most widespread and are currently found on the distributional periphery, while the most recently evolved species occur at the centre of origin and are localised. This would explain the widespread distribution of *V. sepium*, which is regarded as a primitive member of *Vicia*, because of its perennial habit, relatively long peduncle and numerous flowers per peduncle. The latter hypothesis, though, clearly does not explain the wide distribution of *V. sativa*, which is regarded as an advanced species. With two opposing hypotheses being so well represented within *Vicia* it does perhaps act a warning against attempting to draw too firm conclusions about the centre of origin of any genera from the data currently available.

V. faba and *V. ervilia* belong to the earliest domesticated plants of the world. They were domesticated during the Neolithic period (Hopf, 1970; 1986, Zohary and Hopf, 1988; Zohary, 1989) in Western Asia. The earliest finds for *V. ervilia* date back to the epipalaeolithic period at Abu Hureya (ca. 9100-8300 BC). Faba bean is often referred to by ancient writers (Hawtin and Hebblethwaite, 1983). It was most likely domesticated in the region between Afghanistan and the eastern Mediterranean during the period 7000-4000 B.C. (Hanelt, 1972). Cubero (1972) concluded the culture of faba bean spread in four directions from the centre of origin. These were north to central Europe, northwest to Western Europe, west to the Mediterranean, and east to the Far East (India, China and Japan). The *minor* type faba bean was introduced to China in 100 BC (Tao, 1981) and the *major* type in 1200 AD (Hanelt, 1972). Still retaining vestiges of its past, faba bean can loosely be regarded as an incompletely domesticated species, as indicated by the breeding system of the species (which stands between full autogamy and full allogamy), the indeterminate growth habit, and dehiscent pods.

Archaeological records suggest that the common vetch (*V. sativa*) was a weed that spread from the Mediterranean to central Europe with the early crop assemblage of cereals, lentil and bitter vetch (*V. ervilia*). It is not clear whether the domestication of species such as *V. sativa* and *V. ervilia* occurred primarily for fodder or for food grains (Ladizinsky, 1989). The link between animal husbandry, especially ruminant species, and the frequency of certain crops in the archaeological record, particularly during the bronze age, requires further study, and may provide further insights into *Vicia* domestication.

Prehistoric use of plants was more extensive than today, particularly for forage and fodder species. Many species that have been used for millennia were never domesticated, and numerous early domesticates were later abandoned in favour of better performing species. With the expansion of agriculture there was a concomitant increase in diversity at the intraspecific level through adaptation to new habitats (Frankel, 1977). *V. sativa*, *V. villosa* and *V. narbonensis* could be cited as examples. Frankel also predicted that new domesticates would be developed to suit crop production needs in specific ecological and economic niches. Recent developments of *Vicia* crops at ICARDA and in Australia reflect this trend.

Many of the American *Vicia* species have been taxonomically grouped with section *Cracca*, which is the most widely distributed section of the genus. The ecology and molecular evolution of this group are likely to provide the keys to understanding the migration of the genus across Eurasia and America. Subgenus *Vicia* most certainly has its centre of diversity in the Levant, as demonstrated by Maxted (1995), from where many species radiated throughout the Mediterranean and temperate regions. Mountains are today some of the last refuges for the rarer *Vicia* species. An ecogeographic and molecular taxonomic synthesis of *Vicia* and closely related genera, such as *Lathyrus*, may provide answers to the question whether the rare species existing in a mountainous habitat are a significant link in the evolution of *Vicia* and the generation of diversity.

8.4 ECOGEOGRAPHIC DISTRIBUTION

The range of habitats occupied by *Vicia* species is diverse and includes disturbed habitats like field margins and roadsides, as well as more pristine habitats like woodlands and steppes. The species considered more advanced are generally those found in the more disturbed, open communities. The cultivated species have evolved in general from disturbed habitats, they were originally the wild and weedy flora of agricultural fields (Vavilov, 1926). Farming systems have therefore had a great influence on the recent evolution of the genus. Their weedy nature would explain the widespread distribution of many species. Over half of the species existing in North America are introductions from the Old World (Hermann, 1960; Kupicha, 1976).

Cultivation of the faba bean, *V. sativa* and *V. villosa* is widespread, but planting of the other cultivated species is more restricted and largely confined to Southern Europe, North Africa and North West Asia (Tupikova, 1926; Fischer, 1938). The genus contains many restricted endemics, for which only very few sites have been documented or which are bound by specific soil types and climatic regimes (Maxted, 1988; Khattab *et al.*, 1988; Maxted *et al.*, 1989; Maxted, 1995). The ecogeographic distribution of many *Vicia* species, particularly those in subgenus *Vicilla* and those endemic to South America remains poorly understood. The ecogeographic information for these species needs to be reviewed and documented where necessary, as they undoubtedly require active complementary conservation.

The faba bean is grown across the Northern and Southern Temperate zones and at higher altitudes in some sub-tropical regions. The distribution of the faba bean is entirely controlled by humans, as it is not known in the wild. The major faba bean producing countries are China and Ethiopia (1,000,000 and 245,000 ha, respectively; FAO, 1999). In Europe Faba bean is cultivated both in the Mediterranean and in the north. Faba beans were introduced to South America with European colonisation and are mostly grown at higher elevations where *Phaseolus* beans can not be grown. It has been more recently introduction to North America where it cultivation is not of significance and to Australia, where the area has dramatically increased in the past ten years (now 107,000 ha; FAO, 1999).

V. faba is divided into two subspecies, *faba* and *paucijuga* Murat. The most primitive of these and less adapted as a crop, is the relatively small seeded subsp. *paucijuga*. This small plant has a more restricted distribution being confined to Afghanistan, Pakistan and India (Muratova, 1931) and the Terai region of Nepal. Cubero and Suso (1981) used the concept of fertility barriers for defining biological species and concluded that the distinction of *V. faba* into two subspecies was unjustified. They felt that the *paucijuga* forms were probably the most similar to the wild types and more closely related to the supposed extinct progenitor. Subspecies *faba* is divided into varieties according to the size of its seeds. The large seeded faba bean (*V. faba* subsp. *faba* var. *major*), which has a seed weight of more than 1 g, developed in the Southern Mediterranean and China; the small seeded types (var. *minor*, seed weight less than 0.5

g), found in the Ethiopian highlands and Sudan and which have become important in north European agriculture; and the varieties with an intermediate seed size (var. *equina*) developed in the Middle East and North Africa and are the main type grown and eaten in Egypt (Maxted, 1995; Duc, 1997). The crop is known by many names in the English language. Field beans are usually the *minor* and *equina* types. Broadband is the name that is commonly given for the *major* types. The *equina* types are also called horse beans and the *minor* types are sometimes called tick beans. Because of the multiplicity of names, ICARDA has promoted the term faba bean to designate the crop in general.

Among the other cultivated species, *V. sativa*, the common vetch, is widely cultivated for forage and fodder. It is a common pan-temperate and subtropical weed on agricultural and disturbed grounds, with apparently no specific soil requirements. Subspecies *nigra*, the common weedy vetch, is the most extensively distributed. Subspecies *sativa* can also be found on almost all continents. It is the most widely cultivated subspecies and is found throughout Europe, the Mediterranean and Western Asia as a forage and fodder crop. Subspecies *amphicarpa* has a more restricted distribution, it is mainly found in the dryer areas of the Mediterranean and Western Asia and has been introduced in Australia. As its name suggests this subspecies produces, along with the aerial flowers and pods, underground flowers and pods. As such it has potential as a forage in dry areas to provide improved natural recovery following grazing. It prefers sandy or gravelly soil types because it needs loose soil to facilitate the formation of its underground stems (Plitmann, 1973). Subspecies *incisa*, recognised by its incised leaflets and subspecies *macrocarpa*, characterised by its relatively large pods, are also found in the Mediterranean and Western Asia, but both have a more limited distribution than the other subspecies. *Incisa* can be found up to an altitude of 510 m, while *macrocarpa* has been found up to an altitude of 200 m (Maxted, 1995). The latter has particular agricultural interest because of its relatively large pods and seeds.

V. narbonensis L. is a common wild and minor cultivated forage species of Southern Europe, the Mediterranean and Western Asia. It was formerly cultivated more widely on limestone and volcanic soils, often inter-cropped with faba beans (Cubero, pers. comm.) but its cultivation is now restricted to Syria, Turkey and Iraq (Enneking and Maxted, 1995; Enneking, 1995). Currently, cultivars are being registered in Australia (Heazlewood, pers. comm.) and Italy (Venora, pers. comm.). It is one species of a morphologically closely related group of less common species referred to collectively as the *V. narbonensis* complex (Schäfer, 1973; Maxted *et al.*, 1989). This group of seven species is mostly found in field margins, grasslands, as a weed in cultivated fields and in wastelands. They generally prefer calcareous soils, except for *V. narbonensis* var. *jordanica* H. Schäfer and *V. hyaeniscyamus* Mout., which has only been found on basaltic soils. *V. serratifolia* Jacq., a member of the narbonensis group, used to be cultivated as a forage crop in central Europe and parts of France (Clos, 1898). Similarly, *V. johannis* Tamamschjan in Karyagin, a species with a more northern distribution than *V. narbonensis* and thus better cold adaptation, could offer potential as a forage crop. Members of the complex can be found in the countries surrounding the Mediterranean, Europe and Southwest Asia. The largest number of taxa are found in Turkey, Syria and Greece (Bennett and Maxted, 1997).

There are several species cultivated for minor fodder or forage production in *Vicia* sect. *Hypechusa*. Specifically, *V. pannonica* Crantz , Hungarian vetch, is found in Southern Europe and South West Asia, where it grows in disturbed and cultivated habitats up to 2200 m (Maxted, 1995). It is cultivated in the North East Mediterranean. *V. noeana* Reuter ex Boiss. has a restricted distribution in the North East Mediterranean but is cultivated in North West Syria (Ehrman and Maxted, 1990). *V. hybrida* L., hairy yellow vetchling, is widely distributed around the Mediterranean and has been experimented with for forage production in Syria (ICARDA 1994). *V. lutea* L., yellow vetch, has a Mediterranean distribution and has been reported from the Caucasus and Iran (Maxted, 1995).

There are also several species in subgenus *Vicilla* which are cultivated. Since the beginning of this century *V. benghalensis* L., purple vetch, has been cultivated as a cover and green manure crop in the pacific states of the USA and more rarely in South America as a forage crop (Hanelt and Mansfeld, 1986). In Australia the cultivar 'Popany' is very popular for its good biomass production and adaptation to higher rainfall areas. Its natural distribution is the Mediterranean region from where it spread to the Azores and North America (Zeven and Zhukowsky, 1975). *V. villosa* Roth., the hairy vetch, is an annual or biennial species that is distributed throughout Europe, the Mediterranean and Western Asia, and has been naturalized in North America and Japan. It naturally occurs at higher latitudes and altitudes and therefore has high frost resistance (Aarssen *et al.*, 1986; Keatinge *et al.*, 1991), which therefore means that it can be grown in colder environments than common vetch (Townsend 1974). Like many other vetches it is both a noxious weed and a useful fodder or green manure. The seeds of *V. villosa* and *V. benghalensis* contain high levels of the toxic amino acid canavanine which pose a problem for monogastric animals (Enneking *et al.*, 1993), especially when present as contaminants in cereals. *V. ervilia*, bitter vetch, has a mainly Mediterranean distribution. It is cultivated as a minor fodder species throughout the region (Barulina, 1930; Fischer, 1937; Mateo-Box, 1961; Enneking *et al.*, 1995; Esteban, 1997). *V. articulata* Hornem. (Mateo-Box, 1961; Jubete, 1996) also has a Mediterranean distribution and is cultivated in Spain, Greece, Turkey and Syria. It was historically extensively cultivated in Spain earlier this century (Barulina, 1930; Fischer, 1937) but in recent years this area has been reduced, from 160,000 ha in 1960 to virtual disappearance in 1990 (Jubete, 1996). It has recently also been found cultivated as a lentil in Ecuador (Reid and Robertson, pers. comm.). This species is well adapted to infertile acid soils (Mateo-Box, 1961; Bellido, 1994; Jubete, 1996) and competes well against prostrate weeds.

Other cultivated or experimentally cultivated species are described in Hanelt and Mansfeld (1986). Ehrman and Maxted (1989), Maxted (1995) and Bennett and Maxted (1997) have recently published detailed ecogeographic studies of various groups of *Vicia* species. Figure 8.2 shows the natural distribution of the economically useful *Vicia* species based on data supplied by ICARDA, INIA, IPK and CLIMA. *V. faba*) is not included as its distribution is entirely anthropocentric.

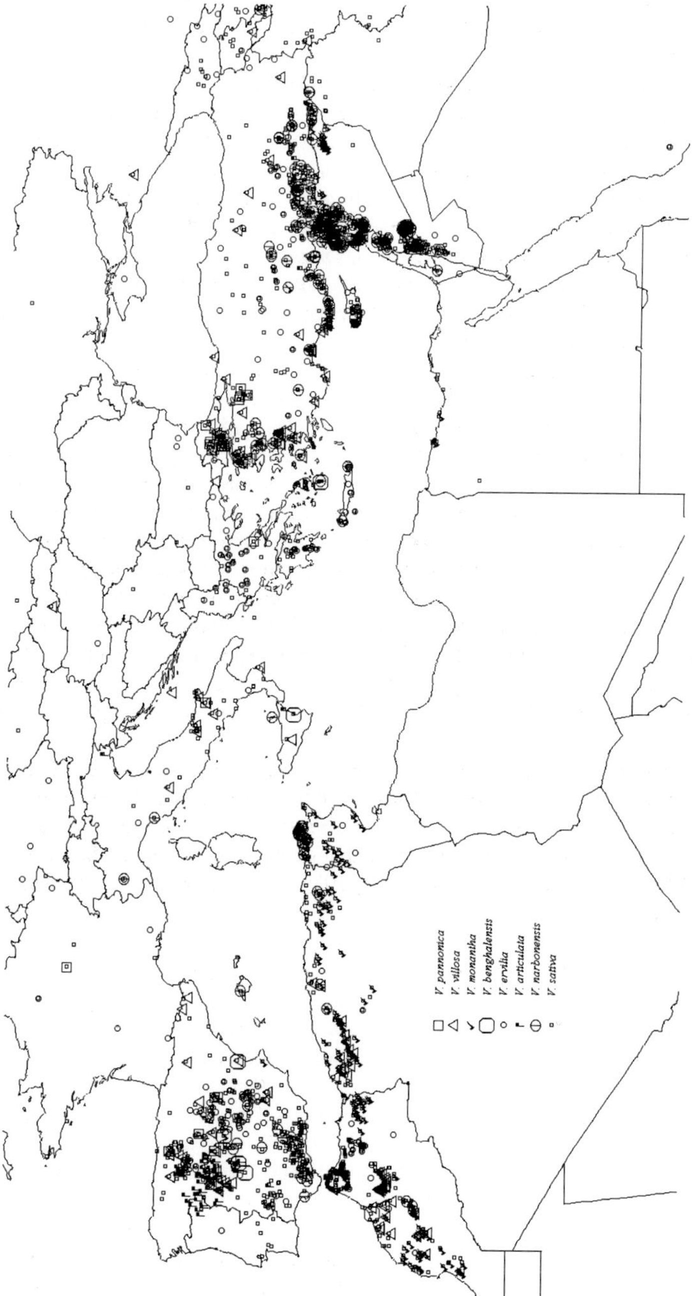

Figure 8.2. Distribution of economically useful *Vicia* species.

8.5 GENETIC DIVERSITY

Within the genus numerous studies have been executed on assessing the genetic diversity, most of them have however focussed on the agriculturally important species of the genus (Tupikova, 1926; Foury, 1950, 1954; Villax, 1963; Kernick, 1978; Abd El Moneim *et al.*, 1990; ICARDA, 1994; Sonnante *et al.*, 1997), and especially on faba bean (e.g. Link *et al.*, 1995; Jaaska, 1997; Torres *et al.*, 1993; Satovic *et al.*, 1996; van de Ven *et al.*, 1990).

Diversity in faba bean is considerable. A study with RAPDs found close relationships between European *V. faba* subsp. *faba* var. *major* and faba bean originating from the Mediterranean region. European *V. faba* subsp. *faba* var. *minor* was found more variable and not as closely related to the Mediterranean forms (Link *et al.*, 1995). Crosses between Mediterranean germplasm and Central European germplasm have proven to provide interesting new basic material for breeding programs (Von Kittlitz *et al.*, 1993).

As a major pulse crop, there has been a significant phenotypic characterization of the genetic diversity in faba bean. Selection in faba bean for resistance for chocolate spot (induced by *Botrytis fabae* Sard.), ascochyta blight (induced by *Ascochyta fabae* Speg.), rust (induced by *Uromyces fabae* (Pers.) de Bary) and stem nematodes (*Ditylenchus dipsaci* (Kühn) Filipjev) at ICARDA have identified useful sources of resistance (Hanounik and Robertson, 1987, 1988). The best sources for chocolate spot resistance have all come from the Colombian-Ecuador mountain region. Several sources of resistance to chocolate spot are also resistant to rust. Screening for resistance for viruses yielded four accessions (all from Afghanistan) with resistance to BLRV, with a long latent period responsible for the resistance and another four accessions with high levels of resistance to BYMV. *Orobanche crenata* Forsk. (an obligate parasitic weed) is the most limiting factor to faba bean production in WANA. The seriousness of this pest is often hidden, because in many cases faba bean is not grown, since complete crop failure can result in areas of severe infestation. The breeding programmes at ICARDA, in collaboration with the Egyptian and Spanish national programmes, developed lines with resistance to this serious pest (Robertson and Saxena, 1993). This has resulted in several lines with good levels of resistance and adaptation to the Mediterranean environment. Drought can be an other major constraint to faba bean production in the Mediterranean region. Research under drought conditions at ICARDA has identified several lines to be more efficient in water use over three seasons (Robertson and Saxena, 1993). Autumn cultivars grown in the Mediterranean region can tolerate frost less than the most resistant European genotype Cote d'Or (Lawes *et al.*, 1983), which tolerates temperatures below -15° C.

The germplasm collection of faba bean has been partially evaluated for its IBPGR/ICARDA descriptor list. Results for 840 lines are summarized in a published catalogue (El-Sherbeeny and Robertson, 1988). The accessions were analysed by country of origin and interesting patterns of diversity by geographic origin were recognized. Egyptian germplasm combined earliness and high yield with a medium seed size. Accessions from Iraq and Syria combined large seed size with long pods and gave the highest seed yield. The countries with the largest seed size (Spain, Iraq, Syria and Turkey) also had the longest pods. The germplasm collection of faba bean was also evaluated for seed protein. Seed protein content varied between 18.0 to 31.0% with a mean of 24.0% (El Sherbeeny and Robertson, 1992). The frequency distribution for protein content displayed a sharp peak about the mean with more than 90% of the accessions lying between 21.0 and 27.0. There were no significant correlations between protein content and seed yield or its components, however, there was moderate variation among countries for protein content, Yugoslavia, had the lowest protein content with 22.5% and Japan had the highest mean of 26.6 %. However, both had comparable ranges. Regional groupings of countries also had similar means and ranges for protein content. Protein content has not been a breeding objective for faba bean; the major activity has been a monitoring of new lines developed to maintain the protein content found in existing cultivars

The faba bean pure line collection was screened for auto-fertility (Robertson and El-Sherbeeny, 1995) which is defined as the ability to self-pollinate in the absence of pollinating insects (Bond and Pope, 1987). A high number of pure line accessions failed to set seeds without tripping and are described as auto-sterile. There were also a large number of accessions with high auto-fertility. There were marked differences among countries of origin for auto-fertility, with Egypt having the highest and the former USSR the lowest auto-fertility. However, most countries expressed a large range, suggesting that breeders should first look in their own material for this trait rather than searching exotic germplasm. In general, there was a higher level of auto-fertility in Middle Eastern countries that may be due to natural selection for non-dependency on pollinators for high yield.

The germplasm collection of faba bean was also evaluated for seed protein. Seed protein content varied between 18.0 to 31.0% with a mean of 24.0% (Robertson and El-Sherbeeny, 1992). Protein content has not been a breeding objective for faba bean; the major activity has been a monitoring of new lines developed to maintain the protein content found in existing cultivars. There are also important anti-nutritional factors in faba bean which affect its utilisation as a human food (Williams *et al.*, 1994). The most important of these are the glucosides, convicine and vicine (Marquardt, 1982). There are lines that have been found in screening in France to be almost free of vicine and convicine.

The basic chromosome numbers in the genus are n = 5, 6 and 7 and the great majority of the species have complements with $2n$ = 10,12 and 14 respectively. (Hanelt and Mettin, 1989). However, plants with these three basic chromosome numbers can exist even within one species as is the case for *V. sativa* (Hollings and Stace, 1974).

For some of the rarer *Vicia* species the genetic diversity is known to be low. In a study on *V. pisiformis* with RAPDs low inter- and intra-population variation was found, although the variation was higher when agronomic characters were observed (Black-Samuelson *et al.*, 1997). However, for many species there is little or no information available on inherent genetic diversity.

The genus contains species which are predominantly inbreeding and some species which are cross-fertilising (Becker-Dillingen,1929; Lechner, 1959; Zhang and Mosjidis, 1995). Even individuals belonging to the same species can have large differences in outcrossing rates, as has been observed in *V. faba* (Abdalla,1977; Porceddu, 1980).

Co-evolution with birds, grazing animals and insects are important factors that have not received sufficient research. With the disappearance of natural habitats, associated fauna also vanishes before there interaction can be observed and understood. Inbreeding endemics may depend on specialised pollinators to generate the necessary diversity to evolve with changing environmental conditions. Interesting examples are the plant-insect interaction associated with extra floral nectaries in *Vicia* subgenus *Vicia* (Koptur, 1979; Koptur, 1986). Ants, attracted to the nectar produced by extra floral nectaries, aid in the protection of the plant against other insect predators, as is particularly seen in *V. sativa* subspecies. Such an interaction can be employed with benefit as an additive component for biological crop protection, especially if plants can be selected for earlier nectar production.

Further research on co-evolutionary aspects could prove interesting. Factors that may find consideration in future co-evolutionary studies are hard-seededness, seed size, shattering, toxicity, palatability, flower size, insect mediated cross pollination, N-fixation and correlations between archaeological remains of animal and plant origin.

8.6 CONSERVATION RESOURCE STATUS

Relatively large *ex situ* seed collections exist of cultivated and wild *Vicia* species (Table 8.3), but there remain numerous gaps in conserved materials. The earliest collections are the accessions collected by Vavilov in the twenties (Vavilov, 1996) and the German expeditions during the Second World War in Albania and Greece (Flitner, 1995). The number of accessions collected over the years is considerable, however gaps still exist, particularly for the South American species. Those species of less immediate utilisation potential have not been systematically conserved *ex situ*. These species have been collected largely by Maxted and co-workers at the University of Southampton, and then the University of Birmingham, UK, in conjunction with IBPGR and ICARDA., and latterly, with ICARDA and CLIMA. All of the collected material has been duplicated at the Genetic Resources Unit at ICARDA, as well as in the country of collection. Detailed information about the current conservation status of *Vicia* species is documented by WIEWS (World Information and Early Warning System on Plant Genetic Resources), which contains information on national PGR holdings (www.fao.org/ag/agp/pgr/wiews/) and SINGER (System-wide Information Network for Genetic Resources), which contains information on CGIAR holdings (http://www.cgiar.org/singer).

Table 8.3. Number of *Vicia* accessions conserved *ex situ* in major collections.

Species	Institute							
	AARI	**ATFC**	**Bari**	**W-6**	**ICARDA**	**INIA**	**IPK**	**VIR**
V. articulata	1	26	19	14	21	83	14	11
V. bengalensis	0	58	49	34	19	0	9	44
V. ervilia	147	254	65	168	324	232	137	548
V. faba	77	1410	1461	564	5497	1064	1507	1261
V. monantha	0	15	19	8	197	0	10	3
V. narbonensis	22	267	71	48	224	17	69	54
V. pannonica	4	35	130	7	136	5	27	54
V. peregrina	12	4	29	29	259	13	9	21
V. sativa	370	724	1726	1178	2840	863	772	2252
V. villosa	35	110	275	177	342	143	23	263
Other *Vicia* species*	214 (87)	276 (39)	306 (57)	357 (35)	1105 (52)	15 (7)	226 (49)	168 (37)
Total	882	3179	4150	2584	11064	2435	2803	4679

* Numbers in brackets indicate the number of other *Vicia* species conserved. AARI = Aegean Agricultural Research Institute, Menemen, Turkey; ATFC = Australian Temperate Field Crop Collection, Horsham, Australia; Bari = Istituto del Germoplasma, Bari, Italy; W-6 = Regional Plant Introduction Station, Washington, USA; ICARDA = International Centre for Agricultural Research in the Dry Areas, Syria; INIA = Instituto Nacional de Investigaciones Agrarias, Madrid, Spain; IPK = Institut für Pflanzengenetik und Kulturpflanzenforschung (IPK), Gatersleben, Germany; VIR =Vavilov Research Institute of Plant Industry, St. Petersburg, Russia.

Although many annual *Vicia* species native to the Mediterranean region have been largely adequately conserved, the germination and growth to fruition of perennial species has proved a major problem at ICARDA and has resulted in the loss of accessions. Maintenance of the genetic integrity and purity of the accessions is an other area that needs attention. The purity of accessions can be difficult to judge in varietal mixtures, as these are frequently found with land races. At the IPK, new *Vicia* accessions are separated into visually homogeneous varieties and can thus be maintained free of admixture from other accessions. This is, of course not always possible, especially for outbreeding and thus continually segregating populations.

Natural out-crossing percentages are unknown for many species in the genus, but can be very high. With high natural out-crossing rates and lack of isolation procedures, the loss of genes through out-crossing and natural selection at the seed multiplication site could be considerable. Faba bean is a partially out-crossing species, with cross pollination the result of insect pollination. Cross pollination has been reported to range from 8-94%, with an average of 35% (Bond and Poulsen, 1983). This is a critical issue for all stages of germplasm activities. Out-crossing necessitates that important decisions be made concerning the method used for collections in the field and the way these collections are rejuvenated.

ICARDA maintains the genetic resources of faba bean in two types of germplasm collections (Robertson, 1985). The original germplasm accessions, usually heterogeneous and heterozygous to some extent, were originally maintained as populations and are now maintained as composite bulks in the land race, or ILB (International Legume Faba Bean) collection. If this is done without pollination control there is danger of the loss of genetic identity resulting from inter-crossing among different accessions. A faba bean pure line collection (BPL accessions, Faba Bean Pure Line collection) has been derived from the ILB collection by a 'pure-breeding' process of single plant progeny rows taking randomly selected single plants to progeny rows in a cyclic manner to obtain uniformity using insect-proofed screen-houses to ensure selfing. The faba bean pure lines (BPLs) have the advantages of : (a) ease of maintenance; (b) repeatability and uniformity of evaluation; (c) reduction in loss due to genetic drift; and (d) uncovering of recessive genes which otherwise might be hidden by heterozygosity. Repeated inbreeding has not led to loss in vigour due to inbreeding depression.

Trait-specific gene-pools (TSGs) were proposed for faba bean by Witcombe (1982), where accessions similar for such traits as maturity, seed size, height, and growth habit are bulked and maintained by growing in isolation to allow inter-crossing. TSG's would allow the reduction of a large number of accessions to a few TSGs without loss of much genetic variability, as variability is not randomly distributed and much germplasm may be duplicated. This method has many advantages of the core collection. There is merit in this approach, however, one limitation is the need for good evaluation data for the material to bulk. Multivariate techniques would be useful to group accessions to be selected for such TSGs.

Burton (1979) proposed that cross-pollinated germplasm accessions be maintained as self-pollinated bulks. This type of collection would contain accessions that are a bulk of homozygous lines, i.e., each line is heterogeneous but homozygous. It should be better than testing a fixed number of pure lines from each open-pollinated accession, in that it might better represent the full range of variability in each original accession. A more refined way to maintain this type of collection would be to self a large number of plants of each accession and take one seed or pod to bulk to produce the next-cycle seed (the Single Seed Descent Method designed by Brim, 1966). This type of collection of selfed composites would maintain the maximum variability with the minimum chance of loss of genes because of inter-crossing among accessions. Also, it would allow the detection of recessive genes in heterogenous populations that are bulks of homozygous genotypes. The ILB collection at ICARDA is evolving into this type of collection as the multiplication phase has been switched to insect-proofed screen-houses because of the impracticality of growing even a small number of accessions in isolation.

About 198 strains of *Rhizobium* collected from nodules of *Vicia* species are conserved by several international research centres (ICARDA, 1999). The vast majority of these strains were collected from nodules of *V. faba* (134) and *V. sativa* (28). Both the conservation of these strains and the collection of new strains that have evolved in a symbiotic relation with other *Vicia* species need more concerted effort.

There has been no attempt to systematically conserve *Vicia* genetic diversity *in situ* in either genetic reserves or on farm, although Maxted (1995) proposed the establishment of reserves in the Syria and Turkey. Undoubtedly, existing genetic reserves (e.g. Ammiad in Eastern Galilee, Israel; Kaz Dag, Aegean Region, Ceylanpinar of Southeastern Turkey, and Amanos, Mersin in Turkey) and other forms of protected areas throughout the range of the genus contain vetch species, but here the conservation is 'passive' (species and genetic diversity is not being monitored and managed) and therefore susceptible to further unobserved genetic erosion. There is an urgent need to make positive steps to establish both reserves for the wild species of *Vicia* and on farm projects to conserve the ancient land races of cultivated vetches.

8.7 CONSERVATION GAPS AND PRIORITIES

There is a need for better information management of current conserved genetic diversity in *Vicia* to permit more effective genetic conservation and to identify more precisely gaps in the current collections. One of the major gaps in the collections is the lack of passport data. At ICARDA, even the country of origin is not known for more than 2000 faba bean accessions, approximately 21% of the entire collection. For many other accessions there is very little information available. Efforts are being extended to rectify as much as possible the lack of passport data availability in order to make their use more efficient.

Priority areas and species for future collection missions can then easily be established. Perhaps one way of doing this would be to ensure the information held in the WIEWS database is comprehensive and combined with that in the SINGER system provides an effective tool for locating genetic diversity. The *Vicia* database of the European Cooperative Programme for Crop Genetic Resources Networks (ECP/GR) held at the Institute of Germplasm, Bari (Maggioni, 1998) could also be used in this manner. For a fully comprehensive system, there is still a need to produce, preferably online, catalogues for *Vicia* cultivars and species along with characterisation and evaluation data.

It is undoubtedly true that the intensification of agriculture around the Mediterranean is leading to serious genetic erosion of *Vicia* diversity (IBPGR, 1985). The clearance of new cultivated land and the associated decline of permanent pastures have lead to the disappearance of habitats for wild climax community species. The introduction of high input farming and new varieties which replace traditional land races for many of the cultivated species is also threatening diversity of weedy and crop species. In Albania, for example, all land races of *V. sativa* have already been lost (Hammer *et al.*, 1996). In South Italy and Sicily many of the land races of *V. faba* (Hammer *et al.*, 1996) and all land races of *V. articulata* have disappeared (Hammer and Perrino, 1995). Collecting expeditions in the 1950's (Whyte, 1958) and 60's focussed mainly on the collection of land races and much crop material is therefore already conserved *ex situ* providing the collection remain extant.

A major gap, however, that needs to be addressed in the faba bean collections available for use in the Mediterranean region is the relatively small number of accessions from China, which is the major producer of faba bean. Other areas that are under-represented are India, Pakistan and Nepal, though Nepal has been just collected in 1995. In the Mediterranean region there is a need for more collection in Algeria, Egypt and Yemen. The major gap in required traits in the faba bean germplasm collections is the lack of adequate resistance sources to *Orobanche crenata* Forsk. Another concern would be to broaden the genetic base of the resistance sources to chocolate spot (*Botrytis fabae*). Also for crops such as *V. articulata*, *V. bengalensis* and its closely related *V. atropurpurea* more germplasm is needed since the total number of accessions available for breeding is relatively low.

Careful analysis of the global *ex situ* collections will reveal the gaps in the collections of land races and a major effort should be focussed on their immediate collection and *ex situ* conservation. Weedy *Vicia* species, which are often associated with traditional farming systems are also disappearing with the changes in land usage. Therefore, attention also needs to be paid to the systematic conservation of these species.

Besides the need to systematically conserve the *Vicia* species associated with cultivated or disturbed land there is a need to conserve those found in climax communities. These species are often perennials and as noted above, the majority of *ex situ* collection effort has focussed on collecting annual species from disturbed habitats. *V. pisiformis* is, for example, a perennial, forest species that is rare and requires urgent conservation attention because of the rapid destruction of its natural habitat (Black-Samuelson *et al.* 1997).

Those species that are not cultivated or are not of immediate exploitation potential are under-represented in *ex situ* collections, but as the majority of countries in the Mediterranean region have signed the Convention on Biological Diversity (CBD, 1992) they are obliged to ensure these species are also conserved. Ehrman and Cocks (1990) recommended that collection efforts should especially be focussed on the species of the drier areas with high population pressures, as those appear to be the most threatened. Some of the wild species might in their own right have agronomic potential if evaluated or be potential gene donors to related *Vicia* and other crops. The genus harbours many potential fodder and grain crops (Kernick, 1978; Hanelt and Mansfeld ,1986; Enneking, 1995), which have not yet been tried and tested, e.g. *V. hyaeniscyamus* (Maxted *et al.*, 1989). It is therefore important that a broad range of species are secured in gene banks from throughout the region. For species with very small *ex situ* conserved gene pools, populations with high diversity need to be identified and sampled more thoroughly for *ex situ* conservation to broaden the genetic base available for plant breeding.

If a national rather than taxonomic view is taken there are certain countries that remain under-collected, notably Albania, Spain, the former Yugoslavia, Turkey, Lebanon, Iran, Afghanistan and many of the Mediterranean islands. Even though some of these countries, particularly Spain and Turkey, have been extensively collected in the past, Ferguson *et al.* (1998) demonstrated, for the wild lentils (*Lens* species), that these countries still do not have an adequate genetic representation of diversity conserved.

An inventory of wild *Vicia* species that occur in national parks would give an indication of the number of *Vicia* species present in protected areas and would assist active *in situ* conservation. Maxted (1995) concluded that the most seriously threatened by extinction species are those restricted to Syria, Lebanon, Turkey and Israel, and the highest concentration of potentially threatened taxa are located in Syria. He recommended the establishment of four genetic reserves for vetch diversity in Syria: Ain Dinar, Al Hasakah, Kessab town, Kessab, Qal'at Al Hosn, Homs and Mimas, Djebel Druze, as well as in the Olimpos Beydaglari National Park,

Belin, Turkey. Unfortunately, these suggestions have been largely ignored by the appropriate national agencies. It was possible to make these detailed recommendation because of the authors extensive study of *Vicia* in the region. Therefore it is likely that other areas of the Mediterranean also warrant the establishment of genetic reserves, possibly in Spain, Greece and the Caucasus, where the species have not been so closely studied.

8.8 Research Needs

Compared with many other forage legume genera *Vicia* is relatively well studied, there have been several ecogeographic studies in recent years, the taxonomy is fairly well understood and there has been a systematic attempt to conserve the genus. However, the ecogeographic characterisation of genetic diversity within *Vicia*, originally proposed by Vavilov (1926) and studied by Tupikova (1926) should be completed. A lot of information on *Vicia* is buried in the multilingual scientific and grey literatures.

Taxonomically, subgenus *Vicilla* still requires further study; especially the South American and East Asian species, which remain poorly understood. Subgenus *Vicia* has been extensively studied and the relationship between the known species are reasonable well understood, apart from the *V. sativa* complex, which is being currently studied (van de Wouw and Maxted, pers. comm.). The recent discoveries of new species closely related to faba bean (Maxted *et al.*, 1989; Maxted, 1993b) suggest that a wild progenitor of faba bean may still be found. This progenitor species would be very interesting to plant breeders, if in fact it was to exist and so the search should be continued.

Associated with the need for fresh collections is the necessity to be able to identify the wild material in the field. Although good traditional keys exist, many conservationists still find these difficult to use and there is an on-going need for more user friendly multi-access keys for *Vicia* and for other legume genera.

Perhaps the major research need is associated with ensuring the utilisation of existing conserved accessions. The diversity present in collections needs to be characterised and evaluated in detail by using both molecular techniques and agro-morphological data. As well as aiding utilisation and maintenance of collections, these data, combined with geographic information systems techniques, could be used to identify gaps in the *ex situ* collections and areas for possible *in situ* conservation. The Mediterranean area, particularly the Fertile Crescent region, is an important centre of diversity for temperate legumes and several other crops groups. There is a serious need to highlight and monitor the amount of genetic erosion in the region.

Biological constraints to faba bean yield include such factors as pollination and seed set (auto-fertility) and flower and pod drop. In some areas susceptibility to frost damage is a major problem. There are exciting new plant types which may alleviate some of the traditional problems of excessive vegetative growth of faba bean with the accompanying large drop of flowers and young pods and the large amount of lodging. The determinate genes and independent vascular supply types of faba bean (eliminating flower and pod shedding) are now available in Mediterranean type germplasm through introgression with land races from this region, offering the possibility for their use in the major producing areas for faba bean. These types need to be put in a more suitable background to allow a positive effect on total productivity.

Many pests limit yields in various regions. While there has been considerable success in finding resistance to fungal pathogens in faba bean, additional sources need to be found to have a broader genetic base for resistance. Another limitation is the need for more sources and combinations of multiple disease and pest resistance as the crop is generally subject to more than one stress at a time.

The amount of genetic erosion within *ex situ* collections is an other area that needs research attention and adequate funding to prevent. The sample size of conserved accessions is generally determined by practical considerations such as available storage, space for regeneration and labour requirements, while genetic considerations are left aside. Many conserved accessions in gene banks have been held over long time periods and it is known that they are not always being regenerated at appropriate intervals (FAO, 1998). Therefore, priority should be given to ensure these materials are tested for viability and, if necessary, regenerated to avoid any unnecessary loss of genetic diversity. It is also important to maintain the genetic integrity and purity of accessions. The problems of loss of genetic integrity might be quantified by comparing the composition of material donated to other collections with that kept at the donor institute.

Before appropriate regeneration can take place there is a need to establish the breeding system of the species. If the species is out crossing then regeneration its self can lead to unnecessary loss of genetic diversity. Therefore it is important to know if a species is an inbreeder or outbreeder. This information is not available for the majority of *Vicia* species. Much can be learned from looking at flower anatomy, diversity of characters within accessions and by screening for molecular markers. It is also important to ensure that conserved accessions are duplicated between at least two gene banks as a safety precaution against accidental loss of collections.

In recent years there has been some interest in reviewing the agronomic potential of previously undomesticated *Vicia* species (e.g. *V. noeana*, *V. hybrida*, *V. sativa* subsp. *macrocarpa*). As many of the wild species have only recently been conserved *ex situ* it is now appropriate to undertake a more systematic review of forage and fodder potential of the entire range of *Vicia* species, both for new species for cultivation but also for genes or gene complexes that might prove of use to agriculture.

REFERENCES

Aarssen, L.W., Hall, I.V. and Jensen, K.I.N., 1986. The biology of Canadian weeds .76. *Vicia angustifolia* L., *Vicia cracca* L., *Vicia sativa* L, *Vicia tetrasperma* (L.) Schreb and *Vicia villosa* Roth. *Canadian Journal of Plant Science*, 66: 711-737.

Abdalla, M.M.F., 1977. Intraspecific unilateral incomparability in *Vicia faba* L. *Theoretical and Applied Genetics*, 50: 227-233.

Abd El Moneim, A.M.; Khair, M.A., and Cocks, P.S., 1990. Growth analysis, herbage and seed yield of certain forage legume species under rainfed conditions. *Journal of Agronomy and Crop Science -Zeitschrift fuer Acker- und Pflanzenbau* 164: 34-41

Baki, A.A.; Teasdale, J.R., and Korcak, R.F., 1997. Nitrogen requirements of fresh-market tomatoes on hairy vetch and black polyethylene mulch. *Horticultural Science,* 32 (2): 217-221

Barulina, H., 1930. *Lentils of the USSR and of other countries - A botanic-agronomical monograph.* Leningrad

Becker-Dillingen, J., 1929. *Handbuch des Hülsenfruchterbaues und Futterbaues. Vol. 3.* Hülsenfruchterbau and Futterbau Handbuch des gesamten Pflanzenbaues einschließlich der Pflanzenzüchtung. Berlin: Paul Parey

Bellido, L., 1994. Grain legumes for animal feed. In: *Neglected crops. 1492 from a different perspective* (eds. Hernandez Bermejo, J. E. and Leon, J.). *Plant production and protection series No 26.* Rome: FAO

Bennett, S.J. and Maxted, N., 1997. An ecogeographic analysis of the *Vicia narbonensis* complex. *Genetic Resources and Crop Evolution*, 44: 411-428.

Black-Samuelsson, S., Eriksson, G., Gustafsson, L. and Gustafsson, P., 1997. RAPD and morphological analysis of the rare plant species *Vicia pisiformis* (Fabaceae) . *Biological Journal of the Linnean Society*, 61: 325-343.

Bond, D.A. and M. Pope. 1987. Proportion of cross-bred and selfed seed obtained from successive generations of winter bean (*Vicia faba* L.) crops. *Journal of Agricultural Science, Cambridge*. 108:103-108.

Bond, D.A. and Poulsen, M.H., 1983. Pollination. In: *The Faba Bean* (ed. Hebblethwaite, P.D.). pp.77-101. London: Butterworth.

Brim, C.A. 1966. A modified pedigree method of selection in soybeans. *Crop Science*, 6:220.

Bueno Perez, A., 1979. *Taxonometria y cariologia en el genero* Vicia. Ph.D. Thesis, Instituto Nacional De Investigaciones Agrarias, Madrid.

Burton, G.W. 1970. Handling cross-pollinated germplasm efficiently. *Crop Science*, 19: 685-690.

Chrtkova-Zertova, A., 1979. *Flora Iranica, Vol. 140 - Papilionaceae 1 -* Vicia *and* Faba. pp. 16-57. Akademische Druck- u. Verlagsanstalt, Graz, Austria.

CBD, 1992. *Convention on Biological Diversity: Text and Annexes*. pp. 1-34. Secretariat of the Convention on Biological Diversity, Montreal.

Clos, D., 1898. Les *Vicia narbonensis* L. et *serratifolia* Jacq. espèces autonomes. *Bulletin de Societe Botanique de France*, 45 (3. Sér. 5): 380-385

Cubero, J.I. 1972. On the evolution of *Vicia faba* L. *Theoretical and Applied Genetics*, 45:48-51.

Cubero, J.I. and M.J. Suso. 1981. Primitive and modern forms of *Vicia faba*. *Kulturpflanze*, 29: 137-145

Davis, P.H. and Plitmann, U., 1970. *Vicia* L. In: *Flora of Turkey and the East Aegean Islands, Vol. 3* (ed. Davis, P.H.). pp. 274-325. Edinburgh University Press, Edinburgh

Duc, G., 1997. Faba bean (*Vicia faba* L.). *Field Crops Research*, 53, 99-109.

Ehrman, T. and Cocks, P.S., 1990. Ecogeography of annual legumes in Syria: distribution patterns. *Journal of Applied Ecology*, 27: 578-579.

Ehrman, T. and Maxted, N., 1989. Ecogeographic survey and collection of Syrian Vicieae and Cicereae (Leguminosae). *FAO/IBPGR Plant Genetic Resources Newsletter*, 77: 1-8.

El-Sherbeeny, M. and L.D. Robertson, 1992. Protein content variation in a pure line faba bean (*Vicia faba* L.) collection. *Journal of Science Food and Agriculture*, 58: 193-196.

Enneking, D., 1995. *The toxicity of Vicia species and their utilisation as grain legumes. Centre for Legumes in Mediterranean Agriculture Occasional Publication* 6. CLIMA, Perth.

Enneking, D., Giles, L.C., Tate, M.E. and Davies, R.L., 1993. L-Canavanine: A natural feed intake inhibitor for pigs (isolation, identification and significance). *Journal of the Science of Food and Agriculture*, 61 (3): 315-325

Enneking, D., Lahlou, A., Noutfia, A. and Bounejmate, M., 1995. A note on *Vicia ervilia* cultivation, utilization and toxicity in Morocco. *Al Awamia*, 89:141-148

Enneking, D. and Maxted, N., 1995. Narbon bean: *Vicia narbonensis* L. (Leguminosae). In: *Evolution of crop plants*, 2nd edition (eds. Smartt, J. and Simmonds, N.W.). pp. 316-321. Longman Group, Harlow, Essex

Erskine, W., Smartt, J. and Muehlbauer, F.J., 1994. Mimicry of lentil and the domestication of common vetch and grass pea. *Economic Botany*, 48: 326-332.

Esteban, J., 1996. El Yero [Bitter vetch]. In: *El cultivo de las Leguminosas de grano en Castilla y León*, Valladolid: Junta de Castilla y León (eds. Franco Jubete, F. and Ramos Monreal, A.).). pp. 161-193. Consejeria de Agricultura y Ganadería.

FAO, 1998. *The state of the world's plant genetic resources for food and agriculture*. pp. 1-510. Food and Agriculture Organisation of the United Nations, Rome.

FAO 1999. *FAO production yearbook vol.52 1998, FAO statistics series no 135.* Food and Agriculture Organisation of the United Nations, Rome.

Ferguson, M.E., Ford-Lloyd, B.V., Robertson, L.D., Maxted, N. and Newbury, H.J., 1998. Mapping of geographical distribution of genetic variation in the genus *Lens* for enhanced conservation of plant genetic diversity. *Molecular Ecology*, 7: 1743-1755.

Fischer, A., 1937. Züchtung und Anbau der Ervilie, *V. ervilia* und der Erve, *V. monantha* als Futterpflanzen. *Züchter*, 9: 286-288.

Fischer, A., 1938. Die geographischen Grundlagen der Züchtungsforschung bei der Gattung *Vicia. Züchter*, 10 (2): 51-56

Flitner, M., 1995) *Sammler, Räuber und Gelehrte. Die politischen Interessen an pflanzengenetischen Ressourcen 1895-1995.* Campus, Frankfurt

Foury, A., 1950. Les légumineuses fourragères au Maroc (premiére partie*). Les Cahiers De La Recherche Agronomique, INRA, Rabat* 3: 25-286

Foury, A., 1954. Les légumineuses fourragères au Maroc (deuxiéme partie). *Les Cahiers De La Recherche Agronomique, INRA, Rabat* 5: 287-658

Jubete, F., 1996. La Algarroba [Single flowered vetch, Spanish lentil]. In: *El cultivo de las Leguminosas de grano en Castilla y León, Valladolid: Junta de Castilla y León* (eds. Franco Jubete, F. and Ramos Monreal, A.). pp. 247-251. Consejeria de Agricultura y Ganadería.

Francis, C.M., Enneking, D. and Abd El Moneim, A., 1998. When and where will vetches have an impact as grain legumes. In: *Paper presented at III Intl. Food Legume Conference, Adelaide 1997*, Adelaide .

Frankel, O.H., 1977. Natural variation and its conservation. In: *Genetic diversity in plants* (eds. Muhammed, A., Aksel, R.and von Borstel, R.C.). pp. 21-44. Plenum Publishing Corporation.

Gams, H., 1924. *Vicia* L. In: *Illustrierte Flora von Mitteleuropa* (ed. Hegi, G.), Vol. 4(3): 1506-1562. Lehmann, München.

Hammer, K., Knüpffer, L., Xhuveli, L. and Perrino, P., 1996. Estimating genetic erosion in landraces - two case studies. *Genetic Resources and Crop Evolution*, 43: 329-336.

Hammer, K. and Perrino, P., 1995. Plant genetic resources in south Italy and Sicily - studies towards *in situ* and on-farm conservation. *Plant Genetic Resources Newsletter*, 103: 19-23.

Hanelt, P., 1972. Die infraspezifische Variabilität von *Vicia faba* L. und ihre Gliederung. *Kulturpflanze*, 20: 75-128.

Hanelt, P. and Mansfeld, R., 1986. *Vicia* L. In: *Verzeichnis landwirtschaftlicher und gärtnerischer Kulturpfllanzen (ohne Zierpflanzen)* (ed. Schultze-Motel, J.), 2nd revised and extended edn., Vol. 2, pp. 615-630. Springer, Berlin.

Hanelt, P. and Mettin, D., 1989. Biosystematics of the genus *Vicia* L., Leguminosae). *Annual Review of Ecology and Systematics*, 20: 199-223.

Hanf, M., 1990. *Ackerunkräuter Europas mit ihren Keimlingen und Samen*. 3rd revised edn. BLV Verlagsgesellschaft, Munich

Hanounik, S.B. and Robertson, L.D., 1987. New sources of resistance in *Vicia faba* to chocolate spot caused by *Botryis fabae*. *Plant Disease*, 72: 696-698.

Hanounik, S.B. and Robertson, L.D., 1988. Resistance in *Vicia faba* germplasm to blight caused by *Ascochyta fabae*. *Plant Disease*, 73: 202-205.

Hawtin, G.C. and Hebblethwaite, P.D.,. 1983. Background and history of faba bean production. In: *The Faba Bean (*Vicia faba *L.): A Basis for Improvement*, (ed. Hebblethwaite, P.D.). Pp. 3-22. Butterworths, London, U.K.

Hermann, F. J., 1960. *Vetches of the United States, native, naturalised and cultivated*, pp. 1-84. USDA Agricultural Handbook No. 168, Washington DC.

Hollings, E. and Stace, C.A., 1974. Karyotype variation and evolution in the *Vicia sativa* aggregate. *New Phytologist*, 73: 195-208.

Hopf, M., 1970. Zur Geschichte der Ackerbohne (*Vicia faba* L.). *Jahrbuch des Römisch-Germanischen Zentralmuseums*, 17: 306-322.

Hopf, M., 1986. Archaeological evidence of the spread and use of some members of the Leguminosae family. In *The origin and domestication of cultivated plants* (ed. Barigozzi, C.). Symposium, 25-27 November 1985. pp. 35-60. Elsevier Amsterdam.

IBPGR, 1985. *Forages for Mediterranean and adjacent arid / semi-arid areas: Report of a working group meeting held at Limossal, Cyprus*. International Board for Plant Genetic Resources, Rome, Italy.

ICARDA, Legume Program 1994. Forage Legume Improvement. In: *ICARDA Legume Program Annual Report 1993*. pp. 139-239. Aleppo, Syria: ICARDA.

ICARDA, 1999. *SGRP database of nitrogen fixing germplasm maintained in CGIAR centres: CIAT, ICARDA, IITA, ILRI and IRRI, version 0.1*. ICARDA, Aleppo.

ILDIS, 1999. *International Legume Database and Information Service. http://www.ildis.org/*

Jaaska, V., 1997. Isoenzyme diversity and phylogenetic affinities in *Vicia* subgenus *Vicia* (Fabaceae). *Genetic Resources and Crop Evolution*, 44: 557-574.

Keatinge, J.D.H., Asghar, A., Roidar Khan, B., Abd El Moneim, A.M., and Ahmad, S., 1991. Germplasm evaluation of annual sown forage legumes under environmental conditions marginal for crop growth in the highlands of West Asia. *Journal of Agronomy and Crop Science -Zeitschrift für Acker und Pflanzenbau*, 166: 48-57.

Kernick, M.D., 1978. *Indigenous arid and semi-arid forage plants of North Africa, the Near and the Middle East.* EMASAR Phase II. Ecological Management of Arid and semi-arid rangelands in Africa, the Near and Middle East No. IV, Rome: FAO.

Khattab, A., Maxted, N. and Bisby, F.A., 1988. Close relatives of the faba bean from Syria : a new species of *Vicia* and notes on *V. hyaeniscyamus* (Leguminosae). *Kew Bulletin*, 43(3): 535-540.

Koptur, S., 1979. Facultative mutualism between weedy vetches bearing extrafloral nectaries and weedy ants in California. *American Journal of Botany*, 66: 1016-1020.

Koptur, S., 1986. Interactions between vetches bearing extrafloral nectaries, their biotic protective agents and herbivores. *American Journal of Botany*, 73: 671-675.

Kupicha, F.K., 1974. *Taxonomic studies in the tribe Vicieae (Leguminosae).* Ph.D. thesis, University of Edinburgh

Kupicha, F.K., 1976. The infrageneric structure of *Vicia*. *Notes from the Royal Botanic Garden Edinburgh*, 34: 287-326.

Kupicha, F.K., 1981. Vicieae. In: *Advances in Legume Systematics* (eds. Polhill, R.M. and Raven, P.M.). pp. 377-381. Royal Botanic Gardens, Kew.

Ladizinsky, G., 1989. Origin and domestication of the Southwest Asian grain legumes. In: *Foraging and Farming* (eds. Harris, D.H. and Hillman, G.C.). pp. 376-389. Unwin Hyman, London.

Lawes, D.A., Bond, D.A. and Poulsen, M.H. , 1983. Classification, origin, breeding methods and objectives. In: *The Faba Bean (*Vicia faba *L.): A Basis for Improvement*, (ed. Hebblethwaite, P.D.). pp. 23-76. Butterworths, London, UK.

Lechner, L., 1959. Wicken-(*Vicia*-) Arten. In: *Handbuch der Pflanzenzüchtung Bd. IV. Züchtung der Futterpflanzen* (eds. Kappert, H. and Rudorf, W.). pp. 52-95. Paul Parey, Berlin.

Link, W., Dixkens, C., Singh, M., Schwall, M. and Melchinger, A.E., 1995. Genetic diversity in European and Mediterranean faba bean germ plasm revealed by RAPD markers. *Theoretical and Applied Genetics*, 90: 27-32.

Linnaeus, C., 1753. *Species Plantarum*, Vol 2. Salvius, Stockholm.

Maggioni, L., 1998. ECP/GR to continue into a sixth phase. *IPGRI Newsletter for Europe*: 14: 7.

Marquardt, R.R. 1982. Favism. In: *Faba bean Improvement: World Crops: Production, Utilization, Description, Vol. 6* (Hawtin, G.C. and Webb, C.). pp. 343-353 Martinus Nijhoff, The Hague, The Netherlands.

Mateo-Box, J.M., 1961. *Leguminosas de grano.* Ed. Salvat, Barcelona.

Maxted, N., 1988. A new species of *Vicia* from South-West Turkey. *Notes From Edinburgh Botanical Gardens*, 45(3): 453-456.

Maxted, N., 1993a. A phenetic investigation of *Vicia* L. subgenus *Vicia* (Leguminosae, Vicieae). *Botanical Journal of the Linnean Society*, 111: 155-182.

Maxted, N., 1993b. Towards a faba bean progenitor! *FABIS*, 31: 3-8.

Maxted, N., 1995. *An ecogeographical study of* Vicia *subgenus* Vicia. Systematic and Ecogeographic Studies on Crop Genepools 8. International Plant Genetic Resources Institute, Rome, Italy.

Maxted, N., Callimassia, M.A. and Bennett, M.D., 1991b. Cytotaxonomic studies of eastern Mediterranean *Vicia* species (Leguminosae). *Plant Systematics and Evolution*, 177: 221-234.

Maxted, N., Khabttab, A. and Bisby, F.A., 1991a. Domesticated legumes and their wild relatives: newly discovered relatives of *Vicia faba* do little to resolve the enigma of its origin. *Botanika Chronika*, 10: 129-159.

Maxted, N., Obari, H. and Tan, A., 1989. Vicieae (Leguminosae): new and interesting endemic species from the eastern Mediterranean. *FAO/IBPGR Plant Genetic Resources Newsletter*, 78/79: 21-25.

Muratova, V.S., 1931. Common beans (*Vicia faba* L.). *Bulletin of Genetics and Plant Breeding, Leningrad)*, 50: 248-298.

Nikiforova, O.D., 1988 . *Dikorastushchie viki Sibiri. (Wild vetches of Siberia)*. Nauka, Novosibirsk, USSR.

Plitmann, U., 1967. *Biosystematical study in the annual species of* Vicia *of the Middle East*. The Hebrew University of Jerusalem, Jerusalem.

Plitmann, U., 1973. Biological flora of Israel. 4: *Vicia sativa* subsp. *amphicarpa* (Dorth.) Aschers. and Graebn. *Israel Journal of Botany*, 22: 178-194.

Porceddu, E., Monti, L.M., Furscianti, L. and Volpi, N., 1980. Analysis of cross pollination in *Vicia faba* L. *Zeitschrift fuer Pflanzenzüchtung*, 84: 313-322.

Potokina, E.K., 1997. *Vicia sativa* L. aggregate (Fabaceae) in the flora of former USSR. *Genetic Resources and Crop Evolution*, 44: 199-209.

Radzhi, A.D., 1971. Conspectus systematis specierum Caucasicarum Generis *Vicia* L. *Novititates Systematic Plant Vascular, (Leningrad)*, 7: 228-240.

Robertson, L.D., 1985. Faba bean germplasm collection, maintenance, evaluation, and use. In: *Faba Beans, Kabuli Chickpeas, and Lentils in the 1980s* (eds. Saxena, M.C. and Verma, S.). pp. 15-21 ICARDA, Aleppo, Syria.

Robertson, L.D. and El-Sherbeeny, M.H., 1988. *Faba Bean Germplasm Catalog: Pure Line Collection*. Pp. 1-140. ICARDA, Aleppo, Syria.

Robertson, L.D. and El-Sherbeeny, M.H., 1995. Autofertility in a pure line faba bean (*Vicia faba* L.) germplasm collection. *Genetic Resources and Crop Evolution*, 42:157-163.

Robertson, L.D. and Saxena, M.C., 1993. Problems and prospects of stress resistance breeding in faba bean. In *Breeding for stress tolerance in cool-season food legumes* (eds. Singh, K.D. and Saxena, M.C.). pp. 37-50. John Wiley and Sons, Chichester, UK.

Satovic, Z., Torres, A. M. and Cubero, J. I., 1996. Genetic mapping of new morphological, isozyme and RAPD markers in *Vicia faba* L. using trisomics. *Theoretical and Applied Genetics*, 93: 1130-1138.

Saxena, M.C., Abd El Moneim, A.M., and Ratinam, M., 1993. Vetches (*Vicia* spp.) and chicklings (*Lathyrus* spp.) in the farming systems in West Asia and North Africa and improvement of these crops at ICARDA. Potential for Vicia and Lathyrus species as new grain and fodder legumes for southern Australia. In: *CLIMA, Proceedings of the Vicia/Lathyrus workshop, held 22-23.9.1992 in Perth, Western Australia*. Perth, Western Australia. Co-operative Research Center for Legumes in Mediterranean Agriculture (CLIMA), Occasional Publication No.1: 1-7.

Schäfer, H.I., 1973. Zur Taxonomie der *Vicia narbonensis* - Gruppe. *Kulturpflanze*, 21: 211-273.

Sonnante, G., Piergiovanni, A.R. and Pignone, D., 1997. Distribution of allozyme variation in *Vicia benghalensis* (Leguminosae) suggests the existence of two genepools. *Plant Systematics and Evolution*, 207: 99-110.

Stankevich, A.K., 1978. K sistematike politipnogo vida *Vicia sativa* L.. *Bulletin VIR*, 81: 3-11.

Stankevich, AK., 1982. On taxonomic position of some sections of the genus *Vicia* L. *Bulletin of Applied Botany, Genetics and Plant Breeding (Leningrad)*, 72: 21-27.

Stebbins, G.L., 1967. Adaptive radiation and trends of evolution in higher plants. *Evolutionary Biology*, 1: 101-140.

Takhtadjan, A., 1969. *Flowering plants: origin and dispersal.* Oliver and Boyd, Edinburgh and London.

Tao, Z.H., 1981. Faba bean production and research in China. *FABIS Newsletter*, 3: 24.

Tate, M.E. and Enneking, D., 1992. A mess of red pottage. *Nature*, 359: 357-358.

Tate, M.E., Rathjen, J., Delaere, I., and Enneking, D., 1999. Covert trade in toxic vetch continues. *Nature*, 400 (6742): 207.

Torres, A.M., Weeden, N.F. and Martin, A., 1993. Linkage among isozyme, RFLP and RAPD markers in *Vicia faba*. *Theoretical and Applied Genetics*, 85: 937-945.

Tournefort, J.P. de, 1694. *Elements de botanique. Tom 1.* l'Imprimerie Royale, Paris.

Townsend, C.C., 1974. *Flora of Iraq, Vol. 3.* Ministry of Agriculture and Agrarian Reform of Iraq, Baghdad.

Tupikova, A., 1926. Botanico-agronomical investigations of annual vetches [Botaniko-agronomiceskoe issledovanie odnoletnych vik]. *Bulletin of Applied Botany, Genetics and Plant Breeding (Leningrad)*, 16(1): 151-246

van de Ven, M., Powell, W., Ramsay, G. and Waugh, R., 1990. Restriction fragment length polymorphisms as genetic markers in *Vicia*. *Heredity*, 65: 329-342.

Vavilov, N. I., 1926. Weeds as progenitors of cultivated plants. *Bulletin of Applied Botany, Genetics and Plant Breeding (Leningrad)*, 16: 195-217.

Vavilov, N. I., 1996. *Five continents* (New Translation). IPGRI, Rome.

Villax, E. J., 1963. La culture des plantes fourrageres (dans la region méditerranéenne occidentale, Maroc- Portugal- Tunisie- Algerie- Espagne- France). *Les Cahiers De La Recherche Agronomique, INRA, Rabat* 17: 1-641.

Von Kittlitz, E., Ibrahim, K.I.M., Ruckenbauer, P. and Robertson, L.D., 1993. Analysis and use of interpool crosses (Mediterranean x central European) in faba beans (*Vicia faba* L.). *Plant Breeding-Zeitschrift Für Pflanzenzüchtung*, 110: 307-314.

Whyte, R.O., 1958. *Plant exploration, collection, and introduction.* FAO, Plant Production Branch, Agriculture Division, Rome.

Williams, P.C., Bhatty, R.S., Deshpande, S.S., Hussein L.A. and Savage, G.P., 1994. Improving nutritional quality of cool season food legumes. In: *Expanding the Production and Use of Cool Season Food Legumes* (Muehlbauer F.J. and Kaiser, W.J.). pp. 113-129. Kluwer, Dordrecht, The Netherlands.

Willis, J.C., 1921. *Age and area; a study in geographic distribution and origin of species.* Cambridge University Press, Cambridge, U.K.

Witcombe, J., 1982. Genetic resources of faba beans. In: *Faba Bean Improvement.* Proceedings of the ICARDA/IFAD Nile Valley Project Conference (Hawtin, G.C. and Webb, C.). pp. 1-13. Martinus Nijhoff, The Hague, The Netherlands.

Zeven, A.C. and Zhukowsky, P.M., 1975. *Dictionary of cultivated plants and their centres of diversity.* CAPD, Wageningen.

Zhang, X. and Mosjidis, J.A., 1995. Breeding systems of several *Vicia* species. *Crop Science*, 35: 1200-1202.

Zohary, D., 1989. Domestication of the Southwest Asian Neolithic crop assemblage of cereals, pulses, and flax: the evidence from the living plants. In: *Foraging and Farming: the Evolution of Plant Exploitation* (eds. Harris, D. R. and Hillman, G. C.). pp. 358-373. Unwin Hyman Ltd, London.

Zohary, D. and Hopf, M., 1988. *Domestication of plants in the old world*, Clarendon, Oxford.

Zohary, D. and Plitmann, U., 1979. Chromosome polymorphism, hybridization and colonization in the *Vicia sativa* Group (Fabaceae). *Plant Systematics and Evolution*, 131: 143-156.

Chapter 9

Grasspea and Chicklings (*Lathyrus* L.)

Ashutosh Sarker, Ali Abd El Moneim and Nigel Maxted

9.1 INTRODUCTION

The genetic diversity of the genus *Lathyrus* is of immense significance, particularly for rain-fed cropping systems of many countries (Campbell *et al.*, 1994) as a resource for the improvement of *L. sativus* L., but also because the genus is largely under-utilised. Several species are cultivated for food, feed, and fodder, as well as for ornamental purposes (Sarker *et al.*, 1997), but there is potential for further exploitation of the *Lathyrus* gene pool. Therefore, the collection, conservation, characterization, study of genetic diversity and utilization of the genus *Lathyrus* is a priority. There is an urgent need actively conserve the genetic diversity of the genus using both *ex-situ* (gene banks) and *in-situ* (natural habitats) techniques. This will permit a critical assessment to be made of the genetic diversity, evolution and genetic erosion of the genus, as well as greatly enhancing further exploitation (Sabanci, 1996).

The genus *Lathyrus* is large and contains about 130 species (ILDIS, 1999), chiefly located in Europe, Asia and North America, extending to temperate South America and tropical East Africa. The genus is primarily located in the Mediterranean and Irano-Turanian regions (Kupicha, 1981). It is adapted to temperate regions but can also be found at high altitudes in tropical Africa. Endemic species are present on all continents, except Australia and Antarctica.

L. sativus L., *L. cicera* L. and *L. ochrus* (L.) DC. are important human food, animal feed and fodder sources. *L. sativus* is widely cultivated for human consumption, as well fodder and green manure. The primary centre of cultivation is in Southern Asia, particularly in Bangladesh, China, India, Nepal, Pakistan and also in Ethiopia (Asthana, 1996), with more limited production in southern Europe and West Asia. It is an important "poor person's insurance crop" because it has relatively good tolerance of water-logging (in the case of flooding), its ability to grow on residual moisture after the end of the rains or in drought and its low production costs (Tadesse, 1997) Figures for grasspea production are not available for all countries, but Campbell (1997) provides a summary of production figures from various sources, see Table 9.1.

Grasspea is nutritionally on a par with other grain legume species, containing up to 30% crude protein (which is high in lysine), about 60% carbohydrate and 0.6% fat (Hartman *et al* 1974). The grasspea is favoured for its ability to mature and produce a yield in times of drought when other crops have failed. The seed, however, may contain 0.1-2.5% of the water soluble non-protein amino acid ODAP (β-N-oxalyl-α,β diaminopropionic acid) or OAP (l-3-oxalylamino-2-amino propionic acid), which have been found to be neurotoxins, the causative agent in the crippling, irreversible neurological disorder, lathyrism (Barrow *et al.*, 1974; Rutter and Percy, 1984; Kaul and Combes, 1986), which leads to paralysis of the lower limbs. These neurotoxins need to be genetically removed if *Lathyrus* is to become an important crop (Abd El Moneim and Cocks, 1993). At present, several grasspea - producing countries are involved in the development of very low or toxin-free *L. sativus* varieties (Malek *et al.*, 1996). Tadesse (1997), for example, reports two promising genotypes, accession 46057, which has a grain yield of 2.729 t/ha and ODAP content of 0.25%, and accession 201513, which has a grain yield of 2.632 t/ha and ODAP content of 0.27%. Additionally, the primary, secondary and tertiary gene pools may play an important role for its improvement. For example, a toxin-free gene has been identified in *L. tingitanus* L. which is being used to develop toxin-free grasspea varieties in China (Zhou and Arora, 1996).

Table 9.1. Production of grasspea in various countries (Campbell, 1997).

Country	Area (1,000 ha)	Production (1,000 t)	Yield (kg/ha)
India	1,500	800	533
Bangladesh	239	174	728
Pakistan	130	45	346
China	20	-	-

L. sativus, *L. cicera* and *L. ochrus* are mainly used as feed legumes. *L. cicera* is cultivated in Greece, Cyprus, Iran, Iraq, Jordan, Spain and Syria and *L. ochrus* in Cyprus, Greece, Syria and Turkey (Saxena *et al.*, 1993). Some other species are used as minor forage or fodder crops, *L. hirsutus* L. is cultivated in southern United States as a fodder species and *L. clymenum* L. was found being cultivated on Kos, Greece in 1999 (Maxted, Pers. Comm.).

Several species within the genus are cultivated as ornamental species, sweet pea (*L. odoratus* L.), everlasting pea (*L. latifolius* L.) and *L. sylvestris* L., and several other species, particularly in section *Lathyrus*, have potential for the development as new horticultural species (Davis, 1970). Due to the potential the genus has as a food, feed and fodder crop, as well as its extensive cultivation of ornamental species it is necessary to collect and conserve all available cultivars, landraces as well as the wild species of the genus *Lathyrus*. Table 9.2 provides a list of those species known to be historically or currently cultivated for agriculture or horticulture.

Many farmers experience members of the genus as noxious weeds (Gams, 1924; Aarssen *et al.*, 1986). *L. annuus* L. and *L. hierosolymitanus* Boiss. are persistent and particularly difficult to control. *L. aphaca* L. can cause problems in cereals: their twining habit can make mechanical harvesting very difficult and may cause lodging of the crop, increasing the incidence of fungal diseases (Aarssen *et al.*, 1986).

The genus is well placed to help meet the increased global demand for animal feed and to provide crops for a diversity of farming systems, particularly when low neurotoxins lines are more widely available. *Lathyrus* conservation has been given priority by the International Board for Plant Genetic Resources (IBPGR 1985) in recent years to prevent genetic erosion and the total extinction of various species of *Lathyrus*. Many national programs and international bodies have launched germplasm collection and conservation activities of this underutilised genus. However, an extensive and systematic approach has not so far been adopted in collection, conservation and evaluation. This deserves further attention in the future. Additionally, it is also necessary to study the genetic diversity of the available collections and utilize them according to human needs.

Table 9.2. Historic or current cultivated of *Lathyrus* species (Kearney, 1993).

Species	Use	Use Status	Location
L. annuus	Pulse, Fodder	Rare	Europe, N. Africa
L. aphaca	Fodder	Rare	India
L. blepharicarpus	Pulse	Historic	Near East
L. cicera	Pulse, Fodder,	Rare	S. Europe, N. Africa
L. clymenum	Pulse	Rare	Greece
L. gorgoni	Fodder	Historic	Middle East
L. hirsutus	Forage	Common	U.S.A.
L. latifolius	Horticulture	Common	Europe
L. ochrus	Pulse, Fodder	Rare	Greece, Middle East
L. odoratus	Horticulture	Common	Widespread
L. pratensis	Forage	Rare	S. Europe, N. Africa
L. rotundifolius	Horticulture	Common	Widespread
L. sativus	Pulse, Forage	Common	Widespread
L. sylvestris	Forage	Rare	S. Europe, N. Africa
L. tingitanus	Fodder	Rare	N. Africa
L . tuberosus	Tubers	Rare	W. Asia

9.2 TAXONOMIC SUMMARY

The genus *Lathyrus* is a member of the legume tribe Vicieae which includes the major legume crops of the classical Mediterranean civilizations, faba bean, pea and lentil. The precise generic boundaries between *Lathyrus* and the related Vicieae genera (*Vicia* L., *Lens* Mill., *Pisum* L. and *Vavilovia* A. Fedorov) have been much debated, the oroboid species appear to form a bridge between *Lathyrus* and *Vicia* (Kupicha, 1981). This taxonomic confusion has led to an abundant and complex synonymy. Kupicha (1983) undertook a comprehensive revision of the genus. She concluded that although the genus was equal in size to *Vicia* it was much easier to comprehend, she divided the genus into 13 sections, a summary of her classification is provided in Table 9.3.

The majority of the cultivated species are placed in *Lathyrus* section *Lathyrus* and therefore this section has received more taxonomic interest. This section had previously been split into section *Cicerula* and *Lathyrus sensu stricto* on the basis of the type of style by Davis (1970) and Czefranov (1971), but Kupicha (1983) concludes this artificial as the distinction is largely a result of the size of the flower and so reduces *Cicerula* to synonymy. The taxonomic relationships within section *Lathyrus* were studied by Yunus (1990), Yunus and Jackson (1991) and Kearney (1993) and the species ecogeography was studied by Baggott (1997). The progenitor of *L. sativus* is unknown, but several Mediterranean species resemble the cultigen morphologically, namely *L. cicera* , *L. marmoratus* Boiss., *L. blepharicarpus* Boiss. and *L. pseudocicera* Pampan.

Table 9.3. Summary of the sectional classification and distribution of *Lathyrus* (Kupicha, 1983).

Section	Species	Geographical Distribution
Orobus	54 species	Europe, W. and E. Asia, N.W. Africa and N and C. America
Lathyrostylis	20 species	C. and S. Europe, W. Asia and N.W. Africa
Orobon	1 species	Anatolia, Caucasia, Crimea and Iran
Lathyrus	33 species (incl. *L. annuus*, *L. blepharicarpus*, *L. cicera*, *L. gorgoni*, *L. hirsutus*, *L. latifolius*, *L. odoratus*, *L. rotundifolius*, *L. sativus*, *L. sylvestris*, *L. tingitanus*, L . *tuberosus*)	Europe, Canaries, W. and C. Asia and N. Africa
Pratensis	6 species (incl. *L. pratensis*)	Europe, W. and C. Asia and N.W. and N. E. Africa
Aphaca	2 species (incl. *L. aphaca*)	Europe, W. and C. Asia and N. Africa
Clymenum	3 species (incl. *L. clymenum*, *L. ochrus*)	Mediterranean
Orobastrum	1 species	Mediterranean, Crimea and Caucasia
Viciopsis	1 species	S. Europe, E. Anatolia and N. Africa
Linearicarpus	7 species	Europe, W. Asia and N. and E. Africa
Nissolia	1 species	Europe, W. Asia and N.W. Africa
Neurolobus	1 species	W. Crete
Notolathyrus	23 species	Temperate S. America and S.E. USA

9.3 CENTRE OF DIVERSITY

Europe, Mediterranean, and West and Southern Asia are the most important centre of current diversity for *Lathyrus*. Secondary centres of diversity exist in South America, North America and Ethiopia, extend into East Africa (Kupicha, 1983). The highest number of species are found in Turkey, then the countries of the Caucasus and Syria (ILDIS, 1999). Further, it has been observed that *L. ochrus* and *L. sativus* are mostly distributed in coastal, lowland sites, while *L.* cicera is the most common species in highland and cold temperate sites. The natural distribution of the *L. sativus* has been completely obscured by cultivation, even in southwest and central Asia (Townsend and Guest, 1974) .

As with *Vicia* elucidation of the centre of origin of the genus is however more problematic. It can be argued that relative species concentrations can be used to indicate the centre of origin, which would therefore suggest a South Eastern Europe and North Western Asia origin. Over half of the *Lathyrus* taxa are endemic to this area. However, due to floristic migration caused by the ice ages Kupicha (1974) concludes that *Lathyrus* evolved in the early tertiary and the centre of origin is likely to have been much further North than is indicated by contemporary concentrations of taxa.

The earliest archaeological remains of *Lathyrus* appear in the Neolithic age in the Balkans and Near East of Bulgaria, Cyprus, Iraq, Iran, and Turkey (Erskine *et al.*, 1994). A single *Lathyrus* seed, presumed to be a field weed, was found in Cayono in Turkey and dated at around 7200 B.C., where bitter vetch was the prevalent pulse (van Zeist, 1972). Compared to the early domesticates, lentil, pea and bitter vetch, *Lathyrus* is only found in small quantities in Turkey, Cyprus, Iraq, Iran and Bulgaria dating back from 6750 to 4770 B. C. However, a different picture appears from late Neolithic age finds at Dimini in Greece (c. 4000-3500 B. C), where grasspea is as frequent as pea and lentil (Kroll, 1979). This increased frequency of grasspea is suggestive of domestication. *Lathyrus* was the chief crop component mixed with lentil (c. 2100-1800), providing stronger evidence of domestication by the Middle Bronze Age (Helback, 1965). It was also found mixed in substantial quantities with other leguminous crops in later finds. *L. clymenum* was cultivated in the Bronze Age on Thera, Crete and Melos in Greece (Sarpaki and Jones, 1990) and *L. ochrus* was possibly cultivated on Knossos, Greece at the same time (Jones, 1992). *L. cicera* is believed to have been domesticated in southwestern Europe by 4000 - 3000 B.C.(Kislev, 1989).

Written records provide very little to knowledge about the origin of grasspea. *Lathuros* is an ancient Greek plant name probably used for a pulse and possibly for *L. sativus* (Westphal, 1974). The Romans also do not mention *Lathyrus*, which reflects little importance or lack of knowledge of the crop. Thus, the archaeological evidence suggests that domestication of *Lathyrus* occurred possibly during the late Neolithic and surely by the Bronze Age. Prior to that time, it was probably a tolerated weed of other pulses (Erskine *et al.*, 1994).

9.4 Ecogeographic Distribution

The range of habitats occupied by *Lathyrus* species is diverse and includes open, disturbed habitats like field margins and roadsides, as well as closed habitats such as woodlands and steppes. The species considered more advanced are generally those found in the more disturbed, open communities. The cultivated species have evolved in general from disturbed habitats, they were originally the wild and weedy flora of agricultural fields (Vavilov, 1926). Farming systems have therefore had a great influence on the recent evolution of the genus. Their weedy nature would explain the widespread distribution of many species.

The genus contains many restricted endemics, for which only very few sites have been documented or which are bound by specific soil types and climatic regimes (Maxted and Goyder, 1988; Ehrman and Maxted, 1990; Maxted *et al.*, 1990; Maxted, 1993; Maxted *et al.*, 1993; Francis *et al.*, 1995; Bennett *et al.*, 1998). The ecogeographic distribution of all but a few *Lathyrus* species is poorly understood, particularly those in section *Notolathyrus* which are endemic to South America. There is a need for a detailed ecogeographic study of the whole genus if it is to be effectively and efficiently conservation and utilised. Baggott and Maxted (in prep.) have started this process with their ecogeographic survey of *Lathyrus* section *Lathyrus*.

9.5 Genetic Diversity

There have been relatively few genetic studies of the genus, with the exception of Yunus (1990) and Kearney (1993), both of which are unpublished doctoral theses and focus on the agriculturally important grasspea and its close relatives in section *Lathyrus*. These have been found to be predominantly self pollinating, with anther dehiscence occurring usually before the flower has fully opened. Inter-specific hybridisation has been successful between *L. sativus* and two other *Lathyrus* species, though the production of successful hybrids remains low. The first successful inter-specific cross was with *L. cicera* (Saw Lwin, 1956; Davies, 1957; 1958). Yunus

(1990) crossed 11 species in section *Lathyrus* with *L. sativus*, and found that *L. cicera* and *L. amphicarpos* gave viable seed. Other species formed pods but these did not form fully developed viable seed. *L. cicera* is thought to be the species closest related to *L. sativus* morphologically (Jackson & Yunus 1984). Plitmann *et al.* (1986) arrived at the same conclusion, based on studies of pollen morphology, karyotype and flavonoid aglycones. Table 9.4 provides a summary of the results of crossing experiments between *Lathyrus* species.

We can apply Harlan and De Wet's gene pool concept to this crossability information for *L. sativus* to elucidate its gene pools. The cultivated and wild races of *L. sativus* are included in the primary gene pool. Townsend and Guest (1974) suggest that the primary gene pool is poorly differentiated in terms of morphological characters, as there are no clear-cut discontinuities between the cultivated and wild forms. Although Smartt (1984) concludes that the white flowered, white seeded varieties are the most highly selected and Jackson and Yunus (1984) suggested that the blue flowered, small speckled seeded forms are primitive. Therefore we can tentatively place the white flowered, white seeded varieties in GP1A and the blue flowered, small speckled seeded forms in GP1B. The secondary gene pool include the other biological species that will cross with some difficulty with the crop species. Therefore in the GP2 we can include: *L. chrysanthus*, *L. gorgoni*, *L. marmoratus* and *L. pseudocicera*, with which *L. sativus* can cross and produce ovules, and possible more remotely *L. amphicarpos*, *L. blepharicarpus*, *L. chloranthus*, *L. cicera*, *L. hierosolymitanus* and *L. hirsutus*, with which *L. sativus* can cross and with which pods are formed. The tertiary gene pool include species that can cross with the original crop species only with use of specialised techniques such as embryo rescue and culture or the use of bridging species and the remaining species of the genus can be considered in the tertiary gene pool (GP3).

Cytogenetic studies in section *Lathyrus* showed that vast majority has the chromosome compliment 2n=2x=14. There is some variation in karyotype, but the majority of chromosomes are sub-metacentric. In *L. sativus*, all seven pairs are sub-metacentric while two cross-compatible species (*L. cicera* and *L. amphicarpos*) have one pair metacentric and six pairs sub-metacentric. This indicates that some chromosome structural differentiation has occurred between genomes of different species. From meiotic studies of interspecific hybrids, it would seem that *L. amphicarpos* is structurally more differentiated from *L. sativus* than is *L. cicera*. In F_1 hybrids of *L. cicera* x *L. sativus* the configurations observed were 6II + 2I and 7II. In the hybrid of *L. amphicarpos* x *L. sativus* multivalents were frequently observed, suggesting that translocation changes had occurred (Yunus, 1990; Yunus and Jackson, 1991).

Proper evaluation, characterization and documentation is an important part of utilizing genetic resources by *Lathyrus* workers. However, in-depth evaluation for phenological, morphological, agronomical and quality characters of available germplasm has yet to be done properly and adequately at the national and global level. The Germplasm Resources, Crop Improvement and Agronomy Committee of the International Network for the Improvement of *Lathyrus sativus* and the Eradication of Lathyrism (INILSEL) proposed a list of 16 descriptors to characterize the *Lathyrus* genetic resources (Campbell, 1994).

Table 9.4. Summary interspecific hybridization in *Lathyrus* (Adapted from Kearney, 1993).

Attempted Cross	Success	Reference
L. amphicarpos x *L. cicera*	Partially fertile hybrids	Khawaja (1985)
L. amphicarpos x *L. sativus*	Partially fertile hybrid [1]; Hybrid produced [2]	Khawaja (1985)[1]; Yunus (1990)[2]
L. annuus L. x *L. hierosolymitanus*	Viable hybrid produced	Yamamoto *et al.* (1986); Hammett *et al.* (1996)
L. annuus L. x *L. odoratus*	No success	Khawaja (1988)
L. annuus L. x *L. sativus*	No success[1]; Pod formed[2]	Yamamoto *et al.* (1986)[1]; Yunus (1990)[2]
L. articulatus L. x *L. clymenum*	Hybrid produced	Burton (1956); Saw Lwin (1956); Davies (1958); Trankovskij (1962)
L. articulatus L. x *L. ochrus*	Hybrid produced	Davies (1958); Trankovskij (1962)
L. basalticus x *L. sativus*	Suspected self	Yunus (1990)
L. belinensis x *L. odoratus*	Hybrids Formed with shrivelled seed	Hammett *et al.* (1994)
L. blepharicarpus x *L. cicera*	No success[1]; ovule formed[2]	Yamamoto *et al.* (1986)[1]; Kearney (1993)[2]
L. blepharicarpus x *L. gorgoni*	No success[1]; hybrid produced[2]	Yamamoto *et al.* (1986)[1]; Kearney (1993)[2]
L. blepharicarpus x *L. marmoratus*	No success; ovule formed[2]	Yamamoto *et al.* (1986)[1]; Kearney (1993)[2]
L. blepharicarpus x *L. pseudocicera*	No success; hybrid produced[2]	Yamamoto *et al.* (1986)[1]; Kearney (1993)[2]
L. blepharicarpus x *L. sativus*	No success[1]; viable seed produced[2]	Yamamoto *et al.* (1986)[1]; Kearney (1993)[2]
L. cassius x *L. sativus*	No success	Yunus (1990)
L. chloranthus x *L. chrysanthus*	Viable hybrid produced	Murray & Hammett (1989)
L. chloranthus x *L. odoratus*	No success[1]; Pods only[2]	Khawaja (1988)[1]; Murray & Hammett (1989)[2]
L. chloranthus x *L. sativus*	Non viable seed produced	Yunus (1990)
L. chrysanthus x *L. chloranthus*	Shrivelled seed formed	Murray & Hammett (1989)
L. chrysanthus x *L. odoratus*	Pods formed only	Murray & Hammett (1989)
L. chrysanthus x *L. sativus*	No success	Yunus (1990)
L. cicera x *L. blepharicarpus*	Pod formed[1]; viable seed produced[2]	Yamamoto *et al.* (1986)[1]; Kearney (1993)[2]
L. cicera x *L. gorgoni*	Pod formed[1]; ovule formed[2]	Yamamoto *et al.* (1986)[1]; Kearney (1993)[2]
L. cicera x *L. hierosolymitanus*	Suspected self	Yamamoto *et al.* (1986)
L. cicera x *L. marmoratus*	Viable hybrid produced	Yamamoto *et al.* (1986); Kearney (1993)
L. cicera x *L. odoratus*	No success	Yamamoto *et al.* (1986)
L. cicera x *L. pseudocicera*	Viable hybrid produced	Yamamoto *et al.* (1986); Kearney (1993)
L. cicera x *L. sativus*	Hybrid produced[1]; Pod[2]; Sterile hybrid produced[3]	Burton (1956)[1]; Davies (1958)[1]; Yunus (1990)[1]; Yamamoto *et al.* (1986)[2]; Saw Lwin (1956)[3]; Kearney (1993)[3]
L. clymenum x *L. articulatus*	Viable hybrid	Davies (1957)
	Sterile plant produced[1]; Pod [2]	
L. clymenum x *L. ochrus*	Suspected self	Saw Lwin (1956)[1]; Davies (1957)[1]; Yamamoto *et al.* (1986)
L. clymenum x *L. tingitanus*		Yamamoto *et al.* (1986)
L. gorgoni x *L. blepharicarpus*	No success[1]; hybrid produced[2]	Yamamoto *et al.* (1986)[1]; Kearney (1993)[2]
L. gorgoni x *L. cicera*	No success; sterile hybrid produced[2]	Yamamoto *et al.* (1986)[1]; Kearney (1993)[2]
L. gorgoni x *L. marmoratus*	No success; sterile hybrid produced[2]	Yamamoto *et al.* (1986)[1]; Kearney (1993)[2]
L. gorgoni x *L. pseudocicera*	Viable hybrid produced	Yamamoto *et al.* (1986); Kearney (1993)
L. gorgoni x *L. sativus*	Suspected self[1]; ovule formed[2]	Yunus (1990)[1]; Kearney (1993)[2]

L. hierosolymitanus x *L. annuus*	Pod produced[1]; fertile hybrid produced[2]	Yamamoto *et al.* (1986)[1] Hammett *et al.* (1996)[2]
L. hierosolymitanus x *L. blepharicarpus*	No success	Yamamoto *et al.* (1986)
L. hierosolymitanus x *L. cicera*	No success	Yamamoto *et al.* (1986)
L. hierosolymitanus x *L. hirsutus*	Non viable seed produced	Yamamoto *et al.* (1986)
L. hierosolymitanus x *L. gorgoni*	No success	Yamamoto *et al.* (1986)
L. hierosolymitanus x *L. marmoratus*	No success	Yamamoto *et al.* (1986)
L. hierosolymitanus x *L .odoratus*	No success	Yamamoto *et al.* (1986)
L. hierosolymitanus x *L. pseudocicera*	Pod produced	Yamamoto *et al.* (1986)
L. hierosolymitanus x *L. tingitanus*	No success	Yamamoto *et al.* (1986)
L. hierosolymitanus x *L. sativus*	No success	Yunus (1990)
L. hirsutus x *L. annuus*	No success	Yamamoto *et al.* (1986)
L. hirsutus x *L. hierosolymitanus*	No success	Yamamoto *et al.* (1986)
L. hirsutus x *L. odoratus*	Viable hybrid produced	Barker (1916); Saw Lwin (1956); Davies (1958); Trankovskij (1962); Khawaja (1988); Yamamoto *et al.* (1986)
L. hirsutus L. x *L. sativus*	No success	Yamamoto *et al.* (1986); Yunus (1990)
L. latifolius x *L. sativus*	Non viable seed produced	Yunus (1990)
L. marmoratus x *L. blepharicarpus*	Viable hybrid produced[1]; viable seed[2]	Yamamoto *et al.* (1986)[1]; Kearney (1993)[2]
L. marmoratus x *L. cicera*	Pod formed only[1]; hybrid produced[2]	Yamamoto *et al.* (1986)[1]; Kearney (1993)[2]
L. marmoratus x *L. gorgoni*	No success[1]; viable seed produced[2]	Yamamoto *et al.* (1986)[1]; Kearney (1993)[2]
L. marmoratus x *L. hierosolymitanus*	No success	Yamamoto *et al.* (1986)
L. marmoratus x *L. pseudocicera*	Pod formed only[1]; sterile hybrid[2]	Yamamoto *et al.* (1986)[1]; Kearney (1993)[2]
L. marmoratus x *L. sativus*	Pod formed only[1]; sterile hybrid[2]	Yamamoto *et al.* (1986)[1]; Yunus (1990)[1]; Kearney (1993)[2]
L. ochrus x *L. clymenum*	Pod formed only	Yamamoto *et al.* (1986)
L. ochrus x *L. sativus*	Pod formed only	Fouzdar & Tandon (1975)
L. odoratus x *L. annuus*	No success	Khawaja (1988)
L. odoratus x *L. belinensis*	Hybrids Formed viable seed	Hammett *et al.* (1994)
L. odoratus x *L. chloranthus*	Hybrids Formed[1]; Shrivelled seed formed[2]	Khawaja (1988)[1]; Murray & Hammett (1989)[2]
L. odoratus x *L. chrysanthus*	Pods formed only	Murray & Hammett (1989)
L. odoratus x *L. hirsutus*	No success[1]; hybrids formed[2]	Khawaja (1988)[1]; Barker (1916)[2]; Davies (1958)[2]
L. odoratus x *L .pratensis*	Viable hybrid[1]; No success[2]	Taylor (1916)[1]; Khawaja (1988)[2]
L. odoratus x *L. sativus*	No success	Yamamoto *et al.*. (1986); Yunus (1990)
L. pratensis x *L. odoratus*	Viable hybrid	Taylor (1916)
L. pseudocicera x *L. blepharicarpus*	Pods formed only[1]; viable seed[2]	Yamamoto *et al.* (1986)[1]; Kearney (1993)[2]
L. pseudocicera x *L .cicera*	Viable hybrid[1]; sterile hybrid[2]	Yamamoto *et al.* (1986)[1]; Kearney (1993)[2]
L. pseudocicera x *L .gorgoni*	No success[1]; ovule formed[2]	Yamamoto *et al.* (1986)[1]; Kearney (1993)[2]
L. pseudocicera x *L. hierosolymitanus*	No success	Yamamoto *et al.* (1986)
L. pseudocicera x *L. marmoratus*	Unviable seed[1]; hybrids formed[2]	Yamamoto *et al.* (1986)[1]; Kearney (1993)[2]
L. pseudocicera x *L. sativus*	No success[1]; Pod formed only[2]; viable seed produced[3]	Yamamoto *et al.* (1986)[1]; Yunus (1990)[2] Kearney (1993)[3]
L. rotundifolius x *L. tuberosus*	Hybrid produced	Marsden-Jones (1919)

L. sativus x *L. amphicarpos*	Pod produced only	Yunus (1990)
L. sativus x *L. annuus*	No success	Yunus (1990)
L. sativus x *L. basalticus*	Pod produced only	Yunus (1990)
L. sativus x *L. blepharicarpus*	No success[1]; ovule formed[2]	Yamamoto *et al.* (1986)[1]; Kearney (1993)[2]
L. sativus x *L. cassius*	No success	Yunus (1990)
L. sativus x *L. chloranthus*	No success	Yunus (1990)
L. sativus x *L. chrysanthus*	No success	Yunus (1990)
L. sativus x *L. cicera*	No success[1]; Pod produced only[2]	Yamamoto *et al.* (1986)[1]; Yunus (1990)[2]; Kearney (1993)[2]
L. sativus x *L. gorgoni*	No success[1]; ovule formed[2]	Yamamoto *et al.* (1986)[1]; Kearney (1993)[2]
L. sativus x *L. hierosolymitanus*	No success[1]; Pod produced only[2]	Yamamoto *et al.* (1986)[1]; Yunus (1990)[2]
L. sativus x *L. hirsutus*	Pod produced only	Yunus (1990)
L. sativus x *L. latifolius*	No success	Yunus (1990)
L. sativus x *L. marmoratus*	No success[1]; ovule formed[2]	Yamamoto *et al.* (1986)[1]; Yunus (1990)[1]; Kearney (1993)[2]
L. sativus x *L. odoratus*	No success	Yamamoto *et al.* (1986); Yunus (1990)
L. sativus x *L. pseudocicera*	No success[1]; ovule formed[2]	Yamamoto *et al.* (1986)[1]; Yunus (1990)[1]; Kearney (1993)[2]
L. sativus x *L. tingitanus*	No success	Yamamoto *et al.* (1986); Yunus (1990)
L. sylvestris L. x *L. latifolius*	Hybrid produced	Davies (1957)
L. tingitanus L. x *L. cicera* L.	No success	Yamamoto *et al.* (1986)
L. tingitanus L. x *L. clymenum* L.	No success	Yamamoto *et al.* (1986)
L. tingitanus L. x *L. gorgoni* Parl.	No success	Yamamoto *et al.* (1986)
L. tingitanus L. x *L. odoratus* L.	No success	Yamamoto *et al.* (1986)
L. tingitanus L. x *L. sativus*	Pod produced only	Yunus (1990)

The GRU, ICARDA uses a standard list of 21 descriptors for *Lathyrus*, many of which are common to that of INILSEL's descriptors. However, we propose that a more detailed list of descriptors which should include: Growth habit, Flower colour, Anthocyanin present, Leaf shape, Leaf width, Days to 50% flowering, Days to 90% podding , Days to 90% maturity, Plant height, Height to the first flower, Leaf length, Pod length, Pod width, Peduncle length, Internode length, Seeds per pod, Plant type, Seed coat colour, Pod shattering at maturity, 1000-seed weight, Downy mildew resistance, Insect resistance, Seed yield in kg/ha, Biomass in kg/ha, Straw yield in kg/ha, Harvest index, b- ODAP concentration in seeds.

A detailed evaluation of 1082 accessions of 30 species was performed in 1992 at ICARDA in a series of augmented nurseries using one systematic check (IFLA 347, *L. sativus*) and two random checks (IFLA 101, *L. ochrus* and IFLA 536, *L. cicera*). Results are presented only for the three economically important species, viz., *L. sativus*, *L. cicera* and *L. ochrus*. Most of the accessions of *L. sativus* and *L. ochrus* had semi-erect growth habit, while *L. cicera* had prostrate and erect plant types (Table 9.5). Almost all accessions of *L. sativus* had violet flower colour while, most accessions of *L. cicera* showed red and *L. ochrus* germplasm had mostly white flower colour. Most accessions of all three species had weak anthocyanin pigmentation. *L. sativus* and *L. cicera* had mostly narrow and *L. ochrus* mostly medium leaf shape. These descriptors are used for taxonomic identification in *Lathyrus*.

The accessions of *L. cicera* were earlier than the *L. cicera* check, IFLA 536, by up to 11 days (Table 9.6). The seeds per pod was higher for the tested accessions, but 1000-seed weight was smaller and the harvest index was similar to the check. The biological yield was reduced, which resulted in a lower seed yield and straw yield for the accessions tested.

The *L. ochrus* accessions were also earlier than the *L. ochrus* check (Table 9.7). The vegetative descriptors were similar to the check for *L. ochrus*. The 1000-seed weight was smaller for the tested entries as was the seeds per pod. However, for *L. ochrus*, the mean of the tested accessions for biological yield was the same as the check, which resulted in similar values for seed yield and straw yield.

The harvest index, biological yield, seed yield and straw yield were much lower for the tested *L. sativus* germplasm accessions than the *L. sativus* check (Table 9.8). Unlike the other two species, the tested accessions of *L. sativus* were late in flowering and maturity than the *L. sativus* check. Resistance to major biotic and abiotic stresses has been noticed during the whole evaluation process. *L. sativus* accessions resistant to powdery mildew caused by *Erysiphe polygoni* DC. and Botrytis blight caused by *Botrytis cinera* Pers. Ex Fr, have been found in ICARDA collections. Resistant sources for other diseases in other species have also been found but are to be confirmed. The broomrape, *Orobanche crenata* Forsk. causes major yield losses of *Lathyrus* species, particularly in the West Asia and North Africa region; *L. ochrus* accessions were of emerged *O. crenata* shoots (Linke *et al.*, 1993). On the other hand, accessions of *L. sativus* and *L. cicera* were highly susceptible to *O. crenata.* Resistant lines of *L. ochrus* to *O. crenata* are IFLAO 84, 94, 95 and 101. Among the species evaluated, *L. cicera* is resistant to cold, whereas *L. sativus* and *L. ochrus* accessions are generally susceptible. However, one accession of *L.* ochrus , IFLAO 109, from Portugal has a high level of cold tolerance (Robertson *et al.*, 1996). This can be explained from the eco-distribution of *Lathyrus* species, where most *L. sativus* and *L. ochrus* accessions are from low altitude, mild winter environments while *L. cicera* is adapted to high altitude, continental environments with severe winters.

Table 9.5. Frequency of morphological characters among *L. cicera*, *L. ochrus* and *L. sativus* germplasm evaluated at ICARDA, Syria.

Descriptors / score	% of *L. cicera*	% of *L. ochrus*	% of *L. sativus*
Growth habit			
Prostrate	24.0	3.4	0.7
Semi-erect	51.0	96.6	96.0
Erect	25.0	0.0	3.3
Flower colour			
White	0.0	53.4	1.5
Cream	1.0	12.1	2.2
Brick	11.5	0.0	0.4
Pink	5.2	0.0	0.0
Violet	4.2	13.8	95.6
Yellow	2.1	0.0	0.0
Blue	0.0	0.0	0.0
Red	76.0	20.7	0.4
Anthocyanin pigmentation			
Weak	83.3	100.0	83.8
Fair	8.3	0.0	9.9
Strong	7.3	0.0	6.2
Very Strong	1.0	0.0	0.0
Leaf shape	99.0	8.6	99.6
Narrow	1.0	91.4	0.0
Medium	0.0	0.0	0.4
Oval			

Table 9.6. Variability for morpho-agronomic traits among 96 *Lathyrus cicera* germplasm accessions evaluated at ICARDA, Syria.

Descriptors	Check Mean (IFLA 536)	Accessions			
		Mean	Min.	Max.	CV(%)
Days to 50% flowering	126.2	123.9	115.0	136.0	3.3
Days to 90% maturity	161.7	163.9	156.0	181.0	2.8
Days to 90% podding	133.7	128.3	122.0	148.0	4.5
Plant height (cm)	36.7	35.4	24.1	49.8	12.5
Height to 1st flower (cm)	10.7	8.1	2.4	13.2	18.5
Seeds per pod	3.61	3.8	2.3	9.6	24.2
Harvest index (%)	36.6	33.8	12.7	52.0	26.1
1000-seed weight (g)	91.8	83.1	13.9	116.7	24.2
Seed yield (kg/ha)	1 237.0	1 120.0	117.0	2 030.0	52.5
Biomass yield (kg/ha)	3 350.0	3 101.0	635.0	4 972.0	37.7
Straw yield (kg/ha)	2 113.0	2 578.0	488.0	3 067.0	26.1

Table 9.7. Variability for morpho-agronomic traits among 58 *Lathyrus ochrus* germplasm accessions evaluated at ICARDA.

Descriptors	Check Mean (IFLA 101)	Accessions			
		Mean	Min.	Max.	CV(%)
Days to 50% flowering	124.3	120.4	115.0	145.0	4.1
Days to 90% maturity	160.0	157.0	149.0	184.0	3.2
Days to 90% podding	128.0	124.0	118.0	154.0	4.6
Plant height (cm)	33.1	34.7	23	48	15.4
Height to 1st flower (cm)	15.8	13.0	7	19	15.5
Seeds per pod	4.8	4.6	3.32	5.7	11.9
Harvest index (%)	38.5	36.2	12.7	48.6	20.1
1000-seed weight (g)	130.2	121.3	57.2	156.3	17.7
Seed yield (kg/ha)	853.0	815.0	105.0	1 454.0	38.0
Biomass yield (kg/ha)	2 214.0	2 221.0	726.0	3 741.0	32.0
Straw yield (kg/ha)	1 362.0	1 406.0	564.0	2 499.0	32.5

Assessment for quality characters in *Lathyrus* at ICARDA has mostly focused on b-N-oxalyl-L-a-b-diaminopropionic acid (b-ODAP), a free amino acid, which causes Lathyrism disease. Special attention has been given to evaluate *Lathyrus* germplasm for b-ODAP content. Results from preliminary screening indicated that none of the *Lathyrus* species lines was free, although the b-ODAP content was very low in several lines; *L. cicera* had low mean b-ODAP content (0.16%) followed by *L. sativus* (0.48%) and *L. ochrus* (0.57%). On average, b-ODAP levels in *L. ochrus* and *L. sativus* were about four to five times higher than those in *L. cicera*.

Although evaluation of *Lathyrus sativus* germplasm have been carried out in Bangladesh, Canada, Ethiopia, India, Nepal and Pakistan, it has been confined mainly to a few agronomic traits only to meet the breeder's immediate needs without detailed characterization.

9.6 CONSERVATION RESOURCE STATUS

Relatively large *ex situ* seed collections exist of cultivated and wild *Lathyrus* species. However, the collections are not comprehensive in terms of species diversity and there remain numerous gaps in conserved materials, particularly for the South American species and those species of less immediate utilisation potential. The most diverse range of species have been collected by Maxted and co-workers, in conjunction with IBPGR and ICARDA. and latterly with CLIMA have engaged in 14 forage legume collection missions to the North Eastern Mediterranean region since 1986. The Genetic Resource Unit (GRU) of ICARDA has made four collection missions with national collaboration in 1981, 1989, 1990 and 1993. All of the material collected by Maxted and co-workers and the GRU is held in the national collection of the country of collection, as well as being duplicated at the GRU, ICARDA. Each of these collecting expeditions reported varying levels of genetic erosion occurred in the genus, especially in species such as *L. ochrus*, *L. gorgoni* and *L. cicera*. Detailed information about the current conservation status of *Lathyrus* species is documented by WIEWS (World Information and Early Warning System on

Plant Genetic Resources), which contains information on national PGR holdings (www.fao.org/ag/agp/pgr/wiews/) and SINGER (System-wide Information Network for Genetic Resources), which contains information on CGIAR holdings (http://www.cgiar.org/singer) and is summarised for the major collections in Table 9.9.

Turkey has the richest diversity of *Lathyrus* species genetic diversity. Davis (1970) reported the presence of 58 species in Turkey, some of them endemic at local or regional level and many of these are held in the gene bank of the national programme. Prior to 1987, Turkish collection missions were targeted on forage grasses, and legume genetic resources and *Lathyrus* species were not given a high priority. Targeted expedition were launched specifically to collect forage legumes in 1987, 1988, 1995, 1996 and 1997 from nine different agricultural regions (Sabanci, 1996) and this material is held at the Aegean Agricultural Research Institute in Menemen and the majority of accessions are duplicated at ICARDA. The number of species collected is over half of those found by Davis, includes a new species, *L. belinensis* Maxted and Goyder, which is closely related to *L. odoratus*, first discovered during the 1987 mission. These expeditions focused on collecting material from areas of Turkey with a Mediterranean climate (Aegean and Southern Turkey) and they did not attempt to systematically collect representative collections from throughout the country. They also concentrated at the lower altitudes favoured by annual species. Therefore, some endemic species, particularly perennial species, were not encountered and are not currently conserved. Undoubtedly, the environment in Turkey is being changed rapidly by human intervention building dams, constructing recreational areas along the coast and overgrazing (Tan, 1998), the flora is obviously suffering genetic erosion as a result and there is a need to give priority to the collection of *Lathyrus* germplasm from throughout the country, particularly in the under-collected areas of the North, central and South East.

ICARDA has the most extensive collection of germplasm for the Mediterranean region. At present ICARDA is concerned with collection and conservation for *Lathyrus* species in the Mediterranean region and other *Lathyrus*-growing areas of the world. ICARDA holds 'in-trust' *Lathyrus* germplasm from more than 45 countries under the auspices of the Food and Agriculture Organization. While the emphasis at ICARDA for genetic resources and improvement of *Lathyrus* is for three species (*L. sativus*, *L. cicera* and *L. ochrus*), a sizeable collection of 44 other species is being maintained (Table 9.10). The majority of accessions of all species of *Lathyrus* held in the ICARDA gene bank, except *L. sativus*, are from the West Asia and North African region. The collections have been collected from cultivated or from naturally occurring populations, found mostly in disturbed habitats such as roadsides, crop fields and orchards. The *L. sativus* accessions in the ICARDA collection are from Ethiopia and the Indian subcontinent and are local landraces. A standard collecting form for passport data developed by GRU, ICARDA is based on IPGRI standards was used for all collection, which means good quality passport data is available for all accessions. Besides, expeditions within the Mediterranean, expeditions from ICARDA have been made to Bangladesh, Ethiopia, India, Nepal and Pakistan, primarily to collect genetic resources of *L. sativus*.

Table 9.8. Variability among 272 *Lathyrus sativus* germplasm accessions evaluated at ICARDA, Syria.

Descriptors	Check Mean (IFLA 347)	Accessions			
		Mean	Min.	Max.	C.V. (%)
Days to 50% flowering	121.1	126.0	119.0	142.0	2.5
Days to 90% maturity	163.3	173.8	145.0	189.0	3.9
Days to 90% podding	125.1	137.5	122	154	4.9
Plant height (cm)	40.8	41.1	5	60	15.7
Height to 1st flower (cm)	7.3	9.2	3	17	18.5
Seeds per pod	3.3	3.1	1.48	6.5	16.4
Harvest index (%)	26.6	19.5	1.9	54.7	43.8
1000-seed weight (g)	104.1	86.8	34.5	225.9	34.3
Seed yield (kg/ha)	729.0	445.0	29.0	1 406.0	65.9
Biomass yield (kg/ha)	2 682.0	2 167.0	516.0	5 200.0	38.2
Straw yield (kg/ha)	1 953.0	1 722.0	440.0	3 861.0	36.2

Table 9.9. Number of *Lathyrus* accessions conserved *ex situ* in major collections.

Species	Institute					
	AARI	ATFC	ICARDA	IPK	IBEAS	W-6
L. annuus	44	6	68	2	0	3
L. chrysanthus	1	1	3	0	0	0
L. cicera	90	141	182	63	785	31
L. clymenum	1	10	2	25	0	20
L. gorgoni	27	6	60	2	0	1
L. hierosolymitanus	22	13	104	1	0	4
L. hirsutus	2	9	17	8	0	16
L. latifolius	0	1	1	13	326	10
L. marmoratus	4	0	33	0	0	0
L. ochrus	1	85	136	46	0	23
L. odoratus	2	3	3	6	0	23
L. pseudocicera	8	1	65	1	0	0
L. sativus	17	572	1627	170	2382	222
Other *Lathyrus* sp.	300	172	698	108	984	111
Total *	519 (32)	1020(42)	3001 (44)	445	4477 (6)	464 (23)

* Numbers in brackets indicate the number of other *Lathyrus* species conserved.

AARI = Aegean Agricultural Research Institute, Menemen, Turkey
ATFC = Australian Temperate Field Crop Collection, Horsham, Australia
IBEAS = IBEAS, Unviersité de Pau et des Pays de l' Adour, Pau, France
W-6 = Regional Plant Introduction Station, Washington, USA
ICARDA = International Centre for Agricultural Research in the Dry Areas, Syria
IPK = Institut für Pflanzengenetik und Kulturpflanzenforschung (IPK), Gatersleben, Germany

Table 9.10. *Lathyrus* species held by ICARDA.

Species	Number of Accessions	Species	Number of Accessions
L. amphicarpos	2	*L. incurvus*	2
L. angulatus	1	*L. latifolius*	1
L. annuus	68	*L. laxiflorus*	12
L. aphaca	253	*L. marmoratus*	33
L. articulatus	104	*L. nissolia*	9
L. aureus	1	*L. occidentalis*	1
L. basalticus	4	*L. ochrus*	136
L. belinensis	1	*L. odoratus*	3
L. blepharicarpus	33	*L. pallescens*	1
L. cassius	8	*L. pratensis*	2
L. chloranthus	1	*L. pseudocicera*	65
L. chrysanthus	3	*L. rotundifolius*	2
L. cicera	182	*L. sativus*	1627
L. cilicicus	10	*L. saxatilis*	2
L. ciliolatus	3	*L. setifolius*	7
L. clymenum	2	*L. sphaericus*	21
L. digitatus	2	*L. stenophyllus*	2
L. gleospermus	2	*L. sylvestris*	1
L. gorgoni	60	*L. tingitanus*	18
L. herticarpus	2	*L. tuberosus*	4
L. hierosolymitanus	104	*L. vinealis*	4
L. hirsutus	17	Unidentified *Lathyrus* species	36
L. inconspicuus	149		

There are also *ex-situ* collections of *Lathyrus* germplasm conserved in Bangladesh, Chile, China, Canada, Ethiopia, Ecuador, France, Germany, India, Nepal, Pakistan, Russia, Syria (ICARDA), USA and UK. Many of these are relatively small collections, however, and do not represent a true sample of the breadth of taxonomic and genetic variability that is found in the genus.

In-situ conservation, whether in a genetic reserves or on farm, has so far not been adopted for *Lathyrus* species, except for an initial attempt in Turkey (Ertug Firat and Tan, 1997) at Kaz Dag (Aegean Anatolia), Amanos, (Southern Turkey) and Ceylan Pínner (in Southeast Turkey). The target species for conservation in these reserves, as in the Ammiad reserve in Eastern Galilee, Israel (Anikster *et al.*, 1997) are not Lathyrus, but they are likely to be present and 'passive' (species and genetic diversity is not being monitored and managed) conserved. Maxted (1995) proposed the establishment of sites for reserves for Vicieae species in the Syria and Turkey, but these ideas have not yet been acted upon. There is an urgent need to make positive steps to establish both reserves for the wild species of *Lathyrus* and on farm projects to conserve the ancient land races of cultivated *Lathyrus* species.

9.7 CONSERVATION GAPS AND PRIORITIES

Although ICARDA has an extensive collection of germplasm from throughout the Mediterranean region, there are still many un- and under-explored areas. National and global efforts should be made for enrichment of *Lathyrus* genetic resources by augmenting collection of cultivars, landraces and wild relatives in countries with rich genetic diversity. The national programs of this region and ICARDA should pay further attention to obtaining adequately sample from throughout the area. The current status of germplasm collection needs to be critically assessed for each country.

It is undoubtedly true that the intensification of agriculture around the Mediterranean is leading to serious genetic erosion of *Lathyrus* diversity (IBPGR, 1985). The clearance of new cultivated land and the associated decline of permanent pastures have lead to the disappearance of habitats for wild climax community species. The introduction of high input farming and new varieties which replace traditional land races for many of the cultivated species is also threatening diversity of weedy and crop *Lathyrus* species.

There are reports of decreased production of both *L. sativus* and *L. cicera* from the mountainous regions Spain where they have been traditionally cultivated. The status of the remaining landraces from throughout the Mediterranean region needs to be assessed and matched against conserved germplasm to clearly establish conservation gaps. Likely under-collected countries are Albania, Spain, the former Yugoslavia, Turkey, Lebanon, Iran, Afghanistan, many of the Mediterranean islands, the Caucasian and Central Asian republics as well as the Russian Federation. Even though some of these countries, particularly Spain and Turkey, have been extensively collected in the past, Ferguson *et al.* (1998) demonstrated, for the wild lentils (*Lens* species), that these countries still do not have an adequate genetic representation of diversity conserved. Intensive collection of the cultivated species needs to be carried out in Chile and Ecuador, where the landraces are subject to genetic erosion by environmental hazards and human intervention.

An inventory of wild *Lathyrus* species that occur in national parks would give an indication of the number of *Lathyrus* species present in protected areas and would assist active *in situ* conservation. Maxted (1995) concluded that the most seriously threatened by extinction species are those restricted to Syria, Lebanon, Turkey and Israel, and the highest concentration of potentially threatened taxa are located in Syria. He recommended the establishment of four genetic reserves for Vicieae diversity in Syria: Ain Dinar, Al Hasakah, Kessab town, Kessab, Qal'at Al Hosn, Homs and Mimas, Djebel Druze, as well as in the Olimpos Beydaglari National Park, Belin, Turkey. It was possible to make these detailed recommendation because of the authors extensive study of *Vicia* in the Eastern Mediterranean. Therefore it is likely that other areas of the Mediterranean also warrant the establishment of genetic reserves, possibly in Spain, Greece and the Caucasus, where the species have not been so closely studied.

To improve the utilization of *Lathyrus*, more emphasis needs to be given to detailed characterization, evaluation and documentation of existing conserved genetic resources. A global data base network should be established under the control of IPGRI for easy access to information. The current ECP/GR (European Cooperative Programme for Genetic Resources) platform includes a database for *Lathyrus*, administered by IBEAS, Pau, France, but their information is focused on six cultivated species only: *L. cicera*, *L. heterophyllus*, *L. latifolius*, *L. sativus*, *L. sylvestris* and *L. tuberosus*. There is a need for a database that includes data on all 130

Lathyrus species. When available this would greatly facilitate all aspects of *Lathyrus* conservation and utilisation.

The search for genes for improving quantitatively inherited traits, such as yield in landraces and weedy and wild relatives, should be emphasized. The rigorous assessment of all available germplasm should be intensified to identify resistant sources against diseases (powdery mildew, Botrytis blight), insects (thrips, aphids), *Orobanche,* and cold. Basic studies need to be undertaken to establish:

- the genetic control of traits such as flower colour, b- ODAP content etc.; out-crossing mechanisms
- linkage maps for *Lathyrus* species
- the reproductive biology in *Lathyrus* species
- the inter-relations between different *Lathyrus* species using genetic and cytogenetic techniques, molecular markers and
- biotechnological approaches; and interspecific hybridization.

In any future development of *Lathyrus* as a safe crop for human consumption, zero or very low levels of neurotoxin (b-ODAP) content will be essential. Efforts should be made to search for toxin-free gene(s) in *Lathyrus* genetic resources for use in conventional breeding programs. Methods have already been developed for the production of transgenic *Lathyrus* plants. Transfer of this technology and the sharing of toxin-free transgenic plants among major *Lathyrus sativus* growing countries needs to be carried out. However, development of zero or low b-ODAP varieties should be coupled with development of appropriate strategies for maintenance of genetic purity through isolation mechanisms.

Another important aspect would be to evaluate the genetic potential of different species adapted to various agro-geographical areas for dry matter/ biological yield/ grain yield in comparison with other legume crops considered as more suitable for replacement on the basis of their promising attributes. There will be an increasing evaluation of the germplasm collections for biochemical and molecular markers to assess the genetic diversity of this genus, and to plan future research in areas such as further *ex-situ* collections, *in-situ* conservation, core collections, etc. The work with wild species will assume greater importance in the near future, as gene transfer between the cultigens and the wild species is improved by new biotechnological techniques, which will drive the need for further collection of wild and weedy species of this important genus.

The International Plant Genetic Resources Institute (IPGRI) jointly established with ICARDA and the Food and Agriculture Organization (FAO) its West Asia and North Africa Plant Genetic Resources Network (WANANET), operating from ICARDA headquarters, Syria, representing 13 countries. The Genetic Resources Unit of ICARDA and the WANANET are working in collaboration with the national programs to systematically fill some of these gaps by the collection and conservation of *Lathyrus* germplasm throughout the region.

9.8 RESEARCH NEEDS

Compared with many other forage legume genera and particularly with *Vicia, Lathyrus* is not well studied, there have been no comprehensive ecogeographic studies, the taxonomy is still poorly understood, even for section *Lathyrus* which contains most of the cultivated species. There is a need for a taxonomic review of the whole genus, particularly including the South American species that are very poorly known. The recent discoveries of new species (Mattatia and Heyn, 1976: Maxted and Goyder, 1988) suggest that there is still *Lathyrus* diversity yet to be discovered.

Associated with the need for fresh collections is the necessity to be able to identify the wild material in the field. Although good traditional keys exist, many conservationists still find these difficult to use and there is an on-going need for more user friendly multi-access keys for *Lathyrus* and for other legume genera.

Perhaps the major research need is associated with ensuring the utilisation of existing conserved accessions. The diversity present in collections needs to be characterised and evaluated in detail by using both molecular techniques and agro-morphological data. As well as aiding utilisation and maintenance of collections, these data, combined with geographic information systems techniques, could be used to identify gaps in the *ex situ* collections and areas for possible *in situ* conservation. The Mediterranean area, particularly the Fertile Crescent region, is an important centre of diversity for temperate legumes and several other crops groups. There is a serious need to highlight and monitor the amount of genetic erosion in the region.

Before appropriate regeneration can take place there is a need to establish the breeding system of the species. If the species is out crossing then regeneration its self can lead to unnecessary loss of genetic diversity. Therefore it is important to know if a species is an inbreeder or outbreeder. This information is not available for the majority of *Lathyrus* species. Much can be learned from looking at flower anatomy, diversity of characters within accessions and by screening for molecular markers. It is also important to ensure that conserved accessions are duplicated between at least two gene banks as a safety precaution against accidental loss of collections.

In recent years there has been some interest in reviewing the agronomic potential of previously undomesticated or minor crop *Lathyrus* species (e.g. *L. ochrus* and *L. hierosolymitanus*). As many of the wild species have only recently been conserved *ex situ* it is now appropriate to undertake a more systematic review of forage and fodder potential of the entire range of *Lathyrus* species, both for new species for crop cultivation and ornamental exploitation, but also for genes or gene complexes that might prove of use to agriculture and horticulture. For example, the recent discovery of *L. belinensis* Maxted and Goyder, which is closely related to *L. odoratus* is a major find for horticulturalists. The sweet pea, *L. odoratus*, has many flowered coloured forms but not yellow. Jones (1986) and Unwin (1987), both comments that the production of a yellow flowered sweet pea is one of the central goals of the contemporary sweet pea breeder. The flower of *L. belinensis* is relatively large (16-28 mm.) compared to other wild *Lathyrus* species and the yellow standard with red veining give the plant aesthetic appeal. Initial attempts to cross *L. odoratus* with *L. belinensis* using traditional hybridisation methods failed (Maxted, unpublished), but embryo rescue techniques have succeeded in producing a hybrid (Hammett *et al.*, 1994) and therefore it seems soon that the extremely marketable yellow sweet pea will soon be available. It is worth noting that *L. belinensis* has a very limited distributions to around Belin village, Antalya, Turkey and that road construction, forestry planting, increased tourism and changes in agricultural practice all threaten the few populations that have been found of this very restricted species.

References

Abd El Moneim, A. M. and Cocks, P.S., 1993. Adaptation and yield stability of selected lines of *Lathyrus* spp. under rainfed conditions in West Asia. *Euphytica*, 66: 89-97.

Aletor, V.A., Abd El Moneim, A.M. and Goodchild, A.V., 1994. Evaluation of the seeds of selected lines of three *Lathyrus* spp. for BOAA, tannins, trypsin-inhibitor activity and certain *in vitro* characteristics. *Journal of Science, Food and Agriculture*, 65: 143-151.

Anikster, Y., Feldman, M. and Horovitz, A., 1997. The Ammiad experiment. In: *Plant genetic conservation: the* in situ *approach* (eds. Maxted, N., Ford-Lloyd, B.V. & Hawkes, J.G.). pp. 239-253. Chapman & Hall, London.

Asthana, A.N., 1996. Grasspea cultivation in problem areas: present approaches. In: Lathyrus *Genetic Resources in Asia* (eds. Arora, R.K., Mathur, P.N., Riley, K.W. and Adham, Y.). pp. 143-148. IPGRI, New Delhi.

Baggott, C., 1997. *Ecogeographic survey of* Lathyrus *section* Lathyrus. Unpublished M.Sc. Thesis, University of Birmingham., UK.

Barker, B.T.P., 1916. Sweet Pea Hybrids. *Gardeners Chronicle Ser 3*, 60: 126.

Barrow, M.V., Simpson, C.F., and Miller, E.J., 1974. Lathyrism: a review. *Quarterly Review of Biology*, 49(2): 101-129.

Bennett, S., Maxted, N. and Sabanci, C.O., 1998. The ecogeography and collection of grain, forage and pasture legumes in south-west Turkey. *Genetic Resources and Crop Evolution*, 45: 253-269.

Campbell, C.G, Mehra R.B., Agarwal, S.K., Chen, Y.Z., Abd El Moneim, A.M., Khawaja, J., Yadov, C.R., Tay, J.W. and Araya, W.A., 1994. Current status and future strategy in breeding grasspea (*Lathyrus sativus*). In: *Expanding the Production and Use of Cool-Season Food Legumes* (eds. Muehlbauer, F.J. and Kaiser, W.J.). pp. 617-630. Kluwer Academic Pub., Dordrecht: The Netherlands.

Chernoff, M., Plitmann, U. and Kislev, M.E., 1992. Seed characters and testa texture in species of the Vicieae: their taxonomic significance. *Israel Journal of Botany*, 41: 167-186.

Czefranova, Z., 1971. Conspectus Systematis Generis *Lathyrus* L.. *Novititas Systematic Plant Vascular, (Leningrad)*, 8: 191-201.

Davies, A.J.S., 1957. Successful crossing in the genus *Lathyrus* through stylar amputation. *Nature* 180: 612.

Davies, A.J.S., 1958. *A cytogenetic study in the genus* Lathyrus. Unpublished Ph.D Thesis. University of Manchester, UK.

Davis, P.H., 1970. *Lathyrus* L. In: *Flora of Turkey and the East Aegean Islands, Vol. 3* (ed. Davis, P.H.). pp. 328-389. Edinburgh University Press, Edinburgh.

Ehrman, T.A.M. and Maxted, N., 1990. Ecogeographic survey and collection of Syrian Vicieae and Cicereae (Leguminosae). *Plant Genetic Resources Newsletter,* 77: 1-8.

Erskine, W., Smartt, J. and Muehlbauer, F.J., 1994. Mimicry of lentil and the domestication of common vetch and grasspea. *Economic Botany*, 48 : 326-332.

Ertug Firat A. and Tan, A., 1997. *In situ* conservation of genetic diversity in Turkey. In: *Plant genetic conservation: the* in situ *approach* (eds. Maxted, N., Ford-Lloyd, B.V. & Hawkes, J.G.). pp. 254-262. Chapman & Hall, London.

Ferguson, M.E., Ford-Lloyd, B.V., Robertson, L.D., Maxted, N. and Newbury, H.J., 1998. Mapping of geographical distribution of genetic variation in the genus *Lens* for enhanced conservation of plant genetic diversity. *Molecular Ecology*, 7: 1743-1755.

Francis, C., Maxted, N., Kyriakakis, S. and Bennett, S., 1995. Crete: observations on the ecogeography of grain, forage and pasture legumes. *CLIMA Occasional Paper No. 11.* Centre for Legumes in Mediterranean Areas, Perth.

Gams, H., 1924. *Lathyrus* L. In: Hegi, G. (Ed). *Illustrierte Flora von Mitteleuropa*, München: Lehmann, Vol. 4(3): 1494-1506.

Hammett, K.R.W., Murray, B.G., Markham, K.R and Hallett, I.C., 1994. Interspecific hybridisation between *Lathyrus odoratus* and *L. belinensis*. *International Journal of Plant Science*, 155(6): 763-771.

Hammett, K.R.W., Murray, B.G., Markham, K.R, Hallett, I.C. and Osterloh, I., 1996. New interspecific hybrids in *Lathyrus* (Leguminosae) : *Lathyrus annuus* x *L. hierosolymitanus*. *Botanical Journal of the Linnean Society*, 122: 89-101.

Hartman, C.P., Divakar, N.G. and Nagaraja Rao, U.N., 1974. A Study on *Lathyrus sativus*. Indian *Journal of Nutrition and Dietitian*, 11: 178-191.

Helback, H., 1965. Isin-Larsan and Horian food remains at Tel Bazmosian in the Dokan valley. *Sumer*, 19: 27-35.

IBPGR, 1985. *Forages for Mediterranean and adjacent arid / semi-arid areas: Report of a working group meeting held at Limassol, Cyprus*. International Board for Plant Genetic Resources, Rome, Italy.

ILDIS, 1999. *International Legume Database and Information Service*. http://www.ildis.org/

Jones, B.R., 1986. *The complete guide to sweet peas*. John Gifford, London.

Jones, G., 1992. Ancient and modern cultivation of *Lalthyrus ochrus* L in the Greek islands. *Annual of the British School of Archaeology at Athens*, 87: 211-217.

Kaul, A.K. and Combes, D., 1986. Lathyrus *and Lathyrism.* Third World Medical Research Foundation, New York.

Kearney, J.P., 1993. *Wild* Lathyrus *species as genetic resources for improvement of grasspea (*L. sativus*)*. Unpublished Ph.D. Thesis, University of Southampton, UK.

Khawaja, H.I.T., 1985. *Cytogenetic studies in the genus* Lathyrus. Unpublished Ph.D. Thesis, University of London, UK.

Khawaja, H.I.T., 1988. A new interspecific *Lathyrus* hybrid to introduce the yellow flower character into the Sweet Pea. *Euphytica*, 37: 69-75.

Kislev, M.E., 1989. Origins of cultivation of *Lathyrus sativus* and *L. cicera* (Fabaceae). *Economic Botany*, 43: 262 - 270.

Kroll, H., 1979. Kulturpflanzen aus Dimini. *Archaeo-Physika*, 8:173-189.

Kupicha, F.K., 1974. *Taxonomic studies in the tribe Vicieae (Leguminosae)*. Unpublished Ph.D. thesis, University of Edinburgh

Kupicha, F.K., 1981. Vicieae. In: *Advances in Legume Systematics* (eds. Polhill, R.M. and Raven, P.M.). pp. 377-381. Royal Botanic Gardens, Kew.

Kupicha, F.K., 1983. The infrageneric structure of *Lathyrus*. *Notes of the Royal Botanic Garden, Edinburgh*, 41: 209-244.

Linke, K.H., Abd El Moneim A.M. and Saxena, M.C., 1993. Variation in resistance of some forage legumes species to *Orobanche crenata*. *Field Crops Research*, 32: 277-285.

Malek, M.A., Sarwar, C.D.M., Sarker, A. and Hassan, M.S., 1996. Status of grasspea research and future strategy in Bangladesh. In: Lathyrus *Genetic Resources in Asia* (eds. Arora, R.K., Mathur, P.N., Riley, K.W. and Adham, Y.). pp. 7-12. IPGRI, New Delhi.

Marsden-Jones, M., 1919. Hybrids of *Lathyrus*. *Journal of the Royal Horticulture Society*, 45: 92-93.

Mattatia, J. and Heyn, C.C., 1976. A new *Lathyrus* from Israel: *L. hirticarpus sp. nov. Israel Journal of Botany*, 25: 216-220.

Maxted, N., 1993. Notes on *Lathyrus hirticarpus* from Syria. *Edinburgh Journal of Botany*, 50(1): 83-87.

Maxted, N. and Goyder, D.J., 1988. A new species of *Lathyrus* sect. *Lathyrus* (Leguminosae - Papilionoideae) from Turkey. *Kew Bulletin*, 43(4): 711-714.

Maxted, N., Khattab, A., Ehrman, T., Bisby, F.A., Obari, H., Sweidan, Y. & Goyder, D.J., 1993e. Important collections of Vicieae from Syria. *Proceedings of the 5th OPTIMA Meeting*, pp. 693-699.

Maxted, N., Obari, H. and Tan, A., 1990. New and interesting Vicieae from the Eastern Mediterranean. *Plant Genetic Resources Newsletter*, 78: 21-26.

Plitmann, U., Heyn, C.C. and Weinberger, H., 1986. Comparative taxonomy of some wild species allied to *Lathyrus sativus*. In: Lathyrus *and Lathyrism* (eds. Kaul, A.K. and Combes, D.). pp. 8-21. Third World Medical Research Foundation, New York.

Robertson, L.D., Singh, K.B., Erskine, W. and Abd El Moneim, A.M., 1996. Useful genetic diversity in germplasm collections of food and forage legumes from West Asia and North Africa. *Genetic Resources and Crop Evolution*, 43: 447-460.

Rutter, J. and Percy, S., 1984. The pulse that maims. New Scientist, 1418: 22-23.

Sabanci, C.O., 1996. *Lathyrus* genetic resources in Turkey. In: Lathyrus *Genetic Resources in Asia* (eds. Arora, R.K., Mathur, P.N., Riley, K.W. and Adham, Y.). pp. 77-86. IPGRI, New Delhi.

Sarker, A., Robertson, L.D., Abd El Moneim, A.M. and Campbell, C.G., 1997. *Lathyrus* species: conserved germplasm resources, priorities for collection and future prospects. In: *Proceedings of International Food Legume Research Conference III*, Adelaide, Australia. (in press).

Saw Lwin, A., 1956. On the cytology of *Lathyrus pratensis*. *Annales Academiae Scientiarum Fennicae, Series A. IV Biologica*, 78: 1-19.

Saxena M.C., Abd El Moneim, A.M. and Raninam, M., 1993. Vetches (*Vicia* spp.) and chicklings (*Lathyrus* spp.) in the farming systems in West Asia and North Africa and improvement of these crops at ICARDA. In: *Potential for* Vicia *and* Lathyrus *species as New Grain and Fodder Legumes for Southern Australia* (eds. Garlinge J.R. and Perry, M.W.). pp. 2 - 9. CLIMA , Perth.

Sarpaki, A. and Jones, G., 1990. Ancient and modern cultivation of *Lathyrus clymenum* L in the Greek islands. *Annual of the British School of Archaeology at Athens*, 85: 363-368.

Syouf, M.Q., 1996. Natural distribution and utilization of *Lathyrus* in Jordan. In: *Lathyrus* Genetic Resources in Asia (eds. Arora, R.K., Mathur, P.N., Riley, K.W. and Adham, Y.). pp. 67-76. IPGRI, New Delhi.

Tadesse, W., 1997. Grasspea (*Lathyrus sativus* L.): The hardy pulse crop. *FABIS*, 40: 1.

Tan, A., 1998. Current status of plant genetic resources conservation in Turkey. In: *The proceedings of international symposium on* in situ *conservation of plant diversity* (eds. Zencirci, N., Kaya, Z., Anikster, Y. & Adams, W.T.). pp. 5-16. Central Research Institute for Field Crops, Ankara, Turkey

Townsend, C.C. and Guest, E., 1974. *Flora of Iraq, Volume 3, Leguminales.* Ministry of

Agriculture and Agrarian Reform, Bagdad, Iraq.

Trankovskij, D.A., 1962. Interspecific hybridisation in the genus *Lathyrus*. Bulletin of Moscow *Nature and Biology Series*, 67: 140-141.

Unwin, C.W.J., 1986. *Sweet peas: their history development culture, 4th Edition*. Silent Books, Cambridge.

Vavilov, N. I., 1926. Weeds as progenitors of cultivated plants. *Bulletin of Applied Botany, Genetics and Plant Breeding*, 16, 195-217.

Westphal, W., 1974. *Pulses in Ethiopia, their taxonomy and agricultural significance*. Haile Sellassie University, Ethiopia, Agricultural University of Wageningen, the Netherlands, Centre for Agricultural Publishing and Documentation, Wageningen, the Netherlands.

Yamamoto, K., Fujiware, T. and Blumenreich, I., 1989. Isozymic variation and interspecific crossability in annual species of the genus *Lathyrus* L. In: Lathyrus *and Lathyrism* (eds. Kaul, A.K. and Combes, D.). pp. 118-21. Third World Medical Research Foundation, New York.

Yunus, A.G., 1990. *Biosystematics of* Lathyrus *section* Lathyrus *with special reference to the grasspea,* L. sativus *L.* Unpublished Ph.D. Thesis, University of Birmingham, UK.

Yunus, A.G. and Jackson, M.T., 1991. The gene pools of the Grasspea (*Lathyrus sativus* L.). *Plant Breeding*, 106: 319-328.

Zeist, W. van , 1972. Palaeobotanical results in the 1970 season at Cayano, Turkey. *Helinium*, 12: 3-19.

Zhou, M. and Arora, R.K., 1996. Conservation and use of underutilized crops in Asia. In: Lathyrus *Genetic Resources in Asia* (eds. Arora, R.K., Mathur, P.N., Riley, K.W. and Adham, Y.). pp. 91-96. IPGRI, New Delhi.

Chapter 10

Peas (*Pisum* L.)

Nigel Maxted and Mike Ambrose

10.1 INTRODUCTION

The garden or field pea is cultivated worldwide in temperate climates, but *Pisum sativum* L. is naturally found in Europe, north-west Asia and extending south to temperate east Africa, while *P. fulvum* Sibth. and Sm. is restricted to the Middle East. The pea has been cultivated for millennia, possibly because of the low levels of toxins in the seed (Liener, 1982) and the relatively high protein content of 25% (Monti 1983). Peas remain today one of the most important temperate pulse, fodder and vegetable crops. Garden peas (*P. sativum* var. *sativum*) are produced primarily for human consumption, field peas (*P. sativum* var. *arvense* (L.) Poiret) for livestock and traditionally as green manure. The pods are also eaten immature as a vegetable (e.g. mangetout, sugar snap peas or snow peas). In a number of developed countries, a significant proportion of the crop is now harvested in an immature state and frozen to make a convenience food.

Peas form an important component of the human diet in temperate regions of North America, the former USSR, Europe, India, China and to a lesser extent in sub-tropical climatic regions (Roy Davies, 1995). They are the forth most important legume pulse, with an estimated world production of approximately 12 Million tonnes for dry peas and 0.7 Mt for green peas, planted on approximately 7 million ha (6.2 m/ha dry peas and 0.8 m/ha green peas) in 1999. The majority are used for pulse production and the remainder largely frozen for convenience food production (FAO, 1999).

The pea is well placed to meet increased global demand for high protein human and animal feed and to act a leguminous break crop in diverse farming systems (Ambrose *et al.*, 1997). It is therefore important to ensure that its genetic resources are conserved and available for further exploitation and crop development. *Pisum sativum* L., is taxonomically isolated from other cultivated legumes and is placed in a small genus with its wild relative, *P. fulvum*. (ILDIS, 1999). However, the diversity of the genus *Pisum* is not currently well understood, conserved or utilised, even though its conservation has been given priority by the International Board for Plant Genetic Resources (IBPGR, 1985) in recent years.

10.2 TAXONOMIC SUMMARY

The genus *Pisum* is a member of the legume tribe Vicieae of subfamily Papilionoideae, along with *Lathyrus* L., *Lens* Mill., *Vavilovia* A. Fedorov and *Vicia* L. The genus is distinguished from related genera by the presence of relatively large, leafy stipules, which are semi-amplexical around the terete stem. The genus is most closely related to the monospecific genus *Vavilovia*, which contains the one dwarf, perennial, alpine species, *V. formosa* (Stev.) A. Fed. (formerly regarded as the *Pisum* species, *P. formosum* (Stev.) Boiss.), an endemic of high altitude areas in the Middle East.

There has been some agreement in the literature over the number of taxa in the genus but much dispute over their rank. Boissier (1872) recognised four annual species *P. elatius* Bieb., *P. humile* Boissier and Noë, *P. fulvum* and *P. formosanum* (Stev.) Alef. (syn. *V. formosa*), the former 3 have remained key taxa within the genus subsequently. Their inter-relationships and taxonomic position was partially clarified by breeding experiments undertaken by Ben-Ze'ev and Zohary (1973).

They found when crossing the putative species, *P. sativum*, *P. elatius*, *P. humile* and *P. fulvum*, that the former three taxa were completely inter-fertile and only *P. fulvum* was more isolated, and thus deserving of specific distinction from *P. sativum*. Therefore, the classification proposed by Davis (1970) which recognised two species and multiple subspecific taxa within *P. sativum* has largely been adopted, see Table 10.1. Davis was unlikely to have been aware of the work of Berger (1928), who previously came to a similar conclusion, but Berger gives all the taxa within *P. sativum* subspecific rank. The classification produced by Davis, produced for a national flora, does not include taxa found outside of the Middle East and so is not comprehensive. For example, the primarily Ethiopian taxon, *P. abyssinicum* A. Br., which possibly on diverse sources of evidence deserves specific rank within the genus (Kloz, 1971; Przybylska *et al.*, 1983; Parzysz and Przybylska, 1984; Ambrose, 1992; Ellis *et al.*, 1998) is excluded.

Smartt (1990) makes the point that widely cultivated species often suffer highly disrupted selection pressures which leads to the production of extreme morphological variants. For example, mangetout has been given specific status as *P. macrocarpum* Ser. ex Schur, but it is known that pod size is under simple genetic control and therefore this rank would appear unwarranted. Lehmann (1954), and Lehmann and Blixt (1984) have taken this to an extreme by proposing an artificial classification of *Pisum* containing: *P. formosum* (Stev.) Boiss., *P. fulvum* Sibth. & Sm., *P. elatius* (M.B.) Stev., *P. abyssinicum* A. Br., *P. syriacum* (Berger) Lehm. and *P. sativum* L.; dividing *P. sativum* into three subspecies, *asiaticum* Gov., *transcaucasicum* Gov. and *sativum*; and splitting *P. sativum* subsp. *sativum* further into 5 convarieties and 101 varieties. This exercise was undertaken to aid identification of the wide range of existing morphological variants, many of which were linked to known highly heritable alleles. Some feel that the erection of so many 'pseudo' taxa based on single gene differences has unnecessarily complicated the taxonomy of *Pisum*. However, Lehmann and Blixt (1984) argue that it provides a rapid tool for describing the range of variability found in the cultivated species and is particularly useful for breeders and researchers. Which ever interpretation there is now an abundant and complex synonymy that needs resolution and it must be clearly stated that the taxonomy of the genus requires urgent clarification!

Table 10.1. Classification of *Pisum* taxa.

Species	Subspecies	Varieties
P. sativum L.	*sativum*	*sativum*
		arvense (L.) Poiret
	elatius (Bieb.) Aschers. & Graebn.	*elatius*
		brevipedunculatum Davis & Meikle
		pumilio Meikle (syn. *P. humile* Boiss. & Noë)
P. abyssinicum A. Br.		
P. fulvum. Sibth. & Sm.		

All *Pisum* taxa are inbreeders with a diploid number of 2n=14 (Ben-Ze'ev and Zohary, 1973). The taxa within *P. sativum* are fully inter-fertile and crosses between *P. sativum* and *P. fulvum* are possible, particularly if *P. sativum* is the female parent (Ben-Ze'ev and Zohary, 1973). A small number of crosses between *Pisum* and *V. formosa* made at the N.I. Vavilov Institute of Plant Industry (St. Peterburg, Russia) have also proven successful, if *P. fulvum* is the female parent, but the two resultant F_1 hybrid plants were not viable (A. Golubev, pers. comm.).

10.3 CENTRE OF DIVERSITY

Associated with the need for a fresh taxonomic treatment of *Pisum* is the need for an ecogeographic survey of existing *Pisum* herbarium and gene bank collections along with a more detailed ecogeographic study of *Pisum* taxa that would incorporate systematically collecting fresh material of the wild taxa. The objective would be to ensure the material conserved accurately reflects the wild and cultivated gene pool. Gentry (1971) made initial field surveys of wild *Pisum* taxa in Sicily, Greece and Turkey, along with collating herbarium specimen passport data, but the collated result were seen as a preliminary survey only. The work of Ben-Ze'ev and Zohary (1973) focussed on inter and intra-specific relationships but they do provide some ecogeographic conclusions. These sources, along with Davis (1970) and data from personal field observation provide the geographical distribution for *Pisum* taxa, see Table 10.2 (country codes follow Hollis and Brummitt, 1992).

It is difficult to provide precise details for the centre of diversity for the crop taxa because their range has been modified significantly by human activities during the millennia it has been cultivated. Its cultivation is mentioned by Greek and Roman writers in the 500 BC (Roy Davies, 1995). Carbonised peas are found during excavations of Neolithic sites from c. 7,000 BC. Zohary and Hopf (1988) conclude that cultivation of peas pre-dates wheat and barley. Vavilov (1949) considered the centre of origin for peas was Ethiopia, the Mediterranean and central Asia, with a secondary centre in the Near East. Archaeological evidence suggests that this was the area of crop domestication (Zohary and Hopf, 1973).

Table 10.2. Geographic distribution of *Pisum* and *Vavilovia* taxa.

Taxon	Geographic Distribution
P. sativum var. *sativum*	Pan-temperate
P. sativum var. *arvense*	Europe and South-west Asia
P. sativum var. *elatius*	ALB, ALG, ARM, AZB, BGR, CYP, EGY, ESP, FRA, GEO, GRC, HUN, ISR, IRQ, IRN, ITA, JOR, LBN, LBY, MOR, PRT, ROM, SYR, TUN, TUR, UKR, YUG
P. sativum var. *brevipedunculatum*	CYP, TUR
P. sativum var. *pumilio*	CYP, EGY, IRN, IRQ, ISR, LEB, JOR, SYR, TUR
P. abyssinicum	ETH, YEM
P. fulvum	ARM(?), CYP, GRC, IRQ, ISR, JOR, LBN, SYR, TUR, YUG
V. formosa	ARM, AZB, GEO, IRN, IRQ, SYR, TUR

10.4 ECOGEOGRAPHIC DISTRIBUTION

There is limited information available for the ecological requirement of Pisum taxa because there is yet to be a detailed ecogeographic study. However, the limited data in the literature (Govorov 1937, Davis, 1970; Ben-Ze'ev and Zohary, 1973;) together with personal field observation is summarised in Table 10.3.

Table 10.3. Ecological distribution of *Pisum* and *Vavilovia* taxa.

Taxon	Cultivation Status	Ecological Preferences
P. sativum var. *sativum*	Crop	Cultivated & spontaneous
P. sativum var. *arvense*	Crop	Cultivated & spontaneous
P. sativum var. *elatius*	Wild	Rocky & grassy slopes, forest & field margins, scrub, ruins, 0 - 1,700m
P. sativum var. *brevipedunculatum*	Wild	Fields, vineyards & roadsides, 700 - 1,000m
P. sativum var. *pumilio*	Wild	Open scrub, rocky slopes & fallow fields, 700 - 1,800m
P. abyssinicum	Crop / Wild	Cultivated & spontaneous
P. fulvum	Wild	Rocky limestone slopes, & dwarf shrubland, 30 -1,500m
V. formosa	Wild	Loose limestone scree, > 1,500m

10.5 GENETIC DIVERSITY

An understanding of gene pool diversity is fundamental to gene pool exploitation. The use of available characterisation data on conserved germplasm along with multivariate analysis on qualitative and quantitative characters (Ambrose, 1992; Amurrio *et al.*, 1995) is helping to identify a core collection for *Pisum*. Along with traditional morphological characterisation, more recent molecular techniques are being used to identify accession groupings (Hoey *et al.*, 1995; Lu *et al.*, 1996). The application of amplified fragment length polymorphism (AFLP) analysis on 56 accessions of the core collection from the John Innes Collection (Ambrose *et al.*, 1997) revealed new insights into the phylogenetic relationship between *Pisum* species and within *P. sativum*. However, this work has shown the need for further systematic sampling of the *Pisum* gene pool before a clear picture of the range of genetic diversity can be interpreted. Early suggestions from a limited number of accessions covering combined and vining cultivars from both European and North American gene pools indicate that the range of genetic diversity is highly restricted and shows clear signs of founder effect. The corollary to this lack of diversity is that we can anticipate that there is much adaptive variation still to be transferred/exploited.

The primary goal of pea breeders remains high yield and resistence to disease and lodging and for those cropped for freezing uniformity of ripening is also important (Roy Davies, 1995). The various root rots and mildews remain a major limiting factor for production (Ali *et al.*, 1994). The linkage map for pea has undergone extensive revision over recent years and has grown to include a range of PCR markers (Ellis *et al.*, 1992, Laucou *et al.*,1998, Weeden *et al.*, 1998). Significant progress has also been made in resolving outstanding issues relating to numbers of linkage groups and establishing clearer relationships between the genetic and cytogenetic maps (Hall *et al.*, 1997a and b). These developments are likely to further assist in crop development in the future. Although the genetic map for *Pisum* remains not fully resolved, it is the most comprehensive of all the grain legumes and the techniques being developed are using peas as their experimental material, and thus rapid progress is expected. Thus far a total of 789 genes have been registered and published (http://www.jic.bbsrc.ac.uk/germplasm/pisum) of which 180 genes appear on the current consensus map (Weeden *et al.* 1998).

10.6 Conservation Resource Status

Current estimates indicate there to be in the region of 76,000 *Pisum* accessions held in *ex-situ* collections (FAO, 1998). Detailed information about the current conservation status of *Pisum* species is documented by WIEWS (World Information and Early Warning System on Plant Genetic Resources), which contains information on national PGR holdings (www.fao.org/ag/agp/pgr/wiews/) and SINGER (System-wide Information Network for Genetic Resources), which contains information on CGIAR holdings (http://www.cgiar.org/singer). Details of the composition of the major collections are provided in Table 10.4.

Table 10.4. Number of *Pisum* accessions conserved *ex situ* in major collections.

Species	Institutes							
	AARI	ATFC	ICARDA	IPK	JI	NGB	W-6	VIR
P. sativum var. *sativum*	10	3683	882	2384	1680	1150	3718	6509
P. sativum var. *arvense*	15	13	ND	ND	ND	ND	58	ND
P. sativum var. *elatius*	8	17	10	15	31	8	51	3
P. sativum var. *brevipedunculatum*	1	ND	ND	ND	ND	ND	ND	ND
P. sativum var. *pumilio*	2	7	1	0	4	2	24	0
P. abyssinicum	4	16	6	41	33	4	17	4
P. fulvum	2	53	31	4	55	10	48	2
V. formosa	0	0	0	0	0	1	0	5
Total	42	3789	930	2444	1803	1175	3916	6523

NB. AARI - Aegean Agricultural Research Institute, Menemen, Turkey; ATFC - Australian Temperate Field Crop Collection, Horsham, Australia; ICARDA - International Centre for Agricultural Research in the Dry Areas, Syria; IPK - Institut für Pflanzengenetik und Kulturpflanzenforschung (IPK), Gatersleben, Germany; JI - John Innes Centre, Norwich, UK; NGB - Nordic Genebank Alnarp, Sweden; W-6 - Regional Plant Introduction Station, Washington, USA; and VIR - Vavilov Research Institute of Plant Industry, St. Petersburg, Russia. ND - Not differentiated.

The initiative within the European Cooperative programme of Genetic Resources (http://www.cgiar.org/ecpgr/platform/index.htm) has lead to the formation of central crop databases for a number of crop species within Europe (Maggioni, 1998). The Grain Legume Working Group is responsible for *Pisum*, though the European *Pisum* Database was initiated earlier (Ambrose *et al.*, 1994). This database currently holds basic passport data on 32,000 accessions from 32 different collections throughout Europe (including ICARDA). Although preliminary analysis of accession for some of the wild taxa shows relatively high levels of putative duplicates (Ambrose *et al.*, 1997; Du Plessis, 1999). Ambrose *et al.* (1997) suspect that the level of duplication across all collections in Europe could be as high as 60-80%. However, this may not be surprising when the rarity of accessions of the wild taxa is considered along with the desire of national breeding programmes to screen wild germplasm for desirable traits. *P. fulvum* accessions are known to have resistance to drought and bruchid beetle attack (Ali *et al.*, 1994), and some accessions are known to produce amphicarpic legume and to be very drought resistant, traits are of great interest to breeders (Ginzberger, Pers. Comm.). There is yet another problem in interpreting current *Pisum ex situ* holdings, because there is no generally accepted classification of *Pisum* taxa being applied by the collections, it is difficult to merge data sets where different taxon and nomenclature concepts have been applied.

As with many legume groups, *in situ* conservation, whether in a genetic reserves or on farm, has so far not yet been specifically adopted for *Pisum* species. The wild taxa are undoubtedly currently conserved in existing conservation areas (national parks, wildlife reserves, etc.), but here they are being 'passively' conserved (species and genetic diversity is not being actively monitored and managed) and thus it is are susceptible to genetic erosion and possible extinction.

10.7 Conservation Gaps and Priorities

Although there are relatively large *ex situ* seed collections of cultivated and wild *Pisum* germplasm, the collections are not comprehensive in terms of taxon or genetic diversity and there remain numerous gaps in conserved materials, particularly for Ethiopian material and *V. formosa*. Maxted and co-workers, in conjunction with IBPGR and ICARDA, and latterly with CLIMA have engaged in 14 legume collection missions to the North-eastern Mediterranean region (centre of primary diversity) since 1986 and *Pisum* germplasm was collected as part of each of these missions. Even though these missions collected diverse forms of *P. sativum* and many accessions of *P. fulvum*, the gene pool remains insufficiently sampled. Both the expeditions led by Maxted and those initiated by the Genetic Resource Unit of ICARDA have reported varying levels of genetic erosion occurred in the genus, especially in landraces of *P. sativum* var. *sativum* and var. *arvense*, and in *P. abyssinicum* in areas of traditional pea production and where *Pisum* germplasm has yet to be sampled. Though having made this point, as the current holdings of *Pisum* are poorly characterised, detailed priorities will need to be revised once the genetic diversity in the gene pool is better understood.

As well as the need for further germplasm collection for *ex situ* conservation, there is a complementary need to make positive steps to establish *in situ* on farm conservation projects, particularly in those countries that retain extensive landrace material such as: Albania, Armenia, Ethiopia, Iran, Morocco, Portugal, Syria, Turkey and Yemen. In terms of the establishment of genetic reserves to conserve the wild species, perhaps the highest priority should be given to ensuring *V. formosa* is effective conserved as it is so difficult to maintain *ex situ*. Therefore, it is suggested that priority is given to establishing a reserve for *V. formosa* in Armenia, East of Yerevan. Existing genetic reserves in Syria and Turkey (Anikster *et al.*, 1997; Ertug Firat and Tan, 1997) are likely to contain *Pisum* taxa, and it is suggested that these reserves should be surveyed for *Pisum* populations. Recorded population should subsequently be routinely monitored to ensure population health.

10.8 RESEARCH NEEDS

Clearly one of the highest priorities for research that will aid *Pisum* conservation and use is the production of a stable, comprehensive classification of the genus. It does appear surprising that for such an important grain legume there is no generally agreed classification. The relatively small number of taxa involved make the lack of such a classification even more inexplicable. Allied to the poor taxonomic and genetic understanding of diversity in the genus, is the lack of ecogeographic data. Therefore, it is recommended that tied to future targeted collecting is a detailed ecogeographic study that will provide the essential baseline data on which to found future effective conservation and utilisation. It is unlikely that one research group or institute will be able to undertake such a study and closer international collaboration is recommended, possible under the aegis of ECP/GR. As recommended, targeted germplasm collections will continue to play an important role in underpinning breeding and research. There is a responsibility to ensure long-term preservation and the activities of the ECP/GR should ensure this is achieved in a more collaborative manner with inter-site safety duplication. The development of an international *Pisum* core collection spread through a number of different institutions around the temperate world could prevent duplication of individual initiatives and investment of time, expertise and resources. As Ambrose *et al.* (1997) conclude the centralisation of *Pisum* information could also be usefully extended to provide a single point access to information on *Pisum* germplasm on a world basis.

Ambrose and Green (1991) state that the greatest challenge for Pisum conservationists is the thorough morphological and molecular characterisation and evaluation of existing collections. This remains a prime research goal. It is well established that unless conserved germplasm is characterised and evaluated it is unlikely to be used in breeding programmes and conserved material with no prospect of utilisation is redundant.

REFERENCES

Ali, S.M., Sharma, B. and Ambrose, M.J., 1994. Current status and future strategy in breeding pea to improve resistence to biotic and abiotic stresses. *Euphytica*, 73: 115-126.

Ambrose, M.J. and Green, F.N., 1991. A review of *Pisum* genetic resources and germplasm utilisation. *Aspects of Applied Biology*, 27:243-251.

Ambrose, M.J., 1992. Making use of exotic germplasm: Improving the odds. In: *1st European conference on grain legumes*. pp. 39-40. European Association for Grain Legume Research, Paris.

Ambrose, M.J., Green, F.N., Winfield, P.J., 1994. Development of a computerised gene list for *Pisum*. In: *Evaluation and exploitation of genetic resources pre-breeding* (eds. Balfourier, F. and Perretant, M.R.). Proceedings of the Genetic Resources Section Meeting of EUCARPIA, 15-18 March 1994, Clermont-Ferrand, France. pp 269-270.

Ambrose, M.J., Weeden, N.F. and Simon, C.J., 1997. *Pisum* spp. conserved resources, priorities for collection and future prospects. *International Food Legume Research Conference III*, Adelaide, Australia, September, 1997.

Amurrio, J.M., de Ron, A.M., and Zeven, A.C., 1995. Numerical taxonomy of Iberian pea landraces based on quantitative and qualitative characters. *Euphytica*, 82:195-205.

Anikster, Y., Feldman, M. and Horovitz, A., 1997. The Ammiad experiment. In: *Plant genetic conservation: the* in situ *approach* (eds. Maxted, N., Ford-Lloyd, B.V. and Hawkes, J.G.). pp. 239-253. Chapman and Hall, London.

Ben-Ze'ev, N. and Zohary, D., 1973. Species relationships in the genus *Pisum* L. *Israel Journal of Botany*, 22: 73-91.

Berger, A., 1928. Classification of Pisum. In: *Vegetables of New York. Part 1 - Peas* (eds. Hedrick, U.P., Hall, F.H., Hawthorne, L.R. and Berger, A.). pp. 1-132. New York State Agricultural Experiment Station Report, Geneva, New York.

Boissier, E., 1872. *Flora Orientalis, Vol. 2.* pp. 622-624. H. Georg, Geneva & Basel.

Davis, P. H., 1970. *Pisum* L. In: *Flora of Turkey and the East Aegean Islands, Vol. 3* (ed. P. H. Davis). pp.370-373. Edinburgh University Press, Edinburgh.

Du Plessis, M., 1999. *Preliminary analysis of the European* Pisum *database*. MSc Thesis, University Of Birmingham, Birmingham, U.K.

Ehrman, T.A.M. and Maxted, N., 1989. Ecogeographic survey and collection of Syrian Vicieae and Cicereae (Leguminosae). *Plant Genetic Resources Newsletter*, 77:1-8.

Ellis, T.H.N., Turner, L., Hellens, R.P., Lee, D., Harker, C.L., Enard, C., Doomoney, C. and Davies, D.R., 1992. Linkage maps in peas. *Genetics*, 130: 649-663.

Ellis, T.H.N., Poyser, S.J., Knox, M.R., Vershinin, A.V. and Ambrose, M.J., 1998. Polymorphism of insertion sites of *Ty1-copia* class retrotransposons and its use for linkage and diversity analysis in peas. *Molecular Genetics*, 260: 9-19.

Ertug Firat, A. and Tan, A., 1997. *In situ* conservation of genetic diversity in Turkey. In: *Plant genetic conservation: the* in situ *approach* (eds. Maxted, N., Ford-Lloyd, B.V. and Hawkes, J.G.). pp. 254-262. Chapman and Hall, London.

FAO, 1998. *The state of the world's plant genetic resources for food and agriculture.* pp. 1-510. Food and Agriculture Organisation of the United Nations, Rome.

FAO, 1999. http://www.fao.org/ *Food and Agricultural Organisation of the United Nations*, Rome.

Gentry, H.S., 1971. *Pisum* Resources: a preliminary survey. *Plant Genetic Resources Newsletter*, 25: 3-13.

Goverov, L.I., 1937. Erbsen. In: *Flora of cultivated plants, Vol. IV* (eds. Vavilov, N.I. and Wulff, E.V.). pp 231-336.

Hall, K.M., Parker, J.S., Ellis, T.H.N., 1998a. The relationship between genetic and cytogenetic maps of pea. I. Standard and translocation karyotypes. *Genome*, 40: 744-754.

Hall, K.L., Parker, J.S., Ellis, T.H.N., Turner, L., Knox, M.R., Hofer, J.M.I., Lu, J., Ferrandiz, C., Hunter, P.J., Taylor, J.D., Baird, K., 1998b. The relationship between genetic and cytogenetic maps of pea. II Physical maps of linkage mapping populations. *Genome*, 40: 775-769.

Hoey, B.K., Crowe, K.R., Jones, V.R., Polans, N.O., 1996. A phylogenetic analysis of *Pisum* based on morphological characters and allozyme and RAPD markers. *Theoretical and Applied Genetics*, 92: 92-100.

Hollis, S. and Brummitt, R.K., 1992. *World geographical scheme for recording plant distributions*. Hunt Institute for Botanical Documentation, Pittsburgh.

IBPGR, 1985. *Forages for Mediterranean and adjacent arid / semi-arid areas: Report of a working group meeting held at Limassol, Cyprus*. International Board for Plant Genetic Resources, Rome, Italy.

ILDIS (1999) http://www.ildis.org/ *International Legume Database and Information Service*, Reading.

Kloz, J., 1971. Serology of the Leguminosae. In: *Chemotaxonomy of the Leguminosae* (eds. Harbourne, J., Boulter, D. and Turner, B.L.). pp. 309-365. Academic Press, London.

Laucou, V., Haurogne, K., Ellis, N. and Rameau, C., 1998. Genetic mapping in pea. 1. RAPD-based genetic linkage map of *Pisum sativum*. *Theoretical and Applied Genetics*, 97: 905-915.

Lehmann, C., 1954. Das morphologische System der Saaterbsen (*Pisum sativum* L. *sens. lat.* Gov. ssp. *sativum*). *Zuercher Chter.*, 24: 316-337.

Lehmann, C. and Blixt, S., 1984. Artificial infraspecific classification in relation to phenotypic manifestation of certain genes in *Pisum*. *Agri. Hort. Genetica, Band*, XLII: 49-74.

Liener, I.E., 1982. Toxic constituents in legumes. In: *Chemistry and biochemistry of legumes* (ed. Arora, S.K.). pp. 217-257. Oxford and IBH Publishing Co., New Delhi.

Lu., J., Knox, M.R., Ambrose, M.J., Brown, J.K., M. Ellis, T.H.N., 1996. Comparative analysis of genetic diversity in pea assessed by RFLP- and PCR- based methods. *Theoretical and Applied Genetics*, 93: 1103-1111.

Maggioni, L., 1998. ECP/GR to continue into a sixth phase. *Newsletter for Europe*, 14: 7.

Matthews, P., Swiecicki, W. and Wingfield, P., 1994. European *Pisum* Catalogue. In: Evaluation and exploitation of genetic resources pre-breeding (eds. F. Balfourier and M.R. Perretant). *Proceedings of the Genetic Resources Section Meeting of EUCARPIA*, 15-18 March 1994, Clermont-Ferrand, France. pp. 271-275.

Matthews, P. and Ambrose, M.J., 1995. Development and use of a 'Core' collection for the John Innes *Pisum* collection. In: *2nd European Grain legume Conference*. pp. 194-195. Copenhagen, Denmark, European Association for Grain Legume Research, Paris.

Monti, L., 1983. Natural and induced variability in peas for protein production. In: *Perspectives for pea and lupins as protein crops*. (eds R. Thompson and R. Casey). pp 23-29. Martinus, Nijhoff, The Hague.

Parazysz, H. and Przybylska, J., 1984. Isoenzyme variation in the genus *Pisum* VI. Electrophoretic analysis of variation in the legumin fraction composition. *Genetica Polonica*, 24(1): 21-38.

Pisum Gene List, (1999). *Pea genetics*. http://www.jic.bbsrc.ac.uk/germplasm/pisum/

Przybylska, J., 1983. Comparative study of seed proteins in the genus Pisum II. Electrophoretic patterns of alcohol dehydrogenase and isocitrate from cotyledons of ungerminated seeds. *Genetica Polonica*, 25(3): 255-259.

Roy Davies, D., 1995. Peas: *Pisum sativum* (Leguminosae - Papiionoideae). In: *Evolution of crop plants* (eds. Smartt, J. & Simmonds, N.W.), 2nd edition. pp. 294-296. Longman Group, Harlow, Essex.

Smartt, J., 1990. *Grain legumes: evolution and genetic resources*. Cambridge University Press, Cambridge.

Weeden, N.F., Ellis, T.H.N., Timmerman-Vaughan, G.M., Sweicicki, W.K., Rosov, S.M., and Berdnikov, V.A., 1998. A consensus linkage map for *Pisum sativum*. Pisum *Genetics*, 30: 1-4.

Westphal, E., 1974. *Pulses in Ethiopia, their taxonomy and agricultural significance*. Centre for Agricultural Publishing and Documentation, Wageningen.

Vavilov, N.I., 1949. The origin, variation, immunity and breeding of cultivated plants. *Chronika Botanica*, 13: 1-54.

Zohary, D. and Hopf, M., 1973. Domestication of pulses in the Old World. *Science*, 182: 887-894.

Chapter 11

Lupins (*Lupinus* L.)

Wallace Cowling

11.1 INTRODUCTION

Lupins belong to one of the most diverse and widespread of flowering plant genera. Wild species of *Lupinus* L. occur naturally in North and South America, the Mediterranean region and northern Africa. The Mediterranean region harbours the genetic resources of the most important agricultural species of lupin, namely: *L. albus* L., *L. angustifolius* L., and *L. luteus* L.

Lupins in the Mediterranean region have been profoundly influenced by the activities of humans, at least since at least Neolithic times (Gladstones, 1974; 1998). Human influence may have threatened the existence of some wild lupins, and enhanced the spread of other wild or semi-domesticated lupin species. The ancient Egyptians, Greeks and Romans used *L. albus* for soil improvement and to precede cereals in crop rotation, and seeds were soaked in brine to de-bitter them prior to consumption (Gladstones, 1970; Hondelmann, 1984; 1996). Seeds of *L. albus* are still a popular snack food in many countries and islands of the Mediterranean region, where they are known by a range of names with Greek, Latin or arabic origins (Table 11.1). The albus lupin has been cultivated in the Mediterranean region for millennia, and current-day landraces show tremendous morphological diversity that will be described in this chapter. In contrast, the wild form of albus lupin, *L. albus* var. *graecus* (Boiss. and Spruner) Gladst., is very rare and restricted to the Balkans and southern Italy (see Section 11.3.1 below).

L. angustifolius was not domesticated in early times, but may have been deliberately or inadvertently spread by early civilisations around the Mediterranean region (Gladstones, 1974; 1998). Wild species with attractive flowers, such as *L. luteus* and *L. pilosus* Murr., were taken to botanical gardens in northern Europe where they were selected as ornamentals (Hanelt, 1960). Many lupin species are very rare in the Mediterranean region, possibly as a result of habitat restriction through clearing or over grazing. One species from northern Africa, *L. somaliensis* Baker, is known only from a single 100-year-old botanical specimen (Gladstones, 1974). Very few accessions of *L. digitatus* Forsk., *L. princei* Harms, or *L. palaestinus* Boiss. exist in world gene banks (Cowling *et al.*, 1998a). Many of these species have lost substantial natural habitat due to human activities, especially in the past 100 years. For these reasons, there are major concerns about the genetic resource conservation status of many lupin species in the Mediterranean region. This chapter examines the taxonomy and genetic diversity of Lupinus in the Mediterranean region, which is one of the major centres of diversity of this genus and the source of the domesticated lupins used in modern agriculture.

Table 11.1. Species names, major synonyms, common names and chromosome numbers of the Mediterranean and North African lupins (reproduced with permission from the International Plant Genetic Resources Institute from Table 1 in Cowling *et al.*, 1998a).

Species	Botanical Synonyms	Vernacular Names	Chromosome No.
Smooth-seeded Mediterranean and North African group			
L. albus L.	*L. termis* Forsk. *L. sativus* Gater. *L. varius* Gaertn. *L. prolifer* Desr. *L. hirsutus* Eichw.	Greek-derivatives: termis, turmas, turmus. Latin derivative: lupino. English: albus lupin, white lupin. French: lupin. German: weisse Lupine. Italian: lupino. Portuguese and Spanish derivatives: altramuz, tremoços, entramoços, chochos. Russian: ljupin belyj. Turkish: aci bakla	
L. albus L. var. *albus*	*L. albus* L. *L. albus* L. subsp. *albus* *L. albus* L. var. *termis* (Forsk.) Alef *L. albus* L. subsp. *termis* (Forsk.) Caruel	As above	50
var. *graecus* (Boiss. & Sprun.) Gladst.	*L. albus* L. subsp. *graecus* (Boiss. et Sprun.) Franco et P.Silva *L. graecus* Boiss. et Sprun. *L. jugoslavicus* Kazimierski et Nowacki *L. vavilovi* Atabek. et Maissurjan *L. albus* ssp. *graecus* (Boiss. & Spruner) Franco et P.Silva	English: wild albus lupin	50
L. angustifolius L.	*L. varius* L. *L. sylvestris* Lam. *L. linifolius* Roth. *L. reticulatus* Desv. *L. leucospermus* Boiss. *L. philistaeus* Boiss. *L. cryptanthus* Shuttlew. ex Campb. *L. opsianthus* Atabek.	English: narrow-leafed lupin, blue lupine. German: schmalblättrige oder blaue Lupine.	40
L. luteus L.	*L. odoratus* hort. ex DC.	English: yellow lupin. German: gelbe Lupine.	52

L. hispanicus Boiss. & Reut.			52
subsp. *hispanicus*			52
subsp. *bicolor* (Merino) Gladst.	*L. luteus* var. *bicolor* Merino *L. hispanicus* var. *bicolor* Merino *L. bicolor* (Merino) Rothm. *L. rothmaleri* Klink. *L. versicolor* Caball.		52
L. micranthus Guss.	*L. hirsutus* L.		52

Rough-seeded Mediterranean and North African group

L. cosentinii Guss.	?*L. semiverticillatus* Desr. in Lam. *L. hirsutus* Black *L. digitatus* Lojac. *L. varius* Caruel *L. pilosus* Black *L. pilosus* subsp. *digitatus* Fiori et Paol.	English: sandplain lupin, Western Australian blue lupin	32
L. digitatus Forsk.	*L. tassilicus* Maire *L. pilosus* subsp. *tassilicus* (Maire) Quezel et Santa		36
L. princei Harms			38
L. pilosus Murr	*L. hirsutus* L. *L. varius* L. *L. varius* L. subsp. *orientalis* Franco et P.Silva		42
L. palaestinus Boiss.			42
L. atlanticus Gladst.		English: Atlas lupin	38
L. somaliensis Baker			?

11.2 TAXONOMIC SUMMARY

Family: *Fabaceae*
Tribe: *Genisteae* (Adanson) Bentham
Genus: *Lupinus* L., from Latin *lupus*, "wolf"
Type species: *L. albus* L.

Two major groups are recognised in the genus *Lupinus* – the geographically separated and isolated New World and Old World species. Several hundred species have been named in the New World where they extend from Alaska to Argentina (Dunn, 1984). DNA evidence confirms that North American species and *L. mutabilis* Sweet from the Andes form a relatively "young" group (Käss and Wink, 1997), whereas lupins from the east coast of South America and the Old World are much older. *Lupinus* most likely originated with the Genisteae and Thermopsideae in tropical Africa (Käss and Wink, 1997) with s several migrations northwards to the Mediterranean (to form the Old World lupins) and across the Atlantic or the Bering Straight (to form the New World lupins) (Gladstones, 1998). Despite such a broad distribution and variable morphology in the genus, the unity of the genus is not questioned and all authors agree in assigning the genus to the tribe *Genisteae*. This chapter focuses on lupin species from the Old World – the Mediterranean and North African species (Table 11.1).

Until the 1970s, the taxonomy of Old World lupin species was in a state of confusion as a result of mistakes and misunderstandings from the early years of Linnaean taxonomy. Based on the major taxonomic revision of Gladstones (1974), twelve species of Mediterranean and African lupin species are recognised, one of which (*L. somaliensis*) may be extinct (Table 11.1). The separation of the Old World lupins into the rough-seeded group and the smooth-seeded group was formalised by Plitmann and Heyn (1984) who named the rough-seeded lupins Section *Scabrispermae*. There is a high degree of genomic similarity in the rough-seeded lupins with high levels of chromosome pairing and gene transfer in interspecific crosses (Gupta *et al.*, 1996). This is in contrast to the smooth-seeded Mediterranean lupins which have major genetic barriers between most species (Plitmann and Heyn, 1984).

Several lupin species continue to be incorrectly named in the literature by one of their botanical synonyms (Table 11.1). For example, *L. termis* Forsk. is used by some authors for the landraces of *L. albus* var. *albus* that are grown in the Nile Valley of Egypt. It is now clear that the Egyptian landraces represent one of several geographical races of *L. albus* var. *albus* in the Mediterranean, North African and Atlantic region (Simpson, 1986a; Simpson, 1986b) (see Section 3.1 below). The wild form of *L. albus* has been regarded as a distinct species by some authors (Table 11.1), but Gladstones (1974) proposed that it be considered simply as a taxonomic variety (*L. albus* var. *graecus*), since it is fully interfertile with *L. albus* var. *albus* and differs mainly in the characters selected during domestication (Gladstones, 1974).

The yellow lupin, *L. luteus*, has the same chromosome number (2n = 52) as *L. hispanicus* Boiss. and Reuter (Table 11.1). Interspecific hybrids have been obtained artificially between these two species (Kazimierski and Kazimierska, 1965; 1970; Swiecicki, 1985).

L. angustifolius is very diverse morphologically with two distinct groups with large or small seeds (Cowling *et al.*, 1998a; Gladstones and Crosbie, 1979). The large-seeded types may have been selected for use as a green manure crop or for grazing by early farmers (Gladstones, 1998). The small-seeded and fine-leafed types were found in "wild" situations and were given various names such as *L. varius* L., *L. linifolius* Roth. and *L. reticulatus* Desv. (Table 11.1), but freely intercrossed with the large-seeded types (Gladstones, 1974). Lee and Gladstones (1979) proposed that *L. varius* be regarded as a synonym under *L. angustifolius* and this was formally accepted in 1983 (Anon, 1983).

L. micranthus Guss. is serologically quite distinct from the other smooth-seeded Mediterranean species (Cristofolini, 1989; Nowacki and Prus-Glowacki, 1971), despite having the same chromosome number (2n = 52) as *L. luteus* and *L. hispanicus* (Table 11.1). The name *L. hirsutus* L. has been misapplied to both *L. micranthus* and *L. pilosus* (Lee and Gladstones,

1979), and the name *L. hirsutus* has now been formally rejected (Anon, 1983) for Mediterranean lupin species.

The taxonomy of the rough-seeded lupins proposed by Gladstones (1974) has been largely accepted (Table 11.1). *L. tassilicus* Maire is now considered a synonym of *L. digitatus* based on herbarium specimens, chromosome numbers and plant growth (Plitmann and Heyn, 1984). Gladstones (1974) discontinued the incorrect use of *L. digitatus* for *L. cosentinii* Guss. He also created the name *L. atlanticus* Gladst. for a rough-seeded lupin species found in Morocco (Gladstones, 1974). It has been shown that *L. princei* has the same chromosome number (2n = 38) as *L. atlanticus*, but the two species are quite distinct and do not intercross (Carstairs *et al.*, 1992). *L. pilosus* also shares the same chromosome number (2n = 42) as *L. palaestinus* but the two species are distinct, forming aborted pods on crossing (Gupta *et al.*, 1996). Recently the name *L. anatolicus* was proposed for a lupin collected in Turkey that resembled *L. pilosus* but had relatively smooth seed coats (Swiecicki *et al.*, 1996). However, some accessions of *L. pilosus* from Syria have a relatively smooth seed coat, sometimes occurring as a mixture with typical rough-seeded types at the same collection site (Buirchell and Cowling, 1998; Clements *et al.*, 1996). The Turkish accession identified as *L. anatolicus* (Swiecicki *et al.*, 1996) had the same chromosome number as *L. pilosus* (2n = 42) and crossed readily with *L. pilosus* to produce viable F1 and F2 plants (Buirchell and Cowling, 1998). It therefore seems premature to suggest that it should be separated from *L. pilosus*.

11.3. Centres of Diversity

The geographical distribution of the Old World lupins in the Mediterranean region and North Africa is described in Hanelt (1960) and Gladstones (1974; 1998), based largely on the collection sites of botanical specimens and published flora of the Mediterranean region. Precise collection site data for lupins in the Mediterranean region were used to construct distribution maps for the smooth-seeded and rough-seeded lupins (Figs. 1A and 1B, respectively). These maps help to physically visualise centres of diversity for the various Mediterranean lupin species, based on actual collection site data.

11.3.1 Centres of Diversity in *L. albus*

The primary centre of diversity of *L. albus* may be inferred from the current-day distribution of *L. albus* var. *graecus* in the southern Balkans and southern Italy. It is always rare, and sometimes occurs as natural hybrids with *L. albus* var. *albus* in southern Greece and on some Aegean islands (Cowling, 1986; Simpson and McGibbon, 1982a). There is one report of a wild *L. albus* var. *graecus* from north-eastern Greece. This accession is hard-seeded and late-flowering with dark blue flowers, shattering pods and small reddish-brown and bitter seeds (Clements and Cowling, 1990a; 1991). *L. albus* var. *graecus* was also found in Calabria in southern Italy in 1983 (Perrino *et al.*, 1984). Although this appears to be an anomalous discovery, the authors speculate that the early Greek settlers took it there. This is the first report of wild *L. albus* var. *graecus* outside of the Balkans.

The restricted geographic range of *L. albus* var. *graecus* suggests that landraces of *L. albus* var. *albus* have been moved around the Mediterranean region since Neolithic times. The current distribution of *L. albus* var. *albus* includes the mid-Atlantic islands, the Canary Islands, Egypt, Ethiopia, Kenya, Greece, Turkey, Syria, Israel, Italy and the Iberian Peninsula (Figure 11.1A). The ancient practice of eating cooked and debittered seeds of *L. albus* var. *albus* as a snack food has survived to the present in both Arabic and western Mediterranean cultures. Agricultural selection has resulted in great genetic diversity in *L. albus* var. *albus* around the Mediterranean region.

Figure 11.1. Map of precise locations in the Mediterranean region of collections of (a) the smooth-seeded and (b) the rough-seeded lupins held in the International Lupin Collection at Agriculture Western Australia, Perth.

Figure 11.1A. Lupin collection sites: smooth-seeded. (For Color Plate version, see p. xviii)

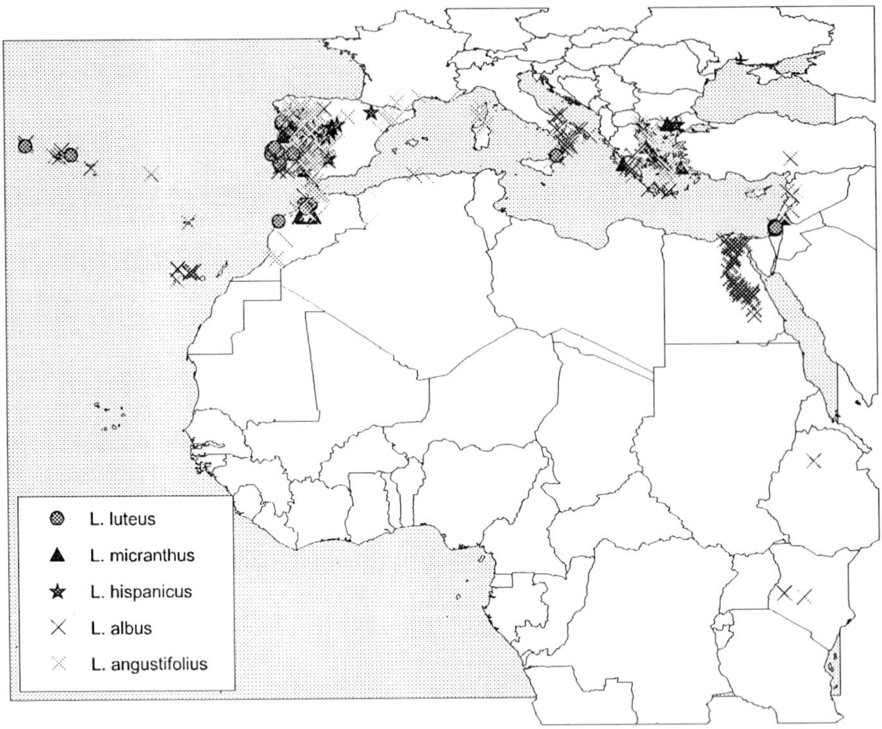

Figure 11.1B. Lupin collection sites: rough-seeded. (For Color Plate version, see p. xviii)

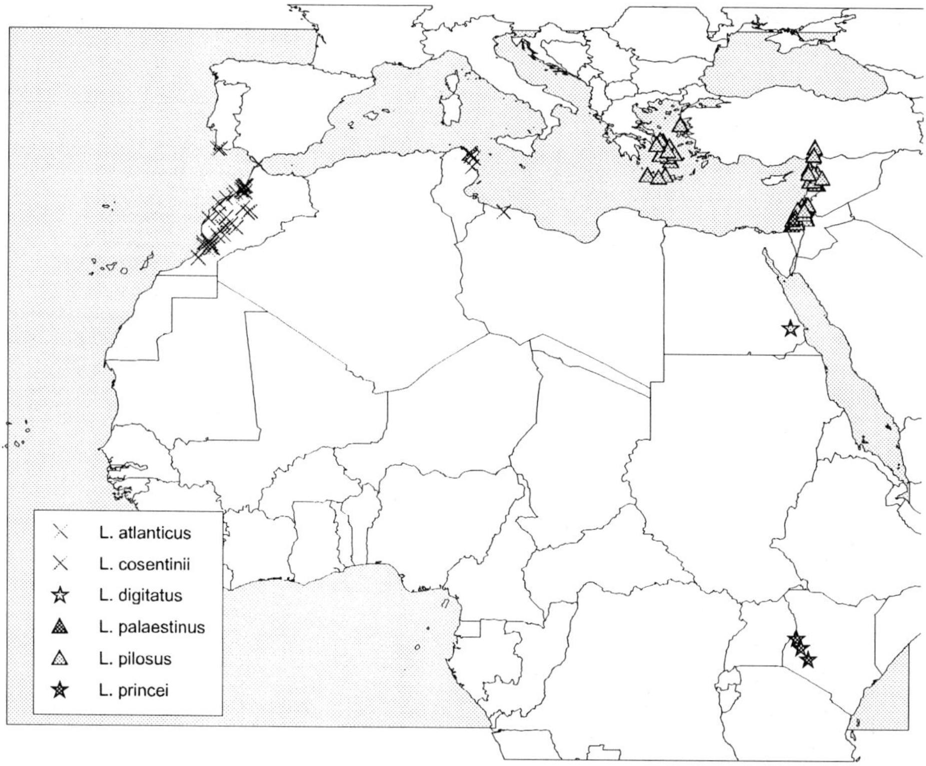

Simpson (1986a; 1986b) used principal component analysis and non-metric multidimensional scaling to characterise four distinct geographical races of *L. albus* in the Mediterranean region:

1. The Iberian race - with two distinct local ecotypes. The first type from Azores Islands, Catalonia, Cadiz and central Portugal was characterised by large leaves, pods and seeds. The second local ecotype from the northern provinces of Palencia and Leon had small leaves, pods and seeds and was relatively early flowering.
2. The Nile Valley race - with strong affinities to southern Iberian populations. This type had smaller and fewer main stem leaves, as well as shorter rosette stages.
3. The Turkish race - quite distinct and homogeneous, characterised by unusually small pods, seeds and leaves, few main stem leaves, few primary pods, more secondary pods and numerous tertiary and higher order level pods. The earliest flowering genotypes are from this region and they also lack a vernalisation response (Simpson, 1984). In addition, there were populations that contained pink flowers, a feature not found in any other group.

Recent collections of *L. albus* var. *albus* from the highlands of Ethiopia also have very small seeds (Buirchell and Cowling, 1998).

4. The Balkan race - both wild and cultivated populations. This race was characterised by short stature and abundant primary and secondary pods. The wild types (var. *graecus*) were distinguished by their shorter stems, deep blue flowers, shattering pods, and coloured seed with impermeable testa.

11.3.2 Centres of Diversity in *L. angustifolius*

L. angustifolius is the most common and widespread lupin the Mediterranean region (Fig. 1A). In an ecogeographical study of *L. angustifolius* in Greece, Clements and Cowling (1994) found that types with very fine leaves and small grey seeds tended to be found in elevated areas of northern Greece and scattered on some islands. In contrast, types with large leaves, pods and seeds were found at lower elevations, as weedy volunteers in pastures, and sometimes on fertile alluvial soil (Clements and Cowling, 1994). This is consistent with the theory that early farmers selected large-seeded types in arable areas for food, forage or green manure (Gladstones, 1998).

In a similar fashion to *L. albus* var. *albus*, genetic diversity in wild *L. angustifolius* may have been expanded through human intervention. The Greek islands seem to be particularly rich in genetic diversity in this species. The island of Naxos in the Aegean Sea, with its ancient history of trade and cultural development, had representatives of 9 out of 13 morphological groups of *L. angustifolius* identified in the Aegean region (Clements and Cowling, 1994). The Aegean islands were used as ports for sea traders from the time of the ancient Greeks, and *L. angustifolius* may have been accidentally or deliberately moved among the various centres of civilisation in the Mediterranean region.

11.3.3 Centres of Diversity in *L. luteus*

Based on the hypothesis that the smooth-seeded lupins originated in the eastern Mediterranean (Gladstones, 1998), *L. luteus* may now be mostly separated from its origins. Wild *L. luteus* is rare in the Middle East and is found mostly on the Iberian Peninsula (Figure 11.1A). It remains to be determined if the current-day eastern and western Mediterranean types are ancestors of an earlier progenitor of *L. luteus* from the eastern region.

A recent collection of wild *L. luteus* from southern Spain had a low branching growth habit identical to *L. micranthus* amongst which it was found (W.A. Cowling, Pers. Obs.). This supports the idea that, while genetically distinct, these lupin species have evolved under similar conditions and in response to similar stimuli (in this case, most likely grazing pressure).

11.3.4 Centres of Diversity in Other Smooth-Seeded Lupins

L. micranthus has a similar distribution pattern to *L. angustifolius* around the perimeter of the Mediterranean Sea, although it is not as common as *L. angustifolius* (Figure 11.1A). It may have been accidentally carried in hay and forage by humans, and appears to be adapted to the very poorest soils and most drought-prone conditions.

L. hispanicus is found mainly on the Iberian Peninsula in Spain and Portugal with *L. hispanicus* ssp. *hispanicus* at intermediate altitudes in southern and central Spain and *L. hispanicus* ssp. *bicolor* at higher altitudes (up to 1,500 m) in north-western Spain and northern Portugal. Botanical reports exist of *L. hispanicus* in Turkey, and Cowling (1986) collected seed of a cream flowered lupin with pure white seeds at two locations in northern Greece. These collections may have belonged to the *L. hispanicus* group and their presence in the eastern Mediterranean adds weight to the hypothesis that the smooth-seeded lupins evolved in the east (Gladstones, 1998).

11.3.5 Centres of Diversity in the Rough-Seeded lupins

The rough-seeded lupins are found in Northern Africa, the eastern Mediterranean, and in southern Portugal and Spain (Figure 11.1B). The present geographical distribution of this closely related group includes the moist Mediterranean coast (for example, *L. cosentinii*, *L. palaestinus* and *L. pilosus*) and the cooler highlands of north and east Africa (for example, *L. atlanticus* in Morocco and *L. princei* in Kenya). One species is a desert ephemeral (*L. digitatus*) (Figure 11.1B).

None of the rough-seeded lupins have extensively colonised the northern Mediterranean region, except for *L. pilosus* in the east and *L. cosentinii* in the west. *L. cosentinii* is found in coastal areas of Morocco, Spain, Portugal, Algeria and Tunisia, and rarely at higher altitudes in Morocco. *L. pilosus* is found in the eastern Mediterranean from southern Greece to Syria and Israel. *L. palaestinus* is found in a small region of southern Israel. *L. atlanticus* is adapted to the higher altitudes of the Atlas Mountains of Morocco. *L. princei* may have become "stranded" in the cool highlands of Kenya and northern Tanzania, as temperatures increased in interglacial periods. *L. digitatus* is recorded over large areas of northern Africa surrounding the Sahara Desert, the Senegal Valley and the Nile Valley (Gladstones, 1974). It is very rarely collected today and is an endangered species.

11.4. ECOGEOGRAPHICAL DISTRIBUTION

Detailed ecogeographical information on lupins in the Mediterranean region is provided in Cowling *et al.* (1998a) and Buirchell and Cowling (1998) and will be summarised briefly here.

The genus *Lupinus* is adapted to a wide range of soil types, with specialisation occurring among the species. The smooth-seeded lupins tend to be found on acidic soils (mean pH < 7.0) in the western Mediterranean and the rough-seeded lupins on alkaline soils (mean pH > 7.0) in the eastern Mediterranean or northern Africa (Cowling *et al.*, 1998a) (Figs. 1A and B). For example, *L. pilosus* is the dominant lupin on the alkaline soils of coastal Syria and Israel. Collection sites in northern Spain and Greece are acidic and contain mostly *L. angustifolius* and *L. hispanicus*. However, there are numerous exceptions to this rule: *L. albus* var. *albus* is found on a range of soils from acid to alkaline. *L. albus* var. *albus* was the main species collected on the highly alkaline soils of the Nile Valley (Cowling *et al.*, 1998a). Simpson and McGibbon (1982a) also found *L. albus* on limestone-derived soils in Greece. There is a trend of increasing soil acidity in lupin occurrence from south to north in Morocco, the Balkans and the Iberian Peninsula (Cowling *et al.*, 1998a). Evidence from *L. pilosus* (Clements *et al.*, 1996) and *L. angustifolius* (Clements and Cowling, 1994) suggests that soil pH at lupin collection sites tends to decrease at higher altitudes.

Some species appear to be adapted to high elevation, although it is difficult to separate adaptation to high elevation from adaptation to acidic soil types. For example, *L. hispanicus* is found mainly on very acid soils at high elevation on the Iberian Peninsula (Cowling *et al.*, 1998a), but *L. angustifolius* is found in the same region and has a much wider adaptation to Mediterranean environments. Adaptation to high elevation requires later flowering to avoid frost damage: *L. angustifolius* collections from higher elevations in the Aegean region had a more pronounced rosette stage, later flowering, smaller seeds and more yield on lateral branches than on the main stem (Clements and Cowling, 1994). These smaller-seeded and finer-leafed types of *L. angustifolius* were previously thought to be different species (Gladstones, 1974), but may simply represent an adaptation to the climate at higher elevations.

Other species collected over a wide range of elevations include *L. atlanticus* which is found at a wide range of altitudes in the Atlas Mountains in Morocco, and *L. pilosus*, found on low elevation coastal sandy soils up to 1100 m on inland mountains (Cowling *et al.*, 1998a). In contrast, *L. luteus* is found close to sea level on the Moroccan and the Iberian peninsulas, and *L. cosentinii* is normally found close to sea level in the same region, with occasional collections at high elevations in Morocco.

Landraces of *L. albus* var. *albus* are widely distributed from sea level at the Mediterranean coastline to over 2400 m in the Kenyan highlands. This demonstrates the power of human intervention on ecogeographical adaptation in *L. albus*. *L. princei* inhabits a niche in the tropical highlands of east Africa similar to *L. mutabilis* in the Andes, and both have been collected above 2000 m (Cowling *et al.*, 1998a). Most lupin species grow in environments with at least 300 mm average annual rainfall, although lower values have been recorded for *L. angustifolius, L. cosentinii* and *L. atlanticus*. The latter two species, along with *L. palaestinus* and presumably *L. digitatus*, occur mainly where rainfall is less than 600 mm. The remaining species are normally found in rainfall regions above 600 mm with some collections of *L. pilosus* at sites with up to 2000 mm annual rainfall (Cowling *et al.*, 1998a). *L. angustifolius* plants collected from higher rainfall areas of the Aegean region tended to be taller at maturity and later flowering when grown at similar latitudes in Western Australia (Clements and Cowling, 1994). Similar associations were found in *L. pilosus* from the eastern Mediterranean, with later flowering and smaller seeded types collected in higher rainfall environments (Clements *et al.*, 1996).

11.5. Genetic Diversity

One measure of genetic diversity of lupins from the Mediterranean region is obtained from field evaluation data at genetic resource centres. Most of the genetic diversity data summarised below is from the International Lupin Collection held at Agriculture Western Australia, Perth, Western Australia, and details on individual accessions may be found in Clements and Cowling (1991). Perth is at a similar latitude (32°S) to the North African coast of the Mediterranean Sea (32°N) and has a similar Mediterranean climate, and therefore provides a relevant measure of phenological development such as flowering time.

11.5.1 Flowering Time

There is large genetic diversity for flowering time in most species when planted in the mild winter/spring growing season in Perth. Most of this diversity is due to variation in vernalisation requirement, as flowering in Mediterranean lupins such as *L. angustifolius* is only mildly influenced by day length (Reader *et al.*, 1995). Some accessions of *L. albus, L. angustifolius, L. cosentinii* and *L. pilosus* flowered in less than 80 days in Perth (Buirchell and Cowling, 1998). This indicates very little vernalisation requirement for flowering, but other accessions of the same species took longer than 130 days to flower and had a strong vernalisation requirement. *L. luteus, L. micranthus* and *L. hispanicus* were the latest flowering Mediterranean species, especially *L. hispanicus* which averaged 143 days to first flower (Buirchell and Cowling, 1998) and had no early flowering types (Cowling *et al.*, 1998a). This species occurs mainly at high elevations in Spain.

Later flowering time in *L. angustifolius* from the Aegean region was associated with cold winter temperatures and high rainfall and altitudes at collection sites (Clements and Cowling, 1994). Flowering time in *L. pilosus* was not correlated with altitude of collection site, but increased as rainfall increased (Clements *et al.*, 1996). Several selections of *L. pilosus* from European botanical gardens were early flowering with little vernalisation requirement, probably as a result of selection as ornamental flowering lupins for spring planting in northern Europe (Clements *et al.*, 1996). *L. princei* had the longest time to flowering of all lupin species. This species grows over long mild seasons in the tropical highlands of Kenya and Tanzania (Cowling *et al.*, 1998a) and flowering time may be controlled by different physiological mechanisms than the Mediterranean lupins. Such late flowering makes it unsuitable for cultivation in a Mediterranean-type climate.

11.5.2 Seed Weight

L. albus wild types and landraces have the widest range of seed weights (Cowling *et al.*, 1998a). Some landraces from Ethiopia had very small seeds (16 g/100 seeds) (Buirchell and Cowling, 1998), and were most likely selected for small seeds by early farmers, although the reason for this is unknown. These small-seeded Ethiopian *L. albus* landraces do readily cross with cultivated varieties (B.J. Buirchell, unpublished data). *L. albus* var. *albus* was undoubtedly selected for large seeds for use as a snack food, resulting in the large-seeded Italian landraces (77 g/100 seeds) (Buirchell and Cowling, 1998).

Seed weight in *L. angustifolius* wild types ranges from 2.9 to 24.4 g/100 seeds (Buirchell and Cowling, 1998). The distribution of seed size in 586 accessions of wild *L. angustifolius* had a distinctively bimodal shape (Cowling *et al.*, 1998a), supporting the earlier measurements of Gladstones and Crosbie (1979). Gladstones (1998) suggests that *L. angustifolius* consists of two types, one truly "wild" with small seeds, later flowering, fine-leaved and short, and the other with large seeds, tall, with larger leaves and earlier flowering. However, the two groups are not sufficiently separated to be regarded as subspecies. The large-seeded group is commonly associated with olive groves or vineyards, and on more fertile soils – and may be the product of human selection for cultivation as a green manure (Gladstones, 1974; Gladstones, 1998).

11.5.3 Pod Set on the Main Stem

There is large genetic diversity among and within lupin species for pod set on the main stem. Wild and landrace *L. albus* and *L. angustifolius* on average set fewer main stem pods (6 and 10 pods, respectively) than *L. luteus* and *L. hispanicus* (17 and 20 pods respectively) (Cowling *et al.*, 1998a). This species difference is also observed in domesticated cultivars, with *L. luteus* cultivars having better main stem pod set than *L. angustifolius* cultivars in Western Australia (W.A. Cowling, unpublished data). The greater number of pods on the main stem of *L. luteus* is accompanied by poorer pod set on the lateral branches than *L. angustifolius*.

There is a wide range of pod set on the main stem of *L. angustifolius* wild types, from 1 to 27 pods (Buirchell and Cowling, 1998), and this diversity is likely to be important in breeding of cultivated varieties.

11.6. Conservation Resource Status

Sweet cultivated lupins have been fully domesticated only in the past 60 years, and are genetically very close to their wild and landrace ancestors in the Mediterranean region. Primary crosses between domesticated types and wild or landrace types continue to play an extremely important role in modern lupin breeding in improving disease and pest resistance, frost tolerance, and adaptation to alkaline soils (Cowling *et al.*, 1998b). The importance of genetic resource conservation of wild and landrace lupin species has been well recognized by lupin breeders since the early years of 'modern' lupin breeding (Cowling et al., 1998b).

Several researchers collected lupins in Mediterranean countries in the 1930s and 1940s (Hackbarth and Troll, 1957). Much loss of lupin genetic resources occurred in Germany during World War II (Hackbarth and Troll, 1957). New collecting missions were made into the Mediterranean region from the late 1960s to the 1980s (Buirchell, 1992; Cowling, 1986; Gladstones, 1973; 1976; Hammer and Perrino, 1995; Lehmann and Hammer, 1983; Mota *et al.*, 1981; 1982; Papineau and Huyghe, 1989, and Simpson and McGibbon, 1982a; 1982b). There were many new introductions to the Australian Lupin Collection between 1968 and 1989 (Clements and Cowling, 1990b; Gladstones and Crosbie, 1979).

The *ex situ* conservation status of Mediterranean lupins is relatively healthy, with more than 18,000 accessions accounted for in germplasm collections around the world (Cowling *et al.*, 1998a). However, there is a high degree of duplication of accessions as a result of seed exchange among institutions, and many accessions are morphological selections from within collection sites. Many of the lesser-known Mediterranean and North African lupin species, such as *L. palaestinus*, *L. digitatus* and *L. princei*, are very poorly represented in world germplasm collections (Cowling *et al.*, 1998a).

The *in situ* conservation status of lupins in the Mediterranean region may be subjectively inferred from the frequency of collection sites of various species in Figures 11.1A and 11.1B. *L. angustifolius* and *L. albus* var. *albus* appear to be common but other species such as *L. digitatus* appear to be under threat. However, these figures do not tell the whole story, as there has been little effort to collect species such as *L. digitatus* and to conserve them in germplasm collections. In the case of *L. albus* var. *albus*, cultivation of old landraces of *L. albus* var. *albus* is diminishing rapidly in some areas due to loss of traditional agricultural practices. For example, *L. albus* var. *albus* populations dramatically declined between two collecting missions to the Azores Islands in 1989 and 1994 (Cowling *et al.*, 1998b). All wild lupins in the Mediterranean region are under threat from overgrazing and increasing human population pressure.

11.7. CONSERVATION GAPS AND PRIORITIES

The development of lupins as crop plants will continue to depend on genetic improvements arising from wild or landrace relatives. Fully domesticated lupin varieties of *L. albus*, *L. luteus* and *L. angustifolius* have been available for less than 60 years, and diseases and pests continue to threaten and reduce cultivation of lupins worldwide (Sweetingham *et al.*, 1998). One of the most important priorities for lupin genetic resource conservation is the identification of disease and pest resistance in wild and landrace relatives. Expanding the range of soils and climates on which lupins may be grown is also a major priority.

Examples of important conservation priorities include:

- Collection and identification of *L. albus* var. *albus* adapted to alkaline soils with high bicarbonate content that will potentially expand the albus industry in Europe and elsewhere;
- Collection and identification of cold-tolerant types of *L. albus* that will benefit the production of winter lupins in northern Europe;
- Collection and identification of anthracnose resistance in *L. angustifolius*, *L. luteus* and *L. albus*. This disease has caused major disruption to lupin industries worldwide in recent years;
- Collection and identification of resistance to fusarium wilt disease in *L. angustifolius*, *L. luteus* and *L. albus*. While this disease appears to be confined to northern Europe at present, it is a major concern to Australian lupin breeders due to the susceptibility of Australian *L. angustifolius* varieties to this disease.
- Collection and breeding of rough-seeded lupins.

A fully domesticated variety of *L. cosentinii* (cv. Erregulla) was bred some years ago (Gladstones, 1982) and *L. atlanticus* is now fully domesticated (Buirchell, 1999). Most domestication genes of *L. pilosus* have now been found (Buirchell, 1999). However, the number of wild genotypes of rough-seeded lupins in genetic resource centres for these species is relatively low (Cowling *et al.*, 1998a). The rough-seeded lupins have great potential to expand the uses of cultivated lupins on alkaline soils. Collection and conservation of rough-seeded lupins is a major priority.

11.8 RESEARCH NEEDS

While lupins remain a small crop by world standards (Cowling *et al.*, 1998a), the needs for research on lupin genetic resources and breeding will always outstrip the availability of funds. As demonstrated in this review, genetic resources research has been strongest in *L. albus* (Simpson,

1986a; 1986b) and *L. angustifolius* (Clements and Cowling, 1994) in Europe and Australia respectively. Similar ecogeographical research has been carried out on *L. pilosus* (Clements *et al.*, 1996) and *L. atlanticus* (Buirchell and Cowling, 1998), and needs to be expanded to other important lupin species such as *L. luteus*.

Lupins have several problems that limit their use in agriculture. *L. angustifolius* has a poor harvest index, which may be the result of the indeterminate pattern of development and inefficient partitioning between reproductive and vegetative growth (Perry *et al.*, 1998). There is a need for further research into lupin development and the possible genetic restriction of branching to achieve better crop yields (Perry *et al.*, 1998). Genetic resources will play a vital part in this research – it is important to understand plant growth and development in wild and landrace relatives of crop plants in order to transfer useful traits to modern cultivars.

The issue of lupin seed quality has not been tackled in this review, and it may be the limiting factor for the expansion of lupins into modern food and feed industries. Sweet lupins have now been accepted in the Food Standards Code in Australia and Great Britain (Cowling *et al.*, 1998a), as a result of low alkaloids and other anti-nutritional factors in lupin seed. Lupins have potential as a human food, including the use of valuable dietary fibre in cotyledonary cell walls (Petterson, 1998). However, lupins have a relatively high hull content compared with other grain legumes (24-25% in *L. angustifolius* and *L. luteus*) (Cowling *et al.*, 1998a) which reduces their value as a legume for human consumption. The high hull content increases cooking time of whole seeds, making them unattractive as a pulse, and increases the cost of extraction of food products such as dietary fibre from cotyledons. High crude fibre is also a nutritional constraint in diets for monogastric animals where it reduces energy content and causes other problems (Edwards and van Barneveld, 1998). Recent research has demonstrated significant genetic variation in hull content in wild and domesticated lupin germplasm (Reader and Dracup, 1997), and lupin genetic resources may play a vital role in improving the quality of lupins for both human and animal consumption in the future.

ACKNOWLEDGEMENTS

I thank the International Plant Genetic Resources Institute, Rome, for permission to reproduce Table 11.1 from Cowling *et al.* (1998a), and Colin Smith of Agriculture Western Australia who constructed the maps reproduced in Figure 11.1, based on collection site data for accessions held in the International Lupin Collection at Agriculture Western Australia, South Perth.

REFERENCES

Anon., 1983. Proposals 474 and 475. *Lupinus varius* L. and *Lupinus hirsutus* L. (*Leguminosae*). *Taxon*, 32: 624.

Buirchell, B.J., 1992. Collecting wild *Lupinus* spp. in Morocco. *Plant Genetic Resources Newsletter*, 90: 36-38.

Buirchell, B.J., 1999. Genetic diversity in rough-seeded lupins. In: *Towards the 21st Century* (ed. Hill, G.D.), Proceedings of the 8th International Lupin Conference, Asilomar, California, p. 290-303. International Lupin Association, Canterbury, New Zealand.

Buirchell, B.J. and W.A. Cowling, 1998. Genetic resources in lupins. In: *Lupins as Crop Plants: Biology, Production and Utilization* (eds. Gladstones, J.S., Atkins, C. and Hamblin, J.). p. 41-66, CAB International, Oxon, UK.

Carstairs, S.A., Buirchell, B.J. and Cowling, W.A., 1992. Chromosome number, size and interspecific crossing ability of three Old World lupins, *Lupinus princei* Harms, *L. atlanticus* Gladstones and *L. digitatus* Forskål, and implications for cyto-systemic relationships among the rough-seeded lupins. *Journal of the Royal Society of Western Australia*, 75: 83-88.

Clements, J.C., Buirchell, B.J. and Cowling, W.A., 1996. Relationship between morphological variation and geographical origin or selection history in *Lupinus pilosus*. *Zeitschrift für Pflanzenzüchtung - Plant Breeding, Berlin*, 115: 16-22.

Clements, J.C. and Cowling, W.A., 1990a. *The Australian Lupin Collection: Passport Data for Wild and Semi-Domesticated Accessions Introduced into Australia to 1990.* Research Report 1/90. Department of Agriculture of Western Australia, Western Australia.

Clements, J.C. and Cowling, W.A., 1990b. Summary of wild and domesticated lupin accessions introduced to Australia from 1974 to 1989. *Australian Plant Introduction Review*, 21: 1-14.

Clements, J.C. and Cowling, W.A., 1991. 1991 *Catalogue of the Australian Lupin Collection: Including Field Evaluation Data for Wild, Semi-Domesticated and Fully Domesticated Accessions. Research Report 3/91.* Department of Agriculture of Western Australia, Western Australia.

Clements, J.C. and Cowling, W.A., 1994. Patterns of morphological diversity in relation to geographical origins of wild *Lupinus angustifolius* from the Aegean region. *Genetic Resources and Crop Evolution*, 41: 109-122.

Cowling, W.A., 1986. Collection of wild *Lupinus* in Greece. *Plant Genetic Resources Newsletter*, 65: 20-22.

Cowling, W.A., Buirchell, B.J. and Tapia, M.E., 1998a. *Lupin (*Lupinus *spp).* Promoting the conservation and use of underutilised and neglected crops. 23. Institute of Plant Genetics and Crop Plant Research, Gatersleben/International Plant Genetic Resources Institute, Rome, Italy.

Cowling, W.A., Huyghe, C. and Swiecicki, W., 1998b. Lupin breeding. In: *Lupins as Crop Plants: Biology, Production and Utilization* (eds. Gladstones, J.S., Atkins, C. and Hamblin, J.). p. 93-120, CAB International, Oxon, UK.

Cristofolini, G., 1989. A serological contribution to the systematics of the genus *Lupinus* (*Fabaceae*). *Plant Systematics and Evolution*, 166: 265-278.

Dunn, D.B., 1984. Cytotaxonomy and distribution of New World lupin species. In: *Proceedings of the 3rd International Lupin Conference* (ed. Anon). pp. 68-85. La Rochelle, France.

Edwards, A.C. and van Barneveld, R.J., 1998. Lupins for livestock and fish. In: Lupins as Crop Plants: Biology, Production and Utilization (eds. Gladstones, J.S., Atkins, C. and Hamblin, J.). pp. 385-409, CAB International, Oxon, UK.

Gladstones, J.S., 1970. Lupins as crop plants. *Field Crops Abstracts*, 23: 123-148.

Gladstones, J.S., 1973. Observations on the environment and ecology of some annual legumes in southern Italy. *Plant Introduction Review, CSIRO (Australia)*, 9: 11-29.

Gladstones, J.S., 1974. *Lupins of the Mediterranean Region and Africa.* Technical Bulletin No. 26. Department of Agriculture of Western Australia, Perth.

Gladstones, J.S., 1976. Observations on the distribution and ecology in Iberia and North Africa of some annual legumes adapted to neutral and acid soils. *Plant Introduction Review, CSIRO (Australia)*, 11: 9-23.

Gladstones, J.S., 1982. Breeding lupins in Western Australia. *Journal of Agriculture - Western Australia (4th Series)*, 23: 73-76.

Gladstones, J.S., 1998. Distribution, origin, taxonomy, history and importance. In: *Lupins as Crop Plants: Biology, Production and Utilization* (eds. Gladstones, J.S., Atkins, C. and Hamblin, J.). pp. 1-39, CAB International, Oxon, UK.

Gladstones, J.S. and Crosbie, G.B., 1979. *Lupin Wild Types Introduced into Western Australia to 1973: Collection Site Data, Preliminary Ratings of Field Characteristics and Disease Reactions, and Measurements of Seed Protein and Oil Contents.* Technical Bulletin No. 43. Department of Agriculture of Western Australia, Perth.

Gupta, S., Buirchell, B.J. and Cowling, W.A., 1996. Interspecific reproductive barriers and genomic similarity among the rough-seeded *Lupinus* species. *Zeitschrift für Pflanzenzüchtung - Plant Breeding, Berlin*, 115: 123-127.

Hackbarth, J. and Troll, H.-J., 1957. Lupinen als Körnerleguminosen und Futterpflanzen [Lupins as grain legumes and fodder plants]. In: *Handbuch der Pflanzenzüchtung IV* (eds. Kappert, H. and Rudorf, W.). pp. 1-51, Verlag Paul Parey, Berlin and Hamburg. 2nd Ed.

Hammer, K. and Perrino, P., 1995. Plant genetic resources in South Italy and Sicily - studies toward and *in situ* and on-farm conservation. *Plant Genetic Resources Newsletter*, 103: 19-23.

Hanelt, P., 1960. *Die Lupinen: Zur Botanik und Geschichte landwirtschaftlich wichtiger Lupinenarten. Die Neue Brehm-Bücherei.* A. Ziemsen Verlag, Wittenberg Lutherstadt.

Hondelmann, W., 1984. The lupin - ancient and modern crop plant. *Theoretical and Applied Genetics*, 68: 1-9.

Hondelmann, W., 1996. *Die Lupine: Geschichte und Evolution einer Kulturpflanze. Landbauforschung Völkenrode Sonderheft 162.* Bundesforschungsanstalt für Landwirtschaft Braunschweig-Völkenrode (FAL), D-38116 Braunschweig, Germany.

Käss, E. and Wink, M., 1997. Molecular phylogeny and phylogeography of *Lupinus* (Leguminosae) inferred from nucleotide sequences of the rbcl gene and its 1+2 regions of rDNA. *Plant Systematics and Evolution*, 208: 139-167.

Kazimierski, T. and Kazimierska, E.M., 1965. Studies on a species hybrid *Lupinus rothmaleri* Klink. X *Lupinus luteus* L. *Genetica Polonica*, 6: 125-140.

Kazimierski, T. and Kazimierska, E.M., 1970. Further studies on the hybrid *Lupinus rothmaleri* Klink. X *Lupinus luteus* L. *Genetica Polonica*, 11: 187-205.

Lee, A.T. and Gladstones, J.S., 1979. (474-475) Proposals to reject *Lupinus varius* and *Lupinus hirsutus* of Europe (Fabaceae). *Taxon*, 28: 616-621.

Lehmann, C.O. and Hammer, K., 1983. Botanical results of a journey to Spain to collect cultivated and wild Leguminosae. [Botanische Ergebnisse einer Reise nach Spanien 1978 zur Sammlung kultivierter und wildwachsender Leguminosen]. *Kulturpflanze*, 31: 185-206.

Mota, M., Gusmão, L. and Bettencourt, E., 1981. *Lupinus* and *Secale* collecting in Portugal. *Plant Genetic Resources Newsletter*, 47: 26-27.

Mota, M., Gusmão, L. and Bettencourt, E., 1982. *Lupinus* and *Secale* collecting in Portugal II. *Plant Genetic Resources Newsletter*, 50: 22-23.

Nowacki, E. and Prus-Glowacki, W., 1971. Differentiation of protein fractions in species and varieties of the genus *Lupinus* with the use of serological methods. *Genetica Polonica*, 12: 245-260.

Papineau, J. and Huyghe, C., 1989. Collecting white lupin in the Azores. *Plant Genetic Resources Newsletter*, 88-89: 77-78.

Perrino, P., Hammer, K. and Hanelt, P., 1984. Collection of land-races of cultivated plants in south Italy 1983. *Kulturpflanze*, 32: 207-216.

Perry, M.W., Dracup, M., Nelson, P., Jarvis, R., Rowland, I. and French, R.J., 1998. Agronomy and farming systems. In: *Lupins as Crop Plants: Biology, Production and Utilization* (eds. Gladstones, J.S., Atkins, C. and Hamblin, J.). pp. 291-338, CAB International, Oxon, UK.

Petterson, D.S., 1998. Composition and food uses of lupins. In: *Lupins as Crop Plants: Biology, Production and Utilization* (eds. Gladstones, J.S., Atkins, C. and Hamblin, J.). pp. 353-384, CAB International, Oxon, UK.

Plitmann, U. and Heyn, C.C., 1984. Old world *Lupinus*: taxonomy, evolutionary relationships, and links with new world species. In: *Proceedings of the 3rd International Lupin Conference* (ed. Anon.). pp. 55-56. La Rochelle, France.

Reader, M. and Dracup, M., 1997. Hull and pod wall weights in lupin. In: *International Food Legume Research Conference III: Program and Papers, Adelaide, South Australia* (ed. Anon.). p. 106., Adelaide, South Australia.

Reader, M., Dracup, M. and Kirby, E.J.M., 1995. Time to flowering in narrow-leafed lupin. *Australian Journal of Agricultural Research*, 46: 1063-1077.

Simpson, M.J.A., 1984. *Geographical variation in* Lupinus albus *L.* PhD Thesis. University of Reading, Reading, UK

Simpson, M.J.A., 1986a. Geographical variation in *Lupinus albus* L. I. Iberia. *Zeitschrift für Pflanzenzüchtung*, 96: 232-240.

Simpson, M.J.A., 1986b. Geographical variation in *Lupinus albus* L. II. Northwest Spain, the Nile Valley, the Balkans and Turkey. *Zeitschrift für Pflanzenzüchtung*, 96: 241-251.

Simpson, M.J.A. and McGibbon, R., 1982a. *Lupinus* collection in Greece and Yugoslavia. *Plant Genetic Resources Newsletter*, 52: 28-30.

Simpson, M.J.A. and McGibbon, R., 1982b. *Lupinus* collection in the Iberian Peninsula. *Plant Genetic Resources Newsletter*, 50: 14-19.

Sweetingham, M.W., Jones, R.A.C. and Brown, A.G.P., 1998. Diseases and pests. In: *Lupins as Crop Plants: Biology, Production and Utilization* (eds. Gladstones, J.S., Atkins, C. and Hamblin, J.). pp. 263-289, CAB International, Oxon, UK.

Swiecicki, W., 1985. Studies on the interspecific hybrid *Lupinus hispanicus* Boiss. et Reut. X *Lupinus luteus* L. *Lupin Newsletter*, 8: 24-25.

Swiecicki, W., Swiecicki, W.K. and Wolko, B., 1996. *Lupinus anatolicus* - a new lupin species of the Old World. *Genetic Resources and Crop Evolution*, 43: 109-117.

Chapter 12

Minor and Under-utilised Legumes

Sarita Bennett, Clive Francis and Bob Reid

12.1 INTRODUCTION

The Leguminosae is the third largest group of flowering plants after the Compositae and Orchidaceae, with 650 genera and 18000 species. A total of 72 genera are found in Europe (Tutin *et al.*, 1980) and there are over 900 legume species in Turkey alone (Cocks, 1993). Yet there are relatively few genera, and species within those genera, that are of economic importance.

In southern Australia, and in many other parts of the world where legumes are important in the agricultural system, there is a drive towards increasing the genetic base of the pasture, forage and crop legumes in use. The traditional species, subterranean clover, a few annual and perennial medics, and more recently lupins, have been grown over much of southern Australia in a ley farming system, regardless of the ecogeographic requirements of the available cultivars and the ecology and climate of the area in which they are sown. With an increased knowledge of the importance of ecogeography in the success of introduced legumes and of the homology between the Mediterranean and other areas of the world with a Mediterranean climate (Bennett, 1999), the potential of expanding the narrow genetic base through the introduction of new species is considerable. A number of relatively new species are already becoming increasingly important, such as Serradella (*Ornithopus* sp.) and Biserrula (*Biserrula pelecinus*) among others, and the potential of others is rapidly becoming apparent.

'The genetic diversity of legume species in the Mediterranean' section of this book describes, in detail, those genera that are of current recognised economic importance - a total of only eight genera. In this chapter, we have attempted to describe some of the other legume species that are of minor economic importance, but whose potential is apparent. As always, with a task such as this, it is impossible to include all the species that are considered to have some minor potential in all parts of the world. Therefore we have included species that are of minor importance in a number of regions and in particular those species whose importance is increasing.

12.2 *Anthyllis* L.

The genus is comprised of 23 species. *A. vulneraria* is the only cultivated species in the genus. *A. cytisoides* L. has been grown occasionally as a browse species. It is a small Mediterranean shrub that is well adapted to grazing pressure (Alados *et al.*, 1997). It has been tested for stabilising the soil (Cerda, 1998).

12.2.1 *Anthyllis vulneraria* L

Common name; kidney vetch

Taxonomic summary: Annual, biennial or perennial. Cross-pollinated. Lower leaves reduced to a terminal leaflet, or imparipinnate, upper leaves imparipinnate or equifoliolate. Heads many-flowered. Calyx inflated at anthesis, constricted at the apex, with 5 unequal teeth and an oblique mouth. Corolla yellow, red, purple, orange, whitish or parti-coloured. Legume 1(-2) seeded.
A very polymorphic species divisible into about 30 infra-specific taxa (many frequently recognised as species), between which many intermediates occur (Tutin *et al.*, 1968). The most common form in the Mediterranean region is ssp. *spruneri* (Boiss)Bornm. (Polunin and Huxley, 1987). GRIN (1997) however, only recognises four sub-species; subsp. *carpatica* (Pant.) Nyman, subsp. *maritima* (Schweigg.) Corb., subsp. *polyphylla* (DC.) Nyman and subsp. *vulneraria* .
Centre of diversity: Euro-Siberian and Mediterranean.
Diversity: A study in Poland on the variability of *A. vulneraria* using electrophoresis, found that there was a relation between variability and geographic region, with a number of population groups identified (Kalinowski and Szweykowski, 1992a and b). Genetic variation in the species has also been reported, by Akeroyd (1988), in Sicily and Italy.
Map: See Figure 12.1.
Ecogeographic distribution: Native throughout Europe from the Atlantic coast to the Baltic region, Turkey, south into North Africa (Algeria, Morocco and Ethiopia), Caucasus, Iran and Iceland.
Habitat: Dry fields, meadows, alpine pastures, hillsides and scree. No specific soil type preference, but more common on calcareous soils, with a pH range of 4.8 to 8.0.
Altitude: Sea level to 3000m, annual rainfall: 440mm to 1360mm, mean annual temperature: 6.6 to 18.6°C. Flowers April to September, flowering earlier in warmer climates.

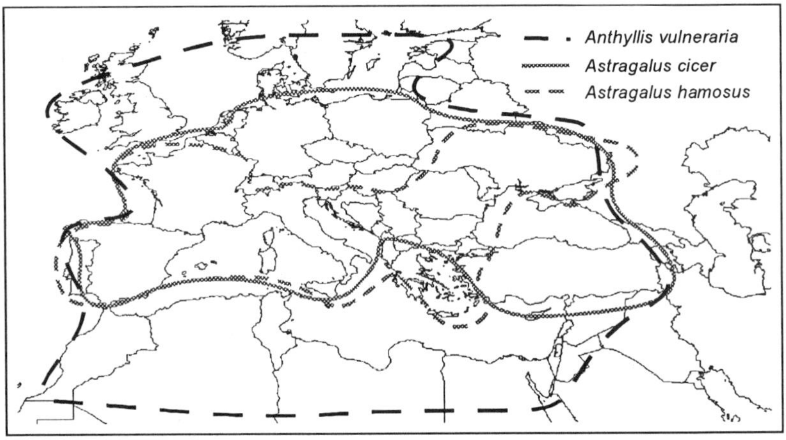

Figure 12.1. Native distribution of *Anthyllis* and *Astragalus* species in the Mediterranean.

Potential value: Cultivars of *A. vulneraria* are reported to show tolerance to high and low pH, and poor and sandy soils (Duke, 1981). It is used for fodder and as a forage crop, being cut anytime from mid-summer to the first frost. *A. vulneraria* is currently an important forage plant in temperate Europe, the Mediterranean region and north Africa where it is commonly fed to sheep and goats (Duke, 1981). May be sown as a substitute for red clover (*Trifolium pratense*) in dry, chalky pastures (Rougemont, 1989). It is also used in folklore medicine where it is considered to be astringent, depurative, diuretic and vulnerary, and is used for dressing wounds.

12.3 *Astragalus* L.

Astragalus is considered to be the largest genera in the Angiosperms; Delimitation of the taxa to date, both at the section and species levels poses considerable problems, with 380 species in Turkey alone (GRIN, 1997). The taxonomy of a number of sections of the genus have been reviewed (Zarre and Podlech, 1996; Gazer, 1993; Podlech, 1990), as well as the relationship between *Astragalus* and *Astracantha* Podlech (Zarre and Podlech, 1997), but a review of the whole genus is badly needed. Many of the species in the genus are spiny or contain the locoweed toxin, swainsonine (Stegelmeier *et al.*, 1999).

12.3.1 *Astragalus cicer* L.

Common name; Cicer milkvetch
Taxonomic summary: Perennial. Cross-pollinated. Ascending or sub-erect plant. Leaflets 10-15 pairs, lanceolate to ovate-lanceolate, with short, adpressed, often sparse hairs on both sides. Peduncles half as long as leaves. Calyx 7-10mm, teeth half as long as tube. Corolla yellow. Legume 10-15mm, ovoid-globose, inflated, with short, black and white hairs.
Centre of Diversity: Euro-siberian. Maximum variation reported from the Mediterranean and sub-mediterranean zones of Europe (Duke, 1981).
Map: See Figure 12.1.
Ecogeographic distribution: Native throughout Europe (rare in south), central and southern Russia, Crimea and Caucasia.
Habitat: scree and steppe, common in both dry and damp meadows. Adapted to a range of soil types, but particularly those that are limestone based.
Altitude: 1900 to 2500m, rainfall: greater than 400mm. Flowers June to July.
Potential value: *A. cicer* is very winter hardy and moderately drought and frost resistant (Duke, 1981). It is valuable forage legume for pastures with cool season grasses, and for erosion control. It is also used in folklore medicine (GRIN, 1997). It shows very low tolerance to acidic soils, but the attributes of drought and defoliation tolerance of this non-bloating species (Townsend, 1993), suggest it is worthy of consideration in neutral to alkaline soils (Hill *et al.*, 1996).

12.3.2 *Astragalus hamosus* L.

Common name: Unknown.
Taxonomic summary: Annual. Leaflets 9-11 pairs, oblong-obovate or truncate, hairy beneath, glabrous above. Peduncles half as long as leaves. Racemes dense with 5-14 flowers. Calyx 5-6mm.

Corolla yellow. Legume 20-50mm, laterally compressed, beak short, with short appressed hairs. Legumes vary in curvature from a half to nearly complete circle, or a horseshoe with uneven sides.
Centre of diversity: Unknown.
Map: See Figure 12.1.
Ecogeographic distribution: Mediterranean area; S. Russia, Crimea; Caucasus; Syrian Desert; N. and N.W. Iran.
Habitat: Fields, grasslands and roadsides. Found on calcareous loam, gravel and sandy clays. Alkaline soils.
Altitude: sea level to 1300m. Flowers March to May.
Potential value: Thought to be a minor pasture plant in Iraq (Townsend, 1974). The pods are sold for medicinal purposes; as an astringent, to promote vomiting and for the expulsion of wind. *A. hamosus* is one of the most widespread potential pasture legumes that is well adapted to alkaline soils. It is often found as a dominant species in rangelands around the Mediterranean and is therefore worthy of further research.

12.4 *Biserrula* L.

A monotypic genus.

12.4.1 *Biserrula pelecinus* L.

Common name: None known
Taxonomic summary: Annual. Leaflets 7-15 pairs, linear-oblong to obovate-orbicular. Calyx campanulate with 5 sub-equal teeth. Corolla blue or pale yellow with blue tip. Legume 10-40mm, brown, dorso-ventrally compressed, indehiscent. Seeds many.
Centre of diversity: Mediterranean.
Map: See Figure 12.2.
Ecogeographic distribution: Mediterranean area of Europe, north Africa and west Asia.
Habitat: sandy fields, waste ground, dry hills, bushy places and track sides.
Altitude: sea level to 700m. Flowers from April to June.
Potential value: Has agronomic potential for acid, duplex soils, and is nodulated by unique root-nodule bacteria (Howieson *et al.*, 1995). Endemic to the Mediterranean region, but is not commonly reported in plant collections (Ewing and Howieson, 1987). Its agronomic attributes include drought tolerance, a prostrate growth, and relatively high productivity and seed yield. It has an advantage over Serradella (*Ornithopus* sp.) as the pods are soft and papery, and are therefore easily harvested using a conventional header (Loi *et al.*, 1997).

12.5 *Hedysarum* L.

Closely allied to *Onobrychis* Adans. The easiest way to separate them is by the shape of the fruit, with most species of *Hedysarum* having two or more segments to the lomentum. In *Onobrychis* species the fruit is constantly one-segmented Tutin *et al.*, 1968).

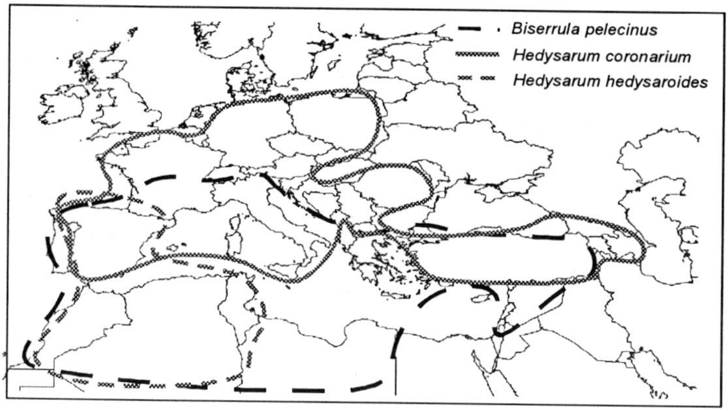

Figure 12.2. Native distribution of *Biserrula* and *Hedysarum* species in the Mediterranean

12.5.1 *Hedysarum coronarium* L.

Common name: Sulla, French honeysuckle, Italian sainfoin.
Taxonomic summary: Biennial, sometimes perennial. Cross-pollinated. Leaflets 3-5 pairs, elliptical to obovate-orbicular, glabrous above, pubescent beneath. Racemes dense, 10 to 35 flowered. Calyx teeth as long as tube. Corolla bright reddish-purple. Legume, 2-4 glabrous segments with small spines.
Centre of diversity: Mediterranean
Diversity: Isozyme analysis of *H. coronarium* by Trifi *et al.* (1989) using ten enzyme systems revealed 3 population types. Cultivars and spontaneous populations both had limited genetic diversity compared to the wild populations, but were very different morphologically.
Map: See Figure 12.2.
Ecogeographic distribution: Native to a restricted area of western mediterranean Europe and north Africa. Has been naturalised elsewhere in southern Europe, and introduced to North America, Australia and India.
Habitat: Rich pastures and cultivated ground, marl and calcareous soils, pH 5.5 to 8.0.
Annual precipitation; 400 mm to 2300mm, mean annual temperature; 5.7 to 29.9°C. Requires mild winters with a minimum temperature above 3°C. Flowers April to May.
Potential value: *H. coronarium* is grown extensively in countries around the Mediterranean for forage, green manure and as a soil improving crop. It has been reported to exhibit tolerance to mild frosts and has shown resistance to viruses (Duke, 1981). Ecotypes from Spain and Italy show greater sensitivity to the cold than those from north Africa. It is comparable in nutritive value to *Trifolium pratense* (red clover) and *Medicago sativa* (alfalfa), but produces so much forage in a short time span that it is necessary to use a combination of grazing and conservation as hay or silage for best cultivation (Kernick, 1978). Martiniello and Ciola (1994) suggest the wild ecotypes of *H.*

coronarium represent an important genetic resource that should be utilised in breeding programmes for developing varieties adapted to mediterranean environments.

12.5.2 *Hedysarum hedysaroides* (L.) Schinz & Thell.

Common name: Alpine French-honeysuckle, alpine sainfoin.
Taxonomic summary: Perennial. Leaflets 3-10 pairs, obtuse. Racemes 15 to 35 (-48) flowered. Calyx teeth ± longer than tube. Corolla reddish-violet, rarely white. Legume with 2-5 segments.
It is known as a widespread and variable species (Davis, 1970) and three subspecies have been recognised; subsp. *hedysaroides*, subsp. *exaltatum* Kerner, and subsp. *arcticum* (Fedtsch.) Ball (Tutin *et al.*, 1968).
Centre of diversity: Unknown.
Map: See Figure 12.2.
Ecogeographic distribution: Native to central Europe, the Balkans and Caucasia.
Habitat: rocky slopes and ledges, grassy places.
Altitude: 2300 to 3100m. Flowers from June to August.
Potential value: *H. hedysaroides* is grown as a fodder crop, and as an ornamental (Sinskaya, 1961)).

12.6 *Lotus* L.

A large number of papers have been published in recent years dealing with the genetics and cyto-taxonomy of Lotus (Tutin *et al.*, 1968). *L. ornithopioides* L. and *L. pedunculatus* Cav. are currently under evaluation in Australia (Cobley *et al.*, 1999; Kelman and Forrester, 1999) as a new pasture species.

12.6.1 *Lotus corniculatus* L.

Common name: bird's-foot trefoil
Taxonomic summary: Perennial. Leaflets lanceolate or oblanceolate. Heads 2 to 7 flowered. Calyx teeth ± equal, sometime curved. Corolla usually yellow. Legume 15 to 30 mm, cylindrical, straight. Seeds many.
Division of the *L. corniculatus* complex has caused much controversy, with some botanists dividing it into many different varieties and sub-species. GRIN (1997) recognises one subspecies (subsp. *frondosus* Freyn) and four varieties; var. *carnosus* Hartm., var. *japonicus* Regel, var. *corniculatus* and var. *hirsutus* W. D. J. Koch. Tutin *et al.* (1968) states that *L. corniculatus* is very variable, and although it may be possible to recognise a number of subspecies it is very confusing because of the species widespread use as a forage crop. Davis (1970) in the Flora of Turkey divides it into 3 varieties; var. *alpinus* Ser; var. *corniculatus* and var. *tenuifolius* L. Each of these varieties are found over all of the *L. corniculatus* distribution but they prefer different habitats. Small *et al.* (1984) attempted to solve the confusion in Turkey, studying herbarium specimens. They found that the traditionally used morphological characters showed a continuous variation with elevation, and suggested that the *L. corniculatus* complex in Turkey may be simply an altitudinally-based ecocline. Despite this they decided to retain the three types which they recognise as three separate species, *L.*

corniculatus (mid-elevation), *L. tenuis* Waldst. and Kit. ex Willd. (lowland) and *L. alpinus* (Ser.) Schleich. ex Ramond (upland), with *L. tenuis* being the most morphologically distinct of the three species.
Centre of diversity: Euro-Siberian, Mediterranean and Irano-Turanian
Map: See Figure 12.3.
Ecogeographical distribution: Native almost throughout Europe and parts of Asia to the Crimea, Caucasus, Iran, India and central Asia. Introduced and naturalised widely elsewhere.
Habitat: Mountain slopes, meadows and dry places, especially by streams. Adapted to a wide range of soils from clays to sandy-loams and pH of 4.5 to 8.2.

Altitude; Sea level to 2500m, annual rainfall; (210-)550mm to 900 (-1910)mm, mean annual temperature; 5.7°C to 23.7°C. Flowers from March to September, requiring a 16 hour day-length for full flowering.
Potential value: *L. corniculatus* is an excellent non-bloating forage which, compared with alfalfa and clover, has softer stems and a higher carbohydrate content (Townsend, 1974). It will grow on poorly drained to drought soils, in acid to mildly alkaline conditions (Duke, 1981). However it is most productive on moderately well-drained, fertile soils. It can tolerate cold temperatures and frosts to -6°C, but is susceptible to high summer temperatures. *L. corniculatus* is thought to be more resistant to pests and diseases than *Trifolium* or *Medicago* species in humid areas (Kernick, 1978), and is moderately salt tolerant, with yield reductions only showing at 6 to 12 pp/1000. Grant and Small (1996) suggest that this complex has vast untapped agricultural potential and that further research on this complex is important in order to clarify further its substantial wild germplasm resource. Kelman *et al.* (1997) have also shown the genetic variation contained within *L. corniculatus* and the potential for selection of improved cultivars.

12.6.2 *Lotus tenuis* Wald. & Kit. Ex Willd. (*L. glaber* Mill.)

Common name: slender trefoil, Narrowleaf trefoil
Taxonomic summary: Perennial. Leaflets linear or linear-lanceolate. Heads 1 to 4(-6) flowered. Calyx teeth equal, usually shorter than tube. Corolla yellow. Legume 15 to 30 mm.
The synonymous use of the illegal *L. glaber* and *L. tenuis* is very confusing. Kirkbride (1995) has proposed the rejection of *L. glaber* and the retention of *L. tenuis*, however the use of either name and *L. corniculatus* ssp. *tenuifolius* L. are found.
Centre of Diversity: Euro-siberian.
Map: See Figure 12.3.
Ecogeographic distribution: Native to most of Europe except the north-east and the extreme north. Introduced to the United States.
Habitat: Damp, heavy soils, river banks and marshes, grassy places. but adapted to dry soils that do not drain well after flooding. Preference for heavy textured clay soils with a pH range of 5.6 to 8.0. Altitude; sea level to 2750m, annual rainfall; 440 to 1160mm, mean annual temperature; 7.0° to 16.9°C. Flowers from April to September.
Potential value: Several cultivars have been developed from *L. tenuis* for dry, poorly draining soils, but all exhibit fine stems and narrow leaves with yields that are generally lower than those of *L. corniculatus*. It has been reported to show tolerance to alkaline soils, salt, mercury, weeds and water-logging. Rogers *et al.* (1997) tested two lines which were relatively salt tolerant.

12.6.3 *Lotus tetragonolobus* L.

Common name: winged-pea, asparagus pea.
Taxonomic summary: Annual. Leaflets obovate to obovate-rhombic. Peduncles shorter to equal length of leaves. Flowers solitary or paired. Calyx teeth 1-2 times as long as tube. Corolla crimson. Legume 30 to 90mm, glabrous.
Centre of diversity: Mediterranean.
Map: See Figure 12.3.
Ecogeographic distribution: Native to the mediterranean area, and north into the Ukraine. Habitat: fields, track sides and macchie. pH range of 6.0 to 7.0.
Annual rainfall; 500 to 600mm, mean annual temperature; 7° to 8°C. Flowers March to June.
Potential value: The pods are edible if picked young, and are cooked whole and served with melted butter. The seeds are roasted and used for a coffee substitute. It is sometimes grown as a culinary herb (Duke, 1981). It is widely cultivated in warm temperate regions of the world and has become naturalised in Europe, however there is very little data available on its production as it is still mainly grown in gardens.

12.6.4 *Lotus uliginosus* Schkuhr.

Common name: big trefoil, greater birds-foot trefoil.
Taxonomic summary: Perennial. Leaflets obovate to obtuse, glaucous beneath. Heads 5 to 12 flowered. Calyx teeth as long as tube. Legume 15 to 35 mm.
Centre of diversity: Euro-siberian
Map: See Figure 12.3.
Ecogeographic distribution: Native to most of Europe, east to western Russia and in north Africa. Habitat: Wet places (marshes and boggy meadows). Shows a preference for acid soils and peat areas with a pH range of 5.5 to 8.2.
Annual rainfall; 350mm to 1360mm, mean annual temperature; 5.9° to 21.3°C. Flowers May to August.
Potential value: *L. uliginosus* is a prominent non-bloating pasture species in the north-west of the United states and in New Zealand. It is superior to *L. corniculatus* as it has a higher protein content, and has been reported to show tolerance to disease, fungus, high pH and water-logging (Duke, 1981). However wide variations in seed set and harvest result in unpredictable yields. It is a possible gene source for *L. corniculatus* (GRIN, 1997) and Grant and Small (1996) suggest it is one of the progenitors of the species. It is moderately salt tolerant (yield is not reduced until 3-6 pp/1000), but shows no resistance to drought. It is one of the few agricultural legumes to persist in marshy, acid soils (Rougemont, 1989).

12.7 *Melilotus* L.

Several species are cultivated locally as fodder, and occur frequently as casuals outside the geographical limits given. Many species smell strongly of coumarin, especially when dry (Tutin *et al.*, 1968). Care must be taken in silage production, as spoiled silage can be poisonous as the coumarin breaks down into substances allied to the rat poison, warfarin, which prevents clotting of the blood (Rougemont, 1989). However, cultivars low in coumarin are now available. *Melilotus* species are considered among the most valuable for honey production (Duke, 1981).

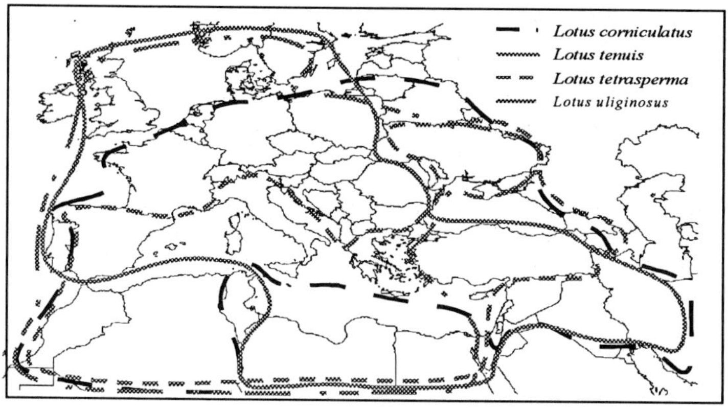

Figure 12.3. Native distribution of *Lotus* species in the Mediterranean.

12.7.1 *Melilotus albus* Medik. (=*M. alba* Medik.)

Common name: white sweet-clover, white melilot, Bokhara clover.
Taxonomic summary: Annual or biennial. Cross-pollinated. Characterised by white flowers. Leaflet narrowly oblong-obovate to suborbicular, serrate. Racemes slender, many flowered. Corolla white. Legume 3-5mm, obovoid, murconate, glabrous, greyish-brown when ripe.
A polymorphic species which Suborov (1951) divided into three subspecies and a large number of ecotypes and forms. Stevenson (1969) collected accessions from around the world and found that they showed great variation in a large number of morphological and agronomic characters, but that the variation was continuous He concluded that *M. albus* is a variable species, but that no clear separation into subspecies is justified.
Centre of diversity: Euro-siberian and Mediterranean.
Map: See Figure 12.4.
Ecogeographic distribution: Native to Europe, west to Asia and south to India. Has been introduced to many temperate parts of the world, including North America and Australia.
Habitat: Disturbed ground and open habitats, often as a weed or ruderal. Adapted to a range of soils from heavy clays to light sands, but is very sensitive to pH preferring a neutral soil.
Altitude; Sea level to 1760m, annual rainfall; 90 to 1600mm, mean annual temperature; 5.7° to 24.3°C. Flowers from May to October.
Potential value: *M. albus* is cultivated for forage and hay, where it is extremely drought resistant. Has been shown to tolerate disease and viruses, frost, heat, poor soil conditions, weeds and water-

logging (Duke, 1981) and is salt tolerant (ILDIS, 1987). However it is very susceptible to many herbicides. It can not stand heavy grazing, but is a useful fodder and green manure crop where clovers persist poorly (Rougemont, 1989). It is valuable in honey production and has medicinal properties where it is considered aromatic, anti-asthmatic, anticoagulant, digestive and vulnerary. With few exceptions the cultivated varieties of *M. albus* belong to the Siberian ecotype which Suborov (1951) called *M. albus* ssp. *borealis*. It has the same rhizobial requirements as *M. polymorpha* L. and *M. sativa* L. (Townsend, 1974).

12.7.2 *Melilotus altissimus* Thuill. (=*M. altissima* Thuill.)

Common name: tall yellow sweet-clover, tall melilot
Taxonomic summary: Biennial or short-lived perennial. Leaflets oblong-ovate, serrate. Racemes many flowered, elongating in fruit. Corolla yellow. Legume 5 to 6 mm, obovoid, acute, pubescent, black when ripe, usually 2-seeded.
Can be distinguished from *M. officinalis* which it resembles by pubescent ovaries and pods, and by the shape of the pods (Stevenson, 1969).
Centre of diversity: Unknown.
Map: See Figure 12.4.
Ecogeographic distribution: Native throughout most of Europe, but is rare in the east and is absent from the islands.
Habitat: Hedges and grassland, damp or saline areas, and as a ruderal. Flowers from June to August.
Potential value: Has potential as a forage (Sinskaya, 1961). *M. altissimus* is not winter-hardy. However forage yields can equal or surpass *M. albus* and *M. officinalis*, and as such may be a good pasture species in areas with milder winters (Stevenson, 1969).

12.7.3 *Melilotus indicus* (L.) All. (=*M. indica* (L.) All.)

Common name: sour-clover, small melilot.
Taxonomic summary: Annual. Self-fertile. Leaflets lanceolate-oblong, serrate. Racemes dense, many flowered. Corolla pale yellow, Legume 1.5 to 3 mm, subglobose, glabrous, whitish-grey when young.
GRIN (1997) recognises two varieties; var. *indicus*, and var. *tommasinii* (Jordan) O. E. Schulz.
Centre of diversity: Hindustani.
Map: See Figure 12.4.
Ecogeographical distribution: Native to the Mediterranean region of Europe, west Asia to India and north Africa. It is widely naturalised elsewhere.
Habitat: : Waste places, scree and disturbed places. Often occurs as a winter weed. Adapted to a range of soils, but prefers well-drained neutral to alkaline soils of medium to high fertility. pH range of 5.0 to 8.2.
Altitude; Sea level to 1750m, annual rainfall; 90mm to 1290mm mean annual temperature; 7.2° to 22.5°C. Flowers April to August.
Potential value: Cultivated as a forage, for hay and silage. It has been shown to be tolerant to high alkaline soils, drought, frosts, insects, viruses and weeds (Duke, 1981). It is also used for reclaiming saline areas and can tolerate salinity levels of 3 to 6 mmhos. *M. indicus* was found to produce

significantly greater dry biomass than *Medicago sativa* under saline conditions (Ashraf *et al.*, 1994), with ecotypic variation for salt tolerance also being exhibited. In a further experiment by Batra and Ghai (1988), *M. indicus* was compared with three *Trifolium* and *Medicago* species, where although it did not produce the highest yields, it was the least affected by saline soils. It has medicinal properties being used as an astringent and narcotic, and the seeds are used for bowel complaints.

12.7.4 *Melilotus officinalis* (L.) Pallas

Common name: yellow sweet-clover, ribbed melilot.
Taxonomic summary: Biennial, very occasionally annual. Cross-pollinated. Leaflets of lower leaves obovate to ovate, the upper ovate-lanceolate, all serrate. Racemes slender, many flowered. Corolla yellow. Legume 3 to 5 mm, transversely rugose, mucronate, glabrous, brown when ripe, usually 1-seeded.
A variable species in which Duke (1981) recognises three subspecies and 16 ecotypes
Centre of diversity: Euro-siberian and Mediterranean.
Map: See Figure 12.4.
Ecogeographic distribution: Native to most of Europe, except the extreme south and the far north. It has been widely introduced and naturalised in many of the former Russian provinces and North America (GRIN, 1997).
Habitat: Disturbed and cultivated ground, often on clay or saline soils, preferring fertile, well-limed and drained soils. Can tolerate sandy conditions. pH range of 4.8 to 8.2, but grows best in neutral conditions.
Altitude; sea level to 4000m, annual rainfall; 310mm to 1600mm, mean annual temperature; 4.9° to 21.8°C. Flowers from May to September.
Potential value: *M. officinalis* is grown for forage, hay and pasture. It is excellent for soil improvement and erosion control as the roots produce copious number of nitrogen-fixing nodules (Duke, 1981). It is good for honey production, and is used in folk medicine as the plant is considered to be aromatic, carminative, styptic and emollient. Cultivars of *M. officinalis* have been reported to be tolerant of alkaline soils, drought, frost, heat, poor soil and saline conditions, insects and weeds. It can tolerate salt levels of up to 3 to 6 mmhos. Fist years growth is considered to be of similar feeding value to alfalfa.

12.8 *Onobrychis* Adans

See discussion under *Hedysarum* for differentiation of the two genera. One of the main centres of the genus is in Anatolia (Turkey), with many problems present between the species. Until a fresh monograph of the genus is prepared, Davis (1968) does not feel that a satisfactory account of the Turkish species can be given. There is even question over whether the two species described below should actually be one species (Cenci and Mizianty, 1997).

Kidambi *et al.* (1990) looked at the variability of 17 *Onobrychis* species and three cultivars of *O. viciifolia* at the esterase enzyme. They found that all the material had some unique banding patterns, but that only four species could actually be uniquely identified. They suggest that three enzyme systems should be used to separate the species.

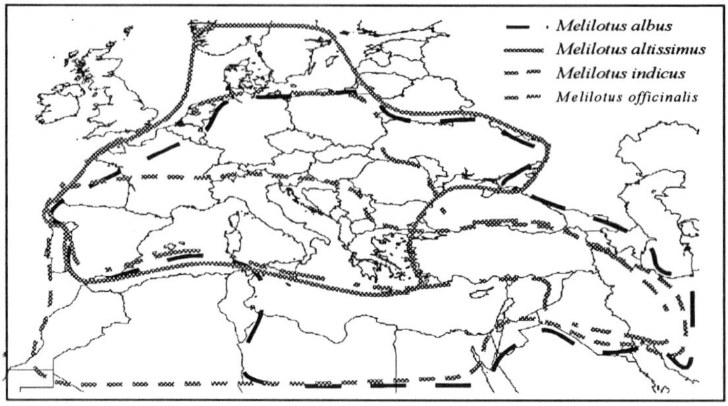

Figure 12.4. Native distribution of *Melilotus* species in the Mediterranean.

12.8.1 *Onobrychis arenaria* (Kit.) DC.

Common name: Sand esparsette.
Taxonomic summary: Perennial. Leaflets usually 3 to 12 pairs, linear-oblong to elliptical. Calyx glabrous or pubescent. Corolla pink with purple veins. Legume 4 to 6 mm, pubescent, margin with 3 to 8 teeth.
There is some confusion over the number of subspecies that this species should be separated into. GRIN (1997) recognise three subspecies; subsp. *arenaria*, subsp. *miniata* Steven and subsp. *sibirica* (Sirj.) Grossh. and Tutin *et al.* (1968) recognises seven subspecies; subsp. *arenaria*, subsp. *taurerica* Hand.-Mazz., subsp. *sibirica*, subsp. *miniata*, subsp. *tommasinii* (Jordan) Ascher. & Graebn., subsp. *lasiostachya* (Boiss.) Hayek and subsp. *cana* (Boiss.) Hayek. Davis (1970) does not recognise *O. arenaria* as occurring in Turkey, but instead states that a closely related species *O. armena* Boiss. & Huet is present instead.
Centre of diversity: Unknown.
Map: See Figure 12.5.
Ecogeographic distribution: Native to central, east and south-east Europe, extending to central France and central Italy.
Flowers June to September.
Potential value: *O. arenaria* is cultivated in parts of eastern Europe as a forage crop. It has the highest protein content of the cultivated *Onobrychis* species (Duke, 1981).

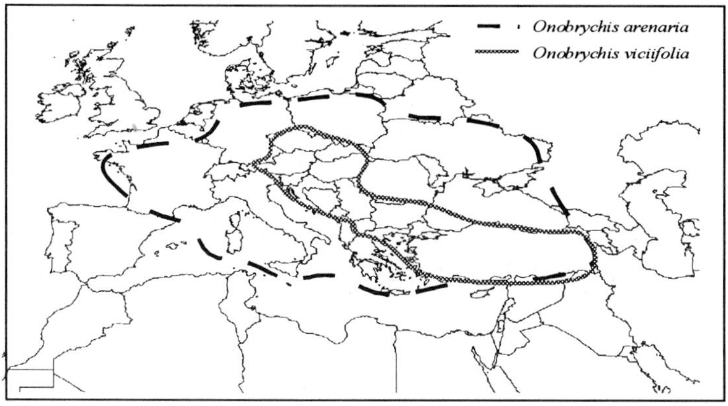

Figure 12.5. Native distribution of *Onobrychis* species in the Mediterranean.

12.8.2 *Onobrychis viciifolia* Scop.

Common name: sainfoin
Taxonomic summary: Perennial. Leaflet 6 to 14 pairs, ovate to oblong, rarely linear. Racemes up to 9 cm in flower. Calyx pubescent, teeth 2 to 3 times as long as tube. Corolla pink with purple veins. Legume 5 to 8 mm, pubescent, margin usually with 6 to 8 teeth.
Centre of diversity: Near East and Euro-Siberian.
Diversity: The genetic resources of sainfoin are hard to preserve as the species is allogamous, self-incompatible, requires insect pollination and its seeds have a low viability of only 5-8 years, even in a cold chamber. Prosperi *et al.* (1994) have isolated some 'genetic pools' using cluster analysis of morphological and agronomic characters in an attempt to preserve the local varieties collected. An altitudinal cline in morphological variation has also been reported in Italy (Cenci and Negri, 1987).
Map: See Figure 12.5.
Ecogeographic distribution: Possibly native to central Europe. The wild distribution of *O. viciifolia* is uncertain as it has long been cultivated and is now widely naturalised in Europe and parts of Asia (Davis, 1970; Tutin *et al.*, 1968).
Habitat: Banks, waste and grassy places, usually in or near cultivation. Adapted to range of soil types, even gravely, shallow, and rocky soils. However it prefers good drainage. pH range of 4.9 to 8.2. Tolerant of frosts to -4.4°C under good snow cover.

Altitude; near sea level to 1700m, annual rainfall; 350mm to 1000mm, mean annual temperature; 5.9° to 18.6°C. Flowers from May to August.
Potential value: *O. viciifolia* is grown for forage in many temperate parts of Europe, Asia and north America, where it is recommended for dryland pastures. It was traditionally one of the most important herbage legumes, but in Europe is being replaced by alfalfa and the clover species

(Rougemont, 1989). It has been shown to contain tolerance to drought, frost, heavy grazing, moderate salt levels and poor soils (Duke, 1981). It is considered to be a short-lived perennial producing its maximum forage yields in the first three to four years (Kernick, 1978). It has a deep root system and is said to aid soil formation on shallow soils overlying chalk (Townsend, 1974). It is also excellent for honey production. Experiments in south-east France have shown that local varieties showed better winter growth and adaptation (Prosperi *et al.*, 1994).

12.9 *Ornithopus* L.

An annual genus of only five species. All species occur in dry, often acid, sandy places, or as ruderals (Tutin *et al.*, 1968). Only one species, *O. micranthus* (Benth.) Arechav., is not found in the Mediterranean and is endemic to South America (Fu and Williams, 1994). Fu *et al.* (1994) have evaluated the taxonomy and agronomic potential of the Mediterranean species.

12.9.1 *Ornithopus compressus* L.

Common name: Yellow serradella.
Taxonomic summary: Annual. Leaflets 7 to 18 pairs, oblong, elliptical or oblong-lanceolate. Heads 3 to 5 flowered. Calyx teeth as least half as long as tube. Corolla yellow. Legume 20 to 50 mm, curved, ± compressed, slightly contracted between segments, segments 5 to 8, oblong, beak curved.
Centre of diversity: Mediterranean.
Map: See Figure 12.6.
Ecogeographic distribution: Mediterranean area and southern Europe.
Habitat: Open ground and pasture. Altitude; Sea level to 300m. Flowers from April to June.
Potential value: Grown as a forage crop. The majority of collected accessions show a prostrate habit that is typical of the species. A few accessions and most current cultivars have a semi-erect habit. This is suitable for hay production and has been selected for and improved through breeding (Fu and Williams, 1994). Genetic variation is also present between sites (Loi *et al.*, 1999).

12.9.2 *Ornithopus sativus* Brot.

Common name: Pink serradella, French serradella.
Taxonomic summary: Annual, mainly self-pollinating. Leaflets 9 to 18 pairs, lanceolate or elliptical to ovate. Heads 2 to 5 flowered. Calyx teeth slightly shorter to equal the tube. Corolla white or pink. Legume 12 to 40 mm, compressed, contracted between segments, segments 3 to 7, elliptical-oblong. Two subspecies are recognised; subsp. *isthmocarpus* (Coss.) Dostal, and subsp. *sativus* (Tutin *et al.*,1968) which are linked by intermediates in central Portugal and western central Spain. Davis (1970) also recognises two subspecies; ssp. *sativus* and ssp. *roseus* (Dufour) Dostal, but following his description ssp. *sativus* has not been confirmed as present in Turkey.
Centre of diversity: Mediterranean.
Map: See Figure 12.6.
Ecogeographic distribution: Native from the Azores to south-west Europe, Morocco and Algeria. It is cultivated in much of Europe and has naturalised locally.

Habitat: Sandy places, and acid soils. pH range 5.5 to 8.2.
Annual rainfall; 5.7mm to 11.9mm, mean annual temperature; 7.0° to 18.3°C. Flowers from April to June.

Potential value: *O. sativus* is grown as a forage and for soil improvement. In Australia it has shown good adaptation to the deep, acid sands of Western Australia where it is becoming increasingly important as a pasture species in rotation with wheat (Hussey *et al.*, 1997). It is suited to cropping in dry, poor soils (Rougemont, 1989).

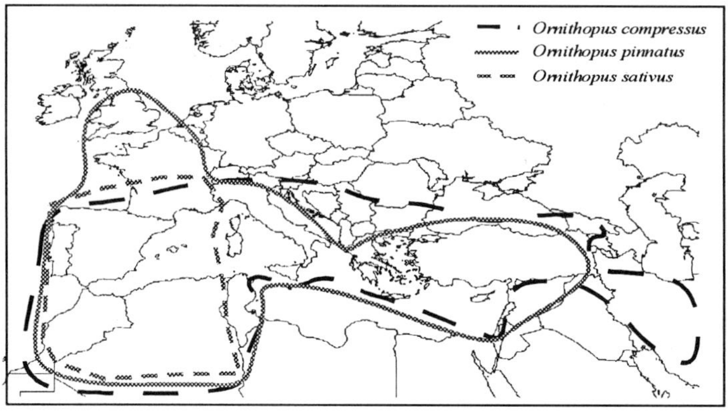

Figure 12.6. Native distribution of *Ornithopus* species in the Mediterranean.

12.10 *Scorpiurus* L.

Both species in the genus are annuals (Tutin et al., 1968). *S. sulcata* L. is not currently cultivated as a pasture plant. However Beale *et al.* (1991) suggest that it should be assessed following pod yields recovered from plants in Morocco, and that collection of the species in Morocco revealed that it was widespread, with no apparent specific requirements for soil type.

12.10.1 *Scorpiurus muricatus* L.

Common name: Caterpillar plant

Taxonomic summary: Annual. Leaves simple. Heads (1-)2 to 5 flowered. Flowers solitary. Calyx campanulate with 5 equal teeth. Corolla yellow or purplish. Legume indehiscent, curved, longitudinally ridged, the ridges smooth, or the outer 4 to 8 tuberculate. Seeds lunate, attenuate at ends.

Often divided into 3 taxa, of varying rank, on a number of morphological characters. However the correlation between these characters is inconsistent, and there does not seem to be any satisfactory geographical separation (Tutin *et al.*, 1968). GRIN (1997) recognise only two varieties; var.

muricatus, and var. *subvillosus* (l.) Fion, of which only var. *subvillosus* occurs in Turkey (Davis, 1968).
Centre of diversity: Unknown
Map: See Figure 12.7.
Ecogeographic distribution: Native to the Mediterranean region, Crimea, Caucasia, Syrian desert, north Iraq, southern Iran and east Africa. It is a rare casual outside of the Mediterranean region.
Habitat: Mainly on rocky slopes, especially limestone and fallow fields.
Altitude; sea level to 850m. Flowers from May to June.
Potential value: It is grown as a minor vegetable. It is considered to be a valuable grazing plant in Iraq (Townsend, 1974).

12.11 *Securigera* DC.

This genus closely resembles *Coronilla* L., but is distinguished by fruit which does not break up into segments.

12.11.1 *Securigera varia* (L.) Lassen

Common name: crown-vetch
Taxonomic summary: Long-lived perennial, cross-pollinated. Leaflets (5-)7 to 12 pairs, oblong or elliptical. Heads (5-)10 to 20 flowered. Calyx campanulate. Corolla white, pink or purple. Legume 20 to 60(-80) mm, segments 3 to 8, oblong, 4 angled.
S. varia, is included in most Floras as *Coronilla varia* L. including the Flora of Turkey (Davis, 1970) and the Flora Europaea (Tutin *et al.*, 1968). It was moved in 1989 by Lassen (1989).
Centre of diversity: Euro-siberian.
Map: See Figure 12.7.
Ecogeographic distribution: Europe, extending to central Russia and south-west Asia, but native limits unclear as it is often cultivated for fodder and naturalised in west and north Europe.
Habitat: Stony slopes and in deciduous woods and scrub. Cultivated ground as a weed. Adapted to a range of soils, but prefers neutral to high pH with a range of 4.8 to 7.8.
Altitude; 250-2150m; annual rainfall; 520 to 1360mm; mean annual temperature; 7.0° to 22.5°C. Flowers from May to August.
Potential value: Primarily grown as a forage crop and compares favourably with alfalfa. It is also useful for erosion control as it has an extensive root system. Has been reported to show tolerance to drought, frost, grazing, low pH, insects and poor soils (Duke, 1981), although Hill *et al.* (1996) found in a trial in New Zealand, that under acid conditions it showed poor seedling vigour and was intolerant of defoliation. However it showed vigorous growth and spread aggressively once established, with little care required.

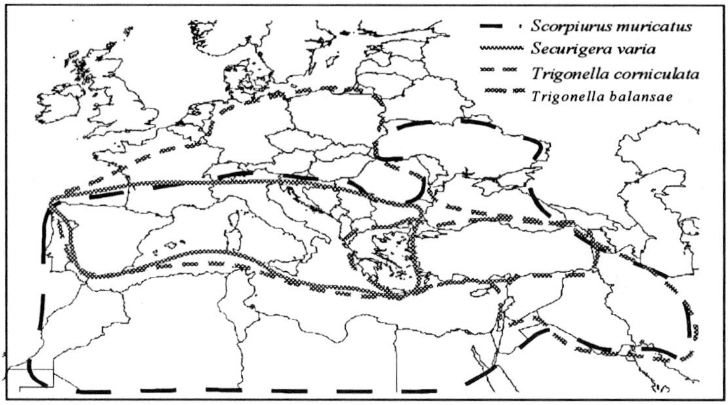

Figure 12.7. Native distribution of *Scorpiurus*, *Securigera* and *Trigonella* species in the Mediterranean.

12.12 *Trigonella* L.

All species are annuals (Tutin *et al.*, 1968). Small *et al.* (1990) have suggested that *Medicago* L. section *Obiculares* belongs in *Trigonella* and Small (1987; 1989) proposed that 32 species of *Trigonella* should belong in *Medicago* according to chemical, floral and seed characters.

12.12.1 *Trigonella balansae* Boiss. & Reuter

Taxonomic summary: Annual. This species is closely allied to *Trigonella corniculata* L. (Davis, 1970), and may be only a subspecies of the above (Tutin *et al.*, 1968). It differs taxonomically in having globose, umbellate flower-heads, pubescent pedicels and unequal calyx teeth. The legume is 2-4mm wide, rounded to sub-acute.
Centre of diversity: Unknown
Map: see Figure 12.7
Ecogeographic distribution: Eastern Europe, possibly also Israel and Jordan
Habitat: Limestone scree, garrigue, cultivated land and ruins.
Altitude; sea level to 800m. Flowers April to June
Potential value: This species has great potential as a pasture plant in the low rainfall areas of southern Australia and evaluation trials are currently in progress.

12.12.2 *Trigonella caerulea* (L.) Ser.

Common name: sweet-trefoil
Taxonomic summary: Annual. Leaflets ovate to oblong, emarginate, denticulate. Racemes globose, dense, many flowered. Peduncles 2 to 5 cm. Calyx teeth equal to tube. Corolla blue or white. Legume 4 to 5 mm, erect, rhomboid-obovate, abruptly contracted to a beak.
Centre of diversity: Mediterranean.
Map: Native distribution is unknown
Ecogeographic distribution: Cultivated for fodder throughout much of Europe, widely naturalised often occurring as a casual, weed or ruderal. It appears to have no indigenous distribution and is probably derived from *T. procumbens* (Besser) Rchb. (Tutin *et al.*, 1968).
Potential value: It is grown for fodder and for flavouring, especially for cheeses (GRIN, 1997).

12.12.3 *Trigonella corniculata* (L.) L.

Common name: None reported.
Taxonomic summary: Annual. Leaflets linear-lanceolate to obovate, obtuse. Racemes ovate-oblong, 8 to 15 flowered, peduncles up to 6cm. Calyx teeth unequal, ± length of tube. Corolla yellow. Legume 10 to 16 mm, pendant, linear, compressed, somewhat curved, glabrous.
Centre of diversity: Unknown.
Map: See Figure 12.7.
Ecogeographic distribution: Mediterranean region, probably introduced to Turkey.
Habitat: Roadsides, waste ground. Widely naturalised on disturbed ground. Flowers from March to April.
Potential value: Grown as a vegetable for human consumption (GRIN, 1997).

12.12.4 *Trigonella foenum-graecum* L.

Common name: fenugreek
Taxonomic summary: Annual. Leaflets obovate to oblong-oblanceolate, denticulate. Flowers solitary or paired, subsessile. Calyx teeth as long as tube. Corolla yellowish-white, tinged with violet at the base. Legume 60 to 110 mm, erect, linear, somewhat curved, glabrous. Beak 20 to 30 mm. All parts of the plant are strongly aromatic.
Centre of diversity: Mediterranean and Near East.
Map: See Figure 12.7.
Ecogeographic distribution: Naturalised in the Mediterranean area, Caucasia, Syrian desert, north Iraq, north, west and central Iran, mostly as escapes from cultivation. Its origin is unknown however, it is now widely distributed (Tutin *et al.*, 1968).
Habitat: Stony slopes, fallow fields, cultivated ground. Grows well on good, well-drained, deep loamy soils, and sandy soils. Does not like clay soils. pH range of 5.5 to 8.2.
Altitude; sea level to 1100m, annual rainfall; 380 to 1530mm, mean annual temperatures; 7.8° to 27.5°C, with a preference for cool, temperate areas without temperature extremes. Flowers from April to July.

Potential value: The plant is widely grown for flavouring, as a food and as a forage crop. The seeds are used to flavour curries and soups, and are eaten in a number of countries around the Mediterranean. As a legume the species is good for soil improvement. Milk from cows fed on *T. foenum-graecum* retains the fenugreek aroma (Duke, 1981). The plant has a number of medicinal properties and is regarded as a tonic, a diuretic and is astringent. It has been reported to be tolerant of disease, drought, high pH, poor soils and salt. *T. foenum-graecum* has potential as a valuable forage in feedlot diets as its silage nutritive value is comparable to mid-bloom alfalfa silage (Mir *et al.*, 1998), plus it has a high dry matter yield (Mir *et al.*, 1993)

12.13 CONSERVATION RESOURCE STATUS (*EX SITU* AND *IN SITU*)

None of the species that have been included in this paper are listed as threatened or even rare (ILDIS, 1997). Specific details are not available, but with the increased interest in the alternative pasture and forage legumes, their collection has increased. Accessions of most of the annual species are held in genebanks belonging to the United States Department of Agriculture in Washington, the *Trifolium* Genetic Resource Centre, Perth Australia, the Australian *Medicago* Genetic Resource Centre in Adelaide, the International Centre for Agricultural Research in Dry Areas (ICARDA) in Syria and the Aegean Agricultural Research Institute (AARI) in Izmir, Turkey amongst others. The problem of collection and storage in *ex-situ* conservation lies, not with the species that are described in this paper, but with the hundreds of other legume species whose potential has not been realised or even tested, as these are the species that will be the least well represented in international gene bank collections. This is where *in-situ* conservation becomes important.

In-situ conservation is the preservation of the whole ecosystem. It seeks to preserve the process of crop evolution though natural and local selection and by hybridisation between wild, weedy and crop plants (Brush, 1995). Unfortunately access to germplasm for the plant breeder is more difficult in *in-situ* reserves. However with the growing availability of techniques for transferring genes between plants that are more-distantly related, the search for useful genes has broadened. This has resulted in an increase in the maintenance of wild crop relatives and to the importance of *in-situ* reserves (Hawtin *et al.*, 1996). Also the amount of inter- and intra-specific diversity that can be conserved in *ex situ* reserves is only a very small proportion of the potentially useful variation.

There are still very few *in situ* conservation genetic reserves (see Chapter 16), and details of any that include the conservation of the species listed in this paper are unavailable. As an example of the kind of projects that need to be set up in all countries around the Mediterranean we provide brief details of the *in situ* project in Turkey. The Aegean Agricultural Research Institute in Izmir, Turkey embarked on a national *in situ* conservation plan within the framework of GEF (Global Environment Facility, World Bank) in 1994. Turkey is considered to be important for *in situ* conservation as it contains the wild relatives of many crop species, plus many other species of economic potential, and along with most Mediterranean countries, is subject to increasing urbanisation, use of modern cultivars and subsequent genetic erosion of landraces and wild species. The project seeks to demonstrate the feasibility of conservation of genetic diversity using an integrated multi-species and multi-site approach (Tan, 1996).

Three areas were chosen as the gene management zones, and within these zones a number of sites will be designated. The zones were chosen to cover the different centres of diversity in Turkey. In each zone a number of target species will be selected and an inventory will be conducted to determine the ecogeographic range and the environmental heterogeneity of each target species (Sebançi, 1998). The three zones are;

a. Kazdag mountains, north-west Turkey (Euro-Siberian, Mediterranean and Irano-Tauranian elements),
b. Ceylanpinar State Farm, south-east Turkey (Mediterranean and Irano-Tauranian elements) and,
c. Anatolian Diagonal, south and central Turkey (Mediterranean, Irano-Tauranian and Euro-Siberian elements). *Vicia* and other forage legumes are likely to be the target species in this zone (Sebançi, 1998).

Unfortunately the details of the species that are to become the target species and a list of the other species that will be conserved within the nominated sites and zones is unavailable, but the potential of such a project can clearly be seen and it is highly recommended that other Mediterranean countries implement such projects.

12.14 Research Needs

The research needs include not only the species and genera listed in this chapter, but many of the other species that are currently of minor use around the Mediterranean climate areas of the world. At this stage the research needs are fairly basic and deal with the potential of the species as a pasture, crop or vegetable species. Initially the ecogeography of the species needs to be determined, followed by its agronomic potential; yield, palatability, flowering time, harvestability of the seed, dormancy mechanisms etc. Also of importance is adaptation to a new environment, and resistance to the pests and diseases that are endemic to the new environment. Current drives in southern Australia include the search for perennials and deep-rooted annuals that are tolerant of water-logging and saline soils. Of paramount importance in the evaluation of any new species is its rhizobial requirements; are they the same as any species currently in cultivation of are they unique to the genera or species in question.

REFERENCES

Akeroyd, J.R., 1988 .Variation in *Anthyllis vulneraria* in Sicily and south Italy. *Notes from the Royal Botanic Garden Edinburgh*, 45: 369-374.

Alados, C.L., Barosso, F.G. and Garcia, L., 1997. Effects of early season defoliation on above-ground growth of *Anthyllis cytisoides*, a Mediterranean browse species. *Journal of Arid Environments*, 37: 269-283.

Ashraf, M., Noor, R., Zafar, Z.U. and Mujahid, M., 1994. Growth and ion distribution in salt stressed *Melilotus indica* (L.) All. and *Medicago sativa* L. *Flora (Jena)*, 189: 207-213.

Batra, L. and Ghai, S.K., 1988. Performance of different forage legumes and *Rhizobium* symbiotic systems under saline conditions. *Indian Journal of Agricultural Sciences*, 58: 350-353.

Beale, P.E., Lahlou, A. and Bounjemate, M., 1991. Distribution of wild annual legume species in Morocco and relationship with soil and climatic factors. *Australian Journal of Agricultural Research*, 42: 1217-1230.

Bennett, S.J., 1999. Using collections to describe ecological relationships. In: *Genetic resources of Mediterranean pasture and forage legumes* (eds. Bennett, S.J and Cocks, P.S.), Proceedings of a workshop, York, Western Australia in October 1996. pp. 41-52. Kluwer Academic Publishers, Dordrecht.

Brush, S.B., 1995 *In situ* conservation in centers of crop diversity. *Crop Science*, 35: 346-354.

Cenci, C.A. and Mizianty, M., 1997. Some morphometric and chemotaxonomic features in the taxonomy of *Onobrychis* (Fabaceae) from Poland. *Fragmenta Floristica et Geobotanica*, 42: 401-404.

Cenci, C.A. and Negri, V., 1987. Variability of the systematic characters in natural populations of *Onobrychis viciifolia* Scop. *Informatore Botanico Italiano*, 19: 422-428.

Cerda, A., 1998. Soil aggregate stability under different Mediterranean vegetation types. *Catena*, 32:73-86.

Cobley, B., Nutt, B.J., Snowball, R., Carr, S.J., Loi, A. and Ewing, M.A., 1999. Development of legumes for use as a phase between extended cropping sequences. *Cooperative Research Centre for Legumes in Mediterranean Agriculture, Annual report for 1998-1999*. pp. 45-46. CLIMA, Perth, Australia.

Cocks, P.S., 1993. *Legumes from the Mediterranean basin: a continuing source of agricultural wealth for southern Australia.* Technical paper No. 1. CLIMA, Perth, Australia.

Davis, P.H., 1970. *Flora of Turkey and the East Aegean Islands. Vol. 3.* Edinburgh University Press, Edinburgh.

Duke, J.A., 1981. *Handbook of legumes of world economic importance*. Plenum Press, New York.

Ewing, M.A and Howieson, J.G., 1987. *A report on the collection of annual* Medicago *and associated* Rhizobium meliloti *from acidic regions of mainland Greece and neighbouring islands of the Aegean Sea.* Western Australian Department of Agriculture, Perth.

Fu, S.M and Williams, W.M., 1994. Description and evaluation of serradella (*Ornithopus* L.) accessions. *New Zealand Journal of Agricultural Research*, 37: 471-479.

Fu, S.M., Hampton, J.G. and Hill, M.J., 1994. An investigation of seed characteristics in serradella (*Ornithopus* L.) accessions. *Plant Varieties and Seeds*, 7: 127-133.

Gazer, M., 1993. Revision of *Astragalus* L. sect. *Sesamei* DC. (Leguminosae). *Sendtnera*, 10: 69-155.

Grant, W.F and Small, E., 1996. The origin of *Lotus corniculatus* (Fabaceae) complex: A synthesis of diverse evidence. *Canadian Journal of Botany*, 74: 975-989.

GRIN, 1999. *Germplasm Resources Information Network*. National Genetic Resources Programme, USDA, USA. Http:\\www.ars-grin.gov/

Hawtin, G., Iwanaga, M. and Hodgkin, T., 1996. Genetic resources in breeding for adaptation. *Euphytica*, 92: 255-266.

Hill, M.J., Mulcahy, C. and Rapp, G.G., 1996. Perennial legumes for the high rainfall zone of eastern Australia. I. Evaluation in single rows and selection of Caucasian clover (*Trifolium ambiguum* M. Bieb.). *Australia Journal of Experimental Agriculture*, 36: 151-163.

Howieson, J.G., Loi, A. and Carr, S.J., 1995. *Biserrula pelecinus* L. - A legume pasture species with potential for acid, duplex soils which is nodulated by unique root-nodule bacteria. Australian *Journal of Agricultural Research*, 46: 997-1009.

Hussey, B.M.J., Keighery, G.J., Cousens, R.D., Dodd, J. and Lloyd, S.G., 1997. *Western weeds. A guide to the weeds of Western Australia.* Plant Protection Society of Western Australia, Perth.

ILDIS, 1999. *International Legume Database Information Service*. University of Reading, UK. Http:\\www.ildis.org/

Kalinowski, A. and Szweykowski, J., 1992a. Variability of three enzymatic systems in populations of *Anthyllis vulneraria* s.l. on the territory of Poland: I. Populations from Pomeranian and Mazur Lake Districts. *Genetica Polonica*, 33: 279-286.

Kalinowski, A. and Szweykowski, J., 1992b. Variability of three enzymatic systems in populations of *Anthyllis vulneraria* s.l. on the territory of Poland: II. Populations from the upland range in south Poland. *Genetica Polonica*, 33: 287-293.

Kelman, W.M. and Forrester, R.I., 1999. Germination and seedling growth at low temperature: Comparisons of *Lotus* species and affects of selection and seed source in *L. pedunculatus* Cav. *Australian Journal of Agricultural Research*, 50: 969-975.

Kelman, W.M., Blumenthal, M.J. and Harris, C.A., 1997. Genetic variation for seasonal herbage yield, growth habit, and condensed tannins in *Lotus pedunculatus* Cav. and *Lotus corniculatus* L. *Australian Journal of Agricultural Research*, 48: 959-968.

Kernick, M.D., 1978. *Ecological management of arid and semi-arid rangelands in Africa and the Near and Middle East. Vol. IV.* Indigenous arid and semi-arid forage plants of North Africa, the Near and Middle East. FAO, Rome.

Kidambi, S.P., Mahan, J.R., Matches, A.G., Burkes, J.J. and Nunna, R.R., 1990. Genetic variability for esterase enzyme in *Onobrychis* spp. *Theoretical and Applied Genetics*, 80: 433-436.

Kirkbride, J.H., 1995. Proposal to reject the name *L. glaber* Mill. (Leguminosae). *Taxon*, 44:423-424.

Lassen, P., 1989. Om Kroniller, Slaktet *Coronilla. Norden. Svenska Botaniska Foreningen*, 83:83-86.

Loi, A., Howieson, J.G., Cocks, P.S. and Carr, S.J., 1999. Genetic variation in populations of two annual Mediterranean annual pasture legumes (*Biserrula pelecinus* L. and *Ornithopus compressus* L.) and associated rhizobia. *Australian Journal of Agricultural Research*, 50: 303-313.

Loi, A., Cocks, P.S., Howieson, J.G. and Carr, S.J., 1997. Morphological characterization of Mediterranean populations of *Biserrula pelecinus* L. *Plant Breeding*, 116: 171-176.

Martiniello, P. and Ciol, A., 1994. The effect of agronomic factors on seed and forage production in perennial legumes sainfoin (*Onobrychis viciifolia* Scop.) and French honeysuckle (*Hedysarum coronarium* L.). *Grass and Forage Science*, 49: 121-129.

Mir, Z., Mir, P.S., Acharya, S.N., Zaman, M.S., Taylor, W.G., Mears, G.J., McAllister, T.A. and Goonewardene, L.A., 1998. Comparison of alfalfa and fenugreek (*Trigonella foenum-graecum*) silages supplemented with grain on performance of growing steers. Canadian *Journal of Animal Science*, 78: 343-349.

Mir, P.S., Mir, Z. and Townley-Smith, L., 1993. Comparison of the nutrient content and in situ degradability of fenugreek (*Trigonella foenum-graecum*) and alfalfa hays. *Canadian Journal of Animal Science*, 73: 993-996.

Podlech, D., 1990. Revision of *Astragalus* L. sect. *Platyglottis* Bunge Leguminosae. *Mitteilungen der Botanischen Staatssammkung Muenchen*, 29: 541-572.

Prosperi, J.M., Demarquet, F., Angevain, M. and Mansat, P., 1994. Agronomic evaluation of local varieties of sainfoin (*Onobrychis sativa* L.) collected in south-east France. *Agronomie*, 14: 285-298.

Rogers, M.E., Noble, C.L. and Pederick, R.J., 1997. Identifying suitable temperate forage legume species for saline areas. *Australian Journal of Experimental Agriculture*, 37: 639-645.

Rougemont De, G.M., 1989. *A field guide to the crops of Britain and Europe*. William Collins and Sons Ltd., London.

Sebançi, C.O., 1999. Plant genetic resources programmes in Turkey with special reference to forage legumes. In: *Genetic resources of pasture and forage legumes* (eds. Bennett, S.J. and Cocks, P.S.), Proceedings of a workshop, York, Western Australia in October 1996. pp. 150-162. Kluwer Academic Press, The Netherlands.

Sinskaya, E.N., 1961. *Flora of cultivated plants of the USSR.* Israel Programme for Russian Translations, Jerusalem.

Small, E., 1989. Polythetic generic separation in tribe *Trifolieae* subtribe *Trigonellinae* Leguminosae. *Canadian Journal of Botany*, 67: 1480-1492.

Small, E., 1987. A taxonomic study of the Medicagoid *Trigonella* Leguminosae. *Canadian Journal of Botany*, 65: 1199-1211.

Small, E., Brookes, B. and Lassen, P., 1990. Circumscription of the genus *Medicago* Leguminosae by seed characters. *Canadian Journal of Botany*, 68: 613-629.

Small, E., Grant W.F. and Crompton, C.W., 1984. A taxonomic study of the *Lotus corniculatus* complex in Turkey. *Canadian Journal of Botany*, 62: 1044-1053.

Stegelmeier, B., James, L.F., Panter, K.E., Ralphs, M.H., Gardner, D.R., Molyneux, R.J. and Pfister, J.A., 1999. The pathogenesis and toxicokinetics of locoweed (*Astragalus* and *Oxytropis* spp.) poisoning in livestock. *Journal of Natural Toxins*, 8: 35-46.

Stevenson, G.A., 1969. An agronomic and taxonomic review of the genus *Melilotus* Mill. *Canadian Journal of Plant Science*, 49: 4-20.

Suvorov, V.V., 1951. *Melilotus* (Tourn.) Adans. In: *Flora of cultivated plants of the USSR. Vol. 13* (ed. Sinskaya, E.N.). p.p.426-627, Israel Programme Russian translation, Jerusalem.

Tan, A., 1996. Current status of plant genetic resources conservation in Turkey. In: *Proceedings of International Symposium on* in situ *Conservation of Plant Genetic Diversity* (eds. Zencirci, N., Kaya, Z., Anikster, Y. & Adams, W.T.), pp. 5-16. Central Research Institute for Field Crops, Ankara, Turkey.

Townsend, C.E., 1993. Breeding, physiology, culture and utilisation of cicer milkvetch (*Astragalus cicer* L.). *Advances in Agronomy*, 49:253-308.

Trifi, F.N., Chati, W.S., Marrakchi, M. and Pernes, J., 1989. Genetic determinism of ten enzymatic systems in *Hedysarum coronarium* L. *Agronomie*, 9: 503-510.

Tutin, T.G., Heywood, V.H., Burges, N.A., Moore, D.M., Valentine, D.H., Walters S.M. and Webb, D.A., 1968. *Flora Europaea. Vol. 2. Rosaceae to Umbelliferae*. Cambridge University Press, Cambridge.

Zarre, M.S. and Podlech, D., 1996. Taxonomic revision of *Astragalus* L. sect. *Hymenostegis* Bunge (Leguminosae). *Sendtnera*, 3:255-312.

Zarre, M.S. and Podlech, D., 1997. Problems in the taxonomy of tragacanthic *Astragalus*. *Sendtnera*, 4: 243-250.

Chapter 13

Root-nodule Bacteria

John G. Howieson

13.1. INTRODUCTION

Probably the most astonishing aspect of the root-nodule bacteria (hereafter rhizobia) of the Mediterranean basin is how little we actually know about them! Given their relative importance to agriculture in that region, as well as in other world regions that share a similar climate, it is surprising that so few comprehensive studies of the ecology of rhizobia in the Mediterranean basin have been undertaken. It is not quite so surprising that some species of rhizobia from the Mediterranean basin have been thoroughly investigated in regions of the world with a Mediterranean-type climate. For legumes of Mediterranean origin to be used successfully in these regions has often required at least a rudimentary understanding of their microsymbiotic partner. Thus, it is from studies both within and outside the region that we have accumulated knowledge of the complex interactions legumes have with rhizobia from the Mediterranean basin.

13.2. RHIZOBIAL TAXONOMY

It is pragmatic to describe species according to their host specificities, as was the original concept of the "cross-inoculation" classification (Fred *et al.* 1932). Today, the root-nodule bacteria are classified according to their genomic DNA or plasmid profiles, their 16SrRNA sequences, their fatty acid constituents or carbohydrate utilisation patterns (Wang *et al.*, 1996). The capacity to fix nitrogen in association with legumes is found in the root nodule bacteria *Rhizobium*, *Bradyrhizobium*, *Sinorhizobium*, as well as the newly described species *Allorhizobium, Mesorhizobium* and in the stem-nodulating bacterium *Azorhizobium* (Young 1996). Whilst the latter is yet to be described from the Mediterranean basin, the former genera are certainly endemic there. Interestingly, the prevalence of the slow-growing genus *Bradyrhizobium* in the Mediterranean basin appears to be relatively low. This is in contrast to the tropics where bradyrhizobia dominate (Norris 1956). It is speculated this differentiation relates to the substantially different geology of the two regions, with the tropical soils often old, weathered and acidic (Uehara 1978) in contrast to the younger, fine textured and alkaline soils of the Mediterranean basin. Bradyrhizobia are generally more acid tolerant than their fast-growing relatives and this is important when attempting to adapt legume symbioses to acid soils (Howieson, 1995).

The lack of specific epithets for many of the rhizobia associated with the minor herbaceous annual legumes is testimony to the paucity of research on these species. The table also noticeably lacks a designation for species nodulating the common woody perennials such as *Cytisus* and *Genista*, although it is possible they could be classified in the group referred to as the "Cowpea miscellany" (Allen and Allen 1981).

The association of the major rhizobial species found in the Mediterranean region with their legume hosts is shown in Table 13.1.

Table 13.1. The common rhizobial species of the Mediterranean basin and the legumes they nodulate.

Rhizobial species	**Genera nodulated**
	Pulse legumes
Rhizobium leguminosarum biovar *viciae*	*Vicia, Pisum, Lens, Lathyrus*
Mesorhizobium ciceri	*Cicer*
Bradyrhizobium sp *(Lupinus)*	*Lupinus*
Bradyrhizobium japonicum	*Glycine*
Rhizobium leguminosarum biovar *phaseoli*	*Phaseolus*
	Herbaceous and shrub legumes
Rhizobium leguminosarum biovar *trifolii*	*Trifolium*
Sinorhizobium meliloti and *S. medicae*	*Medicago*
Mesorhizobium loti	*Lotus*
Rhizobium sp. and *Mesorhizobium* sp.	*Hedysarum, Biserrula, Hymenocarpos* etc.
Bradyrhizobium sp (*Lupinus*)	*Ornithopus*
Bradyrhizobium sp.	*Coronilla, Genista, Cytisus*

13.3. Ecological Studies of Rhizobia in the Mediterranean Basin

It is tempting to assume that, as the Mediterranean basin represents the centre of origin of many of the commercially important legumes and their associated rhizobia, responses to inoculation there are likely to be rare. Somewhat counter-intuitively, the practical reality of the situation is often opposite to this. In many instances where controlled experiments have been conducted, soil-borne rhizobial populations have failed to elicit satisfactory N-fixation in symbiosis with the legume under test. For example, in the eastern Mediterranean comprehensive studies by the microbiologists at ICARDA (International Centre for Agricultural Research in Dry Areas) have revealed that there is in fact a serious widespread deficiency in effective root-nodule bacteria for the regionally important genera *Medicago*, *Cicer* and *Lens*. Despite chickpea having been cultivated in the Anatolian highlands of Turkey for several thousand years, the introduction there of genotypes resistant to the fungal pathogen *Ascochyta rabiei* required inoculation to ensure maximum nitrogen fixation (Keating *et al.* 1995). This was not the consequence of the introduction to the region of a widely divergent chickpea gene pool, as even when local land races of chickpea were evaluated, the indigenous *M. ciceri* were found to be only moderately effective and generally low in number. A similar finding was made with chickpea in northern Syria, where inoculation with an elite strain increased the proportion of nitrogen derived from fixation by 40% (Beck 1992). We learnt from previous studies of *M. ciceri*, using immunological techniques in southern France, that a broad diversity of strains exist in the region but that certain serotypes dominate certain soils (Arsac and Cleyet-Marel 1985).

Just as chickpea were often found to respond to inoculation in the ICARDA studies, medic rhizobia, although numerous in the alkaline soils, were found to be generally poor at N-fixation with 3 of 4 medic species endemic to the West Asian highlands (Materon *et al.* 1995). The annual species *M. aculeata* formed broadly effective associations in this region and whereas *M. rigidula* nodulated effectively with soil rhizobia in northern Syria (Materon and Danso 1991), this was not the case in Anatolia. *M. noeana* and *M. minima* were also ineffectively nodulated. There was no apparent geographical explanation for the presence or absence of effective rhizobia offered by the authors of this study. The question of why the Anatolian soil harbours a large population of ineffective rhizobia for a set of hosts which presumably evolved in parallel with them will only be answered with more detailed research that examines the genetic diversity of the medic rhizobia, in association with their cross-inoculation characteristics.

Further west, on the island of Sardinia, *Hedysarum coronarium* responded strongly to inoculation with a rhizobial strain selected for high nitrogen fixation despite this forage legume having been cultivated for centuries in that region (Sulas *et al.* 1998). This did not arise from experimentation with the legume on a soil type outside its normal sphere of adaptation. Experiments were conducted on both a typical (alluvial, calcareous) and an atypical (acid, sandy) *Hedysarum* soil.

The picture that emerges from these studies is one of complexity, rather than simplicity, in predicting legume interactions with rhizobia endemic to Mediterranean soils. It appears likely that specific rhizobial genotypes may dominate certain soils, but that these genotypes may not necessarily be optimal for N-fixation. Taken together, these particular findings allow us to generate a fundamental hypothesis for rhizobial ecology of the Mediterranean region:

> "Rhizobial genotypes are differentially adapted to soil conditions and may be advantaged both numerically and competitively under conditions that suit them irrespective of the symbiotic reaction with their host."

This tenet may be applied both within species and between species. It is possible to develop a ranking for rhizobial species adaptation to important soil parameters such as pH (e.g. Graham and Parker 1964) and to further rank strains within species according to their acid soil tolerance (e.g. Howieson and Ewing 1986). These rankings are often independent of the symbiotic relationship between legume host and rhizobia. The important concept is that chemical and physical soil conditions, just as the presence or absence of a legume host, engender a major selection pressure on the rhizobial population.

13.4. Implications of Rhizobial Biodiversity for Germplasm Conservation and Improvement

If we accept the hypothesis above, it follows that rhizobial populations will vary genetically and numerically in association with changing edaphic, as well as legume, parameters. Has this hypothesis been substantiated? Few comprehensive studies have been undertaken, although we can piece together a tale from several independent investigations. For simplicity we consider the associations between microsymbionts and legumes of monospecific genera. In doing so we need to be aware that the symbiotic relationship has essentially two important levels:

- the capacity for nodulation, and
- the capacity for high nitrogen fixation.

13.4.1 Monospecific Genera

In quite a novel study of root-nodule bacteria from the Mediterranean basin, Loi *et al.* (1999) investigated variation in the relationship between the herbaceous legume *Biserrula pelecinus* and its rhizobia through a 20m transect across a field in Sardinia, as well as between distant sites in Sardinia. Symbiotic parameters were assessed using legume and microsymbiont partners collected from the same site. Whereas the host ecotypes in the transect showed substantial morphological variation, this was not matched by a consistent pattern of variation in the root-nodule bacteria. Variation in the microsymbiont seemed to be random and unpredictable in terms of the effectiveness of the association for N-fixation. The probability of a rhizobial strain achieving effective N-fixation with a particular host genotype did not decrease as the distance between the point of collection of the host genotype and the rhizobial strains increased. In fact, rhizobial strains from the Greek Islands were often more effective on Sardinian ecotypes than strains collected from within the transect. Plants of *B. pelecinus* were also frequently nodulated by more than one strain of rhizobia *in situ* and in general terms most associations were effective. Investigation of the DNA patterns of the rhizobial strains in this investigation by PCR (Polymerase Chain Reaction) RAPDs (Random Amplified Polymorphic DNA) (de Bruijn 1992; Richardson *et al.* 1995) verified that the test strains were, in fact, discrete genotypes.

In a second monospecific genus, *Hymenocarpus circinnatus*, the host-rhizobial association appeared to be dramatically more specific. Whilst eight rhizobial strains isolated from *H. circinnatus* collected in the Cyclades group of Greek Islands nodulated nine host ecotypes from the same region, only one strain actually fixed nitrogen (WSM1675, Figure 13.1).

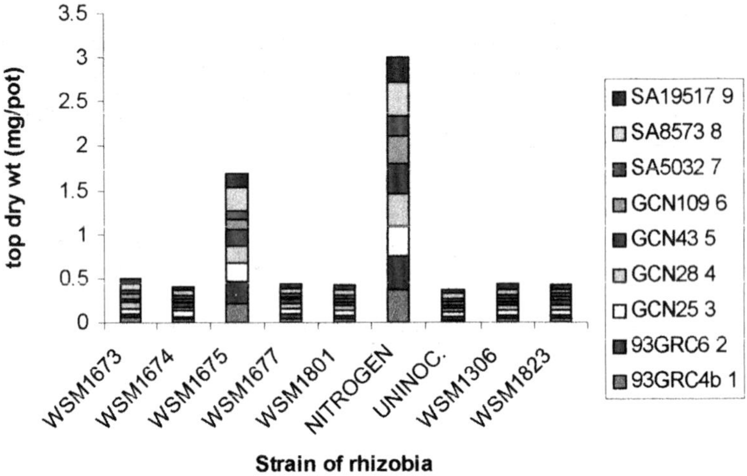

Figure 13.1. The relationship between rhizobial isolates from *H. Circinnatus* and plant genotypes of this legume, with all material collected from the Cyclades group of Greek islands. Only strain WSM1675 effectively fixed nitrogen, despite all strains nodulating all nine hosts (Howieson and Yates, unpublished data). (For Color Plate version, see p. xix)

These two examples indicate that there is a low probability of predicting an optimal symbiotic relationship between legume host and rhizobia based upon regional co-location of the germplasm. This is notwithstanding the fact that we were dealing, in these examples, with mono-specific genera uncomplicated by species interactions.

It became apparent in these studies that within the Mediterranean basin, where large rhizobial biodiversity exists, a legume may in fact be nodulated by multiple strains which vary in their capacity for N-fixation. Unlike legumes introduced to new environments (which have no background rhizobia), the concept of host-rhizobia interactions for N-fixation has a different meaning within the Mediterranean basin, because of simultaneous multiple interactions between legume and microsymbiont.

If we look further, to more complex associations, the picture does not simplify.

13.4.2 Multi Species Interactions with Rhizobia

Table 13.1 indicated that several genera of pulse legume commonly encountered in the Mediterranean regions, including *Vicia*, *Lathyrus*, *Lens* and *Pisum* (see chapters 8,9,10 and 11) are nodulated by the single rhizobial species *Rhizobium leguminosarum* bv. *viciae*. The question that arises here is of seeking evidence for selection on the part of each legume genus for an elite microsymbiotic partner, or more likely a group of partners. Put another way, does R.l. *viciae* diverge to groups which effectively nodulate discrete legume genera, as for the monotypic genera? Or, do we commonly find a mixture of ineffective and effective nodulation? The answer appears to be the latter.

Following an extensive collection of nodule material from these genera in the Cyclades group of Greek Islands, Carr *et al.* (1996) made preliminary observations on the specificity of their microsymbiotic relationships in a field experiment that matched *V. faba* with the 75 resultant rhizobial strains. They found the strains strongly diverged in their effectiveness on *V. faba* according to the host species of origin. For example, 83% of isolates that produced less than 50% of the mean site yield for the experiment were from *Lathyrus* spp. (Table 13.2). At the other end of the scale, the majority of isolates that proved effective for *V. faba* were actually from nodules collected from *Vicia* species.

Table 13.2. Relative yield class of rhizobial isolates (% of site maximum yield) used as inocula for *V. faba* sown on an acidic (pH 4.4 $CaCl_2$) sandy soil at Jingalup, W.A. (Carr *et al.* 1996).

Collection host	Relative yield class (% of maximum site yield)		
	<50	51-75	>75
Vicia sp.	17	68	86
Lathyrus sp.	83	27	14
Unidentified legume	0	5	0

It seems likely that for these pulse legumes, as a general rule, we will eventually be able to group individual *R.l. viciae* strains according to the level of N-fixation achieved in association with each particular genus. It is also probable that from within the divergent groups a low proportion of rhizobial strains will cross the "group" boundaries and be fully effective on the majority of species. Such strains might present themselves as elite genotypes for future commercial manufacture.

A similar question can be posed for the *Trifolium* species. Whilst we are aware of the multitude of clover species endemic to the Mediterranean basin (see chapter 5), we generally have much less knowledge about how their nodule bacteria *Rhizobium leguminosarum* bv. *trifolii* interact within these species groupings. We actually have very little knowledge of the level of biodiversity of this important rhizobial species in the Mediterranean basin. A recent DNA-based study of forty clover rhizobia isolated from 5 clover species growing in a discrete and undisturbed setting on the Tel Hadya Research Station of ICARDA, in northern Syria, revealed a surprising lack of variation (Leonard *et al.* 1999). Nodules were collected at random from the clovers *T. tomentosum*, *T. resupinatum*, *T. stellatum*, *T. purpureum* and *T. angustifolium*. Thirty six of the 40 isolates shared a common *nifHDK* promoter consensus region as revealed by RAPD PCR analysis using the RPO1 primer of Richardson *et al.* (1995).

Thus, the clover species in the sampling region (approximately 5 ha.) appeared to be nodulated by a single rhizobial genotype. Does such lack of biodiversity extend to the greater Mediterranean basin? Little research at the DNA typing level has been conducted but some insights can be drawn from investigations of cross-inoculation responses using rhizobia collected from different geographic regions. Of 12 *R.l. trifolii* strains collected from 6 clover species growing on mildly acid soils on the Mediterranean islands of Sardinia, Corsica, Greece and in Morocco, as well as three Australian commercial strains, all but 3 fixed N optimally with *T. subterraneum*. We might have been tempted to assume that a similar lack of biodiversity was demonstrated here. However further investigations revealed a more variable response when the same strains were evaluated on the species *T. vesiculosum* (Figure 13.2). The effective isolates on *T. vesiculosum* formed only a subset of the effective isolates for *T. subterraneum*, indicating that the latter species is more promiscuous for effective microsymbionts than is the former. More importantly, the symbiotic reaction with *T. vesiculosum* indicated a large biodiversity of clover rhizobia was collected from this transect across the Mediterranean basin.

These results give an indication that the biodiversity of clover rhizobia in the Mediterranean basin may be limited within sites, but diverse between geographically separated sites. Given a wide enough genetic base to experiment with it may be possible to eventually assign general but meaningful species groupings to the clover root-nodule bacteria of the Mediterranean basin. Despite attempts to define such groups, it should also be possible to select strains with a broad host range for effective N-fixation from which to develop elite inoculants.

When evaluating host-strain interactions for N-fixation, there has always been some doubt as to whether strains that give a different phenotypic reaction on the one host are different entities. As an aid to understanding and confirming phenotypic differences displayed between rhizobia, it is now possible to identify rhizobia at the strain level according to their genomic DNA. The PCR technique using multiple primers, or primers with multiple binding sites, has become a powerful tool to aid cross-inoculation and also ecological studies (Hebb *et al.* 1998). When PCR typing (RAPD's) of rhizobia has been used in conjunction with empirical assays, the technique has often confirmed that the phenotypic variability displayed in glasshouse studies represents actual genetic differences between rhizobial strains (Howieson *et al.* 1995; Loi *et al.* 1999). In some investigations in France, genetic differences identified between have been considered sufficiently great to warrant separation of genotypes at the species level (Rome et al. 1996).

Apart from taxonomic studies, how is this phentoypic and genetic information useful? It helps us with our two primary aims –improving legume performance and germplasm conservation.

13.5. Selecting Rhizobia to Improve Legume Performance

Modern agriculture, more so than ever before, is pushing legumes to the boundaries of their agronomic adaptation and demanding optimal performance along the way. A large part of optimising performance is in matching the inoculant rhizobial strain to both the legume host and to specific soil conditions to achieve maximum N-fixation (Howieson *et al.* 1999a). In both the Mediterranean basin and in regions of the world that have exploited Mediterranean legumes, this can only be achieved if rhizobial germplasm is available for evaluation. It is in this regard that the rhizobial germplasm of the Mediterranean is of enormous potential benefit. We have seen in the examples earlier in this chapter that rhizobial biodiversity is rich across the Mediterranean region but that diversity within discrete sites may be limited. The challenge remains to identify niches of particular value and to embark on targeted conservation activities.

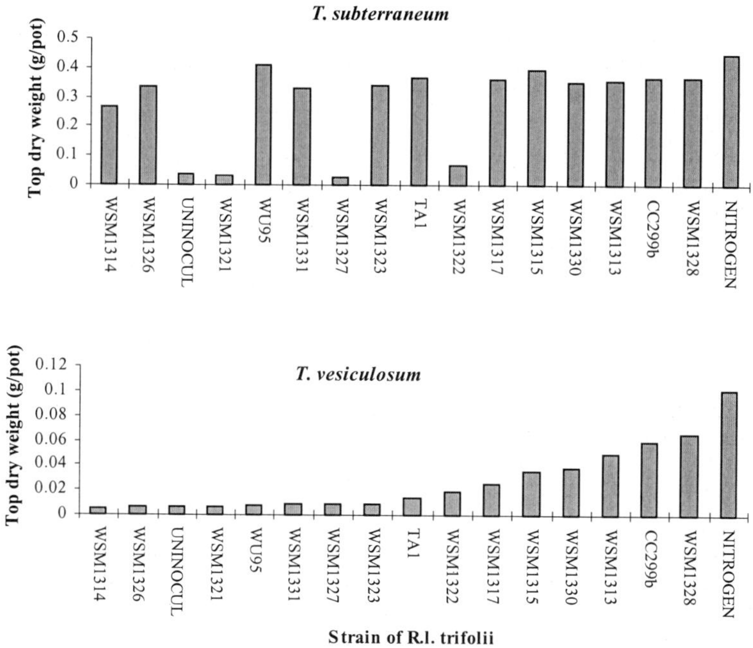

Figure 13.2. The capacity for N-fixation of rhizobia isolated from a range of clover species in the Mediterranean basin with *T. subterraneum* and *T. vesiculosum* (see Howieson and Loi 1994 for strain details)

13.6. CONSERVATION PRIORITIES

Modern studies of rhizobial diversity, particularly those that have used DNA typing, leave us with no doubt that only a minute percentage of the available rhizobial genetic diversity of the Mediterranean basin has been captured in gene banks *ex situ*. Probably the largest and most diverse *ex situ* collection of rhizobia from the Mediterranean basin is held in the WSM (Western Australia Soil Microbiology) collection at the Centre for *Rhizobium* Studies, Perth, Western Australia. This collection holds over three thousand genotypes in a vacuum dried state for both preservation and experimentation. At present, ampoules are provided to *bona fide* researchers free of charge upon request. However, because of the enormity of the available rhizobial biodiversity in the Mediterranean basin, it is futile to attempt to preserve the bulk of this material anywhere except *in situ*.

In this regard, we are fortunate that many of the non-arable lands of the Mediterranean basin represent legume habitats which readily provide an *in situ* repository (Howieson *et al*., 1999b). However we must recognise that a complete reliance on this source of conservation has the following limitations:

- All soil types are not well represented in the non-arable lands,
- We know that pulse and shrub legume biodiversity (and hence rhizobia) are not well captured therein,
- Severe overgrazing of the non-arable lands has already depleted their legume genetic resources in some regions (Robertson and Bounejmate 1999), and
- Arable lands are increasingly being sown to monocultures, which threatens the biodiversity of rhizobia contained therein.

In view of these limitations on *in situ* conservation in non-arable lands, an immediate priority for rhizobial conservation in the Mediterranean basin should be the root-nodule bacteria from the pulse and shrub legumes of the arable lands. An opportunity that should not be overlooked is the potential for the purchase by IPGRI of uninhabited small islands in the Mediterranean Sea carrying a diversity of soil types and legume species. Those currently offered for sale by the Greek Government might satisfy this requirement (S. Samaras, pers. comm., Thessalonika). These islands would provide *in situ* repositories of genes that may provide valuable germplasm for breeding, selection and molecular manipulation in the future.

13.7. Gaps in Knowledge

Few studies have investigated the ecological relationships between rhizobia and legumes from genera containing both annual and perennial species. For example, in the Mediterranean basin, *Vicia*, *Medicago*, *Trifolium* and *Lotus* all contain both annual and perennial species (Pignatti 1982). Within the discrete genera, do rhizobial strains commonly cross from annual to perennial, and if so how frequently and how effectively? We know from several studies that isolates from perennial *Lotus* are often incompatible with annual *Lotus* (Pankhurst, 1980; Lowther *et al.* 1987), however can we say with what frequency cross-over occurs *in situ*? We are also aware that clover rhizobia from the east African perennial species appear symbiotically incompetent with annual clovers from the Mediterranean basin (Norris 1959; Roughley and Date 1986; Lupwayi *et al.* 1997). But within the Mediterranean do strains from the common perennial clovers *T. uniflorum*, *T. repens* and *T. pratense* commonly nodulate the annual clovers and with what outcome? Similar questions may be asked of the *Medicago* and *Vicia* genera. These questions have important commercial implications. For example, there are large parts of southern Australia likely to grow both annual and perennial medics as well as clovers in the next few decades and we need to be able to predict and manage their symbiotic interactions to maximise N-fixation.

A second major knowledge deficiency relates to an understanding of the symbiotic interactions of Mediterranean legumes with indigenous rhizobia associated with legumes in other Mediterranean-type climates. For example, on the western seaboard of both north and south America an unique legume flora has evolved that includes species of *Trifolium* (Aronson *et al.* 1992). Do the Mediterranean clovers interact with rhizobia that nodulate these American species? As well, we understand from early research that the bradyrhizobia from the Western Australian native legumes nodulate effectively with *Phaseolus vulgaris* and *Lupinus cosentinii* but do not associate with either *L. angustifolius* or *Ornithopus* sp. (Lange 1961). This indicates a quite remarkable specificity that needs to be further understood and eventually harnessed. We are also aware that the valuable fodder shrub *Chamaecytisus proliferus*, which is indigenous to the Canary Islands, interacts effectively with native bradyrhizobia from southern Australia, but what of other potentially useful shrub legumes?

Prior to the recent reports of fast-growing rhizobia from the nodules of south-west Australian native legumes *Swainsona* and *Acacia* (Marsudi *et al.* 1999; Yates *et al.* 1999) it was considered that legumes indigenous to this region were nodulated only by bradyrhizobia (Lange and Parker, 1960). Do any of the Mediterranean herbaceous legumes that are nodulated by fast growing rhizobia interact with these newly isolated strains from south-west Australia, or indeed those rhizobia previously reported to nodulate *Acacia* sp. in New South Wales (Barnet et al. 1993)?

A third knowledge gap relates to the taxonomic status of the root-nodule bacteria of the minor herbaceous and shrub legume species of the Mediterranean basin. These species have not been well researched in an agronomic sense, but when they are scientists will need to understand their symbiosis to exploit them optimally in an agricultural setting.

13.8. CONCLUSIONS

Understanding and preserving the biodiversity of rhizobia from the Mediterranean basin is crucial to the continued increase in productivity from legumes cultivated both within and outside the Mediterranean basin. Modern genetic analytical methods indicate that the level of rhizobial biodiversity in this region is much greater than previously supposed. Rhizobial strains vary in both their adaptation to edaphic constraints and in their capacity for nitrogen fixation, yet the variation in these two parameters is unconnected and difficult to predict. Until we understand the factors causing this variation and are in a position to predict where it occurs, preservation of representative legume niches must be considered a high priority to achieve adequate conservation of rhizobial genotypes in the Mediterranean basin. At present, the niches for the shrub and pulse legumes and their rhizobia appear exposed because of the intensification of monoculture agriculture in arable lands. The non-arable lands are also exposed to genetic erosion because of severe overgrazing.

ACKNOWLEDGMENTS

I wish to thank Mr Ron Yates, Ms Tonia Leonard, Ms Jane Malden and Dr Graham O'Hara for contributing to the content of this chapter.

REFERENCES

Allen, O.N. and Allen, E.K., 1981. *The Leguminosae*. University of Wisconsin Press. Macmillan, Madison, London.

Aronson, J., Ovalle, C. and Avendano, J., 1992. Early growth rate and nitrogen fixation potential in 44 legume species grown in two soil types in central Chile. *Forest Ecology Management*, 47: 225-44.

Arsac, J.F. and Cleyet-Marel, J.C., 1985. Serological and ecological studies of *Rhizobium* spp. (*Cicer arietinum*) by immunofluorescence and ELISA technique: Competitive ability for nodule formation between *Rhizobium* strains. *Plant and Soil*, 94: 411-423.

Barnet, Y., Catt, P.C., Jenjaroentham R. and Mann, K., 1993. Fast-growing root-nodule bacteria from Australian *Acacia* spp. In: *New Horizons in Nitrogen Fixation* (eds. Palacios, R., Moira, J. and Newton, W.E.) Kluwer Academic Publishers, Dordrecht.

Beck, D.P., 1992. Yield and nitrogen fixation of Chickpea cultivars in response to inoculation with selected rhizobial strains. *Agronomy Journal*, 84: 510-6.

de Bruijn, F.J., 1992. Use of repetitive (repetitive extragenic palendromic and enterobacterial repetitive intergenic consensus) sequences and the polymerase chain reaction to fingerprint the genomes of *Rhizobium meliloti* isolates and other soil bacteria. *Applied and Environmental Microbiology*, 58: 2180-87.

Carr, S.J., Howieson, J.G., Nutt, B.J. and Yates, R.J., 1996. Specificity in *the Rhizobium leguminosarum bv viciae-Vicia faba* symbiosis. Proceedings of the XI Australian Nitrogen Fixation Conference, 22-27 September, Perth, Western Australia.

Fred, E.B., Baldwin, I.L. and McCoy, E., 1932. *Root Nodule Bacteria and Leguminous Plants*. University of Wisconsin, Madison.

Graham, P.H. and Parker, C.A., 1964. Diagnostic features in the characterisation of root-nodule bacteria of legumes. *Plant and Soil*, 20: 383-396.

Hebb, D.M., Richardson, A.E., Reid R. and Brockwell, J., 1998. PCR as an ecological tool to determine the establishment and persistence of *Rhizobium* strains introduced into the field as seed inoculant. *Australian Journal of Agricultural Research*, 49: 923-34.

Howieson, J.G. and Ewing, M.A., 1986. Acid tolerance in the *Rhizobium meliloti-Medicago* symbiosis. *Australian Journal of Agricultural Research*, 36: 55-64.

Howieson, J.G. and Loi, A., 1994. The distribution and preliminary evaluation of alternative pasture legumes and their associated root-nodule bacteria collected from acidic parts of Greece (Serifos), Morocco, Sardinia and Corsica. *Agricola Mediterranea*, 124: 1-17.

Howieson, J.G., 1995. Characteristics of an ideotype acid tolerant pasture legume symbiosis in Mediterranean agriculture. Proceedings of the Third International Symposium on Plant Soil Interactions at Low pH, *Plant and Soil*, 171: 71-76

Howieson, J.G., Loi, A. and Carr, S.J., 1995. *Biserrula pelecinus L.* -a legume pasture species with potential for acid, duplex soils which is nodulated by unique root-nodule bacteria. *Australian Journal of Agricultural Research*, 46: 997-1007.

Howieson, J.G., Malden, J., Yates, R.J. and O'Hara, G.W., 1999a. Techniques for the selection and development of elite inoculant rhizobial strains in southern Australia. *Symbiosis* (in press).

Howieson, J.G., O'Hara, G.W. and Carr, S.J., 1999b. Changing roles for legumes in Mediterranean agriculture: Developments from an Australian perspective. *Field Crops Research* (in press).

Keating, J.D.H., Materon, L.A., Beck, D.P., Yurtsever, N., Karug, K. and Altuntas, S., 1995. The role of rhizobial biodiversity in legume crop productivity in the West Asian highlands. I. Rationale, methods and overview. *Experimental Agriculture*, 31: 473-483.

Lange, R.T., 1961. Nodule bacteria associated with the indigenous Leguminosae of south-western Australia. *Journal of general Microbiology*, 61: 851-59.

Lange, R.T. and Parker, C.A., 1960. The symbiotic performance of Lupin bacteria under glasshouse and field conditions. *Plant and Soil*, 13: 137-46

Loi, A., Cocks, P.S., Howieson, J.G. and Carr, S.J., 1999. Genetic variation in populations of two Mediterranean annual pasture legumes and associated rhizobia. *Australian Journal of Agricultural Research*, 50: 305-315.

Lowther, W.L., Hay, J.M. and Ryan, L., 1987. Effect of strain of rhizobia, lime and phosphorus on dry matter yield of three *Lotus* species in differing environments in Otago and Southland. *New Zealand Journal of Experimental* Agriculture, 15: 135-42.

Lupwayi, N.Z., Haque, I. And Holl, F.B., 1997. Effectiveness, competitiveness and persistence of inoculant *Rhizobium* for perennial African clovers in a highland vertisol. *Journal of Agricultural Science Cambridge*, 129: 429-37.

Materon, L.A. and Danso, S.K.A., 1991. Nitrogen fixation in two annual *Medicago* legumes, as affected by inoculation and seed density. *Field Crops Research,* 26: 253-262.

Materon, L.A., Keating, J.D.H., Beck, D.P., Yurtsever, N., Karug, K. and Altuntas, S., 1995. The role of rhizobial biodiversity in legume crop productivity in the West Asian highlands. III *Rhizobium meliloti. Experimental Agriculture*, 31: 493-99.

Marsudi, N.D.S., Glenn, A.R. and Dilworth, M.J., 1999. Identification and characterisation of fast- and slow-growing root-nodule bacteria from south-western Australian soils able to nodulate *Acacia saligna. Soil Biol. Biochem.* (in press).

Norris, D.O., 1956. Legumes and the *Rhizobium* symbiosis. *Empire Journal of Experimental Agriculture*, 24: 247-70.

Norris, D.O., 1959. *Rhizobium* affinities of African species of *Trifolium. Empire Journal of Experimental Agriculture*, 27: 89-97.

Pankhurst, C.E., 1980. Studies of rhizobia nodulating *Lotus* species. *Lotus Newsletter*, 11: 3-5.

Pignatti, S., 1982. *Flora d'Italia, Vol III.* Edagricole, Bologna,

Richardson, A.E., Viccars, L.A., Watson, J.M. and Gibson, A.H., 1995. Differentiation of *Rhizobium* strains using the polymerase chain reaction with random and directed primers. *Soil Biology and Biochemistry*, 27: 515-526.

Robertson, L.D. and Bounejmate, M., 1999. Genetic erosion of pasture and forage legumes in the Mediterranean basin. In: *Genetic Resources of Mediterranean Pasture and Forage Legumes* (eds. Bennett, S.J. and Cocks, P.S.). pp. 67-77. Kluwer Academic Publishers, Dordrecht.

Rome, S., Brunel, B., Normand, P., Fernandez, M., and Cleyet-Marel, J.C., 1996. Evidence that two genomic species of *Rhizobium* are associated with *Medicago truncatula. Archives Microbiology*, 165: 285-288.

Roughley, R.J. and Date, R.A., 1986. The effect of strain of *Rhizobium* and temperature on nodulation and early growth of *Trifolium semipilosum. Experimental Agriculture*, 22: 123-131.

Sulas, L., Re, G.A., Loi, A. and Howieson, J.G., 1998. The selection of optimal root-nodule bacteria inoculants increases forage yield of *Hedysarum coronarium* (sulla). *Proc. 17th General Meeting of the European Grassland Federation. Hungary*. May 18-21 1997.

Uehara, G., 1978. Mineralogy of the predominant soils in tropical and sub-tropical regions. In: *Mineral Nutrition of Legumes in Tropical and Subtropical Soils* (eds. Andrew, C.S. and Kamprath, E.J.) CSIRO, Melbourne.

Wang, E.T., Van Berkum, P., Sui, X.H., Beyene, D., Chen, W.X. and Martinez-Romero, E., 1999. Diversity of rhizobia associated with *Amorpha fruiticosa* isolated from Chinese soils and description of *Mesorhizobium amorphae* sp. Nov. *International Journal of Systematic Bacteriology*, 49: 51-65.

Yates, R.J., J.G. Howieson, Gintzburger, G., and O'Hara, G.W., 1999. The interaction of non-indigenous perennial legumes from Mediterranean type climates with root-nodule bacteria in soils of the north-west of Western Australia. *VI International Rangeland Congress*, Townsville. July18-23.

Young, J.P.W., 1996. Phylogeny and taxonomy of rhizobia. *Plant and Soil*, 186: 45-52.

Section 3

Conservation of Legume Species in the Mediterranean

Chapter 14

Ex Situ Conservation of Grain Legumes

Larry D. Robertson and Nigel Maxted

14.1 Introduction

The major food legumes that originated in the Mediterranean and that are important in this region include chickpea (*Cicer arietinum* L.), lentil (*Lens culinaris* L.), and faba bean *(Vicia faba* L.). Lentil is an ancient cool-season field legume that has been traditionally grown in the Middle East, the Indian subcontinent, Ethiopia, North Africa, and to a lesser extent, in southern Europe. It has also been introduced to South and North America. It is usually grown in rotation with cereals and is valued for both its seed for human consumption and straw for animal production. Archeological data has shown that lentil has been cultivated as early as 8000 BC in the Middle East (Hansen and Renfrew 1978). In the dryer regions of the Middle East, North Africa and the Indian subcontinent lentil is one of the principal pulse crops cultivated and is a dietary mainstay of the populations of this region where it is an important source of protein. Faba bean (*Vicia faba L.*) is an old world legume with numerous references by ancient writers (Hawtin and Hebblethwaite 1983). The major producing countries for faba bean are China and Ethiopia. Faba bean is cultivated in Europe in the Mediterranean region and in northern Europe. Throughout North Africa and West Asia faba beans are an important component of the diet, especially of the poorer peoples and has been referred to as the "poor man's meat". Faba beans were introduced to South America with European colonization and are mostly grown at higher elevations where *Phaseolus* beans can not be grown. Faba bean is of a more recent introduction to North America where it is not of significance and to Australia, where the area has dramatically increased in the past ten years. Chickpea is a diploid, self-pollinated, leguminous crop that runs second in area and third in production among the pulses. It is cultivated primarily for its protein-rich seed, and the plant is an efficient symbiotic nitrogen-fixer, playing an important role in farming systems. Two types of chickpea are grown: desi, with angular and coloured sees, primarily grown in South Asia; and kabuli, with large, owl-shaped and beige-coloured seeds, grown in the Mediterranean region, the USA and Latin America.

The Mediterranean region is also the centre of origin and primary diversity for *Lathyrus* species (Zohary and Hopf 1988). In the Mediterranean, *Lathyrus cicera* L. is common in Cyprus, Greece, Iran, Iraq, Spain and Syria, and *Lathyrus ochrus* (L.) DC. in Cyprus and Greece, used mainly for feed and forage (Saxena *et al.* 1993). Lathyrus or chickling (*Lathyrus sativus* L.) has been widely cultivated in South Asia and Ethiopia for over 2500 years (Bell 1989) and is used as a food and feed. It is rich in protein content, around 30 g/ 100 g edible seeds (Aletor *et al.* 1994) and contains a high amount of free L-homoarginine, which can as a precursor of lysine in higher animals (Quereshi *et al.* 1977). One of the drawbacks of *L. sativus*, however, is that its excessive consumption can cause

"Lathyrism", a nervous disorder resulting in an incurable paralysis of the lower limbs of human beings or domestic animals, which is caused by the presence of a free amino acid known as β-*N*-Oxalyl-L-α,β-Diaminopropionic Acid (ODAP) in seed. Agronomically, the species is able to withstand both severe drought as well as water logging. During recent collections in Ethiopia and Bangladesh by the International Centre for Agricultural Research in the Dry Areas (ICARDA), it was noted that most farmers grew *Lathyrus* because of this water logging tolerance. In Ethiopia *Lathyrus* was grown on heavy black clay soils with poor drainage and in Bangladesh the most common cropping system noted was the relay cropping of *Lathyrus* in rice, where *Lathyrus* was planted in the flooded paddy before it was drained and harvested.

Other major grain and food legumes that originated in the Mediterranean region, but which are relatively unimportant in this region are peas (*Pisum sativum* L.),, narrow-leaf lupins (*Lupinus angustifolius*) white lupins (*Lupinus albus*), and yellow lupins (*Lupinus luteus*). Another species of minor world importance is *Trigonella foneum-gracecum* (used as a condiment) which is grown on a small scale both in and outside the Mediterranean region.

14.2 Importance, Production and Utilization

The important food legume crops of the Mediterranean region are faba bean, chickpea, and lentil (Table 14.1, FAO 1999). Chickpea is produced extensively in the Mediterranean region, South Asia, Mexico and the Americas with the largest production in South Asia (76%). Lentil is also produced extensively in these regions, again with the largest production in South Asia (48%). Faba bean is produced extensively in the Mediterranean region, China and Ethiopia with the largest production in China (45%). These species are primarily utilized as food legumes. The majority of the production of both chickpea and lentil in the Mediterranean region is in Turkey and for faba bean in Egypt and Morocco. In general, chickpea and lentil are the most important food legumes in West Asia, while faba bean is more important in North Africa. In the Mediterranean region; chickpea, faba bean and lentil account for the majority of the pulse production (most pea production in Europe is not in the Mediterranean region).

World lentil production doubled in from 1.3 million MT an^{-1} in the triennium 1979-81 to 2.6 million MT an^{-1} in 1991-93, through a 49 % increase in sown area and a 34 % increase in average yield from 598 to 801 kg/ha. This increase came from the three major producing countries, which, collectively, grow 68% of the world's lentil production: 1. India doubled its production, 2. Turkey tripled its production and 3. Canada expanded production tenfold. Production in Asia is concentrated in a band stretching from Turkey in the West to Bangladesh in the East, accounting for 76% of world production. Among other Asian producers, China has recently started releasing lentil data, Bangladesh ranks fourth in production overall. Iran and Nepal have doubled production in the last decade; Syria has maintained its production, while that of Pakistan has been reduced.

Table 14.1. Worldwide area of production (ha) of major food and grain legumes originating from the Mediterranean region (FAO, 1999).

Region/country[1]	Faba bean (dry)	Faba bean (green)	Chickpea	Lentil	Lupins (all)	Pea (dry)	Pea (green)	Total Pulses
North Africa	241700	36950	116175	58650	1200	51550	36600	568135
Nile Valley	452727	255	198721	57981	2718	184950	11500	1680422
West Asia	48910	34810	1539523	976066	0	5935	14953	3098524
South Asia	0	75566	8507925	1619868	95	1038716	210225	27067328
Latin America	207281	30350	146995	35705	26386	82114	74779	7247000
China	1000000	11137	1500	90000	0	750000	112050	3098524
North America	19100	0	0	435640	0	1206580	124942	2533200
Europe	127828	30350	145608	42375	96408	2902339	196286	4780500
Australia	107000	0	265000	82000	1215000	297000	9500	2061000
Other	30500	1550	272163	6015	20060	378807	121550	15903488
World	2185446	190618	11193610	3404300	1361867	6897991	833131	68099493

[1]North Africa=Algeria, Morocco, Libya and Tunisia; Nile Valley=Egypt, Eritrea, Ethiopia and Sudan; West Asia=Turkey, Syria, Lebanon, Syria, Jordan, Iraq, Iran, Israel, and all countries of the Arabian Peninsula; South Asia=Bangladesh, Nepal, India and Pakistan; Latin America=Mexico, all Central and South America and all of the Caribbean; North America=USA and Canada; Europe=all countries on European continent including former Soviet Republics and eastern European countries.

Sixteen percent of lentil production is in North America, where, in addition to Canada, the USA maintains an area of *ca* 55,000 ha and Mexico is a minor producer. Following Asia and N. America, Africa produces approximately 3 % of the lentil crop worldwide, with Ethiopia and Morocco as significant producers and Algeria, Egypt and Tunisia as minor players. In S. America, the producing countries in descending order of importance are Argentina, Chile, Colombia, Peru and Ecuador, but only a total of 50,000 MT is produced annually in the continent. European lentil production is gradually decreasing with Belarus, France, Russia and Spain as the remaining significant producers. In Oceania, New Zealand produces *ca* 10,000 ha annually and Australia may become a significant producer in the future, because of its considerable research interest in the crop. The international export trade in small-seeded, red cotyledon lentil is dominated by Turkey and in the large-seeded, green lentil by Canada and USA.

The largest increases in area and production of faba bean in the period 1979-1993 were for Australia, with an 8-fold increase in area and Morocco with an increase of 33%. In Australia several breeding programmes have been established in the past ten years and the area and production of faba bean is expected to greatly increase in the next ten years. Compared to the other grain legumes, world trade in faba bean is inconsequential, with most produce locally consumed. However, there is expectation that faba bean may become a major export crop in Australia. There is a significant production of green faba bean in the Mediterranean region (Table 14.1). This production is often under-reported because either the crop is harvested first as green pods and then later harvested for dry seeds and reported that way or the vegetable faba bean area is often not reported and there is a significant production of green faba beans in small kitchen gardens.

Chickpea accounts for 12% (or 7.1 million t) of world pulse production. This figure, however, understates the importance of this commodity in specific geographical regions of the world. The area of cultivation is often under-reported because of its secondary status as a companion crop in inter-cropped systems (Jodha and Rao 1987). Compared with significant gains in world pulse production during the last two decades, the chickpea share has declined and the overall declining trend in chickpea area in India and South Asia will likely continue (Kelley and Rao 1994). Production increases in Turkey have come about mainly through area expansion spurred on by strong demand from importers and an attractive export incentive policy of the government. Oceania (mainly Australia) shows a positive trend in area and production; this growth was in response to the relaxation of import restrictions on pulses in India during recent times (Rees *et al.* 1992).

Pea production is heavily concentrated in Europe, North America, South Asia and China (Table 14.1). The large majority of the area in Europe is not in the Mediterranean part of Europe but in the non-Mediterranean climates. The European countries with the largest areas are the Russian Federation, France (non-Mediterranean part), Belarus and Denmark. There is also a large pea production in Australia (FAO 1999). The area of lupin production is reported for all lupins (Table 14.1, FAO 1999). The vast majority of the Lupin area is in Australia (89%), and most of this is in Western Australia. Most of this area is of narrow-leaf lupin (*Lupinus angustifolius*). However, white lupin (*L. albus*) and yellow lupin (*L. luteus*) are grown on small scale, though white lupin has been banned in Western Australia because of sensitivity to anthracnose. In the Mediterranean region white lupins are grown on a limited extent in Morocco and Egypt (FAO 1999). Lupins are also grown to a limited extent in Europe.

14.3 SEED CONSERVATION

The food and grain legume species covered in this paper have orthodox seeds that are easily maintained in active collections at 0° - 4°C at a relative humidity of 20-25% (Robertson *et al.* 1998, Robertson and Erskine 1998, Singh *et al.* 1998). Seeds are dried to seed moisture content of about 6% before storage. A major problem in the conservation of faba bean and to a lesser extent with the *Lupinus* species is the relatively large seed size compared with other crops. This causes a problem in maintenance because of the large space needed to store the collection. The large seed size also results in a low multiplication rate that makes the rate of rejuvenation low.

Clean, healthy and viable seed is need for storage and distribution. Most germplasm collections impose a schedule of seed testing to ensure that only pathogen- and virus-free seed of these crops are kept in collections on a longer-term basis. Also, during multiplication in the field, there is routine monitoring of multiplication plots for viruses and pathogens. It is a normal routine to test all seed lots for viability after a regeneration to ensure seed with a high germination rate is used to conserve these species. The multiplication ratio of the grain and forage legumes is much lower than with the smaller-seeded pasture and forage legumes and can be as low as 10 for the large seeded faba bean accessions.

Wild relatives of chickpea and lentil have orthodox seeds which are easily maintained in the gene banks in active and base collections. Seeds are dried to a moisture content of about 6% before storage in the base and active collections. The base collection is maintained at -20°C and the active collection at 0°C at a relative humidity of 16%. These species have hard seeds and scoring of the seeds is needed to facilitate germination.

At ICARDA and most other international and national genebanks the seed of the cultivated and wild germplasm of these species are kept under a schedule of testing to ensure that only pathogen and virus seed are kept in the collection in the long term maintenance of these genetic resources. Testing for viability when seed is first added to a collection is necessary to ensure that healthy seed that will last is being preserved. Most genebanks have instituted a regular routine monitoring schedule of the viability of seed stored.

14.4 REPRODUCTIVE BIOLOGY

All these food and grain legume species, except faba bean, are normally self-pollinated. Problems are caused in multiplication of the faba bean collection because of its partial out-crossing by insects. Measures that need to be taken in regeneration of faba bean include using insect-proofed screen houses, such as tripping to ensure seed set. (because of autofertility problems that become apparent with selfing), greatly increasing the cost of faba bean germplasm multiplication. A pure line faba bean collection has been developed at ICARDA (BPL, bean pure line). These were developed through a "pre-breeding" process by taking randomly selected single plants to progeny rows in a cyclic manner, using insect-proofed screen houses to ensure selfing (Robertson, 1985).

Lentil is a self-pollinated species with little outcrossing (Wilson and Law 1972) although up to 6.6% outcrossing has been reported (Erskine and Muehlbauer, 1991). Flowers are complete and are borne singly or in multiples on peduncles from the upper part of the plant. Since lentil has low outcrossing, generally accepted at less than 1% (Wilson and Law 1972) it is easily grown for multiplication and rejuvenation in adjacent plots without any measures needed for control of pollination. Because of their seed shattering, the wild *Lens* species are generally grown in plastic or glasshouses for rejuvenation and the plants are bagged at flowering to prevent seed loss. Chickpea is a quantitative long-day plant, but in short days vernalization may replace this requirement in some cultivars (Angus and Moncur 1980). The response to cold also appears to be quantitative (Saxena and Siddique 1980). Flowering begins sooner in warm weather and at long photoperiods (Roberts *et al.* 1980). Only 20-50% of the flowers et pods (Eshel 1968). Temperature is more important than photoperiod in determining the length of the reproductive period.

14.5 COLLECTION TYPES

In contrast with the other temperate legumes, faba bean is partially cross-pollinated by insects (Bond and Poulsen, 1983). The traditional way of maintaining faba bean germplasm is as open-pollinated bulks. The first collection started at ICARDA was the open-pollinated ILB collection which was maintained as populations. The major problem with this type of collection is the loss of genetic identity resulting from inter-crossing among different accessions. There are two ways to reduce the loss of identity of accessions: (1) reduce the rate of inter-crossing among accessions during multiplication through increased distance between accessions or increasing plot size; or (2) reduce generation advance. Witcombe (1982) proposed a system to reduce generation advance using a base collection, foundation seed and active collections. The ILB collection proved difficult to maintain for a large number of accessions and several other types of collections were investigated.

A decision was made at ICARDA to derive a second 'pure-line' collection from the original collection. A set of pure lines of faba bean has been derived from the usually heterogeneous and heterozygous ILB accessions (Robertson, 1985). These were developed through a "pre-breeding" process by taking randomly selected single plants to progeny rows in a cyclic manner, using insect-proofed screen houses to ensure selfing (Figure 14.1). The faba bean pure lines (BPLs) have the advantages of : (a) ease of maintenance; (b) repeatability and uniformity of evaluation; (c) reduction in loss due to genetic drift; and (d) uncovering of recessive genes which otherwise might be hidden by heterozygosity. Repeated inbreeding has not led to loss in vigour due to inbreeding depression.

Trait-specific gene-pools (TSGs) were proposed for faba bean by Witcombe (1982), where accessions similar for such traits as maturity, seed size, height, and growth habit are bulked and maintained by growing in isolation to allow inter-crossing. TSG's would allow the reduction of a large number of accessions to a few TSGs without loss of much genetic variability, as variability is not randomly distributed and much germplasm may be duplicated. This method has many advantages of the core collection. There is merit in this approach, however, one limitation is the need for good evaluation data for the material to bulk. Multivariate techniques would be useful to group accessions to be selected for such TSGs.

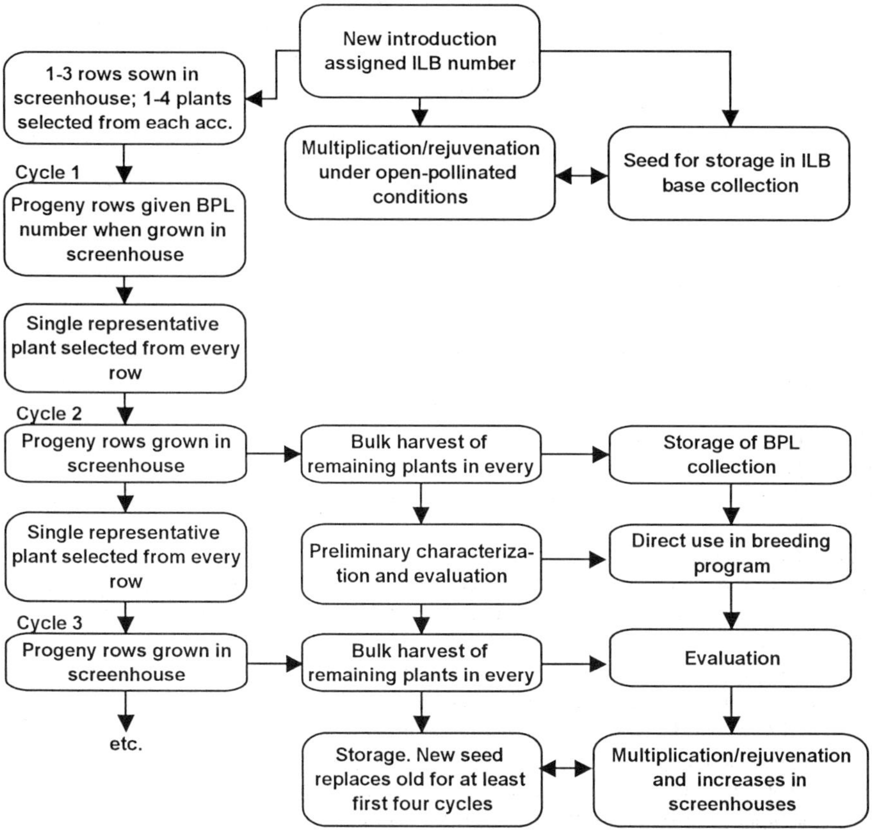

Figure 14.1. Scheme for developing faba bean pure line (BPL) collection at ICARDA.

Burton (1979) proposed that cross-pollinated germplasm accessions be maintained as self-pollinated bulks. This type of collection would contain accessions which are a bulk of homozygous lines, i.e., each line is heterogeneous but homozygous. It should be better than testing a fixed number of pure lines from each open-pollinated accession, in that it might better represent the full range of variability in each original accession. A more refined way to maintain this type of collection would be to self a large number of plants of each accession and take one seed or pod to bulk to produce the next-cycle seed (the Single Seed Descent Method designed by Brim 1966). This type of collection of selfed composites would maintain the maximum variability with the minimum chance of loss of genes because of inter-crossing among accessions. Also, it would allow the detection of recessive genes in heterogenous populations which are bulks of homozygous genotypes. The ILB collection at ICARDA is evolving into this type of collection as the multiplication phase has been switched to insect-proofed screen houses because of the impracticality of growing even a small number of accessions in isolation.

14.6 CONSERVATION STATUS OF COLLECTIONS

The current conservation status of the major food and grain legumes is illustrated in Table 14.2. The largest collections for faba bean and lentil are at ICARDA, with more than double the size of another other collections for lentil and nearly five times the next largest faba bean collection. Both ICARDA and ICRISAT preserve the majority of the chickpea germplasm. For chickpea, the ICARDA collection is only the kabuli type, and at ICRISAT the collection is mostly the desi type and is mostly from South Asia and Iran. The largest collections of lupins are in Australia, where nearly 90% of the production is located (Table 14.1). Though the centre of origin for lupins is in the Mediterranean, the production of them there is negligible and there is relatively little preservation of their germplasm in their centre of origin and diversity. Large pea collections are maintained in the Russian Federation at the Vavilov Institute (VIR) and in Germany. Duplications between these different collections will be discussed in the next section.

Over 87% percent of the food legume accessions at ICARDA are landraces compared to the forage legumes which are mostly naturally occurring populations from the Mediterranean region.. Fifteen percent faba bean accessions held at ICARDA are land races collected by ICARDA and ALAD (Arid Lands Agricultural Development Program). Twenty-one per cent of the lentil accessions held at ICARDA are land races collected by ICARDA and ALAD (Arid Lands Agricultural Development Program, the precursor organization of ICARDA).

14.7 SAFETY AND OTHER DUPLICATION

ICARDA has signed an agreement for the safety duplication of its cultivated lentil collection with the National Bureau of Plant Genetic Resources (NBPGR) in New Delhi, India, and has prepared and stored 91% of its collection there at this time and will store the remainder after regeneration for seed supply and to obtain proper viability for storage. ICARDA has signed an agreement for the safety duplication of its ILB (International Legume Bean) and BPL (Bean Pure Line) faba bean collections and for its forage *Vicia* species with the Federal Institute of Agrobiology (FIA) in Linz, Austria. To date, approximately 1500 accessions of faba bean and 1910 accessions of forage *Vicia* species have been safety duplicated. The chickpea germplasm collection of ICARDA is safety duplicated at the International Crops Research Institute for the Semi-Arid Tropics (ICRISAT). The *Lathyrus* collections at ICARDA are safety duplicated at Station Fédérale de Recherche Agronomique de Changins, in Switzerland. All collections of the National Plant Germplasm System of the United States Department of Agriculture (USDA) are kept at the maintenance and distribution site (Pullman, Washington for the species discussed in this paper) and at a base collection backup at the National Seed Storage Laboratory (Ft. Collins, Colorado). Many accessions in the Australian germplasm collections of these species are duplicates of accessions in international and other national germplasm collections. A large part of the accessions of these species kept at ICARDA and ICRISAT are donations from the USDA collections. The wild germplasm collections of *Lens* and *Cicer* are the same at the USDA facility at Pullman, Washington and ICARDA with the wild Cicer collection also duplicated at ICRISAT in India. Additionally, there is a partial representation of these in the Australian collections.

Table 14.2. International and national collections of the major food and grain legumes originating from the Mediterranean region.

Collection[1]	Faba bean	Chickpea	Lentil	Pea	Narrow-leafed lupins	White lupins
ICARDA	10767	10117	8261	4670	13	76
ICRISAT	0	16992	0	0	0	0
USDA	533	4431	2717	3330	184	320
VIR	1268	1337	2376	6791	671	427
ATFCC	1430	2826	3614	3683	1369	624
IPK and BAZ (DE)	1507	329	384	5049	772	429
CRF-INIA (ES)	1064	493	470	272	556	578
BARI (IT)	2279	357	242	4557	7	121
AARI (TR)	335	1696	577	108	1	0
Total	16569	36525	17822	23795	3565	2454

[1]ICARDA = International Centre for Agricultural Research in the Dry Areas; ICRISAT = International Crop Research Institute for the Semi-Arid Tropics; USDA = United States Department of Agriculture; VIR = Vavilov Research Institute, Russian Federation; ATFCC = Australian Temperate Field Crop Collection; IPK = Institut für Pflanzengenetik und Kulturpflanzenforschung (Germany); BARI = Istituto del Germoplasma, Bari, Italy; and AARI = Aegean Agricultural Research Institute (Turkey).

14.8 Domestication and Centers of Diversity

The immediate ancestor of faba bean is not known, and Schäfer (1973) hypothesized that *V. faba* originated from an extinct ancestor. *Vicia narbonensis* L. (2n=14) has received the greatest attention as the putative ancestor of faba bean. However, with a different karyotype and different chromosome number, *V. narbonensis* has never successfully been crossed with *V. faba*, and therefore cannot be regarded as a direct ancestor of faba bean. *Vicia faba* was most likely domesticated in the region between Afghanistan and the eastern Mediterranean during the period

7000-4000 B.C. (Hanelt 1972). Cubero (1972) concluded the culture of faba bean spread in four directions from the centre of origin. These were north to central Europe, northwest to western Europe, west to the Mediterranean, and east to the Far East (India, China and Japan). The *minor* type faba bean was introduced to China in 100 B.C. (Tao 1981) and the *major* type in 1200 A.D. (Hanelt 1972).

Lentil was domesticated in the Near East Arc in association with the domestication of einkorn, emmer and barley; and later spread to the rest of the world (Vavilov 1926; Ladizinsky 1979). The most likely place for this domestication is the Lidia-Kurdistania region (Cubero 1981; Ladizinsky 1979). Archaeological records of carbonised lentil seeds shows the pattern of dissemination of the crop from its centre of origin (Zohary and Hopf 1988). Lentils spread to Cyprus and south-eastern Europe in the 6th millennium BC and then to Central Europe via the Danube. Lentil spread with wheat and barley to Egypt and was carried to Ethiopia by Hamitic invaders. To the east lentil reached Georgia in the 5th and early 4th millennia BC and reached India and Pakistan around 2000 BC. Lentil reached the Americas in the post-Columbus period. The major factor in the domestication of lentil has been selection pressure for an appropriate phenology (Erskine *et al.* 1989). Barulina (1930) suggested the Hindu Kush-Himalaya border region as the centre of origin of lentil because of the presence of the highest proportion of endemic varieties of the cultivated species. Though the evidence from wild species does not support this hypothesis, this region can be considered a secondary centre of diversity for lentil. This is an area of mountainous topography with many isolated valleys where the effects of genetic drift and gene fixation have led to the high diversity of endemic varieties found by Barulina.

Chickpea most probably originated in the area of present-day southeast Turkey, where three wild species of *Cicer* (*C. bijugum* K.H. Rech., *C. echinospermum* PH. Davis and *C. reticulatum* Ladiz) are found. Seed remnants from Hacillar near Burdur dated to 5450 BC (Helbaek 1970) and the most closely related wild species occur there (Ladizinsky and Alder 1976). De Candolle (1882) traced the origin of chickpea to an area south of the Caucasus and northern Persia (now Iran). Vavilov (1926) designated two primary centres of origin (diversity) - southwest Asia and the Mediterranean - and a secondary one, Ethiopia. In 1951 he recognized four centres of diversity for cultivated chickpea: (I) the Mediterranean, (ii) Central Asia, (iii) the Near East, and (iv) India and a secondary centre in Ethiopia (Vavilov 1951). The earliest record of chickpea in India (Uttar Pradesh) dates from 2000 BC (Chowdury *et al.* 1971). Chickpea spread with human movement toward the west to the Mediterranean basin and south toward the Indian subcontinent via the Silk Route (Afghanistan). Records in Ethiopia date to 1520 BC (van der Maesen 1972). The Spanish and Portuguese introduced chickpea to the New World around 1500 AD, and kabuli is of recent introduction to India (1700 AD). Domestication of chickpea began in the USA and Australia only in the last three decades.

The WANA region is the centre of origin and primary diversity for *Lathyrus* species (Zohary and Hopf 1988). The geographical distribution of the ICARDA *Lathyrus* collection in WANA is shown in Figure 14.1 for those accessions where passport data is available. The accessions are concentrated in Morocco, Algeria, Tunisia, Turkey, Syria and Jordan.

14.9 Geographical Distribution of the Collections

Most accessions of food and forage legumes in the ICARDA collections are from the Mediterranean region. This is to be expected, as the Mediterranean region is the centre of origin and primary diversity for these temperate cool season legumes (Zohary and Hopf 1988). This has led to the successful use of landraces from the collections for direct release as cultivars for the Mediterranean region and beyond (Robertson *et al.* 1996). The geographical distribution of the ICARDA faba bean, chickpea and lentil collections is given in Table 14.3. As mentioned previously, there is a significant duplication of collections for these species between ICARDA the USDA, BARI, and the Australian collections. The Mediterranean countries with large numbers of accessions of faba bean germplasm are Cyprus, Spain, Morocco, Portugal, Syria and Turkey. There are also a large number of accessions from Ethiopia Egypt, Sudan and Canada. With lentil, the largest number of accessions from the Mediterranean region are Egypt, Spain, Greece, Jordan, Morocco, Syria (including a large number of breeding lines of ICARDA), and Turkey. The most significant other representation is that of Iran and India with nearly 2800 accessions. These countries are also highly represented in the USDA and Australian collections. This is because of a USAID project that resulted in a large number of accessions from these countries which has been duplicated at ICARDA and via it to the Australian collection. Other large representations are with Chile, Pakistan and Ethiopia. The chickpea collection has a large representation from Jordan, Morocco, Portugal, Spain, Syria, Tunisia, and Turkey from the Mediterranean region. As with lentil, there is a very large representation of India and Iran with over two thousand accessions at ICARDA (there are 12,000 accessions of chickpea from these two countries in the ICRISAT collection). Again these are mostly from the same USAID project mentioned previously and are heavily duplicated at the USDA and in the Australian collection.

Geographical distributions for *Lathyrus* species in the ICARDA collections in the Mediterranean region are shown in Table 14.4. Almost the entire *Lathyrus* collection of ICARDA is of naturally occurring populations from the Mediterranean region. The collections of *Lathyrus* species are wild forms, found mostly in crop fields and orchards. The *L. sativus* accessions from Ethiopia and the Indian subcontinent are local landraces. There is a significant set of landrace accessions of *L. sativus* from Bangladesh in the ICARDA collection.

14.10 Wild Relatives

The wild progenitors and relatives of lentil and chickpea (annual species) are primarily found in the West Asian region (Cubero 1981, van der Maesen 1987). *Lens culinaris* ssp. *orientalis* (Boiss.) Ponert and *Cicer reticulatum* Lad. are endemic to this region (though the former has probably been distributed to some extent as a weed with the cultigen in the Asian Republics of the former Soviet Union).

Table 14.3. Distribution of the ICARDA grain legume germplasm collections by country of origin.

Country[1]	ILB[2]	BPL	ILC	ILL	Country	ILB	BPL	ILC	ILL
AFG	139	111	891	157	LBN	36	88	28	76
ALB	-	-	-	2	LBY	27	-	2	12
ARG	1	4	-	16	LKA	2	4	-	-
ARM	-	-	4	10	LTU	1	3	-	-
AUS	3	30	4	2	LVA	2	1	-	-
AUT	1	2	-	-	MAR	768	181	225	100
AZE	-	-	22	27	MDA	1	-	9	-
BEL	1	-		2	MEX	11	14	117	69
BGD	35	-	1	81	MWI	-	-	3	-
BGR	14	2	194	51	NGA	-	-	1	-
BLR	-	-	-	1	NLD	30	46	-	1
BOL	-	2	-	1	NOR	-	-	-	1
BRA	-	-	-	3	NPL	97	1	6	402
CAN	291	117	-	6	NZL	-	-	-	1
CHE	2	6	-	-	OMN	4	-	-	-
CHL	-	-	346	353	PAK	35	50	256	271
CHN	390	342	21	13	PAL	17	10	40	2
COL	53	105	1	20	PER	52	47	4	4
CRI	-	-	-	2	POL	42	21		11
CSK	30	-	10	23	PRT	112	15	121	20
CYP	106	251	46	28	ROM	33	14	2	3
DEU	96	78	1	36	RUS	12	7	50	151
DZA	53	41	49	36	SAU	-	-	-	3
ECU	242	241	2	15	SDN	121	44	11	2
EGY	113	127	57	100	SOM	-	-	-	1
ESP	367	782	284	179	SUN	16	16	36	36
ETH	551	683	120	414	SWE	7	12	-	-
FRA	16	15	18	11	SYR	521	178	2095	1338
GBR	86	99	8	1	TJK	31	1	16	67
GEO	-	1	3	2	TKM	7	-	-	3
GRC	34	54	18	119	TUN	39	65	264	22
GTM	-	-	-	2	TUR	225	212	866	430
HUN	15	12	2	26	UKR	16	12	27	17
IDN	1	1	-	-	URY	1	2	-	3
IND	10	10	399	1886	USA	5	8	129	33
IRN	24	18	1736	917	UZB	3	-	74	11
IRQ	77	112	32	27	YEM	15	27	-	67
ITA	85	73	68	13	YUG	23	30	6	25
JOR	22	33	143	390	ZAF	-	3	-	-
JPN	6	12	-	-	UNK	318	792	509	36
KAZ	2	-	3	2	ICA	-	-	733	69
KGZ	3	-	4	-					
					Total	5499	5268	10117	8261

[1]Origin country codes as per ISO (International Standardization Organization) three letter country codes; UNK=Unknown and ICA=ICARDA breeding lines.

[2]ILB=Open pollinated faba bean collection, BPL=Pure line faba bean collection, ILC=Chickpea collection, and ILL=Lentil collection.

Table 14.4. International and national collections of the *Lathyrus* species and *Trigonella foenum-gracecum*.

Collection[1]	*Lathyrus*	*L. sativus*	*L. cicera*	*L. ochrus*	*Trigonella foenum-graecum*
ICARDA	2933	1632	184	136	84
USDA	464	243	33	25	174
VIR	904	688	57	21	154
ATFCC	990	583	142	96	227
IPK (DE)	558	238	68	46	60
CRF-INIA (ES)	307	93	174	0	8
BARI (IT)	190	?	?	?	86
AARI (TR)	600	17	90	1	?
Total	6946	3494	748	325	793

[1]ICARDA = International Centre for Agricultural Research in the Dry Areas; USDA = United States Department of Agriculture; VIR = Vavilov Research Institute, Russian Federation; ATFCC = Australian Temperate Field Crop Collection; IPK = Institut für Pflanzengenetik und Kulturpflanzenforschung (Germany); BARI = Istituto del Germoplasma, Bari, Italy; and AARI = Aegean Agricultural Research Institute (Turkey).

Wild *Cicer* species were scarce in collections before 1970, but now a reasonable number of accessions of annual species are available (Table 14.5) . Availability of perennial species is still rare, and limited information on their potential use is available. Ladizinsky and Adler (1976) proposed grouping *Cicer* species into three gene pools, namely GP1, GP2 and GP3, following the scheme suggested by Harlan and de Wet (1971). Ladizinsky and Adler included *C. reticulatum* in GP1 of chickpea because gene exchange between these species was regular. *C. echinospermum* was placed in GP2 because gene exchange of this with chickpea is impaired because of high sterility of F1 hybrids. All remaining species were placed in GP3. Wild species have been extensively evaluated for resistance to biotic and abiotic stresses. The representation of some wild species of *Cicer* is very limited, especially for *C. cuneatum*, *C. yamashitate* and *C. chorassanicum*. Except for the these species, the annual species of wild *Cicer* are distributed in the Fertile Crescent region (Figure 14.2).

The collections of wild *Lens* are much smaller than those of the cultigens; however, the collections at ICARDA are the largest in the world (Table 14.6). Lentil originated in the 'Fertile Crescent' and its wild relatives are most common in this region (Cubero 1981; Ladizinsky 1979,1993). The progenitor species of lentil, *Lens culinaris* subsp. *orientalis*, is endemic to this region. The wild progenitor has been used also for improving yield levels of the cultigen at ICARDA.. The progenitor of lentil spread through to central Asia as a weed in the cultivated form. However, the genetic diversity found outside the Fertile Crescent in the wild species of lentil is limited (Ferguson *et al.* 1988).

The collections of the *Lathyrus* species at ICARDA are mostly weedy forms, found in crop fields and orchards in WANA.

Table 14.5. Major wild *Cicer* germplasm collections.

Taxa	ICARDA[1]	ICRISAT	USDA	ATFCC	VIR	IPK & BAZ
Annual						
bijugum	42	8	18	39	0	0
chorassanicum	6	1	2	3	0	0
cuneatum	5	1	1	3	0	0
echinospermum	13	1	12	13	0	0
judaicum	76	22	37	3	0	0
pinnatifidum	53	9	21	4	1	3
reticulatum	61	6	19	51	0	0
yamashitae	5	2	3	3	0	0
Total Annuals	261	50	113	119	1	3
Perennial						
acanthophyllum	0	0	3	0	0	0
anatolicum	0	0	6	2	0	0
canariense	0	0	1	0	0	0
flexuosum	0	0	2	0	0	2
grande	0	0	0	0	0	1
macracanthum	0	0	10	0	0	2
microphyllum	0	0	16	0	0	0
montbretii	0	0	8	0	0	0
multijugum	0	0	2	0	0	0
nuristanicum	0	0	4	0	0	0
oxyodon	0	0	2	0	0	0
pungens	0	0	6	0	0	0
songaricum	0	0	3	0	0	1
Total Perennials	**0**	**0**	**176**	**0**	**0**	**6**

[1] Collection abbreviations as in Table 14.2

Table 14.6. Major wild *Lens* germplasm collections.

Taxa	ICARDA[1]	USDA	ATFCC	CRF-INIA	IPK & BAZ
orientialis[2]	217	59	68	0	1
odemensis	57	5	25	0	0
ervoides	133	30	70	0	3
nigricans	47	23	27	21	8
lamottei	8	-	-	4	-
Total	**462**	**117**	**190**	**25**	**12**

[1] Collection abbreviations as in Table 14.2
[2] *Lens culinaris* subsp. *culinaris*

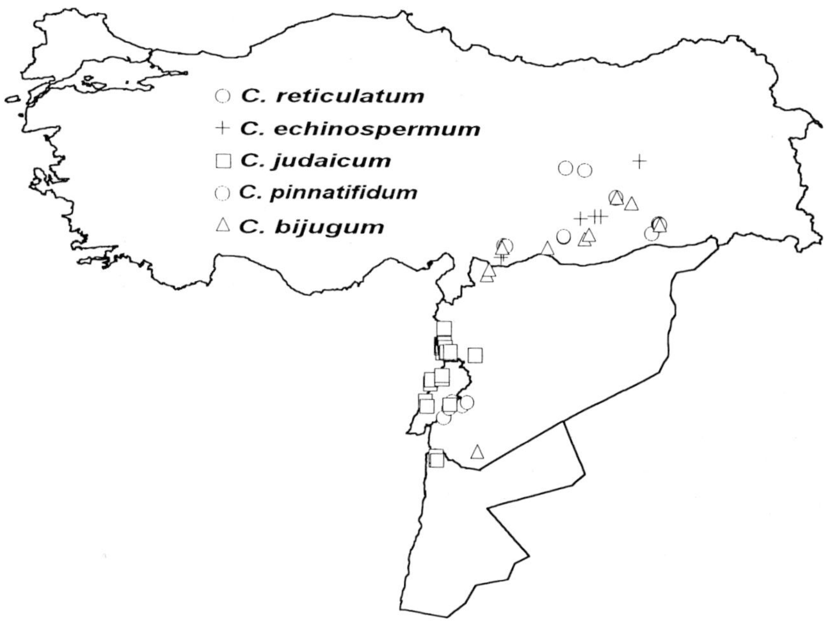

Figure 14.2. Distribution of Near East annual wild *Cicer* species.

14.11 PRIORITIES FOR FUTURE COLLECTIONS, CURRENT GAPS AND PROSPECTS

One of the major gaps in the forage and grain legume collections is the lack of passport data. Even the country of origin is not known for more than 2000 ILB accessions at ICARDA, approximately 21% of the entire collections. For many other accessions there is very little information available. Efforts are being extended to rectify as much as possible the lack of passport data in order to make their use more efficient.

The major geographical gap that needs to be addressed in the faba bean collections is the relatively small number of accessions from China, which is a major producer. This is especially acute for remote areas of China. Other areas that are under-represented are India, Pakistan and Nepal, though Nepal was collected in 1995. In the WANA region there is a need for more collection in Algeria, Egypt and Yemen. The major gap in required traits in faba bean germplasm collections is the lack of adequate resistance sources to *Orobanche crenata*. Another concern would be to broaden the genetic base of the resistance sources to chocolate spot. Because of the importance of chocolate spot and rust resistance with faba bean, there is a crucial need for further collection in the Andean region. Also, there is a need to find faba bean germplasm with virus resistance and heat tolerance.

With lentil there is a general need to increase the collections of wild species, especially *L. culinaris* subsp. *orientalis*. In the cultivated lentil collection of ICARDA there are problems of very incomplete passport data, especially for Iran and India. There are also questions concerning the true origin of many accessions from these countries in the ICARDA collection. A large part of the effort in documentation of genetic resources at ICARDA is being expended to clarify the status of the passport data of the collections in order to make their use more efficient. Within the WANA region, the major gap that needs to be filled is for Algeria and to a lesser extend for Tunisia. Outside this region, the major gap is for Bangladesh. However, a collection mission just completed in April 1995 will fill this gap to a large extent. Specific regions within a country may not be adequately covered. When the passport data is updated this subject will need to be addressed. The major gap in the germplasm collections of lentil (both cultivated and wild) is the lack of accessions with a high biomass. The low biomass restricts the yielding ability of the crop and reduces its competitiveness against weeds. There also is a lack of resistance for pea leaf weevil (*Sitona* sp.) and the parasitic weed broomrape *(Orobanche* sp.).

More accessions of wild annual *Cicer* species need to be collected, particularly the progenitor species, *C. reticulatum*, because they possess genes for resistance to almost all stresses for which evaluations have been made. Introgression of genes from wild species can upgrade the genetic potential for seed yield. Geographic gaps for cultivated chickpea include Russia and CIS, Turkey, Afghanistan, Iraq, and Ethiopia. Known entity gaps include the types of chickpea that have been documented or described as herbarium material but are no longer available as live seed. A limited number of accessions are available for wild Cicer species. Of the 34 perennial *Cicer* species known, live seed is available only for nine species.

There will be an increasing evaluation of the germplasm collections for biochemical and molecular markers to assess the genetic diversity of these crops; and to plan future research in areas such as further *ex situ* collections, *in situ* conservation, core collections, etc. The work with wild species will assume greater importance in the coming years, as gene transfer between the cultigens and the wild species is improved by new biotechnological techniques which will drive the need for further collection of these important species.

REFERENCES

Angus, J.F. and Moncur, M.W., 1980. Photoperiodic and vernalisation effects on phasic development in chickpea. *International Chickpea Newsletter*, 2: 8-9.

Barulina, H., 1930. Lentils of the U.S.S.R. and other countries. *Bulletin of Applied Genetics and Plant Breeding (Leningrad) Supplement*, 40:1-319.

Bond, D.A. and Poulsen, M.H., 1983. Pollination. In: *The Faba Bean (*Vicia faba *L.): A Basis for Improvement* (ed. Hebblethwaite, P.D.). pp. 23-76. Butterworths, London, U.K.

Brim, C.A., 1966. A modified pedigree method of selection in soybeans. *Crop Science*, 6: 220.

Burton, G.W., 1979. Handling cross-pollinated germplasm efficiently. *Crop Science*, 19: 685-690.

Chowdury, K.A., Saraswat, K.S., Hasan, S.N. and Gaur, R.G., 1971. 4000-3500 year old barley, rice and pulses from Atranjikhera. *Science and Culture*, 37: 531-533.

Cubero, J.I., 1972. On the evolution of *Vicia faba* L. *Theoretical and Applied Genetics*, 45: 48-51.

Cubero, J.I., 1981. Origin, taxonomy and domestication. In: *Lentils* (eds. Webb, C. and Hawtin, G.). pp. 15-38. CAB/ICARDA, Slough, England.

De Candolle, A., 1883. *Origine des Plantes Cultivees*. pp. 208-260. Paris.

Erskine, W. and Muehlbauer, F.J., 1991. Allozyme and morphological variability, outcrossing rate and core collection formation in lentil germplasm. *Theoretical and Applied Genetics*, 83: 119-125.

Erskine, W., Rihawe, S. and Capper, B.S., 1990b. Variation in lentil straw quality. *Annual Feed Science Technology*, 28: 61-69.

Eshel, Y., 1968. Flower development and pollen viability of chickpea (*Cicer arietinum* L.). *Israel Journal of Agricultural Research*, 18: 31-33.

FAO, 1993. *FAO Agricultural Production Yearbook 1992: Vol. 46.* FAO, Rome, Italy.

FAO, 1999. *FAO Agricultural Production Yearbook 1998: Vol. 52.* FAO, Rome, Italy.

Ferguson, M.E., Ford-Lloyd, B.V., Robertson, L.D., Maxted, N. and Newbury, H.J, 1988. Mapping the geographical distribution of genetic variation in the genus *Lens* for the enhanced conservation of plant genetic diversity. *Molecular Ecology*, 7: 1743-1755.

Halbaek, H., 1970. The plant husbandry at Hacillar. In: *Excavation at Hacillar* (ed. Mellaart, J.). pp. 189-244. Edinburgh University Press, Gerald Duckworth and Co., London.

Hansen, J. and Renfrew, J.M., 1978. Palaeolithic-Neolithic seed remains at Franchthi cave, Greece. *Nature*, 271: 349-352.

Hanelt, P., 1972. Zur Geschichte des Anbaues von *Vicia faba* L. und ihrer verschiedenen Formen. *Kulturpflanze*, 20: 209-223.

Harlan, J.R. and de Wet, J.M.M., 1971. Toward a rational classification system of cultivated plants. *Taxon*, 20: 509-517.

Hawtin, G.C. and Hebblethwaite, P.D., 1983. Background and history of faba bean production. In: *The Faba Bean (*Vicia faba *L.): A Basis for Improvement* (ed. Hebblethwaite, P.D.). pp. 3-22. Butterworths, London, U.K.

Jodha, N.S. and Subba Rao, K.V., 1987. Chickpea: World importance and distribution. In: *The Chickpea* (eds. Saxena, M.C. and Sindh, K.B.). pp. 1-10. CAB International, Wallingford, UK.

Ladizinsky, G., 1979. The origin of lentil and its wild gene pool. *Euphytica*, 28: 179-187.

Ladizinsky, G., 1993. Wild lentils. *Critical Reviews in Plant Sciences*, 12: 169-184.

Ladizinsky, G. and Adler, A., 1976. Genetic relationships among the annual species of *Cicer* L. *Theoretical and Applied Genetics*, 48: 197-204.

Rees, R.O., Brower, J.B., Mahoney, J.E., Walton, G.H., Brinsmead, R.B., Knights, E.J. and Beech, D.F., 1992. Pea and chickpea production in Australia. In: *Expanding the Production and Use of Cool Season Food Legumes* (eds. Muehlbauer, F.J. and Kaiser, W.J.). pp. 412-425. Kluwer Academic Publisher, Dordrecht, The Netherlands.

Roberts, E.H., Summerfield, J.R., Minchin, F.R. and Hadley, P., 1980. Phenology of chickpeas (*Cicer arietinum*) in contrasting aerial environments. *Experimental Agriculture*, 16: 353-360.

Robertson, L.D., 1985. Faba bean germplasm collection, maintenance, evaluation and use. In: *Faba Beans, Kabuli Chickpeas, and Lentils in the 1980s* (Saxena, M.C. and Varma, S.). pp. 15-21. ICARDA, Aleppo, Syria.

Robertson, L.D., 1997. Faba Bean. In: *Biodiversity in Trust: Conservation and Use of Plant Genetic Resources in CGIAR Centres* (eds. Fuccillo, D., Sears, L. and Stapleton, P.). pp. 168-180. Cambridge University Press, Cambridge, UK.

Robertson, L.D. and Erskine, W., 1997. Lentil. In: *Biodiversity in Trust: Conservation and Use of Plant Genetic Resources in CGIAR Centres* (eds. Fuccillo, D., Sears, L. and Stapleton, P.). pp. 128-138. Cambridge University Press, Cambridge, UK.

Robertson, L.D., Singh, K.B., Erskine, W. and Abd El Moneim, A.M., 1996. Useful genetic diversity in germplasm collections of food and forage legumes from West Asia and North Africa. *Genetics Resources and Crop Evolution*, 43: 447-460.

Saxena, M.C. and Siddique, M.H., 1980. Response of some diverse kabuli chickpea genotypes to vernalisation. *International Chickpea Newsletter*, 2: 7-8.

Singh, K.B., Pundir, R.P.S., Robertson, L.D., van Rheenen, H.A., Singh, U., Kelley, T.J., Rao, P.P., Johansen, C. and Saxena, N.P., 1997. Chickpea. In: *Biodiversity in Trust: Conservation and Use of Plant Genetic Resources in CGIAR Centres* (eds. Fuccillo, D., Sears, L. and Stapleton, P.). pp. 100-113. Cambridge University Press, Cambridge, UK.

Tao, Z.H., 1981. Faba bean production and research in China. *FABIS Newsletter*, 3: 24.

van der Maesen, L.J.G., 1987. Origin, history and taxonomy of chickpea. In: *The Chickpea* (Saxena, M.C. and Singh, K.B). pp. 11-34. CAB/ICARDA, Oxon, England.

van der Maesen, L.J.G., 1972. Cicer *L., A monograph of the genus with special reference to chickpea (*Cicer arietinum *L.), its ecology and cultivation*. Maded. Landbou. Wageningen, 72-10.

Vavilov, N.I., 1951. The origin, variation, immunity and breeding of cultivated plants. *Chronica Botanica*, New York 13-1/6:26-38, 75-78, 151 (1949-50).

Vavilov, N.I., 1992. *Origin and geography of cultivated plants*. (English translation from Russian). pp. 1-498. Cambridge University Press, Cambridge.

Wilson, V.E. and Law, A.G., 1972. Natural crossing in *Lens esculenta* Moench. *Journal of American Society of Horticultural Science*, 97: 142-143.

Witcombe, J., 1982. Genetic resources of faba beans. In: *Faba Bean Improvement: Proceedings of the ICARDA/IFAD Nile Valley Project Conference* (eds. Hawtin, G. and Webb, C.). pp. 1-13 Martinus Nijhoff, The Hague, The Netherlands.

Zohary, D. and Hopf, M., 1988. *Domestication of plants in the old world*. pp. 1-249. Clarendon Press, Oxford.

Chapter 15

Ex Situ Conservation of Forage Legumes

Ruaraidh Sackville Hamilton, Steve Hughes and Nigel Maxted

15.1 INTRODUCTION

This chapter analyses progress and the current status of *ex situ* conservation of 739 species of 27 genera of forage legumes in relation to their natural distribution. Analysis is based on passport data of individual accessions. As such, the scope of the chapter is restricted to those collections for which the authors have been granted access to accession-level passport data in machine-readable format.

For more comprehensive descriptions of the forage legume collections of all genebanks, including taxonomic and geographical breakdowns of their holdings, see Bettencourt, Hazekamp and Perry (1992), Frison and Serwinski (1995), or search online at http://www.cgiar.org/ipgri/ or http://apps2.fao.org/wiews/ for the most current information. Our objective is not to duplicate the work of IPGRI and FAO, but rather to pool all efforts and assess what has been jointly achieved to date on the *ex situ* conservation of forage legumes.

15.2 SCOPE OF CHAPTER

15.2.1 Taxonomic Scope

The circumscription of forage legume species and the distinction between them and their wild relatives are not well defined. Some species, such as *Medicago sativa, Trifolium repens* and *T. subterraneum* are widely cultivated as forages. Others, such as *Vicia sativa* and *Lathyrus cicera*, are dual-use crops, cultivated as grain legumes as well as forages. Yet others, e.g. *Vicia articulata, V. noeana* and *Lathyrus clymenum* are deliberately used locally, but have not been commercialised. Many more are "adventitious" forages, occurring naturally and being grazed or cut for hay, silage or fresh forage without being deliberately sown. Some of these are wild relatives of the few commercialised species. Many of them are also forage species in their own right with known high forage value. Many others are poorly known, with little or no information on forage value or even on their distribution. These species are included at least as wild relatives, and potentially as forage legumes in their own right.

For the purposes of this chapter, forage legumes and their wild relatives are deemed to include all species of all legume genera other than *Astragalus* that have at least one species of high forage value (Table 15.1). Although some species of *Astragalus* (e.g. *A. hamosus* and *A. cicer*) are commercialised forage species, it is considered inappropriate to include all species of the genus because of its large size (1,442 species accepted by ILDIS 1999), poorly understood taxonomy (Polhill, 1981), and the existence of many toxic species. *A. pelecinus* (accepted taxon in ILDIS, but regarded as a synonym for *Biserrula* spp. in Flora Europaea 1999), is included. Other species of *Astragalus* have been treated on an *ad hoc* basis: database records for all accessions of genus were included in analysis of conserved germplasm, but the species were not included in the prior definition and analysis of distribution.

Table 15.1. Genera of forage legumes, with the number of species dealt with in the previous chapter as grain legumes or their wild relatives and therefore excluded from this chapter.

Tribe	Genus	Number of species in genus	Number of species excluded
Cicereae	*Cicer*	31	31
Coronilleae	*Tripodion* (syn. *Hammatolobium*)	3	0
	Coronilla	10	0
	Hippocrepis	30	0
	Ornithopus	7	0
	Scorpiurus	3	0
	Securigera	10	0
Crotalarieae	*Lotononis*	163	0
Galegeae	*Biserrula* (syn. *Astragalus pelecinus*)	2	0
	Galega	6	0
Genisteae	*Lupinus*	460	4
Hedysareae	*Ebenus*	6	0
	Hedysarum	71	0
	Onobrychis	119	0
Loteae	*Anthyllis*	17	0
	Cytisopsis	1	0
	Dorycnium	5	0
	Hymenocarpos	5	0
	Lotus	134	0
	Tetragonolobus	1	0
Trifolieae	*Medicago*	74	0
	Melilotus	19	0
	Ononis	67	0
	Trifolium	222	0
	Trigonella	73	1
Vicieae	*Lathyrus*	128	1
	Lens	4	4
	Pisum	2	2
	Vavilovia	1	0
	Vicia	200	1
Total number of species		1874	44

Species that are also grain legumes or their wild relatives are covered in the previous chapter and therefore excluded from this one. The excluded species comprise all species of *Cicer*, *Lens* and *Pisum*, four of *Lupinus* (*L. albus*, *L. angustifolius*, *L. digitatus* and *L. luteus*), and *Lathyrus sativus*, *Trigonella foenum-graecum* and *Vicia faba*.

15.2.2 Geographical Scope

For the purposes of this chapter, the Mediterranean is defined as all countries with some Mediterranean coastline, together with adjacent countries that have a similar native forage legume flora. This includes Albania, Algeria, Andorra, Armenia, Azerbaijan, Bosnia and Herzegovina, Bulgaria, Croatia, Cyprus, Egypt, France, Gaza Strip, Georgia, Gibraltar, Greece, Iran, Iraq, Israel, Italy, Jordan, Lebanon, Libya, Macedonia, Malta, Monaco, Montenegro, Morocco, Portugal, San Marino, Serbia, Slovenia, Spain, Syria, Tunisia, Turkey, and West Bank.

This definition extends beyond a natural climatic definition of the Mediterranean, since many countries span Mediterranean and non-Mediterranean climates. However, a country-based definition is required because many *ex situ* conserved accessions have their origin defined primarily by country.

15.3.3 Data Scope

Accession-level passport data were standardised and collated into a single database documenting 114,379 accessions of forage legumes as defined above (Table 15.2). The Singer, CSIRO, European Central Crop and the USDA Genetic Resource Information Network (GRIN) database are centralised databases not associated with any one gene bank. The Singer and USDA GRIN database records are unique, but some of the accessions documented in the CSIRO and ECC centralised databases are also documented in databases received from individual genebanks. Duplicate database records were identified and eliminated from these databases. No attempt was made to identify duplicate accessions. No systematic attempt has been made to identify data errors, although obvious errors have been corrected when encountered if there was no ambiguity concerning the correction.

The diversity of data structures from the different sources precludes a comprehensive collation of all data. Instead, a minimal subset of passport data has been extracted: country of origin and latitude/longitude to enable geographical analysis; year of collection to enable historical analysis; and accession status. Accessions were classified as:

1. collected if any collecting data were present (any data on the collecting expedition or on the location or environment of the collecting site; or the documented status of the accession indicates it was collected);
2. a selected line if there were no collecting data but some other passport fields indicating that it was a breeder's line or advanced cultivar; or
3. unknown if there were no ancillary passport data

In addition, full transcription of passport data was attempted for the major collections held in Australia and the USA, with the objective of assessing how well they conform to the FAO multi-crop passport descriptor lists. However, even with these selected collections this objective was not achievable because of variation in the passport descriptors and their states. Analysis is restricted to an assessment of the extent of documentation and characterisation of accessions in these collections.

Table 15.2. Sources of accession-level passport data used for this chapter.

	AARI	AMGRC	ATGRC	Bari	Bulgaria	CSIRO	ECP/GR	IPK	IGER	Israel	Portugal	SINGER	GRIN	Totals
Anthyllis		56					61	18				47	24	206
Astragalus / Biserrula		725	109				12					938	591	2375
Coronilla		84					1					212	24	321
Dorycnium		24					1						4	29
Ebenus		1											6	7
Galega		2											26	28
Hedysarum		129		149								118	47	443
Hippocrepis		139										302	10	451
Hymenocarpus		151										181	332	664
Lathyrus	222	140		332	253	537	4494	466		13	5	2932	274	9668
Lotononis		2										3	1	6
Lotus		599					432	73	113			122	336	1675
Lupinus		13		60	8	832		241		41	148	30	404	1777
Medicago		24499		1161	93	2373	4099	223	121	231	1	8370	4475	45646
Melilotus		91					340	44				122	397	994
Onobrychis		437					217	72	66			801	278	1871
Ononis		15										4	4	23
Ornithopus		155	796				1					6	10	968
Psoralea		6										2	2	10
Scorpiurus		246										472	51	769
Securigera		7										6	42	55
Tetragonolobus		65					11					17	93	186
Trifolium	2982	4240	6524	490		4221	6102	450	1549	308		3502	3098	33466
Trigonella		482		86								550	37	1155
Vicia	801	546		2400	712	1200		1666		541	127	5460	672	14125
Total	**4005**	**32854**	**7429**	**4678**	**1066**	**9163**	**15771**	**3253**	**1849**	**1134**	**281**	**24197**	**11238**	**116918**

15.4 NATURAL DISTRIBUTION

Information on the natural distribution of forage legumes has been extracted from ILDIS (1999). The ILDIS taxonomic database records all countries in which each species has been recorded, together with its status in each country as native, introduced, or uncertain. The current version of the ILDIS database does not distinguish ex Soviet territories and therefore it was not possible to include Armenia, Azerbaijan and Georgia in the geographical analysis.

15.4.1 Global Distribution

The current version of the ILDIS database contains no data on the distribution of 83 (largely *Vicia* and *Lupinus* species) of the 1,830 species in Table 15.1. A further 579 species have unknown native distribution, being recorded only as introduced species or of uncertain status, leaving 1,168 species with documented native distribution. Of these, 429 are recorded as native only to non-Mediterranean countries (Table 15.3), leaving a total of 739 species native to the Mediterranean. With the exception of *Galega, Lupinus* and *Lotononis*, all genera have the majority of their species native to the Mediterranean, and indeed all species of 11 genera are native.

Overall, 354 species, or 30% of those with recorded native distribution, are endemic to the Mediterranean (Table 15.3), Three genera - *Cytisopsis*, *Tetragonolobus* and *Vavilovia* - are exclusively endemic (Table 15.3). *Cytisopsis* and *Tetragonolobus* are endemic to just one country each, respectively Morocco and Greece (Table 15.4). *Vavilovia* is endemic to the eastern Mediterranean: Iran, Iraq, Lebanon, Syria and Turkey. The majority of the species of *Hymenocarpos*, *Ebenus*, *Hippocrepis*, *Ononis*, *Anthyllis* and *Onobrychis* are endemic to the Mediterranean. At the other extreme, *Biserrula*, *Galega* and *Scorpiurus* have no species endemic to the Mediterranean (Table 15.3).

Most species of *Hedysarum* fall into one of two contrasting groups - those that are endemic to the Mediterranean, and those that are not native to any Mediterranean country. In contrast, most species of *Ornithopus*, *Vicia*, *Lathyrus*, *Medicago*, *Trifolium*, *Dorycnium*, *Melilotus*, *Securigera* and *Scorpiurus* tend to have a broad native distribution that spans both Mediterranean and non-Mediterranean countries.

15.4.2 Species-richness of Mediterranean Countries

Turkey, Spain, Greece and Morocco have the largest number of recorded native species (Table 15.4), each with around 40% of the 733 native Mediterranean species. However, the number of species per country depends on the size of the country. A log-log regression of the number of native species per country on the area of the country showed a regression coefficient of 0.30 ($F_{1,22}$=65.8, P <0.001), accounting for 74% of the variation in number of species between countries. That is, on average the number of species per country increased approximately with the cube root of the area of the country, so that number of species increases rapidly with area of small countries, but is less dependent on size of large countries.

Table 15.3. Number of species of forage legume genera recorded as native across the world, in Mediterranean countries, and exclusively in mediterranean countries. Genera are ranked in order of their tendency towards Mediterranean endemism, as the percentage of species that are found only in the Mediterranean. Source: ILDIS (1999).

Genus	Number of Species		
	Global Total	**Total Mediterranean**	**Exclusively Mediterranean**
Cytisopsis	1	1	1
Tetragonolobus	1	1	1
Vavilovia	1	1	1
Hymenocarpos	4	4	3
Ebenus	6	5	4
Tripodion	3	3	2
Hippocrepis	29	28	19
Ononis	66	63	43
Anthyllis	16	16	10
Onobrychis	118	91	68
Hedysarum	69	37	28
Ornithopus	6	6	2
Trigonella	71	44	22
Coronilla	10	10	3
Vicia	111	87	32
Lathyrus	80	61	22
Medicago	70	60	19
Lotus	99	58	25
Trifolium	161	117	37
Dorycnium	5	5	1
Melilotus	19	17	3
Securigera	9	7	1
Lupinus	42	6	2
Lotononis	162	6	5
Biserrula	2	2	0
Galega	5	1	0
Scorpiurus	2	2	0
Total	1168	739	354

Table 15.4. Native distribution of forage legume species in the Mediterranean. The body of the table shows the number of species of each genus recorded as native to each country. Countries are ranked in order of number of native forage legume species. Genera are ranked in order of their tendency towards Mediterranean endemism, as the percentage of the global count of native species-country records that are for Mediterranean countries. Source: ILDIS (1999).

Genus	TR	ES	GR	MA	IN	IT	YU	DZ	FR	LB/ SY	BG	AL	TN	IL / JO	PT	IT/ SD	IQ	LY	CY	EG	IT/ MA	Mediterranean Total	Non-Mediterranean Total	Grand Total
Cytisopsis				1																		1		1
Tetragonolobus			1																			1		1
Vavilovia	1				1					1							1					4		4
Tripodion	2	1	2	2		1	1	2	1	2		1	1	1	1	1		1	1	1	1	23	4	27
Hymenocarpos	1	2	1	3	1	1	1	2	1	1	1	1	2	1	1	1	1	1	1	1	1	26	7	33
Ebenus			1	1	2			1					1					2		1		9	3	12
Ononis	11	36	12	47	6	19	8	33	13	13	4	6	19	12	18	13	6	11	10	8	9	314	115	429
Onobrychis	23	10	12	8	65	7	12	6	9	11	10	6	4	6	1		12	2	3	2		209	79	289
Coronilla	3	5	4	7	3	7	5	5	6	2	1	4	5	2	5	2	1	3	2		2	74	30	104
Hippocrepis	5	16	7	12	3	5	5	8	6	4	4	5	12	5	2	6	1	7	5	5	4	127	55	182
Dorycnium	4	3	4	1	1	3	2	4	3	3	2	3	2	2	3	3	1	2	1			47	22	69
Scorpiurus	1	2	2	2	1	2	1	2	1	1	1	1	2	1	2	2	1	1	1	1	2	30	16	46
Astragalus	1	1	1	1		1	1	1	1	2	1	1	1	2	1	1		1	1	1		20	11	31
Ornithopus	2	5	2	5	1	3	1	5	4	2	1	1	2	2	5	3			1		1	46	26	72
Anthyllis	3	12	6	4	1	4	5	8	4	1	3	4	3	1	2	3		2		1	2	69	41	110
Medicago	41	27	26	23	21	29	23	23	26	32	17	15	22	27	17	19	19	19	24	16	18	484	307	793
Trigonella	18	2	14	2	24	3	3	5	2	16	4	1	4	12		2	13	5	5	9	2	146	96	242
Securigera	5	2	5	1	2	3	4	1	2	4	4	5	1	3			1		2			45	32	77
Trifolium	81	56	75	46	42	65	66	40	49	49	63	59	33	40	41	32	32	20	26	15	20	950	834	1784
Melilotus	10	11	9	8	4	11	11	9	10	6	5	8	8	7	6	7	3	5	5	7	7	157	143	300
Hedysarum	6	4	1	9	19	3	3	9	4	3	2		6	1	2	1	5	1	1	1	1	82	78	160
Lotus	11	22	23	31	8	18	17	23	16	8	8	10	20	10	13	13	3	17	5	19	7	302	294	596
Vicia	40	29	32	28	34	37	34	25	29	25	28	23	16	18	19	23	20	12	16	12	11	511	501	1013
Lathyrus	29	29	29	19	20	30	32	20	25	15	24	26	13	9	19	14	17	11	8	8	10	407	404	811
Galega	1	1	1	1	1	1	1	1	1		1	1										11	17	28
Lupinus	2	3	2	3		2	1	1	1	2		1	2	3	3	2		1	1			30	73	103
Lotononis		1	1	5	1			2		1	1							1		1		14	239	253
Total	301	280	273	270	261	255	237	236	214	204	185	182	179	165	161	148	137	125	119	109	98	4139	3427	7570
†Area of country (1000 km²)	780	499	130	403	1623	301	236	2317	548	198	111	29	155	117	89	24	436	1618	9	998	0.26			

Thus Andorra, Gibraltar and Monaco have few native forage legume species (Table 15.4), but this is because they are small and does not indicate that they are particularly species poor. The desert countries to the south and east of the Mediterranean - Algeria, Egypt, Iran, Iraq and Libya - have few species relative to their large size because of the predominance of unsuitable habitats for forage legumes (i.e. low rainfall). Excluding the desert countries, the size-diversity relationship among the remaining countries is much closer ($F_{1,17}$=92, P <0.001, accounting for 83% of the variation in number of species). The island countries - Cyprus, Gibraltar, Greece, Malta and Sardinia - are the most species-rich relative to their size.

There is little other than size differentiating the species-richness of the remaining mainland countries. Over 96% of the variation in their number of species, which ranges from 2 to 297 (Table 15.4) is accounted for by their size. Turkey, Spain and Morocco have many species, primarily because they are large countries. Only Albania has an unexpectedly large number of species and France few species for their size.

15.4.3 Distribution of Species Within the Mediterranean

To identify patterns of species distributions within the Mediterranean, countries were clustered on the basis of their native forage legume flora, using group average hierarchical clustering (Genstat 1993) based on the Ecological Similarity Index of Dice (1945):

$$S_{ij} = \frac{2N_{ij}}{N_i + N_j}$$

where S_{ij} is the similarity between two countries i and j, N_{ij} is the number of species found native in both countries, and N_i and N_j are the number of species in each country.

Four main regions were identified (Table 15.5; Figure 15.1). Andorra, Gibraltar and Monaco have too few native species - fewer than 15 each (Table 15.4) - to be included in any region.

Table 15.5. A Classification of Mediterranean Countries Into Regions Based on the Species of Forage Legumes Native to Each Country, and Showing, for Each Region, the Total Number of Native Species Present, the Number Present in All Countries in the Region, the Number That Are Present in the Region But Nowhere Else Within the Mediterranean, and the Number That Are Endemic to the Region.

Region	Region / Countries	Number of native forage legume species in the region			
		Total	In all countries	Unique in the Mediterranean	Globally unique
1	Western Mediterranean	442	85	173	151
1a	France, Italy, Spain, Portugal, Sardinia	340	103	39	32
1b	Algeria, Tunisia, Morocco	321	151	81	80
2	North-eastern Mediterranean	402	115	65	39
2a	Greece, Turkey	361	213	26	13
2b	Albania, Yugoslavia, Bulgaria	265	135	21	13
3	Iraq, Iran	281	117	104	76
4	South-eastern Mediterranean	285	41	26	18
4a	Israel, Jordan, Lebanon, Syria, Cyprus, Malta	240	65	10	8
4b	Egypt, Libya	157	77	15	10

Figure 15.1. Dendrogram (Group Average clustering: Genstat 1993) of the similarities of the forage legume floras of Mediterranean countries.

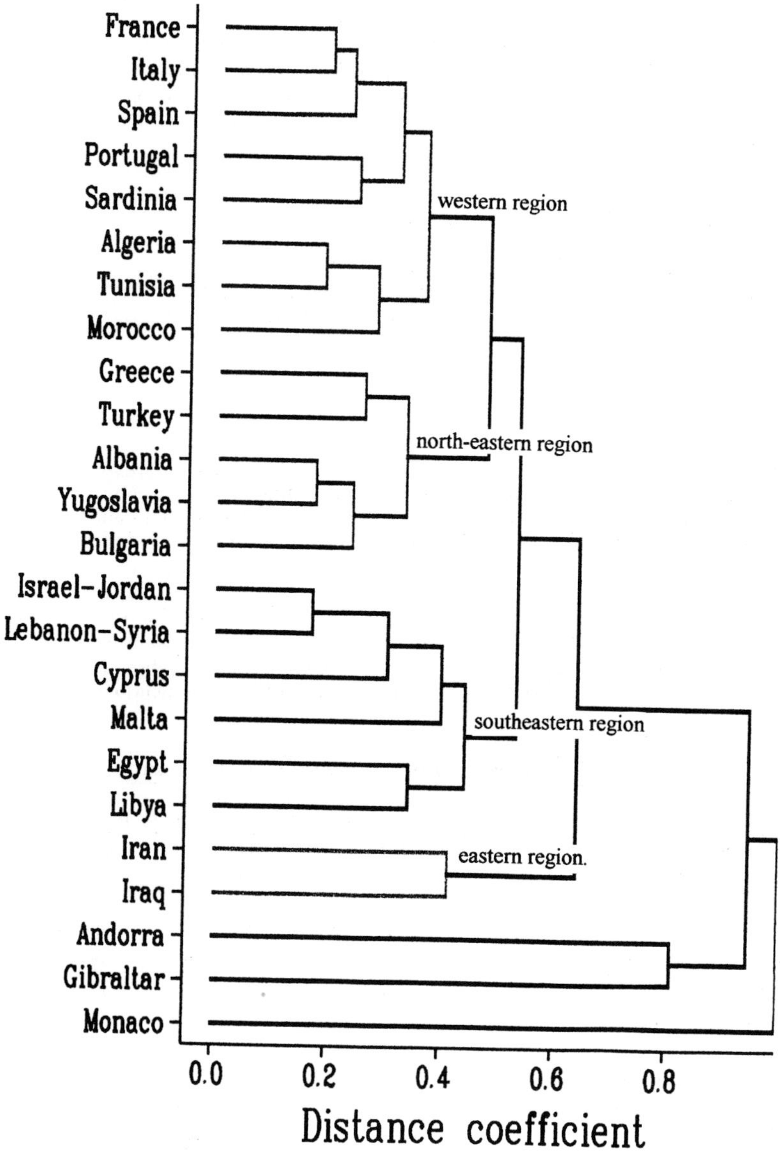

The most distinctive and species-rich region is the western Mediterranean, with 442 species, of which 173 occur nowhere else in the Mediterranean and 151 nowhere else in the world (Table 15.5). This is divided into two sub-regions, west European and west African. The west African subregion is particularly homogeneous and distinctive, with 47% of its species occurring in all countries of the region, and 25% nowhere else.

The north-eastern Mediterranean forms a second region, somewhat less distinctive with 65 species that do not occur anywhere else in the Mediterranean and 39 that are endemic to the region (Table 15.5). The lack of distinctiveness arises because it forms a region of overlapping distributions. It is the north-eastern limit of the distribution of 170 species that are also present further south or west but not further east; the southern limit for 107 species that are also present further east or west but not south; and the north-western limit for 97 species that are also present further south or east but not west. It is divided into two rather similar subregions. Greece and Turkey have the most similar floras, with 213 out of 361 species being present in both countries, but with only 26 species that occur nowhere else in the Mediterranean and only 13 that are endemic to the subregion. Adjacent to Greece and Turkey, the Albania-Yugoslavia-Bulgaria subregion is similar, sharing 224 species with Greece and Turkey.

To the east of the Mediterranean, Iran and Iraq form a distinctive region. A high proportion of the species in the region are found nowhere else in the Mediterranean (104 out of 281: Table 15.5), although 28 of these are principally non-Mediterranean species occurring in other non-Mediterranean countries as well as Iran and Iraq. Of the 177 eastern species that occur elsewhere in the Mediterranean, 95% (168 species) are also native to the north-eastern region. These include 70 that are absent from the western region, of which 38 are endemic to the Mediterranean.

The south-eastern region is distinguished mainly by having few species per country, either because the countries are small (region 4a) or because they are desert countries (region 4b). The region is otherwise not particularly distinctive or homogeneous. Few species are present in all countries of the region, while 90% of the species present are also present in one or more of the adjacent western, north-eastern, or eastern regions.

Of the 354 species that do not occur outside the Mediterranean (Table 15.3), 182 species are endemic to just one country each in the Mediterranean. A further 102 species are endemic to more than one country from just one of the major regions, leaving 70 that are endemic to the Mediterranean but spread across two or three adjacent regions. No endemic Mediterranean species occur in all four major regions. Eighty-six species are widespread throughout the Mediterranean, being recorded as native to all four regions, but all these species also occur elsewhere.

Species of *Ononis*, *Hippocrepis* and *Anthyllis* tend towards a predominantly western distribution, whereas *Onobrychis* and *Trigonella* tend towards an eastern distribution (Table 15.6). Other genera with many species, especially *Trifolium*, *Vicia*, *Lathyrus* and *Medicago*, are widely distributed throughout all regions. For these genera, species differ so greatly in their distribution that the genus is not a useful unit for summarising distribution.

Four species (*Hippocrepis constricta*, *Lotononis platycarpa*, *Medicago laciniata* and *Trigonella anguina*) have a widespread southern distribution, absent from Europe but ranging from Morocco in the west through other north African countries and the mainland countries of the south-eastern region, to Iran and Iraq in the east. Three species (*Trifolium aureum*, *T. medium* and *Vicia cracca*) have a widespread northern distribution ranging from Portugal in the west to Iran and Iraq, but absent from Africa.

Table 15.6. Regional Mediterranean distribution of forage legume species by genus. For each genus, the table shows the number of species present in each ecogeographical (sub)region as defined in Table 15.5, expressed as a percentage of the total number of species of the genus recorded as native in the Mediterranean.

Region		West			North-east			East	South-east		
Subregion		1a	1b	Total	2a	2b	Total	Total	4a	4b	Total
Genus and species number											
Anthyllis	16	81	56	88	38	31	38	6	13	13	19
Astragalus	1	100	100	100	100	100	100	0	100	100	100
Coronilla	10	70	70	90	40	50	50	30	40	30	40
Cytisopsis	1	0	100	100	0	0	0	0	0	0	0
Dorycnium	5	60	80	80	80	80	80	20	60	40	60
Ebenus	5	0	20	20	20	0	20	40	0	40	40
Galega	1	100	100	100	100	100	100	100	0	0	0
Hedysarum	37	16	30	35	16	14	24	57	8	3	8
Hippocrepis	28	61	54	96	25	18	25	11	25	32	32
Hymenocarpos	4	50	100	100	25	25	25	25	25	25	25
Lathyrus	61	57	41	67	62	54	74	36	30	23	38
Lotononis	6	17	83	83	17	17	17	17	17	17	17
Lotus	58	41	60	67	41	33	45	17	22	41	52
Lupinus	6	50	50	67	33	17	33	0	50	17	50
Medicago	60	57	43	63	72	43	78	40	58	35	60
Melilotus	17	71	59	77	65	65	71	24	47	47	53
Onobrychis	91	15	10	20	30	14	34	73	13	3	13
Ononis	63	64	86	97	19	16	22	11	22	19	27
Ornithopus	6	83	100	100	33	17	33	17	33	0	33
Scorpiurus	2	100	100	100	100	50	100	50	100	50	100
Securigera	7	43	29	57	86	71	86	29	57	0	57
Tetragonolobus	1	0	0	0	100	0	100	0	0	0	0
Trifolium	117	60	43	66	77	66	85	40	47	21	50
Trigonella	44	7	11	14	48	9	48	57	46	25	57
Tripodion	3	33	67	67	67	33	67	0	67	33	67
Vavilovia	1	0	0	0	100	0	100	100	100	0	100
Vicia	87	49	38	59	54	40	63	43	33	16	38

15.5 CONSERVATION STATUS

15.5.1 Overall Composition of Collections

Of 114,379 accessions in the database, 88,417, or 77%, are recorded as having been collected from the wild or farmers (Table 15.7). 6% are selected lines, and 17% have no data to indicate whether they were collected or selected. Of the 88,417 known collected accessions, 70,636 or 80% are from the Mediterranean, at least according to data on the country of origin, and a further 4% have unspecified country of origin. 65% of collected accessions have data on latitude and longitude. These data were overlaid on a map of the Mediterranean to classify accessions into Mediterranean or non-Mediterranean origin. For 3% of collected accessions, the country of origin and latitude/longitude data give inconsistent classifications into the region of origin. No attempt has been made to distinguish real errors from rounding errors (in either the digital maps or the latitude/longitude data) that spuriously place border populations on the wrong side of the border of the Mediterranean. A large proportion of the available data record latitude and longitude only to the nearest minute, rather than to the nearest second. This generates rounding errors of up to around 3km, which is especially problematic when analysing coastal or other border accessions. 47,811 accessions may be regarded as being "definitely" from the Mediterranean, in the sense that data on the country of origin and on latitude/longitude are both present and consistent.

Table 15.7. Tabulation of the composition of 114,379 accessions of forage legumes by accession status (collected, selected line, or unknown), region as determined from the country of origin, and region as determined from latitude / longitude data.

	Accession status						
	Collected						
	Region by Latitude / Longitude						
Region by country of origin	Mediterranean	Non-Med	Not specified	Total	Selected Line	Unknown	Total
Mediterranean	47811	2353	20472	70636	908	8568	80112
Non-Med	174	6202	8160	14536	2065	5517	22118
Not specified	97	506	2642	3245	3579	5325	12149
Total	48082	9061	31274	88417	6552	19410	114379

15.5.2 Historical Development

52,040, or 59% of the collected accessions, have known date of collection. Collecting was most intensive, averaging over 2,000 accessions per year, during the late 1970's and throughout the 1980's (Figure 15.2).

Information on the country of collection is nearly complete for accessions known to have been collected since the mid-1970's. Less than 30% of accessions collected before 1965 have associated data on latitude and longitude of the collecting site. This increased to around 75% of accessions collected during the period of most rapid collection. However, there has been surprisingly little recent improvement (to around 85%), despite the advent of Global Positioning Systems that have made it easy now to obtain accurate data.

Figure 15.2. Progress in *ex situ* conservation of forage legumes. Key: thick solid line = total cumulative number of accessions collected. Thin solid line = % accessions collected from the Mediterranean. Thick dotted line = % accessions with defined country of origin. Thin dotted line = % collections with defined latitude/longitude. All percentages are given as 10-year running means.

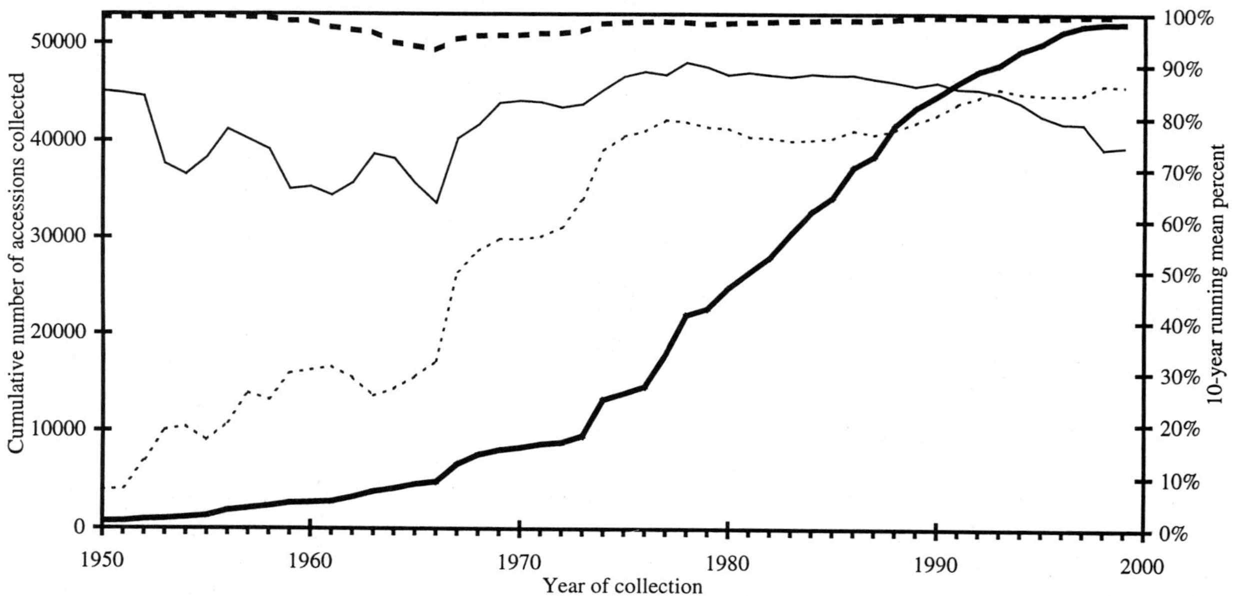

Data from the sources that have been analysed more fully (Table 15.8) demonstrate that for the 51,977 accessions maintained in these Centres, 41,602 or 80% of the accessions have all available passport data transcribed from historical documents and 37,634 or 72% have been characterised. Where possible characterisation was based on International Plant Genetic Resource Institute (IPGRI) forage crop descriptor listings and where available, more genus specific descriptor listings were utilised. For example, characterisation data for annual *Medicago* species maintained at the AMGRC was recorded using IPGRI *Descriptors For Annual Medicago* (IPGRI, 1991).

The bulk of transcription and characterisation efforts have been focussed on five major genera, *Medicago*, *Trifolium*, *Lotus*, *Hedysarum* and *Onobrychis*. These comprised 85% of all accessions documented and 90% of accessions characterised. The AMGRC has passport data complete for 88% of its collection and over 75% is characterised, based predominantly on annual *Medicago*. The ATGRC has 80% of its collection documented and almost 100% characterised, based primarily on *Trifolium subterraneum*. The USDA has completed a review for almost 100% of passport data for *Medicago*, *Trifolium* and *Lotus*, (although *Lotus* has not been uploaded into GRIN) and includes *Hedysarum* and *Onobrychis*. Characterisation has predominantly focussed on cultivated species of these genera, among accessions classified as landrace, cultivated or cultivar (Stephanie Green, *pers comm*).

For these genera, comprehensive data sets do exist in these three Centres, as a result of substantial efforts over the last ten years during which 41% of accessions characterised and 69% of all historical documents were reviewed and transcribed. Emphasis has historically been placed on cultivated species of these genera, considered a high priority by the host Centre. Almost 80% of the collections of these genera are pre 1980, when collecting was at its peak. This priority is a direct reflection on the extent of utilisation by national pasture researchers, the major users of the collections. Within Australia this priority was historically focused on a few annual *Medicago* species and *Trifolium subterranean*.

More recently Australian researchers are placing greater emphasis on the role of alternative species and genera and this is being directly reflected in the current efforts and activities of the two Australian Genetic Resource Centres. For example, in the AMGRC collection of 8,454 accessions of non-*Medicago* species, 62% of all documents transcribed and 100% of all characterisation activities have occurred in the last ten years.

Importantly, while substantial gains have been made in documenting and characterising these collections, Table 15.8 illustrates that further efforts are required to complete the review, documentation and characterisation of 20% of the collection. For the major genera, around 15% require reviewing and documenting. Despite recent efforts, over 50% of the lesser known or under utilised genera require documenting and over 70% or 8,386 accessions have not been characterised. A major outcome from improving the quality, reliability and overall usefulness of the data will be a substantial increase in the effectiveness of use of the collections.

15.5.3 Major Forage Legumes

Just over half of all accessions collected from the Mediterranean belong to one of 15 species - nine species of *Medicago*, four of *Trifolium*, *Lathyrus cicera* and *Vicia sativa* (Table 15.9). These are all widespread species that are also native outside the Mediterranean. In most cases the proportion of Mediterranean to non-Mediterranean accessions of each species matches the proportion of Mediterranean to non-Mediterranean countries where they are native, although there are exceptions. *Vicia sativa* and *Medicago minima*, for example, are both widespread outside the Mediterranean but have been collected mainly from the Mediterranean.

Three species - *Trifolium subterraneum*, *Medicago polymorpha* and *M. truncatula* - have more than 5,000 accessions collected from the Mediterranean (Table 15.9). The nine species of *Medicago* regarded as major forages make up the bulk of all collected accessions of this genus (29,435 out of 38,162).

Table 15.8. Progress in documentation, Characterisation and collecting activities during the last three decades at three major centres: the Australian *Medicago* Genetic Resource Centre (AMGRC), the Australian *Trifolium* Genetic Resource Centre (ATGRC) and the Genetic Resource Centres in the United States Department of Agriculture (USDA).

	Collection Date	Pre 1980			1980s			1990s			Total		
Centre	**Genus**	Documented	Characterised	Collected	Documented	Characterised	Collected	Documented	Characterised	Collected	Documented	Characterised	Collected
AMGRC	*Medicago*	4500	6450	11592	3000	4350	8016	14500	8400	5391	22000	19200	24999
	Other Genera	1000	0	3818	500	0	2001	2500	2500	2635	4000	2500	8454
USDA	*Medicago*												
	+ *Trifolium*												
	+ *Lotus*												
	+ *Hedysarum*												
	+ *Onobrychis*	0	2000	4994	0	4500	1875	8000	1500	1365	8000	8000	8234
	Other Genera	0	0	1363	0	0	651	1500	500	847	1500	500	2861
ATGRC	*Trifolium*	2700	250	3551	860	2300	2060	1690	3984	913	5250	6600	6524
	Other Genera	202	202	206	350	382	385	300	250	314	852	834	905
Total	Major Genera	7200	8700	20137	3860	11150	11951	24190	13950	7669	35250	33800	39757
	Minor Genera	1202	202	5387	850	382	3037	4300	3250	3796	6352	3834	12220
	All Genera	8402	8902	25427	4710	11532	15879	28490	17200	11465	41602	37634	51977

Table 15.9. The native distribution and number of conserved accessions of the 15 species of forage legume that have most accessions collected from the Mediterranean. Species are ranked in order of the preponderance of mediterranean relative to non-Mediterranean accessions.

Species	Number of countries where the species is native		Number of accessions collected			
	Mediterranean	Non-Mediterranean	Mediterranean	Non-Mediterranean	Unknown origin	Total
Trifolium cherleri	20	6	958	6	40	1004
Medicago littoralis	20	6	3161	34	9	3204
M. doliata	16	3	1419	27	18	1464
M. rigidula	21	10	2442	89	2	2533
M. orbicularis	21	11	2426	113	2	2541
M. tornata	13	2	1001	59	6	1066
Vicia sativa	21	35	2382	239	40	2661
M. truncatula	18	7	5545	560	111	6216
T. subterraneum	20	16	5874	747	369	6990
M. minima	21	32	981	146	3	1130
Lathyrus cicera	20	12	807	128	5	940
M. polymorpha	19	18	5750	1237	95	7082
T. resupinatum	12	4	801	206	81	1088
T. repens	20	34	946	732	73	1751
M. sativa	6	9	1424	2538	237	4199
Total			35917	6861	1091	43869
% of all accessions			50.8	47.2	33.6	49.6

Eleven Mediterranean countries each have at least one collected accession of all 15 species (Table 15.10). Of the 35,917 Mediterranean accessions of these 15 species, 37% come from just three countries, Morocco, Turkey and Spain (Table 15.10). The eastern countries tend to be under-represented.

For all 15 species there is at least one country where the species is recorded as native but no accessions have been collected. Conversely, *Vicia sativa* and 8 of the 9 species of *Medicago* have been collected from countries where, according to ILDIS (1999), the species does not occur at all. This suggests that *ex situ* collections could be a useful additional source of data on the distribution of species.

Table 15.10. The number of accessions collected from each country in the Mediterranean, for each of the 15 species of forage legume that have most accessions collected from the region. The format indicates the status of the species in the country as recorded in ILDIS (1999), as follows: bold = native; round parentheses = introduced or uncertain; square parentheses = not recorded as present in the country.

Species	MA	TR	ES	LB/ SY	GR	IT	LY	TN	DZ	CY	IL/ JO	PT	IR	FR	IQ	IT/ SA	BG	YU	EG	MT	GE	AL	AR	AZ
Lathyrus cicera	32	70	317	104	114	7	0	15	12	1	1	124	0	3	6	0	1	0		0		0		
Medicago doliata (syn *M. aculeata*)	533	17	102	95	9	109	[122]	59	198	[33]	25	105	[2]	4	[3]	1		0		2		0		
M. littoralis	358	41	64	30	146	148	1558	176	64	215	197	30	(6)	68	[1]	12	0	1	43	3		0		
M. minima	134	96	38	155	59	87	51	81	77	59	6	1	41	20	6	17	43	4	0	4	[2]	0		
M. orbicularis	135	277	82	360	321	253	26	190	206	57	226	48	78	49	56	26	19	12	0	4	[1]	0		
M. polymorpha	1060	379	440	545	348	276	84	466	481	287	395	180	(242)	69	(238)	188	20	22	6	23		0	[1]	
M. rigidula	44	834	62	629	137	67	0	8	38	25	129	6	175	58	172	16	28	13	0	1		0		
M. sativa	(242)	429	87	38	(24)	114	(6)	(17)	(35)	6	(7)	(10)	137	(172)	(9)	0	(22)	(23)	(16)		[20]	(10)		
M. tornata	615	[2]	44	[20]	[2]	47	122	45	27	3	25	31	[9]	6		0			[3]	0				
M. truncatula	1087	59	115	104	347	258	465	1048	896	521	333	118	(3)	66	(2)	101	[2]	2	7	11				
Trifolium cherleri	56	325	100	72	75	57	0	25	22	114	33	36	4	10	14	2	10	3		0		0		
T. repens	62	349	61	7	63	61		13	10	0	33	83	34	27	1	4	92	45	0	0		1		
T. resupinatum	52	408	(15)	77	37	(23)	0	3	5	2	36	(42)	(58)	(5)	(21)	0	2	(3)	12	0		0		
T. subterraneum	465	392	2055	40	729	863	3	176	24	226	84	493	3	250	0	7	20	41		3		0		
Vicia sativa	316	805	63	365	188	147	0	53	101	77	51	43	62	7	9	0	32	2	11	0	[22]	18	6	[4]
Total number of accessions	5191	4483	3645	2641	2599	2517	2437	2375	2196	1626	1581	1350	854	814	538	374	291	171	98	51	45	29	7	4

15.5.4 Minor Forage Legumes

In addition to the 15 species considered in the previous section, 680 species of forage legume - including 94 Mediterranean endemics and 20 that are not native to the Mediterranean - have been collected from the Mediterranean (Table 15.11). However, disregarding *Astragalus*, which was not assessed, nearly 40% of the species are not species names accepted by ILDIS. These unaccepted species names combine a variety of cases: unidentified species, hybrids, mixtures of species, typing errors, and names regarded as synonyms by ILDIS. 60% of the collected accessions of these minor species are recorded as having been collected from the Mediterranean, including 967 accessions of species that are not native (Table 15.12). 209 accessions of Mediterranean endemics are recorded as having been collected outside the Mediterranean.

The most heavily represented genus of the minor forages is *Trifolium*, in terms of both number of species (with 105 accepted species and 60 "species" not recognised in ILDIS) and number of accessions (16,422, or 37% of all collected accessions of minor forage legumes).

In terms of both number of species (Table 15.13) and number of accessions collected (Table 15.14), Turkey far outranks other countries, with over 9,000 accessions of 326 species. However, this is partly a result of its large size. Relative to their size, Cyprus and Malta have been most intensively collected. Other eastern Mediterranean countries (Lebanon, Syria, Israel, Jordan, and Greece) are also relatively well collected, in terms of absolute numbers of accessions and species, and relative to their size. Further east, Iran has had a comparatively large number of species collected relative to the number of accessions.

A further 264 Mediterranean endemics and 100 species that are native to both Mediterranean and non-Mediterranean countries are not recorded as having been collected at all, or at best are represented by a single accession (Table 15.15). 181 of these are native to just a single Mediterranean country, but the remainder are more widespread. Iran in the east, and Morocco, Algeria and Spain in the west have the highest proportion of species that have not been collected from anywhere in the Mediterranean (Table 15.16).

15.5.5 Distribution Within Mediterranean Countries

All previous analysis of geographical distribution is based on countries. Finer analysis of distribution within countries is possible for the 48,082 accessions collected with latitude/longitude data that place their origin within the Mediterranean. A grid of equal-area 2500km2 cells was defined and accessions were classified according to the cell from which they were collected. This also carries the advantage of avoiding the bias, inherent in country-based analyses, that large countries tend to have more species and more collected accessions.

Space precludes separate treatment of individual taxa. Figure 15.3 shows the combined number of accessions of all forage legume species collected per cell of the grid. The greatest concentration of accessions collected from a single 2500km^2 cell is 1,176 from the north west of Jordan. There is a strong concentration of collecting sites along the coastal belt. The most marked "hot spot" of collecting is along the Turkish-Syrian border (part of the Fertile Crescent). Other marked "hot spots" occur further along the coasts of Turkey and Syria and into Lebanon, Jordan and Israel; also along the coasts of Morocco and Tunisia in the west; and in many central and eastern islands. The only marked inland hot spot is along the Spanish-Portuguese border.

The number of species collected per 2500km^2 shows a broadly similar pattern (Figure 15.4). One cell in Syria has 110 species collected; two others along the Syrian-Turkish border have over 90 species each. Twenty-four cells, mostly in the same region of Syria and the border with Turkey each have more than 50 species per cell.

Table 15.11. The Number of species of forage legume (other than the most 15 intensively collected shown in table 9) collected from the Mediterranean. Accessions are classified by their native distribution given by ILDIS - as global (present both in the Mediterranean and elsewhere), or native only in the Mediterranean, or native only outside the Mediterranean - and by the region in which they were collected as defined by their country of origin.

Genus	**Distribution**			**Species not recognised by ILDIS**	**Total**
	Global	**Mediterranean only**	**Non Mediterranean only**		
Anthyllis	2	1		3	6
†(*Astragalus*)					(88)
Biserrula				1	1
Coronilla	4	2		9	15
Dorycnium	3				3
Ebenus	1	1		2	4
Galega				2	2
Hedysarum	1	4	1	4	10
Hippocrepis	6	1		3	10
Hymenocarpos	1			2	3
Lathyrus	29	7	2	15	53
Lotus	17	7		14	38
Lupinus	4	2	2	6	14
Medicago	27	12	2	23	64
Melilotus	8	1	1	8	18
Onobrychis	8	15	3	13	39
Ononis	6	1		4	11
Ornithopus	4	1		2	7
Scorpiurus	2			1	3
Securigera	4				4
Tetragonolobus				5	5
Trifolium	71	26	8	60	165
Trigonella	14	4		22	40
Vicia	33	9	1	34	77
Total	245	94	20	321	680

† Status in ILDIS not determined

Table 15.12. Number of collected accessions, for all forage legume species collected at least once from the Mediterranean but excluding the 15 most intensively collected shown in Table 9. Accessions are classified by their native distribution given by ILDIS - as global (present both in the Mediterranean and elsewhere), or native only in the Mediterranean, or native only outside the Mediterranean - and by the region in which they were collected according to country of origin.

Native Distribution of Species	**Global**			**Mediterranean only**			**Outside Mediterranean Only**			**Species not recognized by ILDIS (1999)**			
Region of origin of accessions	**Mediterranean**	**Non-Mediteranean**	**Not specified**	**Mediterranean**	**Non-Mediteranean**	**Not specified**	**Mediterranean**	**Non-Mediteranean**	**Not specified**	**Mediterranean**	**Non-Mediteranean**	**Not specified**	**Total**
Anthyllis	47	28	29	1						57	1	9	172
†(*Astragalus*)										(1523)	(459)	(3)	(1985)
Biserrula										165	1	20	186
Coronilla	218	5	1	3						81		1	309
Dorycnium	14	8	3									1	26
Ebenus	1			4						2			7
Galega										2	21		23
Hedysarum	47	5		183	5	2	1	13		8	16		280
Hippocrepis	325			5						119		2	451
Hymenocarpos	151									181			332
Lathyrus	1975	323	5	358	31		935	840	10	324	75	4	4880
Lotononis								1			1		2
Lotus	433	316	102	20	5	2		30	2	180	89	31	1210
Lupinus	191	8	1	269		1	5	182	17	86	244	18	1022
Medicago	4803	826	79	886	7	14	2	129	3	1585	368	25	8727
Melilotus	122	241	37	1			1	4	1	145	32	24	608
Onobrychis	260	75	5	213	141	6	4	37	1	240	47	5	1034
Ononis	13	1	1	1						6	1		23
Ornithopus	799	42	89	2						13	1		946
Scorpiurus	732	8	1							27			768
Securigera	45	24	1										70
Tetragonolobus										66	13	14	93
Trifolium	10065	1336	1252	1179	17	52	18	639	33	1311	407	113	16422
Trigonella	305	11	1	16	3	1		5		641	36	6	1025
Vicia	2401	286	98	216		1	1	15	5	681	212	22	3938
Total	22947	3543	1705	3357	209	79	967	1895	72	7443	2024	298	44539

† Status in ILDIS not determined

Table 15.13. The number of forage species per genus collected at least once from each country in the Mediterranean, for all forage legume species other than the 15 most intensively collected species shown in Table 15.7. Countries are ranked in order of the density of collected accessions (see Table 15.14).

Genus	CY	MT	IL/ JO	LB/ SY	GR	TN	TR	IT- SA	PT	BG	MA	IT	ES	FR	AL	YU	GE	IQ	AR	IN	DZ	AZ	LY	EG	Total
Anthyllis			1		3	2	3		1	1	3	2	1	1		1				1	1				21
†(*Astragalus*)		1	18	11	4	10	15		3	6	10	2	15	3		2		1		53	9		2	1	166
Biserrula			1	1	1	1	1		1		1	1	1												9
Coronilla			1	4	6	2	3		3	2	3	2	2					1		3	1		1	4	38
Dorycnium					3		1			1		1		1		1									8
Ebenus						1	2				1									1					5
Galega							2																		2
Hedysarum	1			2	1	5	1		1	1	3	2	2							1	2				22
Hippocrepis			2	2	3	6	3			2	5							2		1	7		2	3	38
Hymenocarpos	1		3	3	1	2	2	1		1		1				1		2		2			1	1	22
Lathyrus	3	1	9	27	7	5	35		6	9	12	9	9	7	2	4	6	6	4	7	7	7		1	183
Lotus	2		15	4	11	7	12		7	4	11	9	14	10	2	4		1	1	4	4		1		123
Lupinus			3	3	4	2	1		7	1	6	1	7	2		1	1						1	1	41
Medicago	18	7	22	26	25	20	34	8	16	15	27	25	19	25	1	6	2	11		13	16		9	9	354
Melilotus			3	4	1	4	6		6	4	7	5	4	1		2		1	1	5	4			2	60
Onobrychis			4	5	5	1	15			6	3	3	5	4		2	3	4		18	1		2	1	82
Ononis	1		4		5	1	1				3							1			1				17
Ornithopus	1			1	2	2	1	1	5	1	6	2	6	4											32
Psoralea			1				1		1		1				1										5
Scorpiurus	1		3	3	1	3	3	1	2		2	2	2	1				2		1	2		1	2	32
Securigera			1	1	1		3			2		1								2					11
Tetragonolobus	1		2	1		2					2		3	2		1							1		15
Trifolium	25		89	49	63	22	107	2	35	48	30	40	31	35	11	32	2	17	1	29	24		8	3	703
Trigonella			13	13	3	5	27			2	2	2	3			1		7		15	3		4	3	103
Vicia	5	1	14	32	12	6	47		16	21	14	22	15	10	4	1	9	8	9	9	11	8	3	2	279
Grand Total	59	10	209	192	162	109	326	13	110	127	152	132	139	106	21	59	23	64	16	165	93	15	36	33	2371

† Status in ILDIS not determined

Table 15.14. The number of accessions collected from each country in the Mediterranean, for all forage legume species other than the 15 most intensively collected species shown in Table 15.9. Countries are ranked in order of the number of accessions collected relative to the size of the country.

Genus	CY	MA	IL/ JO	LB / SY	GR	TN	TR	IT- SA	PT	BG	MA	IT	ES	FR	AL	YU	GE	IQ	AR	IN	DZ	AZ	LY	EG	Total
Anthyllis			1		7	21	6		1	3	22	3	19	2		7				1	12				105
†(*Astragalus*)		1	239	320	42	114	166		6	24	128	2	31	6		2		10		278	139		7	8	1523
Biserrula			9	30	64	2	9		1		29	11	10												165
Coronilla			41	64	16	33	43		5	2	32	2	2					6		6	43		1	6	302
Dorycnium					7		1			1		2		2		1									14
Ebenus						2	2				2									1					7
Galega							2																		2
Hedysarum	1			3	2	155	2		1	3	17	42	9							1	3				239
Hippocrepis			87	74	16	82	22			5	50							8		13	68		3	21	449
Hymenocarpos	2		83	111	58	2	40	2		1		3				2		11		11			5	1	332
Lathyrus	52	1	47	491	98	216	652		156	55	119	123	215	1239	18	7	14	16	5	29	21	13		1	3588
Lotus	2		45	17	50	127	170		15	19	64	22	37	17	2	14		1	1	10	19		1		633
Lupinus			38	18	43	7	9		101	2	91	1	235	2		1	1						1	1	551
Medicago	637	9	687	828	872	440	715	251	96	95	600	647	278	167	2	26	4	103		135	314		336	34	7276
Melilotus			5	17	1	36	60		8	10	34	9	6	1		3		4	1	11	59			4	269
Onobrychis			103	146	8	2	165			24	6	35	47	7		3	12	4		128	18		7	2	717
Ononis	1		5		6	1	1				4							1			1				20
Ornithopus	2			4	186	22	13	4	149	5	105	76	197	51											814
Psoralea			1				1		1		1				1										5
Scorpiurus	2		67	87	25	169	31	1	18		192	2	27	1				5		4	110		8	10	759
Securigera			1	8	1		12			10		1								12					45
Tetragonolobus	1		18	1		29					5		7	3		1							1		66
Trifolium	144		1424	1474	781	259	5626	3	233	338	507	315	450	77	17	283	18	143	1	188	275		12	5	12573
Trigonella			188	243	55	42	183			3	22	2	4			1		15		136	46		7	15	962
Vicia	20	1	84	754	171	59	1183		112	156	188	155	102	21	27	1	35	23	14	45	124	15	5	4	3299
Total accessions	86	12	3170	4690	2509	1820	9114	261	903	756	2218	1453	1676	1596	67	352	84	350	22	1009	1252	28	394	112	34715
Total accessions per 1000km²	96	46	27	24	19	12	12	11	10	6.8	5.5	4.8	3.4	2.9	2.3	1.5	1.2	0.8	0.73	0.62	0.54	0.33	0.24	0.11	

† Status in ILDIS not determined

Table 15.15. The number of species documented by ILDIS (1999) as being native to at least one country in the Mediterranean but which have not been collected at all or are represented by only one accession collected from the Mediterranean. Species are classified into those that are native only in the Mediterranean and those that are also found elsewhere.

	Number of accessions collected from the Mediterranean			
	No Collected Accessions		One Collected Accession	
Distribution	**Global**	**Mediterranean only**	**Global**	**Mediterranean only**
Anthyllis	3	9	1	1
Coronilla	2	1	1	1
Cytisopsis		1		
Dorycnium	1	1		
Ebenus		3	1	
Galega	1			
Hedysarum	6	24		
Hippocrepis	3	18		
Hymenocarpos		3		
Lathyrus	3	11	7	
Lotononis	1	5		
Lotus	9	17	2	4
Medicago	3	6	1	2
Melilotus	1	2	1	1
Onobrychis	9	49	2	3
Ononis	12	42	2	1
Ornithopus		1		
Securigera	1	1	1	
Tetragonolobus		1		
Trifolium	4	11	3	4
Trigonella	7	16	1	2
Tripodion	1	2		
Vavilovia		1		
Vicia	7	19	3	1
Total	74	244	26	20

Table 15.16. The number of forage legume species recorded as native in each Mediterranean country (Source: ILDIS 1999) but represented in *ex situ* collections by either 0 or 1 accessions collected from anywhere in the region.

Genus	IN	MA	DZ	ES	TN	IQ	EG	YU	GR	TR	IT	LB / SY	BG	LY	FR	PT	AL	IL / JO	MT	IT-SA	CY
Anthyllis		3	6	10	2			3	4	1	2		1	1	2	1	2		1	2	
Coronilla		4	2	2	2			2	1		3			1	2	2	1	1			1
Cytisopsis		1																			
Dorycnium			1						1	1			1								
Ebenus	2						1		1					1							
Galega	1	1	1	1				1	1	1	1		1		1		1				
Hedysarum	19	5	5	1	2	5		1		4	1	2	1		1						
Hippocrepis	1	8	4	11	7		2	2	2	1	2	1	1	4	2		2	1	1	2	1
Hymenocarpos		3	1	1	1											1					
Lathyrus	2	4	3	4	2	3		10	6	4	5		7		3	2	5		1		
Lotononis	1	5	2	1			1		1			1	1	1							
Lotus	3	12	6	5	4	2	7	5	5	1	4	3	2	3	3	2	2	2	1	1	
Medicago	5	1	2	1	2	4	1			7	1	4		2	2			2			1
Melilotus		2	3	3	2		1	2	2	2	2	1	1	1	2	1	2	1	1	2	
Onobrychis	46	5	3	6	2	8	1	4	4	13	1	8	3		4	1		3			
Ononis	2	41	27	30	13	2	5	2	6	5	13	7	1	6	8	13	1	7	5	7	5
Ornithopus		1																			
Securigera			1		1			1	2	1		1	1				2	1			1
Tetragonolobus									1												
Trifolium	2	5	4	2	3	1		11	9	4	8		5		2	1	7		2	1	
Trigonella	13		1	1		6	5	2	6	8	1	6	2		1		1	2	1		1
Tripodion		2	2	1	1		1	1	2	2	1	2		1	1	1	1	1	1	1	1
Vavilovia	1					1				1		1									
Vicia	11	8	2	2	2	3		5	3	6	4	1	4		2			1		1	1
Total	109	111	76	82	46	35	25	52	57	62	49	38	32	21	36	25	27	22	14	17	12
% native species	42	41	32	29	26	26	23	22	21	21	19	19	17	17	17	16	15	13	14	11	10

Figure 15.3. Distribution of collecting sites of 48082 accessions of forage legumes collected with latitude/longitude data from Mediterranean countries. The map shows the number of accessions collected per cell in a grid of 50km x 50km cells. Albers equal-area conic projection, with central meridian at 20°E and standard parallels at 30°N and 46°N. Map prepared with ArcView® and ArcInfo® GIS, and reproduced with permission of ESRI™ (UK). (For Color Plate version, see p. xx)

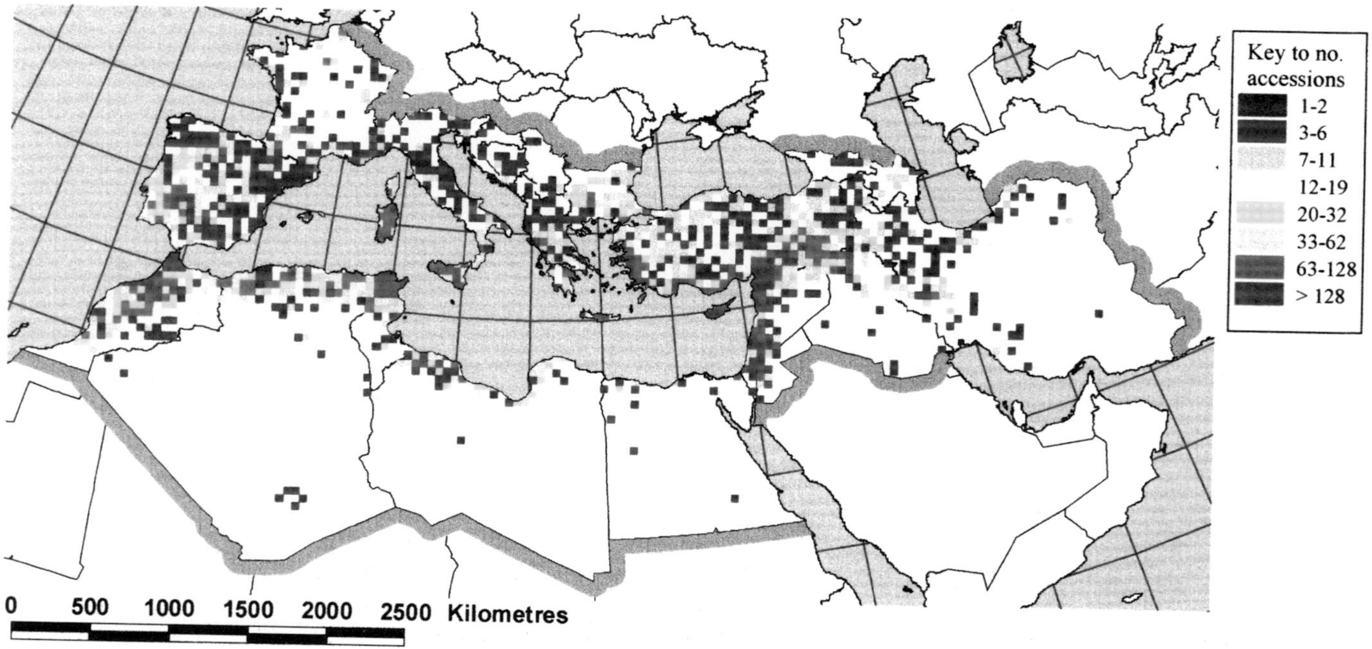

Figure 15.4. Distribution of collecting sites of 48082 accessions of forage legumes collected with latitude/longitude data from Mediterranean countries. The map shows the number of species that have been sampled at least once per cell in a grid of 50km x 50km cells. Albers equal-area conic projection, with central meridian at 20°E and standard parallels at 30°N and 46°N. Map prepared with ArcView® and ArcInfo® GIS, and reproduced with permission of ESRI™ (UK). (For Color Plate version, see p. xxi)

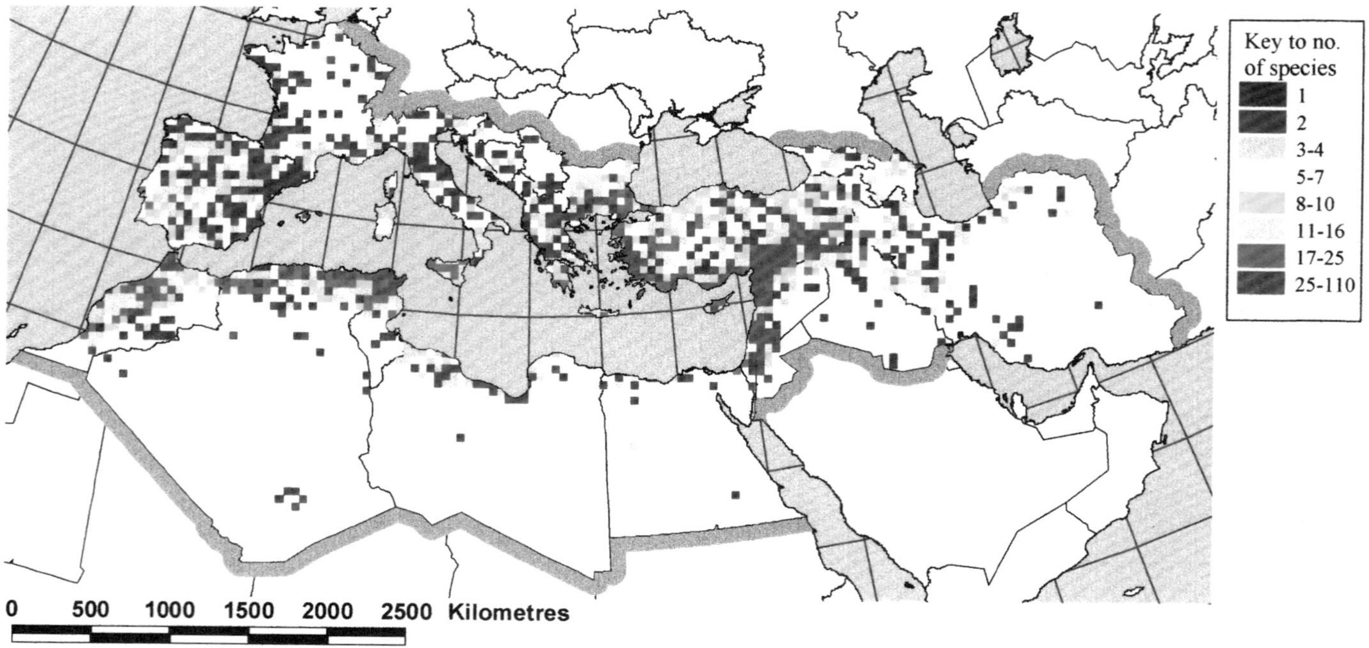

15.6 DISCUSSION

The availability and accuracy of data and the diversity of data formats, especially for the early collections, are major factors limiting the efficacy of the analysis presented here. Nevertheless, within these constraints, some clear messages are apparent.

First is the difficulty of conserving the genetic diversity of such a diverse group of species *ex situ*. Although over 70,000 accessions have been collected from the Mediterranean, even the most intensively collected species, *Trifolium subterraneum*, is represented by only 5,874 accessions of Mediterranean origin, and only 10 species are represented by more than 1,000 Mediterranean accessions each. This is far less than for the main species of most other crop groups and means that for the majority of forage legume species their gene pool is under-represented in *ex situ* collections. 318 forage legume species have not been conserved *ex situ* at all, and 44 others are represented by only one Mediterranean accession. No accessions have been conserved *ex situ* for any species of *Cytisopsis*, *Lotononis*, *Tripodion* or *Vavilovia*. The reason for this omission of these four mono-specific perennial genera is unclear, but they certainly do have utilisation potential and should be collected in the future. Seed and plants of *Vavilovia formosa* (the wild perennial *Pisum* relative) were collected from Azerbaijan by Maxted *et al.* (1989) but the seed could not be germinated and the plant could not be induced to flower in the range of environments possible at the University of Southampton.

The task of conserving *ex situ* an adequate representation of the gene pool of all 739 forage legume species native to the Mediterranean is clearly problematic. *Ex situ* conservation must focus primarily on an economically important subset of species, as has occurred *de facto*. However, the corollary is that integration of *in situ* with *ex situ* conservation is of exceptionally high priority, in order to achieve what cannot be achieved by *ex situ* conservation alone.

The second message is that the distribution of collecting activities within the Mediterranean does not well match the distribution of species diversity of forage legumes, with several marked gaps. Two distinctive regions - the African west Mediterranean, and the northeast Mediterranean - have been relatively well sampled. However, two further regions - the European west Mediterranean (especially France and northern Italy), and Iran and Iraq to the east - are highly distinctive but relatively poorly sampled. Few accessions have been collected from the eastern region. In the European west region, even in Portugal and Spain from which many accessions have been collected, few species have been collected. The former Yugoslavian republics also form a conspicuous gap in collecting.

However, it must be remembered that simple counts of numbers of species and numbers of accessions collected may in certain cases be misleading. Ferguson *et al.* (1998) showed that wild lentils (*Lens* species) had been comparatively over-sampled from Turkey and under-sampled from Uzbekistan. However, molecular studies indicated that the Uzbek material was genetically uniform whereas the Turkish material was extremely genetically variable, thus justifying yet more collecting of Turkish material. The most desirable sampling pattern therefore depends not only on taxonomic diversity (genera and species) but also on patterns of genetic diversity within species. For most genera of forage legumes the appropriate genetic diversity studies have not been undertaken and in their absence relative estimates of taxonomic diversity are the only practical means of comparing *ex situ* conservation priorities.

Additionally, the health and safety of the 114,379 forage legumes currently conserved *ex situ* has not been defined within the scope of this chapter. It is well recognised that the conservation status of individual accessions depends on how well the principal procedures governing *ex situ* conservation are maintained. The maintenance of viability, vigour and genetic integrity are becoming increasingly expensive operations and many genebanks are facing declining resources and are being forced to modify their operations for greater efficiency. It is imperative that a reduction in resources be accompanied by a change in procedures and policies.

ACKNOWLEDGEMENTS

The authors gratefully acknowledge all those that have contributed data: Stephanie Green and Dave Stout (USDA for GRIN forage legume passport and characterisation data); Richard Snowball (Australian Trifolium Genetic Resource Centre for passport and characterisation data); Daniel Combes (Pau, France: ECCDB for *Lathyrus*); Vincent Gensollen (Montpellier, France: ECCDB for perennial *Medicago*); Morten Hulden (Nordic Gene Bank, Norway: ECCDB for *Trifolium pratense*); Lajos Horváth (Tápiószele, Hungary: ECCDB for minor perennial forage legumes); Francisco Gonzalez Lopez (Badajoz, Spain: ECCDB for *Trifolium subterraneum* and annual *Medicago*); Gaetano Laghetti (Istituto del Germoplasma, Bari, Italy); Siyka Stoyanova (Institute for Plant Genetic Resources "K. Malkov", Bulgaria; Cafer Sabanci (Aegean Agricultural Research Institute, Turkey); Eliseu Bettencourt (Portugal) and the CGIAR data sets were obtained from the System-wide Information Network for Genetic Resources (SINGER). We also thank ILDIS for supplying data on the distribution of forage legumes.

REFERENCES

Bettencourt, E., Hazekamp, Th. and Perry, M.C., 1992. *Forages: Legumes, Grasses, Browse plants and others. Directories of Germplasm Collections.* pp. 356. International Plant Genetic Resources Institute, Rome, Italy.

Dice, L.R., 1945. Measures of the amount of ecological association between species. *Ecology*, 26: 297-302.

ESRI ArcAtlas (1998) http://www.esri.com/data/catalog/esri/esri_aa.html

Ferguson, M.E., Ford-Lloyd, B.V., Robertson, L.D., Maxted, N. and Newbury, H.J., 1998. Mapping of geographical distribution of genetic variation in the genus *Lens* for enhanced conservation of plant genetic diversity. *Molecular Ecology*, 7: 1743-1755.

Frison, E.A. and Serwinski, J., 1995. *Directory of European Institutions Holding Crop Genetic Resources Collections, Fourth Edition.* International Plant Genetic Resources Institute, Rome, Italy

Gensollen, V., 1998. The European Central Perennial *Medicago* Database. In: *Report of a Working Group on Forages. 6th Meeting, 6-8 March 1997, Beitostølen, Norway* (eds. Maggioni, L., Marum, P., Sackville Hamilton, N.R., Thomas, I., Gass, T. and Lipman E.). pp. 30. International Plant Genetic Resources Institute, Rome, Italy.

Genstat 5 Committee, 1993. *Genstat 5 Release 3 Reference Manual.* Clarendon Press, Oxford, UK.

Horváth, L., 1998. The European *Bromus*, *Trifolium pratense*, and other perennial forages databases. In: *Report of a Working Group on Forages. 6th Meeting, 6-8 March 1997, Beitostølen, Norway* (eds. Maggioni, L., Marum, P., Sackville Hamilton, N.R., Thomas, I., Gass, T. and Lipman E.). pp. 37-40. International Plant Genetic Resources Institute, Rome, Italy.

ILDIS, 1999. *The International Legume Database and Information Service*. http://www.ildis.org/

IBPGR, 1991. Descriptors for annual *Medicago* / Descriptors pour *Medicago* annuelles. International Board for Plant Genetic Resources, Rome, Italy.

Flora Europaea, 1999. http://www.rbge.org.uk/forms/fe.html

Maxted, N., Bisby, F.A., Forde, M. and Munyenyembe, P., 1989. *Wild forage legume collection in the Caucasus region of the Soviet Union.* Unpublished report. IPGRI, Rome.

Polhill, R.M., 1981. Galegeae. In: *Advances in Legume Systematics, Part 1* (eds. Polhill, R.M. and Raven, P.H.). pp. 357-363. Royal Botanic Gardens, Kew.

Chapter 16

In Situ Conservation

Nigel Maxted, Ayfer Tan, Ahmed Amri and Jan Valkoun

16.1 INTRODUCTION

The Mediterranean Basin, especially the coastline and the semi-arid and arid rangelands, have in the recent past and are currently suffering acutely from human activities that result in genetic erosion. As discussed in earlier chapters, these activities include: increasing industrialization, pollution, urbanization, inadvertent fires, increasing tourism, improper soil use and management, inappropriate animal production systems, and the general over-exploitation of natural resources. Soil erosion, loss of arable land, scarcity of water resources, loss of genetic diversity in terms of taxa and genetic variation are becoming a major concern to the sustainability of agriculture in the region. In fact, most of the national resources are already over-exploited in this region and urgent action is needed to slow the speed of degradation of the ecosystems and to stop desertification. For example, the desertified areas of rangeland, rainfed agriculture and irrigated land currently extend over 90%, 83% and 36% of their total respective areas and are increasing, showing the fragility of the predominant ecosystems in the region (ICARDA 1997).

However, the region still has a high concentration of 24,000 ± 600 vascular plants species (Greuter, 1991), a large proportion of which, about 50%, are endemics (Quézel, 1985; Greuter, 1991; Heywood, 1995), but many (12.5%) are threatened (Walters and Gillett, 1998). Several of these endemic species can be considered as the only vegetation barriers to the expansion of the desertification. This continuous threat to the biological diversity of the region at all levels (genetic, species, community, landscape and ecosystem) means that there is now a critical requirement to rethink human exploitation into more sustainable terms and take conservation action to protection the valuable ecosystem resources of the region.

Much of the current effort to preserve legume genetic resources has been and remains, as is seen in the previous chapters, *ex situ* seed conservation, because this method is generally regarded as technically straight-forward and relatively inexpensive. Here populations are sampled and seed is removed from its original site for storage in remote gene banks. Substantial efforts has been exerted at the international level and by many countries to collect and preserve the genetic diversity of PGRFA species. But these efforts, commendable as they are, remain far from being systematic in terms of genetic representation, focusing on the major crop species, and are unlikely to ever be comprehensive in the genetic sense. The ICARDA gene bank is, for example, conserving in its long term storage facility a total of 120,000 accessions of cereals, food legume and forage legumes species but this still no accurately reflect the diversity of legumes found in the region. Although an effective method of conservation for leguminous species, *ex situ* techniques do have several inherent disadvantages (Maxted *et al.*, 1997a):

- Freezes evolutionary development, especially that which is related to pest and disease resistance

- Genetic diversity may be lost with each regeneration cycle (though individual cycles in legume species can be extended to periods of 20-50 years or more)
- Restricted to a single target taxon per accession (no conservation of associated species, habitat or ecosystem found in the same location along with the target taxon)
- Problems storing seeds of 'recalcitrant' species, particularly of non-forage legume species

Each of these criticisms of *ex situ* seed conservation techniques do not apply to *in situ* conservation, where the germplasm is conserved at the location where it is found. Therefore, as there is no single 'ideal' conservation technique that can be applied in isolation, each technique having its own advantages and disadvantages, Maxted *et al.* (1997a) emphasised the need to apply multiple techniques to conserve any particular gene pool in a complementary manner. Therefore, when attempting to conserve legume diversity, a combination of both *ex situ* and *in situ* techniques should be applied in a complementary manner. In fact, the Convention on Biological Diversity (CBD, 1992) and the Global Plan of Action (FAO, 1996) places particular emphasis on the application of *in situ* techniques over *ex situ*, because of the ancillary benefits for associated species, habitat or ecosystem conservation in the areas where target taxa for genetic conservation are being conserved.

16.2 *In Situ* Conservation of Biodiversity

Much more scientific research has been focused in the past on the scientific basis for *ex situ* rather than *in situ* genetic conservation (Maxted *et al.*, 1997a). Hawkes (1991) observed that *in situ* techniques remain very much in their infancy, however, these techniques provide a 'natural' mean of conservation allowing populations to continue evolving within the environments where they have developed their distinctive traits. As was introduced in Chapter 1, the definition of *in situ* conservation includes two distinct conservation techniques for conserving wild and cultivated species. To avoid confusion Maxted *et al.*, (1997b) proposed the following working definitions of the two distinct conservation techniques:

> "*Genetic Reserve Conservation* - the location, management and monitoring of genetic diversity in natural wild populations within defined areas designated for active, long-term conservation.
>
> *On Farm Conservation* - the sustainable management of genetic diversity of locally developed traditional crop varieties with associated wild and weedy species or forms by farmers within traditional agricultural, horticultural or agri-silvicultural cultivation systems."

Unfortunately, neither of these techniques have been widely applied for temperate legume species. Although there has been much discussion of these two techniques in recent years, there still remain very few quotable practical examples of their application. It is also true that many of the projects referred to as *in situ* are wrongly ascribed, if the definition from the CBD is applied strictly. As Hoyt (1988) concludes the majority of existing nature reserves or natural parks were established to conserve animals or protect aesthetically beautiful landscapes rather than plant conservation.

Although, as is stressed by Maxted *et al.* (1997b) and in the introduction to this text, legume diversity is currently being conserved widely in wasteland, field margins, primary forest, national parks, other protected areas and existing ecologically based conservation projects in the Mediterranean, their existence there is coincidental and passive. In each of these cases the existence of any particular species is not the result of active conservation management. Legume populations conserved in this way that are not actively monitored and managed are more vulnerable to extinction, i.e. any deleterious environmental trend would be unlikely to be noted and counter measures adopted.

With the level of genetic erosion and habitat destruction or degradation that is present in the Mediterranean region, active *in situ* conservation is urgently required to promote the sustainability of legume taxa and the maintenance of the natural or artificial (e.g. agricultural) ecosystems which contain them. Therefore there is a need for active habitat monitoring, management and protection for legume (as well as other biodiversity) conservation. One of the most comprehensive *in situ* conservation project yet developed in the Mediterranean region can be found in Turkey where both genetic reserve and on farm techniques are already being applied to legume conservation. A similar project on *in situ* conservation of several wild relatives of Mediterranean legumes and fruit trees has recently been implemented for other Near East countries.

16.3 GENETIC *IN SITU* CONSERVATION PROJECTS

16.3.1 Genetic Reserve Conservation of Wild Crop Relatives in Turkey

In 1993 a Global Environment Facility project was initiated to conserve *in situ* the wild relatives of globally significant crop species (legumes, cereals etc.) at selected pilot sites in Turkey (Anonymous, 1996, 1998a; Firat and Tan, 1997; Tan, 1998). The project demonstrates the feasibility of a comprehensive and integrated multi-species approach to conservation of genetic diversity. Genetic reserve sites (referred to as Gene Management Zones by the project) were selected where there was a high concentration of the globally significant crop species rather than for their legume composition alone (Anonymous, 1996; Tan and Ulubelda, 1998). However, legume species presence was a factor considered when selecting sites and each sites selected had a significant amount of legume diversity present. The ecological heterogeneity, the ability to manage the site and ease of access to the site for monitoring and management; were each taken in to account in site selection. Conservation of genetic diversity within the legume genera *Lens, Cicer, Vicia* was specifically targeted. In practice the project had five components that built sequentially one upon each other:

1. survey potential sites to locate those containing unique and rich germplasm,
2. establishing and managing genetic reserves at the richest sites,
3. develop a database management system which permits integration with the existing institutional *ex situ* databases,
4. facilitate institutional collaboration,
5. develop a national plan for *in situ* conservation of genetic diversity in Turkey.

The survey and inventory studies involve the identification of several sites containing legume populations that could then be designated for permanent protection and management. The candidate genetic reserve sites have already been determined for the *Lens, Pisum* and *Vicia* species (Kitiki and Tan, 1996, Karagoz, 1998) from Ceylanpinar (Urfa province) and the Southern part of the Anatolian Diagonal (namely the Amanus, Bolkar and Aladag Mountains in Içel and Adana provinces). The decision over the location of the final selection of sites will be made following the result of population diversity analysis of target species within the potential sites. The project led to the establishment of a National Plan for *In situ* Conservation of Genetic Diversity in Turkey, which includes a mechanism for setting up the priorities and developing future plans to conserve the genetic diversity in Turkey. The genetic reserves have been designated for all target species. The database, which contains the project information is linked to GIS and *ex situ* conservation activities (Anonymous, 1998b; Tan, 1998; Tan and Tan, 2000).

16.3.2 On Farm Conservation of Legume Crops in Morocco and Turkey

In 1995, IPGRI together with national programs in nine countries, including Morocco and Turkey from the Mediterranean Region, formulated a global project to strengthen the scientific basis of *in situ* conservation of agro-biodiversity. The nine partner countries were selected because each lies within a region of primary diversity for crop genetic resources with world-wide importance and each has traditional farming communities which maintain diverse plant genetic resources. The countries all have national programmes organized to conserve crop genetic resources *ex situ* and each has a strong interest in developing a complementary on farm project to conserve agro-biodiversity.

Major objectives, hypotheses and preliminary activities for the project were decided upon by IPGRI and National Partners at the First Participants Meeting held in Rome in July 1995. The three main objectives of the project are:

- to support the development of a framework of knowledge on farmer decision-making processes that influence *in situ* conservation of agricultural biodiversity,
- to strengthen national institutions for the planning and implementation of conservation programmes for agricultural biodiversity
- to broaden the use of agricultural biodiversity and the participation in its conservation by farming communities and other groups.

As a result of the meeting, the national partners formulated four key research questions, that are fundamental to the maintenance of genetic diversity on farm:

- to determine and understand the situations in which local cultivars are maintained by farmers,
- to identify the key factors which affect farmer decisions to maintain local cultivars,
- to understand how farmer decision-making affects the amount of genetic variation within crop populations over time, and
- to find ways to assist the continued selection of local cultivars or cultivars that conserve local germplasm.

Within Morocco three agro-ecological zones have been selected for establishment of the on farm project: Demnate / Tanante region, in the province of Azilal in the high Atlas, the valley of Ziz-Tafilalte in the oasis area, and Taza in the Rif and Pre-Rif Mountain area. Priority crops to be included for study in this project are barley, durum wheat and bread wheat, along with the faba bean and alfalfa. These crops were selected on the basis of their economic importance, the appreciation of landraces by local farmers, as indicated by adaptation for abiotic stresses, grain and straw quality, and special uses in preparing local dishes, the availability of scientific data showing the existence of genetic diversity of these crops, and the extent of threat to their diversity by the extension of new released cultivars. The project is implemented within the national PGR programme, involving National Agricultural Research institutions, as well as Non-Governmental Organisations and the extension branch of the Ministry of Agriculture at the local level. More information on the Moroccan element of the project and the IPGRI In Situ Project as a whole can be found at www.cgiar.org/ipgri/.

In turkey the North-west transitional zone has been selected to maintain legume diversity of chickpeas, lentils, faba bean and phaseolus on farm. This zone is particularly rich in terms of diversity containing diverse ecological sites and being influenced by Mediterranean and continental climates, as well as including a range of topographic conditions. But importantly the region also has broad variation in land races cultivated by traditional farmers that remain highly valued by the local communities.

16.3.3 *In Situ* Conservation of Crops and Wild Crop Relatives in the Near East Area

The modern territories of the Near-East (including Iraq, Jordan, Lebanon, Palestine, Syria, South East Turkey and Southern Iran) encompass the region known as the 'Fertile Crescent'; considered one of the four nuclear centres of agricultural origin, where numerous species (notably wheat, barley, lentil, pea and vetch) originated 10,000 years ago and where the wild relatives of these crops and local populations are still found (Hawkes, 1983). Many fruit trees such as almond, olive and pistachio also dominate the traditional ecosystems (Harlan 1975). The wild relatives of many other fruit trees and many forage and rangeland species are also found in this region. This diversity is constantly threatened by the rapid spread of uniform new cultivars, the destruction of natural habitats, land clearing, overuse and overgrazing of species, deforestation, changing of farming systems, civil strife and environmental degradation. Further, the agricultural activity is less attractive to the younger generations and subsequently, the use of the landraces and the related local knowledge of primitive species is rapidly decreasing. The decrease in diversity has lead to loss of soil and to the decrease of air and water quality and subsequently increased poverty and is leading to the migration of rural populations to urban areas.

ICARDA has been involved in a number of research activities related to *in situ* conservation of agro-biodiversity in Central and West Asia and North African Arab countries mainly on aspect of rangeland improvement, *in situ* conservation of wild cereals and legumes progenitors. A new project financed by the Global Environment Facility (GEF), through UNDP, on the conservation and sustainable use of dryland agro-biodiversity in Jordan, Lebanon, Palestine Authority and Syria has been launched in June 1999 for a period of five years. This project was to focus on the application of genetic reserve and on-farm mechanism and techniques. It will promote the conservation and preservation of important wild relatives and landraces of agricultural species and thus ensure sustainable use of agro-biodiversity in the region for the new millennium. The project has five components (GEF/UNDP/ICARDA/IPGRI/ACSAD, 1999):

- conduct eco-geographic surveys of crop target species,
- promote alternative land use practices at project sites to conserve and sustainably use agro-biodiversity,
- increase national capacity to provide training in *in-situ* and on-farm conservation techniques,
- provide suggestions/options for modifying existing legislation and land use rights where necessary and in the national interest to promote the conservation and sustainable use of agro-biodiversity, and
- monitor the impacts of project activities for lessons learned and adaptive project management.

ICARDA is the executing institution of the regional component, in cooperation with IPGRI, ACSAD and the national PGR programmes of the countries represented. The project strategy is to develop community driven *in situ* and on-farm agro-biodiversity conservation initiatives in representative, targeted areas of global agro-biodiversity significance. These community-driven initiatives will be supported by national, legislative, social and economic policies adapted to agro-biodiversity conservation developed during the project process. The project and partners will benefit from institutional capacity strengthening through personnel training and regional networking. Awareness promotion is a priority at all levels of the project. The involvement of land users as primary participants is fundamental to agro-biodiversity management. Innovative approaches to *in situ* and on-farm conservation will be developed alongside appropriate resource management, which will at the same time, maintain the productive capacity of the resources and secure the economic viability of the communities.

The long-term, global objective of the project is to ensure the continuous availability of agro-biodiversity in Jordan, Lebanon, the Palestinian Authority and Syria that is essential to the sustainable development of agriculture in the region, as well as to global food security and production. As a result of the project GEF/UNDP/ICARDA/IPGRI/ACSAD (1999) concludes that the following goals will be achieved:

- information on the distribution and abundance of genetic diversity in the target areas, and the socioeconomic circumstances of communities and land users in the target areas, will be available in national and regional databases, integrated within GIS, for use in long-term monitoring and impact analysis,
- wild relatives of agricultural species will be conserved in identified locations through the adoption of community based habitat management and modified or alternative land use practices,
- alternative income-earning opportunities based on the above conservation of wild species, and alternative land use practices, will have been identified and adopted by communities.
- land races of target species will be conserved on-farm through the full participation of farmers in the selection of adapted land races and mixtures,

- cultivated target species will have been integrated into existing or alternative crop rotations that promote the maintenance of soil fertility,
- field gene-banks of vulnerable fruit tree wild relatives and perennial rangeland species will have been established, providing seed and root stock to communities both within and beyond the target sites,
- community managed pasture nurseries will have been established on common land, providing seed and rootstock;
- both the field gene-banks and the community managed pasture nurseries will provide another source of income to the communities in which they are established, through the sale of seed and root stock;
- national agricultural, environmental and land use policies and legislation, and their implications for agro-biodiversity conservation, will have been analysed; where necessary, alternative policy and legislative options will have been developed and, where feasible, tested at the project sites; and, where appropriate and in the national interest, these options will have been presented to the appropriate authorities for legislative reform.
- interventions developed within the project target sites will have been communicated to other communities and land users through organized field visits and demonstrations of the management techniques and uses of field gene-bank, pasture nurseries, water harvesting and soil conservation structures, and the sale and exchange of locally adapted root stocks and seed;
- public awareness of the importance of biodiversity conservation and environmental degradation will have been increased through public open days, media events, dissemination of information packets and school syllabus material;
- national capacities in the conservation and use of agro-biodiversity will have been strengthened through the graduate and short-term training of national staff provided during the project;
- regional cooperation and the exchange of experiences in the conservation of agro-biodiversity will have been strengthened through the project;
- options for extending the project activities to other sites within the countries involved, and to other countries in the region, will have been identified;
- lessons learned during the project will have been disseminated internationally through publications, a project web-site, and other international media.

The taxonomic focus of the project will be ten target crops (or crop groups) of global or regional significance and their wild relatives (see Table 16.1), all of which originated from the Near East region and are subject to severe genetic erosion. The selection of the target areas, where *in situ* activities would be focused, was undertaken with the aim to capture the maximum genetic diversity of the target crops in the minimal number of areas possible. The target areas were therefore selected based on the presence of target species, the representativeness of major ecosystems and the suitable conditions of work (including willingness of local communities and potential of impact). Two target areas were selected per country and one or more sites were chosen in each of target site to include the diversity of environment. More detailed information on the target areas is provided in Table 16.2.

Although this project promises to yield very interesting results both in terms of actual plant genetic diversity conserved and made available for sustainable utilisation, they also are likely to generate methodologies and techniques that can be applied in a much broader PGR context, as is the IPGRI *in situ* conservation of Agro-biodiversity global project.

Table 16.1. Legume taxa found at target site locations in Jordan, Lebanon, Syria and Palestinian authority (from GEF/UNDP/ICARDA/IPGRI/ACSAD, 1999).

	Jordan		**Lebanon**		**Palestinian Authority**		**Syria**	
Taxon	Ajloun	Muwaqqar	Aarsal	Baalabek	Jennin	Hebron	Al-haffe	Sweida
Lathyrus spp.	✓		✓	✓	✓	✓	✓	✓
Lens spp.	✓		✓	✓				✓
Medicago spp.	✓	✓	✓	✓	✓	✓	✓	✓
Trifolium spp.	✓		✓	✓	✓	✓	✓	✓
Vicia spp.	✓	✓	✓	✓	✓	✓	✓	✓

Each of the projects detailed above has attempted to conserve *in situ* the genetic diversity of legume species in the Mediterranean region, but the point should be re-emphasised here that other than these projects there remains relatively few active *in situ* projects in the region. Though project plans have been proposed, such the *in situ* conservation of *Brassica* species in Sicily (Gustafsson, 1999), these plans have yet to be implemented.

16.4 ECOLOGICAL CONSERVATION *IN SITU* PROJECTS

The previous sections of this chapter have highlighted the limited number of *in situ* genetic conservation projects, which focus on particular taxa and attempt to conserve the full range of genetic (allelic) variation. In contrast ecological conservation activities focus on communities, individual taxon or population survival and extinction are a major concern, but are seen in the context of overall community health. This form of conservation is exemplified by the 'Man and the Biosphere' programme of UNESCO (1996), which established a network of biosphere reserves, representing distinct biomes and ecosystems throughout the world, including the Mediterranean region. The clear emphasis is on conservation of ecosystems: individual species should be conserved as part of an ecosystems rather than on a species by species basis. Reserves have been sited in areas of high species concentration or endemism. The selection criteria has been extensively reviewed for Germany, as an exemplar state, by Weidenhammer (1996).

Table 16.2. Target site characteristics in Jordan, Lebanon, Syria and Palestinian Authority (from GEF/UNDP/ICARDA/IPGRI/ACSAD, 1999).

Country	Target Sites	Site Characteristics
Jordan	Ajloun Lat. 32 20N Long. 35 45E	It is a mountainous area with steep slopes and valleys, located 75 km north of Amman. It has a sub-humid Mediterranean climate and 80% of the soils are shallow. The vegetation cover is mainly indigenous forest of *Pinus* and *Quercus* and here wild species of pistachio, apricot and almond are found. Wild relatives of cereals and forage legume species are also found in undisturbed area and in agricultural landscapes. Overgrazing and land reclamation are major threats to biodiversity. Two existing natural reserves are located in the region.
	Muwaqqar Lat. 31 49N Long. 36 06E	It is a dry area located in the plateau and hills south of Amman representing the steep zone of Jordan. Soils are highly calcareous and eroded by wind and water. Open grazing and barley growing are the predominant land uses. Supplementary irrigated orchards planted to olive tree are being developed. Wild barley, wild species of *Vicia* and *Lathyrus* and local varieties of olive, grapes, figs and almonds are all found. Overgrazing is the major threat to biodiversity.
Lebanon	Aarsal Lat. 34 05N Long. 36 17E	It is part of Anti-Lebanon mountain range with climate ranging from arid to semi-arid. Soils are predominately calcareous and alluvial soils are found in the valleys. The area is used for open grazing and to grow barley and wheat. The planting of grapes and cherries is progressing. Wild relatives of cereals, legumes and fruit trees and many forage legume species are found in this region.
	Baalabek Lat. 34 00N Long. 36 12E	The area is made up of flat plateau rising steeply on one side to an elevation of 1500 m. It has a semi-arid climate and the soils are highly calcareous. Dryland farming of field crops predominates. The flora has >500 species of which many are endemic. Wild relatives of cereals, legumes and fruit trees are still found in this region. Habitat fragmentation and overgrazing are threatening wild relatives while the landraces are replaced by improved cultivars.
Palestinian Authority	Jennin Lat. 32 28N Long. 35 18E	The region is hilly with a gradient down to Jordan Valley covering a climatic gradient from semi-arid to arid. Soils are alluvial and dark rendzina with some basaltic pockets that are lost through overgrazing. Natural reserves exist in this region. Cereals, food legumes, vegetables and olive trees are the basis of agriculture. Wild relatives of cereals, legumes and forage legume species are found but they are threatened by habitat destruction and overgrazing.
	Hebron Lat. 31 32N Long. 35 06E	This site includes the mountain slopes of Hebron and the hills in the south and east parts. The climate is a Mediterranean semi-arid type. In the mountain terra rosa soils predominate while alluvial soils exist in plains and valleys. Landraces of cereals food and feed legumes and fruit trees are found as well as many of their wild relatives. Overgrazing is the major threat to the agro-biodiversity.
Syria	Al-haffa Lat. 35 35N Long. 36 04E	This region extends from an altitude of 500 m to 1000 m at Slenfe mountain. The climate is humid and sub-humid with Mediterranean influences. In local forests wild species of fruit trees predominate. In cultivated areas, landraces of cereals, food legumes and fruit trees are still used. Deforestation, overgrazing and expansion of olive and citrus plantations are threatening biodiversity.
	Sweida Lat. 32 43N Long. 36 33E	This area is predominantly a plateau with a climate ranging from sub-humid to arid. The soils are of basaltic origin. Dryland farming prevails with cultivation of cereals, food legumes and forages. New plantations of apple tree and grapes are expanding rapidly. It is a unique area for biodiversity with >900 native species, some of importance as wild relatives of cereals, food legumes and oleaster and pistachio. Overgrazing, expansion of plantation of apple trees and destruction of habitats are threatening biodiversity.

The concept of biosphere reserves is relatively recent but they have proved to be valuable for biodiversity and ecosystem conservation. Although large exclusionary reserves, which remove land from productive use and ignore the needs of local communities have often proved unpopular locally and may increase the exploitation pressure on surrounding areas. Maxted *et al.* (1997a) has stressed that in many cases individual species can not be conserved *in situ* without conserving the communities in which they naturally occur, so the distinction between genetic and ecological conservation is in practice blurred. The distinction remains, however, that when undertaking genetic conservation, population and genetic diversity levels will be regularly monitored, while this may not be the case in ecological conservation, and so in the later the existence of individual species is more precarious. Experience has also shown that in the Mediterranean region where biosphere reserves are not actively managed they have a higher probability of fire destruction, therefore some level of management, commonly grazing, is essential.

Although biosphere reserves may primarily focus their conservation activities on preservation of endemic and threatened species or habitats and ecosystem rather than PGRFA each of these reserves undoubtedly contains numerous PGRFA species. Therefore it is appropriate here to provide a brief review of ecological based conservation in the Mediterranean region. There is a reasonably comprehensive network of National Parks and other large protected areas in the Mediterranean region (Akeroyd, 1999). The European Union's Special Areas of Conservation covers many of the key areas in EU Member States and the Council of Europe's Emerald Network could have a similar function for non-EU States. Spain has a particularly comprehensive coverage of National Parks, the Sierra Nevada National Park in Andalucia contains large tracts of species rich mountain ranges, with 30-40% plant species endemism. Though the picture of adequate reserve coverage declines to the eastern or southern Mediterranean (Akeroyd, 1999). This points is illustrated by considering the 1997 UN List of National Parks and Protected Areas (IUCN, 1998) for the Mediterranean region summarised in Table 16.3 and Figure 1.2 (of Chapter 1), which shows the location of the 68 UNESCO Man and Biosphere reserves (MAB, 2000) found within the Mediterranean region. IUCN (1994) define the six Protected Area Categories, used in the 1997 UN List of National Parks and Protected Areas and thus in Table 16.3, as follows:

Category I: Strict Nature Reserve/Wilderness Area. To protect nature and maintain natural processes in an undisturbed state in order to have ecologically representative examples of the natural environment for scientific study, environmental monitoring, education and maintenance of genetic resources in a dynamic and evolutionary state. There are two sub-categories: Ia includes protected areas managed mainly for scientific research and monitoring; Ib includes protected areas managed mainly for wilderness protection, subsistence and recreation.

Category II: National Park. To protect outstanding natural and scenic areas of national or international significance for scientific, educational and recreational use. These are relatively large areas not materially altered by human activity and where extractive resource uses are not allowed.

Category III: National Monument/Natural Landmark. To protect and preserve nationally significant natural features because of their special interest or unique characteristics. These are relatively small areas focussed on the protection of specific features.

Table 16.3. UN List of National Parks and Protected Areas summary data for Mediterranean region countries. (IUCN, 1998).

Country		Protected Area Categories																						Totals		
		I						II			III			IV			V			VI						
		Ia			Ib																					
Name	Area (km²)	No.	Area (km²)	%	No.	Area (km²)	%	No.	Area (km²)	%	No.	Area (km²)	%	No.	Area (km²)	%	No.	Area (km²)	%	No.	Area (km²)	%	No.	Area (km²)	%
Albania	28,750	4	145	0.5	0	0	0.0	11	207	0.7	4	34	0.1	26	433	1.5	3	24	0.1	4	182	0.6	52	1,028	0.5
Algeria	2,381,745	4	376	0.1	0	0	0.0	9	57,240	2.4	0	0	0.0	4	492	0.1	1	800	0.1	0	0	0.0	18	58,909	2.5
Armenia	29,800	4	634	2.1	0	0	0.0	1	1,500	5.0	0	0	0.0	0	0	0.0	0	0	0.0	0	0	0.0	5	2,134	7.2
Azerbaijan	86,600	14	1,920	2.2	0	0	0.0	0	0	0.0	0	0	0.0	20	2,856	3.3	0	0	0.0	0	0	0.0	34	4,777	5.5
Bulgaria	110,910	28	651	0.6	0	0	0.0	3	2,212	2.0	2	44	0.1	28	571	0.5	2	1,463	1.3	0	0	0.0	127	5,001	4.5
Croatia	56,538	6	222	0.4	0	0	0.0	6	655	1.2	0	0	0.0	1	52	0.1	13	2,780	5.1	0	0	0.0	26	3,964	7.0
Cyprus	9,250	0	0	0.0	0	0	0.0	1	91	1.0	0	0	0.0	6	681	7.4	3	9	0.1	0	0	0.0	10	782	8.5
Egypt	1,000,250	3	370	0.1	0	0	0.0	1	615	0.1	0	0	0.0	11	6,953	0.7	0	0	0.0	3	1,178	0.1	18	9,116	0.9
France	543,965	6	395	0.1	4	2,162	0.4	5	2,613	0.5	23	16	0.1	1,241	4,315	0.8	40	66,150	12.2	2	118	0.1	1,323	73,241	13.5
Georgia	69,700	14	1,680	2.4	0	0	0.0	2	265	0.4	0	0	0.0	0	0	0.0	0	0	0.0	0	0	0.0	16	1,953	2.8
Greece	131,985	1	5	0.1	0	0	0.0	16	2,720	2.1	15	164	0.1	33	1,296	1.0	23	509	0.4	0	0	0.0	88	4,696	3.6
Iran	1,648,000	0	0	0.0	0	0	0.0	7	10,753	0.7	5	63	0.1	27	30,518	1.9	39	41,697	2.5	0	0	0.0	78	83,066	5.0
Iraq	438,445	0	0	0.0	0	0	0.0	0	0	0.0	0	0	0.0	8	5	0.1	0	0	0.0	0	0	0.0	8	5	0.1
Israel	20,770	0	0	0.0	0	0	0.0	1	30	0.2	0	0	0.0	14	3,132	15.1	1	84	0.5	0	0	0.0	16	3,257	15.7
Italy	301,245	8	7	0.0	0	0	0.0	10	3,705	1.2	14	15	0.0	298	2,735	0.9	93	15,587	5.2	0	0	0.0	423	22,051	7.3
Jordan	96,000	1	12	0.1	0	0	0.0	0	0	0.0	0	0	0.0	5	869	0.9	5	2,099	2.2	0	0	0.0	11	2,980	3.1
Lebanon	10,400	0	0	0.0	0	0	0.0	2	43	0.4	0	0	0.0	1	5	0.1	0	0	0.0	0	0	0.0	3	48	0.5
Libya	1,759,540	0	0	0.0	0	0	0.0	4	510	0.0	0	0	0.0	4	1,220	0.1	0	0	0.0	0	0	0.0	8	1,730	0.1
Macedonia	25,713	0	0	0.0	0	0	0.0	3	1,083	4.2	7	547	2.1	4	89	0.3	2	84	0.3	0	0	0.0	16	1,803	7.0
Malta	316	0	0	0.0	0	0	0.0	0	0	0.0	0	0	0.0	0	0	0.0	1	2	0.8	0	0	0.0	1	2	0.8
Morocco	458,730	1	65	0.1	1	26	0.1	0	0	0.0	0	0	0.0	7	2,381	0.5	2	698	0.2	0	0	0.0	11	3,174	0.7

Portugal	92,390	2	36	0.1	1	97	0.1	1	211	0.2	10	23	0.0	23	1,104	1.2	17	4,543	4.9	0	0	0.0	53	6,026	6.5
Romania	237,500	12	607	0.3	0	0	0.0	15	9,781	4.1	5	2	0.0	11	364	0.2	8	126	0.1	0	0	0.0	52	10,896	4.6
Slovenia	20,251	0	0	0.0	0	0	0.0	1	838	4.1	0	0	0.0	0	0	0.0	13	310	1.5	0	0	0.0	14	1,148	5.7
Spain	504,880	0	0	0.0	2	36	0.1	11	2,202	0.4	0	0	0.0	136	17,256	3.4	154	22,876	4.5	0	0	0.0	303	42,401	8.4
Syria	185,680	0	0	0.0	0	0	0.0	0	0	0.0	0	0	0.0	0	0	0.0	0	0	0.0	0	0	0.0	0	0	0.0
Tunisia	164,150	0	0	0.0	0	0	0.0	6	444	0.2	0	0	0.0	1	1	0.1	0	0	0.0	0	0	0.0	7	445	0.3
Turkey	779,450	4	209	0.1	0	0	0.0	19	3,959	0.5	2	2	0.1	17	4,593	0.6	9	1,034	0.1	14	2,115	0.3	78	11,967	1.5
Yugoslavia	102,173	0	0	0.0	1	10	0.1	6	1,421	1.4	10	17	0.1	31	878	0.9	29	1,041	1.0	0	0	0.0	104	3,390	3.3
Totals	**11,295,128**	**112**	**7,336**	**0.1**	**9**	**2,333**	**<.1**	**141**	**103,100**	**0.9**	**97**	**929**	**<.1**	**1,957**	**82,801**	**0.7**	**458**	**161,918**	**1.4**	**23**	**3,595**	**<.1**	**2,903**	**359,992**	**3.2**

Category IV: Habitat/Species Management Area. To assure the natural conditions necessary to protect nationally significant species, groups of species, biotic communities or physical features of the environment where these may require specific human manipulation for their perpetuation. Controlled harvesting of some resources may be permitted.

Category V: Protected Landscapes and Seascapes. To maintain nationally significant natural landscapes that are characteristic of the harmonious interaction of man and land while providing opportunities for public enjoyment through recreation and tourism . These are mixed cultural/natural landscapes of high scenic value where traditional land uses are maintained.

Category VI: Managed Resource Protected Area. This is a new category designed to include areas that ensure long-term protection and maintenance of biodiversity while providing a sustainable flow of natural products and services to meet community needs. They are intended to be relatively large and predominantly unmodified natural systems where traditional and sustainable resource uses are encouraged.

16.5 A GROWING NEED FOR *IN SITU* PLANT GENETIC CONSERVATION

Post UNCED and the ratification of the CBD there is a growing interest in the application of *in situ* conservation techniques. For legumes in the Mediterranean region there is an urgent need for conservation of PGRFA diversity within profitable and evolving farming systems through full participation and the involvement of the local communities. This is the only appropriate model as *in situ* conservation of biodiversity over large areas depends on the active support of the land users and on the generation of appropriate technologies through participatory approaches to increase productivity, add value to the generated products and to allow the producers enjoy the benefits of their products. The achievement of this goal may be enhanced through five categories of intervention:

- **Participatory approaches:** These will involve the use and development of appropriate technologies which involve farmers directly and consider the importance of local knowledge and the traditional and multiple uses. This approach may include participatory plant breeding, the development of local seed and seedling supply at the local community levels, as well as participatory regeneration and collection. The genetic manipulation will focus on correcting major deficiencies in the local varieties. Participatory approaches may also be used on the management of shared natural resources such as, water and rangeland. Many agricultural practices including contour and ridge planting, water harvesting and alley cropping. Low cost and low input agricultural packages need to be developed to allow an increase in productivity without affecting the biodiversity negatively.
- **Better *In Situ* Conservation Targeting:** The existing *in situ* conservation projects in the Mediterranean and West Asian countries have arisen in an *ad hoc* manner. There is a need for a more systematic approach, aiming on farm and genetic reserve activities at a more efficient coverage that can truly be seen to complement current *ex situ* activities. The siting of new genetic reserve and on farm projects specifically for legume species is discussed in the various Chapters of Section 2 of this text and in Chapter 20. Perhaps there is a coordination role here for the regional agencies with an interest in PGRFA

conservation notably ICARDA, IPGRI and ACSAD in terms of gaps analysis and *in situ* conservation promotion.

- **Economic incentives:** The landraces of many crops and many wild species have many inherent quality traits that are well known to the local communities, but only to few of the other end users. The price of food legume local varieties is at least 30% more than the grain of improved varieties in many countries of the region. At the local level, due to intermediaries, farmers are not taking full advantage of the value of their products. The improvement of marketability through the development of markets for local landraces, the advertisement of their products as biological and originating from organic organization of farmers, are important elements to increase the benefits of farmers while preserving diversity. Added value could also be achieved through processing and the marketing of processed products.
- **Policy initiatives and government incentives:** Changes in policy and property rights have altered land use practices in most of the Mediterranean and West Asian countries in ways that do not appear to be sustainable. The majority of these policies have been established at the national and international levels and therefore do not consider the specificity of the local communities and their traditional organizations. The local agricultural extension services are focusing their efforts on promoting intensive agriculture while in these areas special attention should be given to low input agriculture, local knowledge and preservation of biodiversity. Government incentives and public investments are well justified if social (poverty reduction, decreasing of population migration) and environmental benefits are considered, in additional to increasing productivity of the biodiversity. Due to the importance of old traditional structures in the management of natural resources, policy and property rights must be linked to ensure an income increase of the farmer without jeopardizing the biodiversity.
- **Increasing public awareness:** Public awareness on the importance of biodiversity should constitute a major action at all levels from individuals, communities, national, regional and international levels. This can be achieved through activities, such as specialized courses at school, frequent talks with local communities at any gathering of peoples (mosque, fairs, etc.) training, use of different media supports etc. The tools to be used should provide easy understanding of biodiversity importance to a given audience. Local non-governmental organizations can help a lot in this respect.

The Mediterranean and West Asian region is an area of major importance for world biodiversity. The countries of the region should give ample attention to the genetic reserves and on farm conservation to preserve the landraces of crops and their wild relatives, particularly of the most highly threatened species. This can only partially be achieved through better management of natural resources, there is also a need for more efficient international, regional and national conservation programmes. Together the international agencies working together with the NARSs can achieve the ultimate goal of conserving the rich agro-biodiversity for the needs of the future generations.

16.6 CONCLUSION

In reality the sheer number and ecogeographic diversity of temperate legume species, particularly the forage species, must make *in situ* the preferred conservation option. However, although *in situ* conservation techniques offer the most practical alternative for legume conservation, we have comparatively little practical experience of its implementation. There is a real need to develop the science of both genetic reserve and on *in situ* conservation, while ensuring that its implementation is sensitive to the needs of local communities and is also linked to the requirements of the germplasm user community. One means of establishing better protocols would be to instigate project case studies, such as those discussed outlined, so that we all could learn from their experience and thus refine our methodologies.

The *in situ* conservation projects already established have been largely funded by the Global Environment Facility and have tended to focus their activities on a small number of countries and not necessarily those with the highest levels of biodiversity or legume diversity. There is a need for more systematic funding of *in situ* conservation projects throughout the region and there is a need for more regionally based or national agencies to take the lead in the development of future projects.

There have been suggestions made for site to locate novel genetic reserves for certain legume species in the region, unfortunately these suggestions have yet to be implemented by national agencies. There is a need for individual countries to actually designating and managing genetic reserves to conserve their native diversity and particular legume diversity. The location of priority sites for general legume diversity would follow from ecogeographic studies and this further underscores the need for these studies.

On farm conservation is the most recently developed conservation technique, although it has effectively been practised by farmers in the region for millennia. On farm conservation is particularly important for the conservation of grain legume species, where land races are under immediate threat of replacement with higher yielding varieties. The on farm conservation of forage species is of lower priority, because of the difficulty of conserving the diversity and of number of species involved, and the fact that these species are less subject to genetic erosion.

The distinction is made above between passive and active conservation and it is recognised that legume diversity is often being passively conserved in existing protected areas throughout the region. Our emphasis as genetic conservationists is to link conservation to use. Therefore if the diversity conserved in this way is ever to be of potential use to the germplasm user community, we need to encourage the managers of these protected areas to produce inventories of the wild legume species contain in their reserves and promote the use of the germplasm.

REFERENCES

Akeroyd, J., 1999. Conserving the Mediterranean flora: a way forward. *Plant Talk*, 18: 23-28.

Anonymous, 1996. *Genetik Cestiligin in situ, Yerinde, Muhafazasi Projesi: Gen Muhafaza Zonlari*. Workshop 28-29 Mart 1996. [*In Situ* Conservation of Genetic Diversity Project. Workshop on Gene Management Zones, 28-29 March 1996] ETAE No: 92. Menemen, Turkey.

Anonymous, 1998a. *Final Report of* In Situ *Conservation of Genetic Diversity Project in Turkey*. MARA, Turkey.

Anonymous, 1998b. *Information Technology Management Plan (ITMP), Database Design and GIS Applications Development Project. Project Atlas*. MARA, Ankara, Turkey.

Convention on Biological Diversity, 1992. *Convention on Biological Diversity: Text and Annexes*. pp. 1-34. Secretariat of the Convention on Biological Diversity, Montreal.

GEF/UNDP/ICARDA/IPGRI/ACSAD, 1999. *Conservation and Sustainable use of Dryland Agro-biodiversity in Jordan, Lebanon, Syria and Palestinian Authority*. Unpublished Project Document. GEF/UNDP/ICARDA/IPGRI/ACSAD, Aleppo, Syria.

Food and Agriculture Organisation of the United Nations, 1996. *The conservation and sustainable utilization of plant genetic resources for food and agriculture: Global Plan of Action*. FAO, Rome.

Firat, A.E. and Tan, A., 1995. Turkey maintains its pivotal role in global genetic resources. *Diversity*, 11: 61-63.

Firat, A.E. and Tan, A., 1997. *In situ* conservation of genetic diversity in Turkey. In: *Plant genetic Conservation: the* in situ *approach* (eds. Maxted, N, Ford-Lloyd, B.V. and Hawkes, J.G.). pp. 254-262. Chapman and Hall. London.

Greuter, W., 1991. Botanical diversity, endemism, rarity, and extinction in the Mediterranean area: an analysis based on the published volumes of Med-Checklist. *Botanica Chronika,* 10: 63-79.

Gustafsson, M., 1999.Brassica oleracea and its wild relatives. Diversity and in situ conservation. *Botanica Lithuanica, Supplementum* 2: 53-60.

Harlan J.R., 1995. *The living fields: our agricultural heritage*. pp. 1-271. Cambridge University Press, Cambridge U.K..

Hawkes, J.G., 1993. *The diversity of crop plants*. Harvard University Press, Cambridge, Mass.

Hawkes, J.G., 1991. International workshop on dynamic *in situ* conservation of wild relatives of major cultivated plants: summary of final discussion and recommendations. *Israel Journal of Botany*, 40, 529-536.

Heywood, V.H., 1995.The Mediterranean flora in the context of world diversity. *Ecologia Mediterranea*, 21: 11-18.

Hoyt, E., 1988. *Conserving the wild relatives of crops*. IBPGR / IUCN / WWF, Rome, Italy.

ICARDA, 1997. *Conservation, Management and Sustainable Use of Dryland Biodiversity within Priority Agro-Ecosystems of the Near East. Highlights of the Proceeding of the Workshop*. 54 pp. ICARDA, Aleppo.

IUCN, 1994. *Guidelines for Protected Area Management Categories*. WCMC and IUCN Commission on National Parks and Protected Areas, Gland.

IUCN, 1998. United Nations list of protected areas. IUCN, Gland.

Karagoz, A., 1998. *In situ* conservation of plant genetic resources in the Ceylanpinar State Farm. In: *The Proceedings of International Symposium on* In Situ *Conservation of Plant Genetic Diversity* (eds. Zencirci, N., Kaya, Z., Anikster, Y. and Adams, W.T.). pp. 87-93. Central Research Institute for Field Crops, Ankara, Turkey.

Kitiki, A. and Tan, A., 1998. Vegetation survey in Southern part of Anatolian Diagonal and possible gene management zones for wild crop relatives. In: *The Proceedings of International Symposium on* In Situ *Conservation of Plant Genetic Diversity* (eds. Zencirci, N., Kaya, Z., Anikster, Y. and Adams, W.T.). pp. 129-134. Central Research Institute for Field Crops, Ankara, Turkey.

MAB, 2000. www.unesco.org/mab/brlist.htm UNESCO, Paris.

Maxted, N., Ford-Lloyd, B.V. and Hawkes, J.G., 1997a. Complementary Conservation Strategies. In: *Plant genetic conservation: the* in situ *approach* (eds. Maxted, N, Ford-Lloyd, B.V. and Hawkes, J.G.). pp. 20-55. Chapman & Hall, London.

Maxted, N., Hawkes, J.G. Ford-Lloyd, B.V. and Williams, J.T., 1997b. A Practical Model for *In Situ* Genetic Conservation. In: *Plant genetic conservation: the* in situ *approach* (eds. Maxted, N, Ford-Lloyd, B.V. and Hawkes, J.G.). pp. 339-367. Chapman & Hall, London.

Quézel, P., 1985. Definition of the Mediterranean region and the origin of its flora. In: *Plant conservation in the Mediterranean area* (ed. Gómez-Campo, C.) W. Junk, Dordrecht.

Tan, A., 1998. Current status of plant genetic resources conservation in Turkey. In: *The Proceedings of International Symposium on* In Situ *Conservation of Plant Genetic Diversity* (eds. Zencirci, N., Kaya, Z., Anikster, Y. and Adams, W.T.). pp. 5-16. Central Research Institute for Field Crops, Ankara, Turkey.

Tan, A. and Tan, A.S., 2000. *In situ* conservation of wild species: the case of Turkey. *SAT21 Conference*, Kuala Lumpur, Malaysia.

Tan, A. and Ulubelda, M., 1998. Selection criteria and planing of gene management zones (GMZs) for in situ conservation. In: *The Proceedings of International Symposium on* In Situ *Conservation of Plant Genetic Diversity* (eds. Zencirci, N., Kaya, Z., Anikster, Y. and Adams, W.T.). pp. 363-373. Central Research Institute for Field Crops, Ankara, Turkey.

UNESCO, 1996. *Biosphere reserves: the Seville Strategy and the Statutory Framework of the World Network*. UNESCO, Paris.

Weidenhammer, S. 1996. *Criteria for designation and evaluation of UNESCO biosphere reserves in Germany*. German National Committee for the UNESCO Man and the Biosphere Programme, Bonn.

Section 4

Utilisation of Legumes in the Mediterranean

Chapter 17

Utilisation Of Grain Legume Diversity

William Erskine, Kaddambot Siddique, Tanveer Khan and Wallace Cowling

17.1 INTRODUCTION

The most important cool season grain legumes with a Mediterranean origin are chickpea (*Cicer arietinum* L.), faba bean (*Vicia faba* L.), grasspea (*Lathyrus sativus* L.), lentil (*Lens culinaris* subsp. *culinaris* Medikus), lupins (*Lupinus* sp.) and field pea (*Pisum sativum* L.). Their association with *Rhizobium* and the resulting biological nitrogen fixation reduces the demand for inorganic nitrogen fertiliser in the cropping system and produces high protein seed. As a result, they are widely grown as break crops in cereal-based systems. The uses of these grain legumes in the diet and as feed are highly diverse. Equally diverse, as the preceding chapters illustrate, is the Mediterranean environment, within which the ancestors of the various grain legume crops are found in different, often allopathic, ecological habitats. This diversity in habitat has led to contrasting adaptation among species and to a suite of different legume options for farming systems. Partly as a consequence, the cool-season grain legume crops have spread widely from their Mediterranean origins. Some crops, such as lupins, are of major importance in other areas with a Mediterranean climate such as Southern Australia. Others are important in markedly different climates such as chickpea and lentil in the Indian sub-continent and faba bean in China. The field pea is a gift from the Near East to the rest of the world, as dry pea is now barely consumed in West Asia.

Despite the diversity, the genetic resources of these grain legumes are under-exploited, and this aspect is briefly covered in this chapter.

17.2 CHICKPEA

17.2.1 Uses in Food and Feed

Several reviews of the chemical composition and nutritional quality of chickpea have been published (Williams and Singh, 1987, Bressani and Elias, 1988). Chickpea is a good source of carbohydrate and protein, with protein ranging from 13 - 31%, but it is generally deficient in the sulphur-containing amino acids. It is relatively free from anti-nutritional factors, has high protein digestibility, and contains more phosphorus and calcium than other pulses. Although primarily used for human consumption, chickpea may be a useful protein and energy source for ruminant and monogastric animals, and the stubble after harvest has good forage value.

The uses of chickpea in the human diet are highly diverse and major preparations have been reviewed (Pushpamma and Geervani, 1987, Jambunathan and Singh, 1990). The Desi chickpea is mostly consumed in the Indian subcontinent in the form of whole seed, *dhal* (decorticated split cotyledons) or as flour (*besan*). Kabuli chickpea is mostly consumed as whole seed around the Mediterranean Sea in a ground paste (*hommos-biteheneh)* and fried pattie *(falafel)*. Traditional processing practices such as soaking, sprouting, fermenting, boiling, roasting, parching and frying are still used, while canning of kabuli chickpea and dehydration of immature green chickpeas are becoming more popular.

17.2.2 Exploitation of Genetic Diversity in Breeding

The utilisation of chickpea germplasm gained momentum with the establishment of the crop improvement programs at the International Crops Research Institute for the Semi-Arid Tropics (ICRISAT) in 1972 and the International Center for Agricultural Research in Dry Areas (ICARDA) in 1977. Both centres have collected, conserved and evaluated large germplasm collections for botanical and agronomic traits and reactions to stresses (Singh *et al.*, 1983, Pundir *et al.*, 1988, Singh *et al.*, 1991). Some germplasm accessions were directly released as cultivars by national programs while others are used as donor parents (Singh *et al.*, 1997, Siddique *et al.*, 1997).

Among more than 19,000 accessions screened at ICARDA, 14 were identified as resistant to ascochyta blight (*Ascochyta rabiei* (Pass.) Lab.), a major limitation to chickpea production in West Asia and North Africa (WANA) (Singh and Reddy, 1993). Use of these resistant genotypes in national breeding programs resulted in yield increases up to 75% (Nene and Reddy, 1987, ICARDA, 1994). A total of 5,781 accessions were screened for freezing tolerance at ICARDA and 16 were rated as resistant to cold, a major stress for winter-sown chickpea in WANA (Singh *et al.*, 1989, Singh, 1993). Traditionally, ascochyta blight and cold damage were avoided by sowing in spring. Combined resistance to ascochyta blight and cold allows the replacement of spring sowing by winter sowing, increasing the efficiency of use of winter rainfall and yield.

Screening for drought resistance at ICRISAT and ICARDA has identified several lines with drought escape or drought resistance (van Rheenen, 1991, Saxena, 1987, Singh *et al.*, 1996). Sources of resistance to fusarium wilt caused by *Fusarium oxysporum* Schlecht. emend Snyd. and Hans. f.sp. *ciceri* (Padwick) Snyd. and Hans. have been identified (Singh *et al.*, 1997). Tolerance to pod borer (*Helicoverpa* sp.) and leaf miner (*Liriomyza cicerina* Rondani) have also been identified (Reed *et al.*, 1987, Lateef and Pimbert, 1990, Singh and Weigand, 1996). Collections of wild annual *Cicer* species have been characterised (Pundir *et al.*, 1988, Robertson *et al.*, 1995). The introgression of genes from C. *echinospermum* Davis and C. *reticulatum* Lad. into the cultigen produced several F_7 lines which out-yielded the locally adapted parent by up to 39% (Singh and Ocampo, 1997). Resistance to the cyst nematode (*Heterodera ciceri* Volvas, Greco and Di Vito) has been transferred from C. *reticulatum* to the cultigen (ICARDA, 1997).

17.2.3 Research Needs for Utilisation

Although wild *Cicer* species have been extensively evaluated, their exploitation in breeding programs has been limited to yield and cyst nematode resistance. The wild progenitor of chickpea, *C. reticulatum* shows resistance to several other stresses (Singh *et al.*, 1997), and the introgression of additional genes from wild to cultivated is merited. Molecular markers for resistance to fusarium wilt, ascochyta blight, cold tolerance and *Helicoverpa* resistance are under development (Muehlbauer *et al.*, 1990, Kahl *et al.*, 1994, ICARDA, 1994, Lawlor *et al.*, 1998, Mayer *et al.*, 1997) and their practical application in breeding programs deserves greater attention. Germplasm enhancement through the transfer of desirable genes to an adapted background genotype to develop new and efficient plant types with improved biomass and seed yield potential needs future emphasis. The introgression of genes from incompatible *Cicer* species, especially *C. bijugum* Rech., *C. judaicum* Boiss. and *C. pinnatifidum* Jaub. and Sp., is highly desirable and ovule rescue and *in vitro* culture may be employed to this end. Chickpea has been transformed using recombinant DNA techniques and tissue culture procedures (Schroeder *et al.*, 1997) opening the way to introduce a variety of alien genes. Further effort is required to identify desirable genes in either cultivated or wild backgrounds for transfer to adapted cultivars by conventional and/or biotechnological approaches.

17.3 Faba Bean

17.3.1 Uses in Food and Feed

A review of the chemical composition and nutritive quality of faba bean is to be found in Hulse (1994). The protein content of faba bean ranges from 19 to 39 %; however seeds are low in sulphur amino acids, as in the other pulses. Faba bean contains anti-nutritional factors, the most important of which are the glycosides, convicine and vicine. The glycosides can cause a haemolytic anemia, which is sex-linked and is often fatal when young males first eat faba bean (Marquadt, 1982). The main cause of the disease has been traced to a deficiency in glucose-6-phosphate dehydrogenase.

For human consumption, faba beans are commonly consumed as a green salad and vegetable, in addition to preparations of dry seed (Hawtin and Sears 1993). The dry seeds of faba beans are used whole in dishes such as *cous-cous* and *foul medames* or as a flour in *falafel*. In China, faba bean seeds are important for processing and used in many food and industrial products. The main use of faba bean seed in Europe is as an animal feed protein supplement.

17.3.2 Exploitation of Genetic Diversity in Breeding

Despite being among the most cold-tolerant major grain legumes, frost resistance is a limiting factor to faba bean production in many regions. Frost tolerance varies widely in faba bean germplasm, reaching a maximum in the French cultivar Côte d'Or which can tolerate -25 °C without snow protection (Picard *et al.*, 1985).

Among biotic stresses to faba bean the diseases, chocolate spot, caused by *Botrytis fabae* Sard. and *B. cinerea* Pers. Ex Fr., Ascochyta blight (*Ascochyta fabae* Speg.), rust (*Uromyces fabae* (Pers.) de Bary) and foot and root rot (*Fusarium* and *Phoma* sp.) limit yields. Resistance to the major pathogens in faba bean was reviewed by Porta-Puglia *et al.* (1994). Resistance to *Botrytis* was found in accessions from South America and is now widely used in breeding programs. Sources of resistance have been reported for *Ascochyta fabae* and rust. Resistance has also been reported for the fungal pathogens *Fusarium*, *Uromyces* and *Peronospora.*

A wide range of viruses may also cause significant yield loss in faba bean. Many are spread by aphids (especially by *Aphis craccivora* Koch and *Acyrthosiphon pisum* Harris) though *Aphis fabae* Scop. causes more damage by direct feeding. The parasitic angiosperm broomrape (*Orobanche* spp., particularly *O. crenata* Forsk.) causes serious losses and restricts the cultivation of faba bean in the Mediterranean region. Resistance has been reported for the plant parasite *O. crenata* and the viruses BYMV (bean yellow mosaic virus) and BLRV (bean leaf roll virus). Resistance is also known to aphids (Bond and Lowe, 1975). Cultivars Ascot, Rastatt and derived lines, for example, carry partial resistance to *A. fabae*.

There are adequate existing sources of variation for the anti-nutritional factors which limit the use of faba beans in some circumstances. Alleles at either of two different loci prevent tannin synthesis. Similarly, single recessive alleles greatly reduce the levels of vicine and convicine in the seed (Ramsay and Griffiths, 1993).

17.3.3 Research Needs for Utilisation

The requirement for adequate water supply restricts the cultivation of faba beans within dry rain-fed areas, but makes it particularly suited to temperate regions, cool and rainy seasons in warmer areas, high altitudes in sub-tropical regions or irrigated production systems. Few studies have been made on sources of drought tolerance in faba bean and sources of improved water-use efficiency are needed (Bond *et al.*, 1994).

Salinity is an increasing problem because of the extensive use of irrigation for faba bean production. The few studies on salt tolerance in faba bean (reviewed by Bond *et al.*, 1994) suggest that some variation exists for this trait and that further germplasm evaluation is merited.

Faba beans are genetically isolated from other taxa. Although related species contain a range of useful characters to supplement those present in existing germplasm within *V. faba* (*V. narbonensis* L., for example, is both strongly resistant to *A. fabae* and drought), this variation cannot be accessed through sexual crosses. Despite improvements to embryo and ovule culture methodology (Lazaridou *et al.*, 1993), interspecific hybrids are still lacking. A recent advance in the protoplast culture of faba beans opens a new route to interspecific hybrids (Tegeder *et al.*, 1995), but some doubt must remain as to the prospects for the further use of any interspecific hybrids as the genetic distance between *V. faba* and related species is large.

Given this isolation, gene transformation technology to introduce alien genes is especially important for faba bean. However, it still remains recalcitrant (Schroeder *et al.*, 1997). In the absence of a transformation system for *V. faba*, *V. narbonensis* has been used as a model to test applications of transformation technology. An efficient transformation system based on *Agrobacterium*-mediated transformation of epicotyl explants was used to introduce the Brazil nut 2S globulin into *V. narbonensis* (Saalbach *et al.*, 1995).

17.4 GRASSPEA

17.4.1 Uses in Food and Feed

The *Lathyrus* species with greatest economic importance are *L. sativus* L., *L. cicera* L. and *L. ochrus* (L.) DC. *L. sativus*, also called grasspea, khesari, sabberi or common chickling, is grown widely in Bangladesh, China, Ethiopia, India, Nepal and Pakistan for green feed, hay and grain production for both animal and human consumption. Some countries (e.g. India) have banned its cultivation because it contains the neurotoxin β-N-oxalyl-L-α-β-diaminopropionic acid (ODAP). This toxin has been implicated as the cause of a paralysis of the lower limbs known as lathyrism when the seed of *L. sativus* is consumed as a major part of the diet over 3-4 months (Ross and Spencer, 1989). In common with many grain legumes, *Lathyrus* seed contains trypsin and chymotrypsin inhibitors, but is high in protein (25-30%), rich in lysine (Sardana, 1992) and is palatable and nutritious. *Lathyrus sativus* is mainly consumed as a split product *dhal* in South Asia, whereas elsewhere the boiled pods and roasted seeds are also consumed. Flour made from the seed is used as a component of a batter to make snacks or flat bread. Most of the ODAP toxin is removed when the seed is soaked and boiled, but it is not lost when the seed is ground into flour and cooked (Pushpamma, 1989).

Lathyrus cicera and *L. ochrus* are grown primarily as stock feed for both forage and grain. *Lathyrus cicera*, commonly called dwarf chickling, is grown in West Asia, North Africa and South America. It is used for human consumption to a small extent in South America. *Lathyrus ochrus*, also known as ochrus chickling or Cyprus vetch, is grown in southern Europe.

As with most legumes, these three species improve soil fertility and increase the yield and protein content of succeeding cereal crops. They are under consideration as green manure crops in Australia and Canada, because of their large biomass (Biederbeck *et al.*, 1993, Siddique *et al.*, 1996).

17.4.2 Exploitation of Genetic Diversity in Breeding

Although previously neglected, several institutions have recently made substantial progress in collecting, conserving and characterising important *Lathyrus* species (Sarker *et al.*, 1997, Campbell *et al.*, 1994, Cambell, 1997, Arora *et al.*, 1996, Hanbury *et al.* 1995). These collections are being widely utilised to develop cultivars with a low neurotoxin content in the seed and good biomass and grain production. There are cultivars registered in Australia, Bangladesh, Canada and India (Mehra *et al.*, 1996, Malek *et al.*, 1997, Campbell *et al.*, 1994, Hanbury and Siddique, unpublished data). Hybridisation has been used to transfer useful traits from germplasm accessions to locally adapted genotypes (Campbell *et al.*, 1994). *In vitro* culture has produced useful somaclonal variants in *L. sativus* with low to nil ODAP content and cultivars with improved adaptation and seed yield from this source are being considered for commercial release in India (Mehta and Santha, 1996).

17.4.3 Future Utilisation

Future exploitation of *Lathyrus* sp. for human consumption, particularly *L. sativus*, will depend entirely on the elimination or reduction of the neurotoxin from the seed. The ODAP concentration differs widely across environments, genotypes and species, with *L. cicera* showing the lowest mean ODAP content in the seed followed by *L. sativus* and *L. ochrus* (Hanbury *et al.*, 1995, Siddique *et al.*, 1996, Robertson and Abd El Moneim, 1996). Other antinutritional factors such as tannin and trypsin inhibitors require more emphasis. Early growth and increased final biomass production also deserves attention because of an increasing use for forage and hay production. Resistance to the diseases powdery mildew (*Erysiphe polygoni* DC) and downy mildew (*Peronospora lathi-pulustris* Gaumann) and various insect pests will be required in some environments.

Primary, secondary and tertiary gene pools will increasingly contribute to the improvement of *Lathyrus* sp. Yunus and Jackson (1991) showed that interspecific hybridisation is possible. A toxin-free gene has been identified in *L. tingitanus* that is being used to develop toxin-free *L. sativus* cultivars in China (Zhou and Arora, 1996). Similarly, zero ODAP *L. ciliolatus* has been crossed with *L. sativus* at ICARDA resulting in progeny with very low ODAP in the *L. sativus* background (Abd El Moneim, pers. comm.). Clearly biotechnology combined with interspecific hybridisation should play a major role in reducing the ODAP content and improving biomass and seed yields.

17.5 LENTIL

17.5.1 Uses in Food and Feed

Lentil is a popular human food. In West Asia the straw is also important as a feed for small ruminants. The nutrient content, protein concentration and amino acid composition of lentil seed is similar to that of other grain legumes (Savage, 1988)). Lentil contains a number of anti-nutritional factors, but these are unimportant in human diets because they are heat labile and deactivated by cooking prior to eating. This, together with the short cooking time of lentil in comparison to most other pulses, makes lentil very suitable for human consumption. The chief uses of lentil are summarized in Robertson and Erskine (1997) and the most common lentil-based foods such as dhal in South Asia and as soup and mujadarah in West Asia are detailed in Hawtin and Sears (1993).

17.5.2 Exploitation of Genetic Diversity in Breeding

The major abiotic limiting factors to lentil production are low moisture availability and high temperature stress in spring, and, at high elevations, cold temperatures in winter. Among biotic stresses, the diseases rust, vascular wilt and Ascochyta blight, caused by *Uromyces fabae* (Pers.) Schroet., *Fusarium oxysporum* Schlecht. emend. Snyd. and Hans. *f.*sp. *lentis* and *Ascochyta lentis* Bond. and Vassil., respectively, are the key pathogens of lentil. Other constraints to productivity include agronomic problems of pod dehiscence and lodging in lentil. Adequate variability for these traits is known within the genetic resources of the crop, allowing manipulation through plant breeding (Erskine *et al.*, 1994).

Improvement of the autogamous lentil started with the collection and selection of land races. Hybridisation to recombine the above traits has been followed by breeding with pedigree, modified bulk and single seed descent methods (Muehlbauer and Slinkard, 1981).

The wild *Lens* collection has been evaluated for agronomic characters and resistance to key stresses. No striking variation has been found for grain or straw yield within wild Lens. High levels of winter-hardiness were found within *L. culinaris* subsp. *orientalis* (Boiss.) Ponert (Hamdi *et al.*, 1996). The wild subspecies of *Lens* were found to be more drought resistant than the cultigen with low relative reduction in yield with drought stress (Hamdi and Erskine, 1997). Resistance to vascular wilt and Ascochyta blight has also been found in wild *Lens* (Bayaa *et al.*, 1994, 1995).

With the existing variability, prospects exist for substantial gains in sustainable production of lentils arising from the integration of a change in agronomic practice with a new cultivar as the following examples show:

- In highland West Asia, lentil is currently spring-sown. However, sowing in late autumn/early winter coupled with the use of a winter-hardy cultivar gives yield advantages of approximately 50% (Sakar *et al.*, 1988). The incorporation of winter-hardiness from both wild and cultivated sources can be expected to augment the area and production in this agro-ecological zone.

- In some Mediterranean areas, such as South Syria, lentil production has been abandoned because of vascular wilt disease. New cultivars with wilt resistance in Lebanon and Syria offer improved prospects for such areas. The continued spread of lentil harvest mechanization technology, exploiting genetic variation in standing ability and plant height, will also contribute to increased lentil production in lowland West Asia in the future.

- In South Asia, large areas are left fallow over winter following the harvest of late paddy rice. For such a niche farmers need early-maturing, disease-resistant varieties with late sowing potential. As an extra-early maturing cultivars with combined resistance to diseases become available, a major expansion of lentil production in the rice-based cultivation system is expected.

17.5.3 Research Needs for Utilisation

Among limitations to the crop, several important traits, discussed below, are not currently addressable by breeding because of insufficient genetic variation. The crop has a low biomass yield, restricting its potential in high input systems and its competitiveness against weeds. The variability for biomass within the cultivated lentil and wild *Lens* is limited. Wide crossing of lentil with *Vicia* sp. using embryo rescue should be explored for increased biomass. Lentils are commonly subject to drought; genetic means to alleviate drought are required to supplement drought escape through early flowering. In the Mediterranean region, unsolved key problems are the control of the pea leaf weevil (*Sitona crinitus* Herbst) and the parasitic weed broomrape (*Orobanche* sp.). Genetic variation in reaction has not been found, precluding host plant resistance as a method of control. Crosses have been made between incompatible *Lens* sp. using embryo rescue (Ladizinsky *et al.*, 1985, Ahmed *et al.*, 1995), but scope exists to improve the efficiency of inter-specific hybridisation in *Lens* to increase inter-specific gene transfer. Alien genes, to overcome some of the above limitations, may now be introduced into lentil with the recent development of a transformation system (Barton *et al.*, 1997).

17.6 LUPINS

17.6.1 Uses in Food and Feed

Lupins have a long history of utilisation starting with the ancient Greeks who grew the albus lupin (*L. albus* L.) and soaked the seed to debitter them prior to consumption by stock or humans (Gladstones, 1970; Hondelmann, 1984, 1996). Today the debittered seeds of *L. albus* are an important snack food in the Mediterranean region. A parallel development occurred in South America with the Andean lupin (*L. mutabilis* Sweet) and an early Spanish colonialist remarked on the similarity of the Andean lupin to and 'the lupins we have in Spain' (Hondelmann, 1984).

However, the potential of sweet lupins in food and feed was only realised with the advent of the first sweet cultivars in the 1930s in Germany. Sweet lupin seed is high in protein and dietary fibre, and has low levels of antinutritional factors (Petterson *et al.*, 1997). Many lupin food products are now available commercially in Australia (Nottage and Wallace, 1994, Petterson *et al.*, 1997). Lupin seed is a high quality alternative to soybean in fermented foods (Kyle, 1994, Petterson, 1998, Petterson *et al.*, 1997).

The main end-use for over 1 million tonnes of Australian sweet lupins (*L. angustifolius* L.) produced annually, is as feed for monogastric and ruminant animals (ABARE, 1994). Van Barneveld and Hughes (1994) highlight the potential for use of Australian sweet lupins in the diets of pigs and poultry, and they are also used for dairy and beef cattle (Hough and Jacobs, 1994), sheep (Murray, 1994) and fish (Jenkins *et al.*, 1994).

17.6.2 Exploitation of Genetic Diversity in Breeding

Exploitation of natural genetic diversity in lupins has resulted in major breeding improvements over the past 50 years (Cowling *et al.*, 1998). Often, large improvements have been obtained by matching the phenotype to the environment. For example, the *Ku* gene in *L. angustifolius* removed the vernalisation requirement for flowering, adapting the crop to the short season environment of South-western Australia (Gladstones and Hill, 1969). Similarly, thermoneutral types of *L. luteus* L. with early maturity have improved yields in northern Europe. In France and the UK, there are large improvements in the yield of *L. albus* L. with dwarf, determinate, and vernalising types which are autumn-sown and resistant to freezing damage (Huyghe *et al.*, 1994).

Valuable traits are often found in wild and landrace relatives of lupins. For example, landraces of *L. albus* with: frost tolerance were found in Italy (J. Papineau and C. Huyghe, 1993 unpublished results), with resistance to Pleiocheata root rot in the Azores (Sweetingham *et al.*, 1994) and with resistance to anthracnose in Ethiopia (B.J. Buirchell, H. Yang, and M.W. Sweetingham, Western Australia, 1997, unpublished data). Resistance to grey leaf spot and anthracnose and frost tolerance were found in crosses of *L. angustifolius* with a Portuguese wild type (Wells and Forbes, 1982). Resistance to Phomopsis stem blight in *L. angustifolius* was found in Mediterranean accessions (Cowling *et al.*, 1987). Crosses between *L. angustifolius* and wild types from Spain and Morocco resulted in Phomopsis-resistant cultivars (Gladstones, 1989, 1994). Other crosses with wild *L. angustifolius* resulted in partial resistant to brown spot and root rot caused by *Pleiochaeta setosa* (Kirchn.) Hughes (Cowling *et al.*, 1997).

In the yellow lupin (*L. luteus*) wild relatives have also been important. Resistance to Fusarium wilt, controlled by a dominant gene *Fus1*, was found in wild Portuguese *L. luteus* (Lamberts, 1952; Hackbarth and Troll, 1956). This resistance gene has been widely used in Germany and Poland (Cowling *et al.*, 1998).

Mutagenesis has also been used in lupin breeding (Micke and Swiecicki, 1988) but has produced few cultivars. In the Ukraine, mutants from an *L. albus* landrace from Georgia in the 1950s contributed to the release of two cultivars (Golovchenko, 1982) and in Moscow an early maturing mutant of *L. albus* led to a low alkaloid cultivar (Vavilov and Gataulina, 1984).

17.6.3 Future Prospects and Research Needs

As lupins were fully domesticated relatively recently, they have a narrower cultivated gene pool than other grain legumes. Consequently, broadening the genetic base of cultivated lupins through wide crosses will continue to be a major feature of breeding. Research is now needed to identify the best wild parents for crossing. The greatest challenge to lupin breeding in the 1990s is to improve resistance to anthracnose disease, because a highly virulent strain has spread in seed around the world since the late 1970s.

The prospects for improving the adaptation of lupins to modern agricultural systems through breeding are immense. New plant types are being explored to improve harvest index and harvestability. Lupin architecture is significantly modified by two major genetic systems: restriction of branching (determinacy) and dwarfism (Huyghe *et al.*, 1994). Additionally, breeding and research needs for lupins include:

- abiotic stress tolerances such as lime and frost in *L. albus* and frost in *L. luteus*,
- improved yield potential in sweet domesticated lupins,
- Increased proportion of seed in lupin pods,
- higher digestible energy in lupin feed,
- stable incorporation through transformation of herbicide tolerance and virus resistance,
- development of doubled haploid system.

17.7. FIELD PEA

17.7.1 Uses in food and feed

Field pea is one of the oldest domesticated food plants. It also produces one of the most versatile food items. The dry seed contains 19 - 27% protein, 1 - 2% fat and 30 - 50% soluble carbohydrate, plus it is relatively free of anti-nutritional factors (Petterson *et al.*, 1997). As a vegetable, its immature seed is canned, frozen or cooked fresh, as well as the whole pod of a variant, snow peas, are cooked or eaten raw in salads. Dry seed is used whole, split as dhal or ground into flour. Pea sprouts and re-hydrated seed for canning are other uses. Peas are commonly used in pig and poultry rations as well as for feeding ruminants, particularly as a forage crop (Petterson *et al.*, 1997). As a grain legume, it also improves the yield and protein content of the following cereal crop through residual nitrogen (Rowland *et al.*, 1994).

17.7.2 Exploitation of Genetic Diversity in Breeding

P. sativum has been divided into six sub-species, that may have originated in the Near East (Zeven and Zhukovsky, 1975). Ecotypes of Turkey and Syria are believed to be the precursors of the domesticated types (Ben-Ze'ev and Zohary, 1973). A distinct species *P. fulvum* Sibth. and Sm., partially cross-fertile with *P. sativum*, is a source of useful variation for disease and insect pest resistance (Hardie *et al.*, 1995; Wroth, 1996).

The pea genome is an immense source of variation (Marx, 1985). Although early breeding efforts did not fully exploit the wealth of genetic information available in the species, major breeding programs in Australasia, Europe and North America may provide the opportunity for rare gene combinations to occur. Pedigree and bulk-pedigree methods, early generation yield testing as well as single seed descent have been variously used. The pedigree system is extensively used in Europe, where earlier generations are used to discard plants with colored flowers; Its effectiveness appears to be limited due to the low efficiency of selection in early generations (Stelling and Ebmeyer, 1990), partly due to heterosis (Sarawat *et al.*, 1994). Bulk-pedigree method is used to breed for high biomass in the low rainfall areas of Western Australia (Khan and French, 1996). Although male sterility is known (Myers and Gritton, 1988), it has not been used to develop outcrossing populations.

Much recent attention has focussed on restructuring the pea plant using *afila (af)* gene that converts leaflets into tendrils (Marx, 1977). Another gene that reduces the size of stipule (*st*) and produces an almost leafless plant (afafstst) is now considered to be of little practical value. The semi-leafless plants (afafStSt) intertwine with neighboring plants providing mutual support. This character was initially claimed to improve standing ability, reduce disease severity and improve drought resistance. Although, recent studies cast doubt on some of these assertions (Eteve, 1985; Heath and Hebblethwaite, 1992; Martin *et al.*, 1992) considerable standing ability can be achieved when the *af* gene is combined with improved stem strength and high branching ability. A significant number of semi-leafless cultivars are now grown worldwide. Stelling (1994), however, cautioned that the semi-leafless plant is inherently yield limited and urged breeders to continue selecting for high yielding normal leaf type plants with an acceptable level of standing ability.

In a review of the use of disease and pest resistance genes, Muehlbauer (1987) indicated that the greatest challenge facing breeders is resistance to black spot (*Mycosphaerella pinodes* (Berk. and Blox.) Vestergr.). Resistance to black spot appears to be quantitative with a low heritability (Wroth, 1996) and breeding efforts so far have not been very effective. The search for more diverse sources of resistance and large-scale recombination programs are in progress in Australia and New Zealand. Timmerman *et al.* (1993, 1994) who developed molecular markers for other diseases, are now studying similar possibilities for black spot resistance.

Jaronowski and Micke (1985) reviewing the enrichment of the pea gene pool through mutagenesis described it as a well functioning tool for breeding. More recently, genetic transformation has added increased diversity to the pea genome (Schroeder *et al.*, 1993, 1995, 1997).

17.7.3 Future Prospects and Research Needs

Pea as a field crop is now receiving unprecedented attention from breeders and there is an ever increasing demand for a wide variety of genes to extend adaptation to new areas and improve quality and disease and insect pest resistance. There is an urgent need to undertake germplasm screening similar to that done on other grain legumes at the International Institutes in order to support the breeding programs. There is also a need to explore and collect more primitive forms from areas such as Afghanistan and Ethiopia.

REFERENCES

ABARE, 1994. *Grains Statistics: A Database for the Grains Research and Development Corporation.* Australian Bureau of Agricultural and Resource Economics, Canberra, Australia.

Ahmad M., Fautrier, A.G., McNeil, D.L., Burritt, D.J. and Hill, G.D., 1995. Attempts to overcome post-fertilization barrier in interspecific crosses in the genus Lens. *Plant Breeding*, 114: 558-560.

Arora, R.K., Mathur, P.N., Riley, K.W. and Adham, Y., 1996. *Lathyrus* genetic resources in Asia. In: *Proceedings of a regional workshop*, Raipur, India 27-29 December, 1995. pp 167. Indira Gandhi Agricultural University, Raipur, Raipur, India. IPGRI Office for South Asia, New Delhi, India.

Barneveld, van, R.J. and Hughes, R.J., 1994. The nutritive value of lupins for pigs and poultry. In: *Proceedings of the First Australian Lupin Technical Symposium* (eds. Dracup, M. and Palta, J.). pp. 49-57. Department of Agriculture Western Australia, South Perth.

Barton, J.E., Klyne, A., Tennakoon, D., Francis, C. and Hamblin, J., 1997. Development of a system for gene transfer to lentils. In: *International Food Legume Research Conference III B Program and Posters.* Adelaide, Australia (Abstract). pp 85.

Bayaa, B., Erskine, W. and Hamdi, A., 1994. Response of wild lentil to *Ascochyta fabae* f.sp. *lentis* from Syria. *Genetic Resources and Crop Evolution*, 41: 61-65.

Bayaa, B., Erskine, W. and Hamdi, A., 1995. Evaluation of a wild lentil collection for resistance to vascular wilt. *Genetic Resources and Crop Evolution*, 42: 231-235.

Biederbeck, V.O., Bouman, O.T., Looman, J., Slinkard, A.E., Bailey, L.D., Rice, W.A. and Janzen, H.H., 1993. Productivity of four annual legumes as green manure crop in dryland cropping systems. *Journal of Agronomy*, 85: 103-143.

Ben-Ze'ev, N. and Zohary, D. 1973. Studies in relationship in the one *Pisum* L. I. *Journal of Botany*, 22: 73-91.

Bond, D.A. and Lowe, H.J.B., 1975. Tests for resistance to *Aphis fabae* in field beans (*Vicia faba* L) *Annals of Applied Biology*, 81: 21-32.

Bond, D.A., Jellis, G.J., Rowland, C.G., Le Guen, J., Robertson, L.D., Khalil, S.A. and Li-Juan, L., 1994. Present status and future strategy in breeding faba beans (*Vicia faba* L.) for resistance to biotic and abiotic stresses. In: *Expanding the Production and Use of Cool-Season Food Legumes* (eds. Muehlbauer F.J. and Kaiser, W.J.). pp 592-616 Kluwer Academic Publishers, the Netherlands.

Bressani, R. and Elias, L.G., 1988. Seed quality and nutritional goals in pea, lentil, faba bean and chickpea breeding. In: *World crops: Cool season food legumes* (ed. Summerfield R.J.). pp. 381-404. Kluwer Academic Publishers, Dordrecht, Netherlands.

Campbell, G.C.,1997. *Grasspea.* Lathyrus sativus *L.* Promoting the conservation and use of underutilized and neglected crops. 18. Institute of Plant Genetics and Crop Plant research, Gatersleben. International Plant Genetic Resources Institute Rome, Italy.

Campbell, G.C., Mehra, R.B., Agrawal, S.K., Chen, Y.Z., Abd El Moneim, A.M., Khawaja, H.I.T, Yadav, C.R., Tay, J.U. and Araya, W.A., 1994. Current status and future breeding strategy in grasspea (*Lathyrus sativus* L.). *Euphytica*, 73: 167-75.

Cowling, W.A., Hamblin, J., Wood, P.R. and Gladstones, J.S., 1987. Resistance to Phomopsis stem blight in *Lupinus angustifolius* L. *Crop Science*, 27: 648-652.

Cowling, W.A., Huyghe, C. and Swiecicki, W., 1998. Lupin breeding. In: *Lupins as Crop Plants: Biology, Production and Utilisation* (eds. Gladstones, J.S., Hamblin, J. and Atkins, C.). CABI, Wallingford.

Cowling, W.A., Sweetingham, M.W., Diepeveen, D. and Cullis, B.R.,1997. Heritability of resistance to brown spot and root rot of narrow-leafed lupins caused by *Pleiochaeta setosa* (Kirchn.) Hughes in field experiments. *Plant Breeding*, 116: 341-345.

Erskine, W., Tufail, M., Russell, A., Tyagi, M.C., Rahman, M.M. and Saxena, M.C., 1994. Current and future strategies in breeding lentil for resistance to biotic and abiotic stresses. *Euphytica*, 73: 127-135.

Eteve, G., 1985. Breeding for cold tolerance and winter hardiness in pea. In: *The Pea Crop, A Basis for Improvement* (eds. Hebblethwaite, P.D., Heath, M.C. and Dawkins, T.C.K.). pp 131-136. Butterworths, London.

Gladstones, J.S., 1970. Lupins as crop plants. *Field Crop Abstracts*, 23: 123-148.

Gladstones, J.S., 1989. Breeding Phomopsis-resistant lupins. *Journal of Agriculture Western Australia (4th Series)*, 30: 3-7.

Gladstones, J.S., 1994. An historical view of lupins in Australia. In: *Proceedings of the First Australian Lupin Technical Symposium* (eds. Dracup, M. and Palta, J.). pp. 1-38 Department of Agriculture Western Australia, South Perth.

Gladstones, J.S. and Hill, G.D., 1969. Selection for economic characters in *Lupinus angustifolius and L. digitatus*. 2. Time of flowering. *Australian. Journal of Experimental Agriculture and Animal Husbandry*, 9: 213-220.

Golovchenko, V.I., 1982. Mutagenesis in selection of alkaloidless varieties of white and yellow lupine. In: *Proceedings of the Second International Lupin Conference*. pp. 43-45. Publicaciones Agrarias, Ministry of Agriculture, Madrid, Spain.

Hackbarth, J. and Troll, H.-J., 1956. Lupins as grain legumes and fodder plants. In: *Handbook of Plant Breeding, Part IV* [in German] (eds. Kappert, H. and Rudorf, W.). pp. 1-51. 2nd edn. Verlag Paul Parey, Berlin and Hamburg.

Hamdi, A., and Erskine, W., 1997. Reaction of wild species of the genus *Lens* to drought. *Euphytica*, 91: 173-179.

Hamdi, A., Küsmenoglu, I. and Erskine, W.,1996. Sources of winter hardiness in wild lentil. *Genetic Resources and Crop Evolution*, 43: 63-67.

Hanbury, C.D., Sarker, A., Siddique, K.H.M. and Perry, M.W., 1995. *Evaluation of* Lathyrus *germplasm in a Mediterranean type environment in south-western Australia. Occasional Publication No.8.* Centre for Legumes in Mediterranean Agriculture, Perth.

Hardie, D.C., Baker, G.J. and Marshall, D.R., 1995. Field screening of Pisum accessions to evaluate their susceptibility to the pea weevil (Coleoptera: Bruchidae). *Euphytica*, 84: 155-161.

Hawtin, L. and Sears, L., 1993. *Legume cookbook: Cooking with chickpeas, faba beans and lentils.* ICARDA, Aleppo, Syria

Heath, M.C. and Hebblethwaite, P.D., 1992. Agronomy and physiology of leafed, leafless and semi-leafless peas. II. Plant density, standing ability and disease considerations. In: *Proceedings of First European Conference on Grain Legumes*. pp. 215-216. Anger, France.

Hondelmann, W., 1984. The lupin - ancient and modern crop plant. *Theoretical and Applied Genetics*, 68: 1-9.

Hondelmann, W., 1996. *The Lupins* [in German]. Landbauforschung Völkenrode, Sonderheft 162. Bundesforschungsanstalt für Landwirtschaft Braunschweig-Völkenrode (FAL), D-38116 Braunschweig, Germany.

Hough, G.M. and Jacobs, J.L., 1994. The use of lupins as a feed for dairy and beef cattle. In: *Proceedings of the First Australian Lupin Technical Symposium* (eds. Dracup, M. and Palta, J.). pp. 58-66. Department of Agriculture Western Australia, South Perth.

Hulse, J.H. 1994. Nature, composition and utilization of food legumes. In: *Expanding the Production and Use of Cool-Season Food Legumes* (eds. Muehlbauer, F.J. and Kaiser, W.J.) pp. 77-97.Kluwer Academic Publishers, the Netherlands.

Huyghe, C., Julier, B., Harzic, N. and Papineau, J., 1994. Breeding of *Lupinus albus*: New architectures for a further domestication. In: *Advances in Lupin Research* (eds. Neves Martins, J.M. and Beirão da Costa, L.). pp. 25-41. ISA Press, Lisbon,

ICARDA, 1994. *Legume Program Annual Report for 1993.* ICARDA, Aleppo, Syria.

ICARDA, 1997. ICARDA Annual Report 1996. ICARDA, Aleppo, Syria.

Jambunathan, R. and Singh, U., 1990. Present status and Prospects for utilisation of chickpea. In: *Chickpea in the Nineties* (van Rheenen, H.A. and Saxena, M.C.). pp. 41-46. International Crops Research Institute for the Semi-Arid Tropics, Pantancheru, India.

Jaronowski. J. and Micke, A.,1985. Mutation breeding in peas. *Mutation Breeding Review*, 2: 1-

23.

Jenkins, G.I., Waters, S.P., Hoxey, M.J. and Petterson, D.S., 1994. Lupin seed (*Lupinus angustifolius*) as an alternative to soybean meal in the diet of juvenile snapper (*Pagrus auratus*). In: *Proceedings of the First Australian Lupin Technical Symposium* (eds. Dracup, M. and Palta, J.). pp. 74-78. Department of Agriculture Western Australia, South Perth.

Kahl, G., Kaemmer, D., Weising, K., Kost, S., Weigand, F. and Saxena, M.C., 1994. The potential of gene technology and genome analysis for cool season food legume crops: theory and practice. In: *Expanding the Production and Use of Cool Season Food Legumes* (eds. Muehlbauer, F.J. and Kaiser, W.J.). pp. 705-725 Kluwer Academic Publishers, Dordrecht, The Netherlands.

Khan, T.N. and French, R.J., 1996. Breeding field peas for Western Australia: progress and problems. *Pisum Genetics*, 28: 5-12.

Kyle, W.S.A., 1994. The current and potential uses of lupins for human food. In: *Proceedings of the First Australian Lupin Technical Symposium* (eds. Dracup, M. and Palta, J.). pp. 89-97. Department of Agriculture Western Australia, South Perth.

Ladizinsky G, Cohen, D. and Muehlbauer, F.J., 1985. Hybridization in genus *Lens* by means of embryo culture. *Theoretical and Applied Genetics*, 70: 97-101.

Lamberts, H., 1952. Resistance to *Fusarium oxysporum* in yellow lupins. *Euphytica*, 1: 199-200.

Lateef, S.S. and Pimbert, M.P., 1990. The search for host-plant resistance to *Helicoverpa armigera* in chickpea and pigeonpea at ICRISAT In: *Summary proceedings of the First Consultative Group Meeting on the Host Selection Behavior of* Helicoverpa armigera, 5-7 March 1990, ICRISAT Centre India. pp 14-18 International Crops Research Institute for the Semi-Arid Tropics, Pantancheru, India.

Lawlor, H.J., Siddique, K.H.M., Sedgley, R.H. and Thurling, N., 1998. Improving cold tolerance and insect resistance in chickpea and the use of AFLPs for identification of molecular markers for these traits. *Acta Horticulturae*, (in press).

Lazaridou, T.B., Roupakias, D.G. and Economou, A.S. , 1993. Embryo rescue in *Vicia faba* and *V. narbonensis*. *Plant Cell Tissue Organ Culture*, 33: 297-301.

Malek, M.A., Afzal, A., Rahman, A.M. and Salahuddin, A.B.A., 1997. Lathyrus: evidence of impact on cropping in harsh environments. In: *Proceedings of the International Food Legume Research Conference* (ed. Knight, R.). Adelaide September 22-26, 1997. (in press).

Marquadt, R.R. 1982. Favism in Faba bean improvement. *World Crops*, 343-353.

Martin, I., Tenorio, J.L. and Ayerbe, L., 1992. Yield and evapotranspiration of leafed and semi-leafless peas with different plant populations under drought conditions. In: *Proceedings of First European Conference on Grain Legumes*. pp 211-212. Anger, France.

Marx, G., 1977. Classification, genetics and breeding. In: *The physiology of the garden pea* (Sutcliffe, J.F. and Pate, J.S.). pp. 21-43. Academic Press, London.

Marx, G., 1985. The pea genome: a source of immense variation. In: *The pea crop, A basis for improvement* (Hebblethwaite, P.D., Heath, M.C. and Dawkins, T.C.K.). pp. 45-54. Butterworths, London,

Mayer, M.S., Tullu, A., Simon, C.J., Kumar, J., Kaiser, W.J., Kraft, J.M. and Muelbauer, F.J., 1997. Development of a DNA marker for Fusarium wilt resistance in chickpea. *Crop Science*, 37: 1625-1629.

Mehra, R.B., Raju, D.B. and Himabindu, K., 1996. Evaluation and utilisation of *Lathyrus sativus* collection in India. In: Lathyrus *genetic resources in Asia* (eds. Arora, R.K., Mathur, P.N., Riley, K.W. and Adham, Y.). pp. 37-43. Indira Gandhi Agricultural University, Raipur, Raipur, India. IPGRI Office for South Asia, New Delhi, India.

Mehta, S.L. and Santha, I.M., 1996. Plant biotechnology for development of non-toxic strains of *Lathyrus sativus*. In: Lathyrus *genetic resources in Asia* (eds. Arora, R.K., Mathur, P.N., Riley, K.W. and Adham, Y.). pp. 128-138. Indira Gandhi Agricultural University, Raipur, Raipur, India. IPGRI Office for South Asia, New Delhi, India.

Micke, A. and Swiecicki, W., 1988. Induced mutations in lupins. In: *Proceedings of the Fifth*

International Lupin Conference (ed. Twardowski, T.). pp. 110-127. PWRiL, Poznan, Poland,

Muehlbauer, F. J., 1987. Incorporation of useful characters from germplasm resources into cultivars of food legumes. In: *Legume Genetic Resources for Semi-Arid Temperate Environment, Proceedings of an international workshop* (eds. Smith, A. and Robertson, L.D.). pp. 363-393. ICARDA, Aleppo, Syria.

Muehlbauer, F.J., and Slinkard, A.E., 1981. Genetics and breeding methodology. In: *Lentils* (eds. Webb, C. and Hawtin, G.). pp 69-90. Commonwealth Agricultural Bureaux, Slough, U.K.

Muehlbauer, F.J., Simon, C.J., Spaeth, S.C. and Haddad, N.I., 1990. Genetic improvement of chickpea: key factors to be considered for a breakthrough in productivity. In: *Chickpea in the Nineties* (eds. van Rheenen, H.A. and Saxena, M.C.). pp 209-216. International Crops Research Institute for the Semi-Arid Tropics, Patancheru, India.

Murray, P.J., 1994. The use of lupins as a feed for sheep. In: *Proceedings of the First Australian Lupin Technical Symposium* (eds. Dracup, M. and Palta, J.). pp. 67-73. Department of Agriculture Western Australia, South Perth.

Myers, J. R. and Gritton, E.T., 1988. Genetic male sterility in the pea *(Pisum sativum* L.): I. Inheritance, allelism and linkage. *Euphytica*, 38: 165-174.

Nene, Y.L. and Reddy, M.V., 1987. Chickpea diseases and their control. In: *The Chickpea*. (eds. Saxena, M.C. and Singh, K.B.). pp 283-318 CAB International, Wallingford, United Kingdom.

Nottage, E. and Wallace, R.G., 1994. From theory to reality, the developing of commercial lupin loaves and biscuits. In: *Proceedings of the First Australian Lupin Technical Symposium* (eds. Dracup, M. and Palta, J.). pp. 248-251. Department of Agriculture Western Australia, South Perth.

Petterson, D.S. 1998. Composition and food uses of lupins. In: *Lupins as Crop Plants: Biology: Production and Utilization* (eds. Gladstones, J.S., Hamblin, J. and Atkins, C.) CABI, London.

Petterson, D.S., Sipsas, S. and Mackintosh, J.B., 1997. *The Chemical Composition and Nutritive Value of Australian Grain Legumes (2nd Edition)*. Grains Research and Development Corporation, Canberra, Australia.

Picard, J., Duc, G. and Pelletier, R., 1985. Cote D'OR, a highly frost resistant population of *Vicia faba*. *FABIS*, 13: 11-20.

Porta-Puglia, A, Bernier, C.C. and Jellis, G.J., 1994. Screening techniques and sources of resistance to foliar diseases by fungi and bacteria in cool season food legumes. *Euphytica*, 73: 11-25.

Pundir, R.P.S., Reddy, K.N. and Mengesha, M.H., 1988. *ICRISAT Chickpea Germplasm Catalog: evaluation and analysis.* International Crops Research Institute for the Semi-Arid Tropics, Patancheru, India.

Pushpamma, P. and Geervani, P., 1987. Utilisation of chickpea. In: *The Chickpea* (eds. Saxena M.C. and Singh K.B.). pp 357-368. CAB International, Wallingford, U.K.

Pushpamma, P., 1989. Post-production processing of *Lathyrus sativus* in India. In: *The grasspea-threat and promise* (ed. Spencer, P.S.). pp.198-204. New York, Third World Medical Research Foundation.

Ramsay, G. and Griffiths, D.W., 1993. The genetics of vicine and convicine synthesis in faba beans. In. *Recent Advances of Research in Antinutritional Factors in Legume Seeds* (eds. van der Poel, A.F.B., Huisman, J. and Saini, H.S.). pp. 397-400. Proceedings 2nd International Workshop on Antinutritional Factors in Legume Seeds, Wageningen.

Reed, W., Cardona, C., Sithanantham, S. and Lateef, S.S., 1987. Chickpea insect pests and their control. In: *The Chickpea* (eds. Saxena, M.C. and Singh, K.B.). pp 283-318 CAB International, Wallingford, United Kingdom.

Rheenen, van H.A., 1991. Production aspects and prospects of chickpea. In: *Proceedings of the consultancy meeting on uses of grain legumes* (ed. Jambunathan, R.) pp 31-35 ICRISAT, Pantancheru, India.

Robertson, L.D. and Abd El Monein, A.M., 1996. *Lathyrus* germplasm collection, conservation and utilisation for crop improvement at ICARDA. In: Lathyrus *genetic resources in Asia* (eds. Arora, R.K., Mathur, P.N., Riley, K.W. and Adham, Y.). pp. 97-111. Indira Gandhi Agricultural University, Raipur, Raipur, India. IPGRI Office for South Asia, New Delhi, India.

Robertson, L.D. and Erskine, W., 1997. Lentil. In: *Biodiversity in Trust: Conservation and use of plant genetic resources in CGIAR Centers* (Fuccillo, D., Sears, L. and Stapleton, P.). pp 128-138. Cambridge University Press, U.K.

Robertson, L.D., Singh, K.B. and Ocampo, B., 1995. *A catalogue of annual Cicer species*. ICARDA, Aleppo, Syria.

Ross, S.M. and Spencer, P.S., 1989. B-N-oxalylamino-L-alanine: potential mechanism of neurotoxicity. In: *The grasspea-threat and promise* (ed. Spencer, P.S.). pp 97-108. New York: Third World Medical Research Foundation.

Rowland, I., Mason, M., Pritchard, I. and French, R.,1994. Effect of field peas and wheat on the yield and protein of subsequent wheat crops grown at several rates of applied nitrogen. *Australian Journal of Experimental Agriculture*, 34: 641-646.

Saalbach, I., Waddell, D. and Pickardt, T., 1995. Stable expression of the sulphur-rich 2S albumin gene in transgenic *Vicia narbonensis* increases the methionine content of seeds. *Journal of Plant Physiology*, 145: 674-681.

Sakar, D., Durutan, N. and Meyveci, K., 1988. Factors which limit the productivity of cool season food legumes in Turkey. In: *World Crops: Cool Season Food Legumes* (ed. Summerfield, R.J.). pp. 137-146 Kluwer, the Netherlands. Heterosis for yield and related characters in pea.

Sarawat, P., Stoddard, F.L., Marshall, D.R. and Ali, S., 1994. Heterosis for yield and related characters in pea. *Euphytica*, 80: 39-48.

Sardana, V., 1992. Grain legumes in human nutrition. *Indian Farming*, 42(7): 17-20.

Sarker, A., Robertson, L.D. and Campbell, C.G., 1997. *Lathyrus* spp.: Conserved resources, priorities for collection and future prospects. In: *Proceedings of the International Food Legume Research Conference* (ed. Knight, R.). Adelaide September 22-26, 1997.

Savage, G.P., 1988. The composition and nutritive value of lentils (*Lens culinaris*). *Nutrition Abstracts and Reviews (Series A)*, 58(5): 320-343

Saxena, N.P., 1987. Screening for adaptation to drought case studies with Chickpea and pigeonpea. In: *Adaptation of chickpea and Pigeonpea to Abiotic Stresses* (eds. Saxena, N.P. and Johansen, C.). pp 63-76, Proceedings of the consultants workshop, 19-21 December 1984, ICRISAT Centre, India. ICRISAT, Pantancheru, India.

Schroeder, H., Schotz, A.H. Wardley-Richardson, T., Spencer, D. and Higgins, T.J.V., 1993. Transformation and regeneration of two cultivars of pea (*Pisum sativum* L.). *Plant Physiology*, 101: 751-757.

Schroeder, H., Barton, J.E.,Tabe, L.M., Molvig, L., Grant, J.E., Jones, M. and Higgins, T.J.V., 1997. Gene technology for improved weed, insect, and disease control and for seed protein quality. In: *Proceedings of the International Food Legume Research Conference* (ed. Knight, R.). Adelaide September 22-26, 1997.

Schroeder, H., Gollasch, S., Moore, A., Tabe, L.M., Craig, S., Hardie, D.C., Chrispeels, M.J., Spencer, D. and Higgins, T.J., 1995. Bean α-amylase inhibitor confers resistance to the pea *weevil (Bruchus pisorum*) in transgenic peas (*Pisum sativum* L.). *Plant Physiology*, 107: 1233-1239.

Siddique, K.H.M., Brinsmead, R.B., Knight, R., Paull, J.G. and Rose, I.A., 1997. Adaptation of chickpea (*Cicer arietinum* L.) and faba bean (*Vicia faba* L.) to Australia. In: *Proceedings of the International Food Legume Research Conference* (ed. Knight, R.). Adelaide September 22-26, 1997.

Siddique, K.H.M., Loss, S.P., Herwig, S.P. and Wilson, J.M., 1996. Growth, yield and neurotoxin (ODAP) concentration of three Lathyrus species in mediterranean-type environments of Western Australia. *Australian Journal of Experimental Agriculture*, 36: 209-18.

Singh, K.B. and Ocampo, B., 1997. Exploitation of wild *Cicer* species. *Theoretical and Applied*

Genetics, 95: 418-423.

Singh, K.B. and Reddy, M.V., 1993. Resistance to six races of *Ascochyta rabiei* in the world germplasm collection of chickpea. *Crop Science*, 33: 186-189.

Singh, K.B. and Weigand, S., 1994. Identification of resistant sources in *Cicer* species to *Liriomyza cicerina*. *Genetic Resources and Crop Evolution*, 41: 75-79.

Singh, K.B., 1993. Problems and prospects of stress resistance breeding in chickpea. In: *Breeding for Stress Tolerance in Cool Season Food Legumes* (eds. Singh, K.B. and Saxena, M.C.). pp 17-35. John Wiley and Sons, Chichester, U.K.

Singh, K.B., Holly, L. and Bejiga, G., 1991. *A catalog of kabuli chickpea germplasm: An Evaluation Report of Winter Sown Kabuli Chickpea, Land Races, Breeding Lines and Wild* Cicer *Species*. ICARDA, Aleppo, Syria.

Singh, K.B., Omar, M., Saxena, M.C. and Johansen, C., 1996. Registration of FLIP 87-59C, a drought-tolerant chickpea germplasm line. *Crop Science*, 36: 472.

Singh, K.B., Pundir, R.P.S., Robertson, L.D., van Rheenen, H.A., Singh, U., Kelly, T.J., Parthasarathy Rao, P., Johansen C. and Saxena, N.P., 1997. Chickpea. In: *Biodiversity in Trust* (eds. Fuccillo, D., Sears, L. and Stapleton, P.) pp. 100-113. Cambridge University Press, U.K.

Singh, K.B., Malhotra, R.S. and Witcombe, J.R., 1983. *Kabuli chickpea germplasm catalog*. ICARDA, Aleppo, Syria.

Singh, K.B., Malhotra, R.S. and Saxena, M.C., 1989. Chickpea evaluation for cold tolerance under field conditions. *Crop Science*, 29: 282-285.

Stelling, D. and Ebmeyer, E., 1990. Selection in early generations of dried peas, Pisum sativum L. I. Values of heritability and efficiency of indirect selection. *Plant Breeding*, 105: 169-179.

Stelling, D., 1994. Performance of morphologically divergent plant types in dried peas (*Pisum sativum* L.). *Journal of Agricultural Science (Camb.)*, 123: 357-361.

Sweetingham, M.W., Henderson, J., Cowling, W.A. and Buirchell, B.J., 1994. Glasshouse screening of Lupinus species for resistance to brown spot and Pleiochaeta root rot. In: *Proceedings of the First Australian Lupin Technical Symposium* (eds. Dracup, M. and Palta, J.). pp. 313. Department of Agriculture Western Australia, South Perth.

Tegeder, M., Gebhardt, D. and Schieder, O., 1995. Thiadizuron-induced plant regeneration from protoplasts of *Vicia faba* cv. Mythos. *Plant Cell Reports*, 15: 164-169.

Timmerman, G.M., Frew, T.J., Miller, A.L., Weeden, N.F. and Jermyn, W.A., 1993. Linkage mapping of *sbm*-1, a gene conferring resistance to pea seed-borne mosaic virus, using molecular markers in Pisum sativum. *Theoretical and Applied Genetics*, 85: 609-615.

Timmerman, G.M., Frew, T.J., Weeden, N.F., Miller, A.L. and Goulden, D.S., 1994. Linkage analysis of *er*-1, a recessive *Pisum sativum* gene for resistance to powdery mildew fungus (*Erysiphe pisi*, D. C.). *Theoretical and Applied Genetics*, 88:1050-1055.

Vavilov, P.P. and Gataulina, G.G. 1984. Growth and development pattern of different spring forms of Lupinus albus. In: *Proceedings of the Third International Lupin Conference*. pp. 562-565. International Lupin Association, La Rochelle, France.

Wells, H.D. and Forbes, I.,1982. Genetics and breeding of Lupinus for winter production for the south of the United States of America. In: *Agricultural and Nutritional Aspects of Lupines: Proceedings of the First International Lupine Workshop* (eds. Gross, R. and Bunting, E.S.). pp 77-89. GTZ, Eschborn, Germany.

Williams, P.C. and Singh, U., 1987. The chickpea - Nutritional quality in breeding programs. In: *The Chickpea* (eds. Saxena, M.C. and Singh, K.B.). CAB International, Wallingford, U.K.

Wroth, J., 1996. *Host-pathogen relationship of the Ascochyta blight (*Mycosphaerella pinodes *(Berk. and Blox.) Vestergr.) disease of pea (*Pisum sativum *L).* PhD thesis, University of Western Australia, Perth.

Yunus, A.G. and Jackson, M.T., 1991. The gene pools of the grasspea (*Lathyrus sativus* L.). *Plant Breeding*, 106: 319-328.

Zeven, A. C. and Zhukovsky, P.M., 1975. *Dictionary of cultivated plants and their centres of diversity*. Centre for Agricultural Publishing and Documentation, Wageningen, the

Netherlands.
Zhou, M. and Arora, R.K., 1996. Conservation and use of underutilised crops in Asia. In: Lathyrus *genetic resources in Asia* (eds. Arora, R.K., Mathur, P.N., Riley, K.W., and Adham, Y). pp. 91-95. Indira Gandhi Agricultural University, Raipur, Raipur, India. IPGRI Office for South Asia, New Delhi, India.

Chapter 18

Utilisation of Forage and Pasture Legume Diversity

Kenneth A. Street, Larry D. Robertson and Nigel Maxted

18.1 INTRODUCTION

It is open to question whether there is any difference between pasture and forage legumes. In the literature the terms are often used interchangeably and thus the distinction has become somewhat blurred. For the purpose of this discussion the term ''forages'' will apply to plant species that are cultivated and harvested for animal feed and are utilized or conserved within or at the end of the growing season. *Vicia sativa* (vetches), *Lathyrus* species and *Medicago sativa* are legumes that are used in this way. By contrast we will define 'pastures' as systems used for grazing that remain established beyond one growing season. These include natural pastures and those that are established within the context of an agricultural system and rely on natural regeneration for annual production. Both types of pastures generally involve a mixture of species. A good quality pasture, however, will usually have a high legume component. For example, in southern Australia, commercial varieties of *Trifolium* spp. and *Medicago* spp. are purposely introduced into a farming system, which relies on the legume component to regenerate, from dormant seeds, during the fallow phase of a cereal cropping / animal production enterprise (Puckridge and French, 1983).

It is clear, even from the above definitions that any discussion on utilizing forage or pasture legumes must include a discussion on the dominant farming system into which they are introduced. This chapter is not intended as an exhaustive monograph on the subject of farming systems or forage legumes. Rather the discussion will be limited to the different ways forage and pasture legumes have been used in dry-land farming systems that have evolved in Mediterranean-type environments of southern Australia and in the West Asian and North African (WANA) countries. However, many of the concepts discussed here could be applied to farming systems employed in other climatic zones.

18.2 MEDITERRANEAN-TYPE FARMING SYSTEMS

The actual farming systems employed by Mediterranean farmers has shaped much of what we now regard as the Mediterranean ecosystem. Dryland farming systems that have evolved around the Mediterranean basin over the past 2000 years have a delicate balance between cropping and livestock systems involving cereals and grazed fallows, tree crop plantations and the utilization of rangelands for animal husbandry. Cereals are the most important subsistence and cash crops in the Mediterranean basin. It must be recognised that the Mediterranean has numerous diverse farming systems, but broadly speaking two types of cereal production systems can be identified: "wheat-based" in wetter areas and "barley based" in the drier areas. The transition zone dividing the two occurs between the 300 – 450 mm/year rainfall isohytes (Jones, 1990).

Rainfall above 350mm / year produces reliable wheat yields. In these areas wheat is often grown in rotation with grain legume crops such as lentils, chickpeas and faba beans. For example, in the Syrian province of Idleb where the mean annual rainfall is approximately 460mm, Nordblom (1987) identified 24 different cereal - legume crop rotations. In these regions animal production is usually a secondary enterprise with grazing limited to fallows, crop residues and natural pasture growing in non-arable land. Thus, the main focus of farmers in the higher rainfall regions is commercial production on larger land holdings or subsistence crop production on small holdings.

The reliability and size of wheat yields fall sharply as the mean rainfall decreases below the

350 isohyte and thus barley, which is more stress resistant, replaces wheat as the principle cereal. However, below 300mm per year, even barley does not give reliable yields (Craufurd *et al.*, 1991). In this situation small ruminate production, which does not depend directly on adequate and reliable rainfall, provides more security to the farmer. Thus, as the climate becomes more arid, farming systems become more focused on small ruminant production.

Traditionally, animal husbandry in the regions relied on natural pastures and crop residues that were grazed on a nomadic or semi-nomadic basis. However, over the last 50-80 years natural pastures have been placed under enormous pressure as rising human populations have demanded an extension of cropping areas and an uncontrolled increase in stock numbers. As a result, natural pastures are being over-exploited and are unable to deliver adequate dry matter to sustain production. Thus, increasingly barley is being used to supplement natural pastures in the form of green pasture, stubble, stored grain and straw. The growing demand for animal feed has led to reduced fallow, leading to continuous barley cropping with a subsequent reduction in yields which is clearly unsustainable (Cooper and Bailey 1990). Already in many parts of the region there are widespread signs of wind and water erosion in both arable and grazing lands. Various strategies have been proposed to alleviate this situation. One that has received considerable attention is to replace fallows with legumes and improve natural pastures with legumes. This approach has been successfully employed in the broad acre (highly mechanized large landholder farming systems) cereal farming systems of southern Australian (Puckridge and French 1983).

18.3 Role of Forage and Pasture Legumes in Mediterranean Cereal Production Systems

The brief history of cereal farming in southern Australia is a useful example to illustrate how legume pastures have been successfully utilized to drastically improved production in the system as a whole. The evolution of cereal cropping in southern Australia occurred in three recognizable phases. From 1870 to 1900, wheat yields steadily declined from 860 kg/ha to 460 Kg/ha (Figure 18.1). During this phase the natural Savannah woodland was both grazed and cleared for cropping. Australian soils are ancient and weathered and therefor low in nitrogen and phosphorous and trace elements. As a consequence, the soils were rapidly depleted of their natural fertility, contributing substantially to the initial reduction in yields.

The second phase saw average wheat yields increase by almost 75% as superphosphate fertilizers and bare fallow came into use. Although bare fallowing was thought to benefit subsequent crops through water conservation, it was found that the benefits actually accrued when nitrogen was released through the mineralization of organic matter. However, the initial increase slowed as organic-matter levels became depleted and soil structure was destroyed. Despite the use of improved wheat varieties, from 1921-1950 average yields only increased from about 750-860 kg/ha.

During the third phase a rapid increase in cereal yields and general productivity occurred following the wide spread integration of annual *Medicago* spp. and *Trifolium* spp. into cereal crop rotations. This system was called ley farming. By 1967 more than 20 million hectares were under sown pasture, most of it within a ley farming system (Macindoe, 1975). Donald (1963) noted that throughout 1980 livestock production increased in proportion to the increase in the area of improved pastures. During this period bare fallow decreased by 30% and crop yields increased by up to 27 % in West Australia and 68% in New South Wales while the protein content of wheat grown in southern New South Wales increased by 2%. During the early nineties, areas under pasture increased to almost 30 million hectares but declined to 20 million hectares towards the end of the decade as the world price for wool plummeted and farming enterprises shifted towards more intensive cropping. Even though the emphasis has swung heavily away from animal production, ley legume pastures will still play an important role in south Australian farming systems because of the beneficial effects legumes have on the system productivity as a whole.

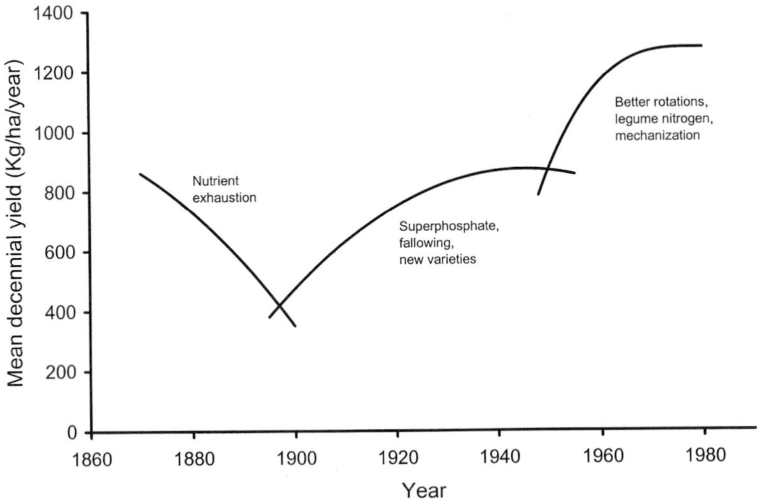

Figure 18.1. Australian wheat yields from 1860 to 1980. Each point indicates the unweighted mean of annual yields in the proceeding decade (Donald, 1981).

18.4 AUSTRALIAN LEY-FARMING

Although ley farming in Australia is still evolving to embrace shifts in the economic and physical environment, the principal remains the same as when it was first widely adopted. In simple terms, it entails replacing bare fallows with legume pastures or in some cases replacing cereals in alternate years. This is not essentially a new idea. Legume crops have been used in rotation with non-legume crops for centuries in Europe to maintain soil fertility. What is unique about the Australian system is that it relies on the pasture legume to regenerate from a soil seed bank after the cropping phase is complete. Thus once the legume component is established a rotation cycle will involve one or more years of a cereal crop followed by one or more years of pasture. In a broad acre system where farms sizes can reach tens of thousands of hectares this system greatly reduces the cost of resowing the pasture after the cereal phase. This factor alone contributed greatly to the widespread adoption of ley farming in Australia.

The legume species used and how the rotation is structured is tremendously varied depending upon the farmers objectives which are usually modified by economic, edaphic and climatic constraints, the need to control grassy weeds in the cropping phase and the number of animals in the system. For example, farmers in the south-west of western Australia on acid soils over clay with an assured winter rainfall may grow wheat for 1 year followed by subterranean clover (*Trifolium subterraneum*) for 1 year (Watson *et al.*, 1977). By contrast South Australian farmers on neutral to alkaline loams over clay with an assured winter rain will use the following rotation with each phase lasting 1 year: Medic (*Medicago* spp.): Fallow - Wheat - Medic - Barley (Webber *et al.*, 1976). For stock production and control of annual ryegrass in paddocks managed as a wheat: lupin rotation in acid sands, French serradella (*Ornithopus sativus*) is established and is grazed for a 2 – 4 year period in conjunction with non-selective herbicides (Webber *et al.*, 1976).

The successful adoption and refinement of ley farming in Australia has been supported by well-funded research. In general this research has focused on the utilization and improvement on either species naturalized in Australia or wild genotypes originating from elsewhere. The next section examines some of the species that have been utilized in Australian ley-farming.

18.4.1 Species Used in Australian Ley Farming

Subterranean clover (*Trifolium subterraneum, T. yanninicum, and T. brachycalycinum*) is by far the most important pasture legume on acid soils in southern Australia and by 1965 was growing on at least 20 million hectares (Francis and Millington, 1965). On alkaline soils annual *Medicago* species including *M. truncatula, M. polymorpha, M. minima, M. praecox, M. laciniata, M. littoralis, M. tornata, M. rugosa and M. scutellata* are the most important pasture legumes. Both the medics and subterranean clovers are not native to Australia. It is highly likely that they were introduced, non-intentionally, into Australia in the 1800s with the first waves of European settlement (Gladstones, 1966). Because most of these settlers arrived on British ships, Gladstone (1966) argued convincingly that the both the medics and subterranean clovers must have originated predominantly from England, France and Spain. Cocks and Matheison (1980) point out that this hypothesis is strongly supported by the natural distribution of these species. For example, England France and Spain are inhabited predominately by *T. subterraneum*. Further, of the 12 medic species native to France, Britain and the Low Countries, 10 have become naturalized in Australia.

Of the 26 commercial cultivars of subterranean clover grown in Australia, most were bred or selected from naturalized material. Only five originated from material collected outside of Australia: three from Greece, one from Italy and one from Turkey. Important varieties from this collected material include 'York', 'Larissa', and 'Rosedale'. By contrast, the naturalized medics did not provide good breeding material because their pods tend to be spiny and accumulate in the sheep's fleece, thereby reducing its value. Of the 31 commercial varieties of annual medics grown in Australia, only 5 are derived wholly from naturalized Australian material. The rest largely originate from the Mediterranean basin. Important varieties from this region include 'Sephi' a *M. truncatula* cultivar from Galilee, 'Zodiac' a *M. murex* cultivar collected in Sardinia, and 'Harbinger' a *M. littoralis* cultivar collected in Iran. Accessions collected from other Mediterranean environments have also been useful for variety release, for example, 'Santiago', a cultivar of *M. polymorpha* collected in Chile.

Table 18.1 lists the different pasture species and the number of commercial cultivars used in Australia. However, as Cocks and Matheison (1980) speculate, perhaps seven cultivars in all contributed to 90 percent of pasture development: 4 subclovers and 3 medics. Despite this, new varieties are still being developed that are adapted to previously unoccupied niches in the physical environment and to the continually evolving farming systems. In addition, new varieties are being developed that have greater utility or superior productivity. For example, the Commonwealth Scientific and Industrial Research Organization (CSIRO) in Australia has successfully transferred a gene for a sulphur-rich sunflower protein to several varieties of Lucerne and subterranean clover. They estimate that sheep grazing these genetically engineered pastures have the potential to produce 10% more wool.

18.5 How Legumes Benefit a Farming System

Increases in both cereal and animal yields after including a legume component into a farming system have been well documented in a large volume of studies. As we would expect, the impact that the legume has is dependent upon a variety of factors including species involved, management practices, and environmental conditions. However, in general the impact of the legume component on cereal yields is almost always significant when compared to a continuous cereal-cropping regime, even a number of years after the legume phase. Table 18.2 details a study in which 3 different wheat: medic rotations were employed on heavy soil in the south west of Western Australia.

In this study the beneficial effects of a medic pasture on both wheat yield and protein content are clearly demonstrated even 3 years after the legume phase had ended. Similar results have been obtained for crop legumes. For example, in a one year field pea (*Pisum sativum*): one year wheat rotation grown in the eastern wheatbelt of Western Australia the crop: legume rotation yielded 0.52 t/ha more wheat than the continuous cereal regime (Gartell, 1998). While it is clear that legumes have a beneficial effect on cereal yield and quality and on animal production, the reason why is not so clear-cut. Possible beneficial effects will now be discussed.

Table 18.1. Pasture and forage species that have been commercialized in Australia.

Species	Common name	Number of varieties
Clovers		
Trifolium repens	White clover	7
Trifolium pratense	Red clover	4
Trifolium fragiferum	Strawberry clover	3
Trifolium subterraneum		
var. *subterraneum*	Sub clover	20
var. *yanninicum*	Sub clover	4
var. brachycalycinum	Sub clover	4
Trifolium hirtum	Rose clover	4
Trifolium cherleri	Cupped clover	3
Trifolium ambiguum	Caucasian clover	6
Trifolium purpureum	Purple clover	1
Trifolium semipilosum var *glabrecens*	Kenya white clover	1
Trifolium balansae	Balansa clover	2
Trifolium resupinatum	Persian clover	2
Lucerne		
Medicago sativa	Lucerne	19
Medicago falcata x Medicago sativa	Lucerne	2
Annual medics		
Medicago truncatula var *truncatula*	Barrel medic	13
Medicago littoralis	Strand medic	2
Medicago rugosa	Gamma medic	3
Medicago tornata	Disc medic	4
Medicago polymorpha	Bur medic	3
Medicago scutellata	Snail medic	4
Medicago murex	Murex medic	2
Medicago sphaerocarpos	Sphere medic	
Serradella		
Ornithopus compressus	Yellow serradella	8
Ornithopus pinnatus	Slender serradella	1
Ornithopus sativus	French serradella	1
Bisserula		
Biserrula pelecinus	Biserrula	1
Vetch		
Vicia sativa subsp. *sativa*	Common vetch	4
Vicia villosa subsp. *dasycarpa*	Wooly pod vetch	1
Vicia benghalensis	Purple vetch	1
Astragalus		
Astragalus hamosa	Milk vetch	1
Onobrychis		
Onobrychis viciifolia	sanfoin	1

Table 18.2. Medic pastures in rotation with wheat: average grain yields and grain protein levels for 1985-89 (I.C. Rowland; cited in Gartell, 1998).

Rotation	Yield (t/ha)	% protein
Continuous wheat	1.00	9.1
1 year wheat: 1 year medic	1.45	10.9
3 years medic: 1 year wheat	1.62	11.9
3 year medic: 3 years wheat		
- 1st wheat crop	1.4	11.7
- 2nd wheat crop	1.32	10.1
- 3rd wheat crop	1.24	9.9

18.5.1 Nitrogen fixation

Legumes are generally regarded as beneficial to crop rotations due to their ability to fix nitrogen and thereby improve soil fertility. There is no simple way to measure the amount of nitrogen fixed by pastures and how much of it eventually becomes available to the cereal crop. Thus, it is difficult to compare studies because different approaches have been used to quantify the addition of nitrogen to the system. Perhaps the simplest method that has been used is to directly measure the amount of nitrogen fixed in the legume dry matter. In a large study, levels of N_2 fixation were quantified for 540 pastures containing subterranean clover, annual medic, Lucerne or white clover growing in different farming systems over a wide range of rainfall zones in Victoria and New South Wales, Australia. It was found that the efficiency with which N was fixed was generally similar regardless of the legume species or the environment in which they were grown. On average, 20 to 25 kg shoot N was fixed for every ton of legume shoot dry matter. They proposed that these values may provide useful, predictive guides to the likely inputs of fixed N through pasture herbage (Peoples *et al.*, http://life.csu.edu.au/agronomy/papers/73/73.html).

It is misleading to use total N fixed as a measure of increase in soil fertility because some it will be lost due to a variety of factors such as leaching and removal in agricultural products. Perhaps a better estimate of fertility build up is a gross measurement of changes in total soil nitrogen. Many studies that used this approach have been conducted in Australia under a variety of contrasting conditions. They clearly show a steady but slow build up of total N under a legume pasture that tends towards an equilibria after at least 2 decades (e.g. Russell, 1960). Table 18.3 shows the results of selected long-term studies.

Table 18.3. Mean annual nitrogen increments in some southern Australian soils under grass / clover based pastures. (Adapted from Rowland and Halse, 1980).

Location	Soil Type	Length of Study (yrs)	Depth (cm)	Increment Kg/ha/year
South West Western Australia	Sand over clay	40	0-12	38
New South Wales	Podsolic	17	0-15	48
South West Western Australia	Sand over sandy clay	27	0-15	35
South West Western Australia	Sandy loam	20	0-15	63
South Australia	Solonetz	35	0-15	52

In addition to increased N fertility, legume pastures also lead to an increase in soil organic matter. This in turn leads to an increased cation exchange and a subsequent reduction in the soil pH as H ions are released. In some cases this change in pH is associated with an enhanced availability of trace elements such as copper zinc and iron which may be advantageous under infertile conditions (Williams and David, 1976).

18.5.2 Soil Structure

One of the dangers associated with continuous cereal cropping is that the soil structure deteriorates causing a decrease in water penetration and retention. The breakdown in structure also leads to wind and water erosion. Pasture legumes can dramatically improve soil structural conditions in the following ways:

1. an overall increase in porosity;
2. an increase in aggregate stability;
3. a decrease in bulk cohesive strength and
4. an increase in water holding capacity.

Figure 18.2 shows the changes that occur in the percentage of water stable aggregates under different rotations after land in the south east of Western Australia was cleared of native vegetation. What this figure clearly shows is that soil structure steadily improved under a legume dominant pasture and sharply declined under cereal cropping.

18.5.3 Weed Control

Another way legume dominant pastures can benefit cereal yields in a rotation system is by providing an opportunity for grassy weed control during the legume phase. A combination of timely grazing management and strategic herbicide application can greatly reduce the weed seed bank that will germinate in the cropping phase. Table 18.4 demonstrates how effective weed control during the pasture phase is at improving wheat yields during the cropping phase.

Although the higher wheat yields realized in the weed control treatments cannot be attributed only to reduced weed competition in the 1982 crop it does demonstrate the effectiveness of using a legume pasture to control weeds.

18.5.4 Control of Pests and Diseases

In the trial outlined above it is also possible that some of the effect was due to the control of diseases and pests that either build up in the wheat phase or are harbored by grassy species during the pasture phase. Numerous studies in Australia have shown that successive cropping with cereals causes a rapid build up of pests and diseases and therefor a decline in yield. Introducing a pasture phase into the cropping program has been an effective way of combatting cereal pests. For example, in Australia the fungal disease Take-all (caused by *Gaeumannomyecs gramis)* and cereal cyst nematode (caused by *Hererodera avenae)* are serious wheat diseases that can be controlled by breaking their life cycles with a pasture phase (CSIRO, 1977). However, since pastures will tend towards grass dominance if not carefully managed, there is still a risk that diseases can be harbored in the pasture phase. Thus, to act as an effective disease break for a cereal crop a legume dominant pasture must be maintained.

Figure 18.2. Changes in water-stable aggregation of soil under different rotations since land was cleared for cropping, soil type: Wongan loamy sand. (Rowland and Halse, 1980).

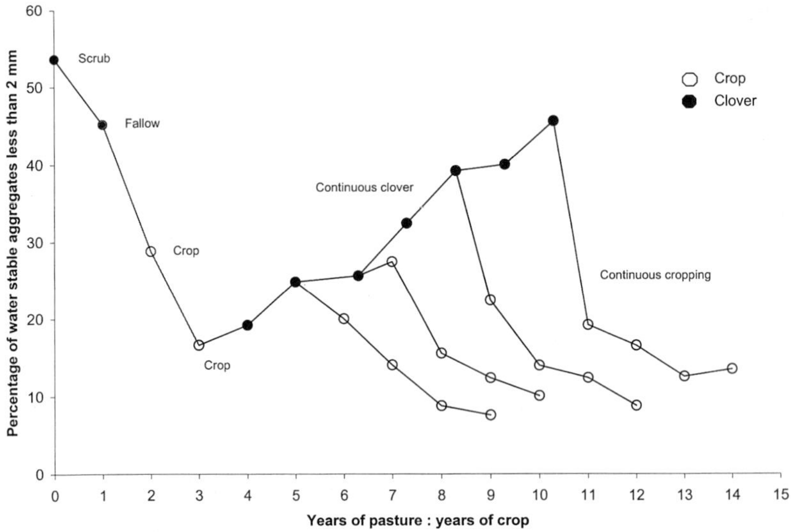

Table 18.4. Weed control in the year before the wheat crop: grain and protein yields of wheat in 1982 on red-brown earth in southern Western Australia when different weed control treatments were applied (Thorn; cited in Gartell, 1998).

	Wheat grain (1982 crop)		**Weediness**
1981 treatments	**Yield t/ha**	**% protein**	**(1982 crop)**
Subclover based pasture - no weed control	1.62	12.1	Much grass few *Emex* spp.
Subclover based pasture Broad leaf herbicide	2.05	12.1	Some grass few *Emex* spp.
Subclover based pasture Grass herbicide	2.64	13.4	No grass most *Emex* spp.
Subclover based pasture Broad leaf + grass herbicide	2.91	13.8	No grass no *Emex* spp.
Wheat, Broad leaf + grass herbicide	1.97	11.4	Little grass few *Emex* spp.

18.5.5 Animal Productivity

Donald (1963) estimated that half the increase in wool production in Australia from 1947 to 1963 was due to the impact that legume pastures had on sheep numbers and fleece weights. Numerous grazing studies have conclusively shown that animals grazing pastures with a high legume content out perform those grazing non-legume pastures. For example, Christian and Shaw (1952) compared the live weight gains of steers grazing (2 steers/ha) a *Chloris gayana / Medicago sativa* pasture and a *Chloris gayana* pasture without the *Medicago* in southeast

Queensland, Australia. Steers on the improved pasture gained 179 kg/year while those on the non-legume pasture gained only 90 kg/year. In northeast Syria phosphate fertilizer was applied to natural pastures which resulted in a 4-fold increase in the legume component. On the improved pasture sheep gained weight faster and required less supplementary feeding (Osman *et al.*, 1991).

18.6 THE ROLE OF PASTURE AND FORAGE LEGUMES IN WANA

18.6.1 Fallow Replacement with Legumes

In the lower rainfall areas of the WANA region where farming is more focused on animal production and barley, fallow replacement with forage legumes has been identified as one way to arrest the unsustainable trend towards continuous barley cropping (e.g. Jones, 1990). Indeed, the outstanding success of the Australian ley farming system provides a workable model that could be adapted to the region. While the Australian emphasis is on clover and medic pastures, in the WANA region, research and development has focused on *Lathyrus* and vetch species as forages grown in rotation with barley (e.g. Tully, 1984). Jones (1990) identifies at least four possible modes of utilization:

1. Green grazing to fatten lambs or increase milk production in spring. This method has the advantage of low labour costs but does not yield any grain.
2. Haymaking, which conserves good quality feed for winter. However, the yield is low, grain yield is nil and the labour costs are high (Thomson and Oglah, 1988).
3. Mature harvesting that conserves high quality feed for winter and seed for subsequent planting but occurs at a time when labour is in peak demand and so harvesting costs are high.
4. Standing hay in which the mature crop is grazed. This has the advantage of low labour costs but does not conserve the grain.

In a more novel approach, Christiansen *et al.* (1994a) proposed a barley: medic system managed as a mixture. The aim here is to establish a self regenerating legume component which adds nutritional value to the barley stubble and has the advantage of adding nitrogen and organic matter to the soil. In a trial to test the feasibility of this system, Christiansen *et al.* (1994b) demonstrated that enough seed of *M. rotata* and *M. rigidula* was produced under a barley crop to enable regeneration at a useful density the following year.

A considerable body of research in the WANA region has demonstrated that rotations including pasture or forage legumes give higher yields. This has been shown with a number of species, including *Vicia sativa* subsp. *amphicarpa* (Street, 1999), *Vicia sativa* (Rashed, 1986), *Medicago* species (Abd El Moneim and Cocks, 1986), *Lathyrus ochrus, Vicia dasycarpa, V. ervilia* (Papastylianou, 1987), *Pisum sativum* (Keating *et al.*, 1988) and *M. truncatula, M. orbicularis, M. polymorpha,* and *M. aculeata* (Zeghida, *et al.*, 1994). In all of these trials, an improvement in soil fertility and or barley yield or pasture productivity was demonstrated Christiansen *et al.*, (1994a). However, most of these trials were conducted in the artificial environment of a research station. Perhaps more realistic trials are those that are conducted on working farms in which animals are part of the system.

One such on farm trial conducted in northwest Syria in a 270 mm annual rainfall barley growing area compared the yields of continuous barley with barley - fallow, barley - common vetch (*Vicia sativa*) and barley - chickling (*Lathyrus sativus*) rotations. Feed yields from both rotation phases was greatest with the barley - vetch and barley - chickling rotations while the barely - vetch rotation gave the highest yield of metabolisable energy. Further, total crude protein was twice as high for the legume rotations as for the barley - barley and barley - fallow treatments. Table 18.5 shows a 7-year by 8-farm mean dry matter production, totalled over both phases of the rotation (Christiansen *et al.*, 1994a).

Table 18.5. Mean dry matter production totalled over both phases of the rotation. Each mean is over 7 years and 8 farms for each rotation (Adapted from Christiansen *et al.*, 1994a).

Rotation	Dry matter yield (t/ha)
Barley - Fallow	2.91
Barley - Barley	4.83
Barley - Vetch	5.02
Barley - Chickling	5.32

The authors of the above trial point out that grazing lambs with forage legumes fits well with the forage calendar because it provides high quality forage for weaning and relieves the pressure from natural pastures at a time when they are flowering and setting seed.

18.6.2 Improved Pasture in Marginal Land

Traditionally most of the non-arable land within and adjacent to cropping zones in the Mediterranean basin, have been used for animal grazing. This land constitutes 30 - 60 % of the total land area in some countries but it is often intensively grazed resulting in erosion and low productivity. In most cases animals grazed exclusively on marginal lands need supplementary feeding. However, legume species abound in these marginal lands. Osman *et al.* (1990) lists 64 legume species that are common in the WANA region. The potential exists to improve the productivity of these naturally occurring legumes, and thus the pasture as a whole, by applying phosphate fertilizer and managing grazing. In one experiment, legume seed reserves increase from 2.7 g / m^2 to 17 .5 g / m^2, the percentage of legumes in the pasture increased from 17 to 46% and total dry matter increased from 200 kg/ha to 1000 kg/ha after natural pastures in north east Syria were fertilized with phosphate (Osman *et al.*, 1991; Osman *et al.*, 1994). In a similar experiment the carrying capacity of phosphate fertilized marginal lands increased from 1.65 to 5.31 sheep per hectare. However, this experiment, relied on a controlled stocking rate, something that is not easy to attain, particularly on communal land. However, the technical potential to improve marginal lands with phosphate exists and awaits a suitable socio-economic context in which to be adopted.

Despite research results that demonstrate increased yields, enhanced soil fertility and improved soil structure, fallow replacement with legumes and pasture improvement have been poorly adopted in the WANA region. Arndt and Dalrymple (1977) eloquently put it this way "The capacity of scientists to generate new (agricultural) technology outstrips our understanding of the social and economic implications of technological change". Clearly the issue is not one of technological innovation alone. As Cocks and Gintzburger point out, adoption of improved practices in the Mediterranean region are constrained by many socioeconomic factors. For example, small farm sizes, lack of credit, low access to information, poor education and unclear land tenure will all conspire to limit a farmer's capacity to adopt new ideas. For example, in the WANA region most marginal land is communal. Who then should pay for and apply the phosphate fertilizer and who should police the grazing intensity so that an improvement in natural pastures can be realized? At this stage there are no practicable answers.

The constraints to adoption are often purely economic and/or technical. This is emphasized by the farmers who participated in the on-farm legume - barley rotation trial cited above. They commented that the practice was unlikely to be adopted until inexpensive and efficient mechanized harvesting is available to avoid the high cost of manual labour. In this case it was cheaper for them to buy barley feed than pay for the labour to harvest the vetch or chickling. The question then becomes who will invest the time and money to develop these technologies. Clearly, a small landowner simply does not have the resources to invest in this type of technological development. Furthermore, small farm size exposes farmers to high risk when experimenting with new technologies or adopting biologically sustainable land care techniques. For example, a farmer with less than ten hectares cannot afford to have a three year pasture legume phase in a rotation to improve the soil organic matter content on his land. His limited access to affordable credit makes it difficult to adopt changes that may cause a loss of income in the initial stages. Clearly many of the solutions to improved farming practices in the WANA region lie with the social scientists and policy makers.

However, considerable scope still exists to improve the attractiveness of forage legumes by overcoming biological constraints to adoption such as drought and cold tolerance, disease resistance, improved harvest index and reduced anti-nutritional factors.

18.7 YIELD

Obviously, a genotype that consistently yields well in a variety of environments is more desirable than a genotype with a narrow adaptation and poor yield. Thus, a high biomass under a range of conditions is a goal of many breeding programs for pasture or forage species. However, it is not as simple as just choosing high yielding genotypes from a genetic resource collection. Consistent yield, is correlated with a whole variety of factors such as flowering time, disease and pest resistance and resistance to drought and cold. Further, a high yielding genotype that does not lend itself to harvesting because it lodges or the seed pod shatters will have little chances of adoption. In a pasture context, issues of long term persistence will often be more important than breeding for yield per se. For example, a medic or subclover genotype with low seed dormancy will not persist well in a rotation where the cropping phase is longer than one year (Taylor, 1981; Maller *et al.*, 1991). Finally, a high yielding genotype that has low digestibility or has anti-nutritional factors that make it unpalatable will also be of limited value. Thus, although the overall aim in pasture and forage legume improvement is to develop high yielding varieties, "high yielding" must be seen in the broader context of adaptation to the physical environment that it is intended for and utility within the farming system that it is intended for. Therefor, it is perhaps misleading to present tables showing yields of potential forage or pasture legumes because they are usually generated in a single year in the artificial context of a plant breeding institute and have little relation to the genotype's potential for wide spread adoption.

However for the sake of comparison, Table 18.6 presents the results of an advanced yield trial for a variety of potential fodder legumes adapted to low rainfall areas. The results are for elite germplasm that was selected in a screening trial in which a large number of wild accessions of the same species were grown out and selected for performance. Performance was judged on factors such as yield potential, end use agronomic traits, pest and disease resistance and quality aspects. The trial was conducted in northeast Syria in an area that received 233 mm of rain.

Timing of biomass accumulation and seed yield will determine how the species would be utilized. For example, *V. sativa*, *V. ervilia*, and *L. ochrus* have high seed yields and could be used as dual purpose crops for seed and straw while those with lower seed yields could be used for hay or green grazing (Abd El-Moneim, 1999). *V. villosa* subsp. *dasycarpa* is particularly suited for this purpose because it produces its biomass earlier in the season and so would be good for grazing in early spring when there is a shortage of feed.

The forage-breeding program at ICARDA has been successful in selecting forage varieties with potential for high biomass and seed yield in the WANA region (Table 18.7). Twelve varieties of *Vicia* and *Lathyrus* have been released based on the elite lines selected from within the ICARDA germplasm collection. Two varieties have also been released outside the WANA region based on these selections. One in Australia ('Chalus' a *Lathyrus cicera* cultivar) and one in Pakistan ('Kuhak-96', a *V. villosa* subsp. *dasycarpa* cultivar).

Table 18.6. Means for biological and grain yields of elite lines of forage legumes grown in northeast Syria at 233 mm of rainfall in heavy alkaline soils. (Abd El-Moneim, 1999).

Species	Biological yield (kg/ha)	St. error	Grain yield (kg/ha)	St. error	No. of Accessions
Vicia narbonensis	4200	250	900	87	25
V. villosa subsp. *dasycarpa*	5000	390	300	60	25
V. ervilia	2000	160	500	55	25
V. sativa	8200	510	1500	130	36
V. palaestina	3800	290	900	110	25
Lathyrus sativus	2800	NA	310	30	25
L. cicera	3500	390	460	60	25
L. ochrus	1500	210	700	85	25

18.8 NUTRITIVE VALUE

The level of animal production can be manipulated through pasture quality, which is essentially determined by the nutritive value of the pasture. The amount eaten or voluntary intake, its digestibility, its energy, mineral, protein, and vitamin content and the absence of harmful substances determine the nutritive value of pasture. Digestibility and protein content are particularly important features of a pasture quality. Highly digestible feed is more rapidly broken down in the rumen, hence it has a faster rate of passage and the animal can increase its intake of the particular pasture (Jones and Lazenby, 1988 cited in Scott, 1995). Protein content determines the amount of protein available to the animal to convert into animal protein. This will determine growth rate, milk and wool yield. Thus breeding or selection of genotypes with a high digestibility and protein contents will dramatically increase potential animal production (McPhee *et al.*, 1998).

Considerable variation in protein content and digestibility occurs between and within species. Further, these factors also vary as the plant matures. Abd El Moneim *et al* (1990) established that as *Vicia* and *Lathyrus* species matured the digestibility and crude protein contents progressively declined after 50 –100 % podding. This is demonstrated in a further study that compared feed quality of hay and straw for different species and genotypes within species (Table 18.8).

What this trial like others of its kind shows is that sufficient variation in feed quality within species exists to exploit in breeding programs. Further, nutritive values differ between species depending upon when they are harvested or grazed. This will influence management decisions such as whether to make hay or harvest seed and graze the straw.

18.8.1 Antinutritional Factors

Although legumes provide both animal and human food, particularly protein, they also contain a variety of constituents that are either toxic or reduce protein digestion and metabolism. Table 18.9 lists some of these anti-nutritional factors that occur in either the seed or foliage of forage or pasture legumes.

Antinutritional factors can have devastating effects on animal production. For example, the fertility of some Australian ewe flocks fell from 80% to 8 % within a few years of grazing the Dwalingup variety of subclover (Bennetts *et al.*, 1946, cited in Cocks *et al.*, 1980). Both medics and clovers contain varying amount of oestrogens. The amounts of the oestrogen compounds produced are thought to be influenced by phosphorus supply and diseases (Cocks *et al.*, 1980). In another example, high levels of β-N-oxalylamino-L-alanine and β-N-oxalyl-α,β-diaminopropronic acid in the grain of Lathyrus spp. can cause a degenerative muscle disease called Lathyrism.

Table 18.7. Forage variety release in the West Asia and North African region based on the ICARDA germplasm collection.

Country	Year	Name	Parentage	Special Features
Australia	1998	(*Lathyrus cicera*) Chalus	IFLLC-1279	n.a.
Cyprus	1998	(*Vicia narbonensis*) acc. 568	n.a.	Drylands
Jordan	1994	(*L. ochrus*) IFLLO-185	IFLLO-185	High yield and resistant to Orobanche
Jordan	1994	(*V. villosa ssp. dasycarpa*) IFLVD 683	IFLVD 683	Tolerant to Orobanche and high biomass
Jordan	1994	(*V. sativa*) IFLVS - 715	IFLVS - 715	High yield and resistant to Orobanche
Lebanon	1997	(*L. cicera*) Jaboulah	IFLLC-492	High yield, drought tolerant
Lebanon	1997	(*V. ervilia*) Amara	IFLVE-2520	High yield and cold tolerant
Lebanon	1997	(*V. sativa*) Baraka	IFLVS - 715	High yield, cold tolerant and non-shattering
Morocco	1990	(*V. sativa*) ILFVS-1812	ILFVS-1812	High yield, erect, tolerance to Orobanche and early
Morocco	1992	(*V. villosa subsp. dasycarpa*) IVLVD-2053	IVLVD-2053	High yield, cold tolerant and resistant to Orobanche
Morocco	1994	(*V. narbonensis*) IFLVN-2387	IFLVN-2387	High yield and early maturing
Morocco	1994	(*V. narbonensis*) IFLVN-2391	IFLVN-2391	High yield and early maturing
Morocco	1994	(*V. sativa*) IFLVS-709	IFLVS-709	High yield and non-shattering
Pakistan	1997	(*V. villosa ssp. dasycarpa*) Kuhak-96	IFLVD-683	High yield and cold tolerant

The process to develop varieties low in anti-nutritional factors involves screening large numbers of germplasm accessions in the hope of finding variants low in the undesirable constituents. For example, in an important find Abd El-Moneim (1999) located *Lathyrus* variants low in β-N-oxalylamino-L-alanine and β-N-oxalyl-α,β-diaminopropronic acid. This is significant because these species have the ability to grow in low rainfall areas and produce sizeable quantities of grain, which are palatable to both humans and animals. Now the potential exists to introduce the low neurotoxin variants into breeding programs. This example demonstrates the importance of screening a diverse genetic base for variation.

Table 18.8. Mean and range of protein content and digestibilities for hay and straw of *Vicia* and *Lathyrus* subsp. The means are for 25 lines that were grown in northeast Syria in 1997. The location received 434 mm rainfall. (Abd El-Moneim, 1999).

Species	Hay		Straw	
	Protein %	DOMD%	Protein %	DOMD%
V. villosa subsp. *dasycarpa*	22	72	11	61
Range	17-24	60-76	7-13	45-72
V. ervilia	19	81	15	69
Range	15-23	72-83	12-17	51-78
V. sativa	20	77	50	61
Range	18-22	72-82	45-55	47-68
V. palaestina	17	70	10	44
Range	13-19	58-79	5-14	32-57
Lathyrus sativus	22	76	13	63
Range	20-25	66-82	8-15	55-75
L. cicera	23	75	15	61
Range	21-26	66-80	10-18	47-68

Table 18.9. Some anti-nutritional factors found in pasture and forage legumes.

Compound	Effect	Species
β-N-oxalylamino-L-alanine	neurotoxin	*Lathyrus*
β-N-oxalyl-α,β-diaminopropronic acid	neurotoxin	*Lathyrus*
Heamagglutinins	Protein metabolism	Vicia species
Oestrogens	infertility	*Medicago* and *Trifolium*
Tannins	Protein metabolism	*Lathyrus, Vicia*

18.9 HARVESTIBILITY

The widespread adoption of a pasture or forage species will also depend on pragmatic factors such as whether the variety is easily and economically harvested. For example, although the on-farm trail cited above showed that replacing bare fallowing with vetch was more productive, the villagers thought that the practice would not be widely adopted until an economic way of harvesting the vetch was found (Thomson and Oglah, 1988). A variety of factors affect how easily a species may be harvested such as seed retention, susceptibility to lodging, stem thickness and the height above the ground at which the seed pods are produced on the plant. These factors are particularly important for forage legumes that are harvested for grain or hay.

Although many of the constraints may be overcome with appropriate harvesting technology or management practices, seed retention is a major limitation, particularly for *Vicia* and *Lathyrus* species whose pods shatters at maturity. This causes two problems, firstly, reduction in grain yield and secondly, build up of seed in the soil that causes weediness in the cereal crops. Fortunately, *Vicia sativa* mutants have been isolated whose pods do not shatter (Abd El-Moneim, 1993). Further more, this trait has a simple inheritance mechanism (Abd El-Moneim, 1999). These mutants have been used to produce eight superior lines, which only shatter at a rate of 2-5% and have other desirable traits compared to 70-80% in the original breeding lines.

The situation is different for pasture legumes because many of the traits that make plants easy to harvest such as, erect growth habit and seed retention are traits that reduce its persistence in a pasture context. For example, one of the traits that contribute to the persistence of subterranean clovers in a ley-farming system is its ability to place its pods underground where they are safe from grazing animals. For medics, short internodes and petioles, which result in a prostrate growth habit, were shown to contribute significantly to their persistence under grazing (Cocks, 1992a). In these cases the attributes that favoured persistence do not favour easy harvesting of either the stem for forage or the seed. Thus, commercial seed production for medics and clover pastures relies on specially developed machinery. In Australia the agricultural sector is well serviced by research and development and farmers have sufficient money to invest in expensive machinery. In the WANA region however, the adoption of medic leys has been hampered by a lack of appropriate and inexpensive machinery to harvest and thresh the seedpods.

18.10 ADAPTATION TO ENVIRONMENT AND AGRICULTURAL ECOSYSTEMS

For a variety to perform well in a given environment it must be adapted to the local conditions. Although the emphasis on physiological adaptation differs somewhat between pasture and forage cultivars in a rainfed Mediterranean environment, a well adapted genotype of either must: a. finish their life cycle in a given time frame that is dictated by moisture availability and the onset of summer drought , b. withstand the local winter temperature extremes, c. be able to perform well under the local edaphic constraints and d. have mechanisms for coping with local pests and diseases. A typical plant-breeding program must address all of these issues.

18.10.1 Phenology: the Importance of Timing

The time from germination to setting of mature seeds has obvious ecological significance. In unpredictable environments the 'decision' of when to produce seed is a trade-off between setting as many seeds as possible by delaying reproduction in favour of accumulating biomass or producing fruits as fast as possible in anticipation of sudden death (Zeide, 1978). Annuals are an example of the second option. Even so, within this strategy, flowering time represents a tradeoff. For example, the timing and volume of winter rainfall will determine the effective growing season for Mediterranean annuals (Cooper *et al.*, 1987). Thus we expect well adapted genotypes to flower late enough to accumulate adequate biomass but early enough so that grain filling is not be impeded by severe moisture stress as the soil dries out.

Thus for wild germplasm it is not surprising to find a correlation between length of season and flowering time. This has been documented in studies where the different ecotypes within a species have been compared (e.g. Smith *et al.*, 1995; Piano *et al.*, 1996). In a study examining the reproductive patterns of annual legumes in Syria, Ehrman and Cocks (1996) demonstrated that there was a well defined shift from late to early flowering along an aridity gradient stretching from the mesic coast to the arid interior. A shift in flowering time has also been observed for within site elevation (Cocks, 1992b) and altitude above sea level (Loi *et al.*, 1997). In both cases it was hypothesised that soil moisture availability was the governing factor. In the study by Ehrman and Cocks (1996) the ease with which many species alter their flowering schedule in response to aridity indicates that considerable variation exists for this trait in many species and that it is under fairly simple genetic control (review by Rathke and Lacey, 1985).

Thus the time at which reproductive events occur is critical if the pasture or forage variety is to be adapted to local conditions. For example, in high rainfall environments maximum seed and herbage yield would be achieved if the cultivar flowered later. However, in a dry environment early flowering varieties will ensure that the switch from vegetative to reproductive development will occur in time to allow seed filling while there is still adequate moisture in the soil. For example, the use of early cultivars in Western Australia trebled the area of land suitable for subterranean clover (Donald, 1970). In a breeding program it is essential therefor, to screen large numbers of genotypes to understand the level of variation for flowering and pod setting times so that appropriate genotypes can be selected for the target environment.

18.10.2 Cold Tolerance

Frost is one of the most limiting factors to winter growing annuals in Mediterranean environments. Since damage caused by frost may in many cases can lead to total crop failure (Saxena *et al.*,1993), a major objective of breeding programs is to introduce cold tolerance into desirable genotypes. Once again breeders must look to natural populations for exploitable variation. A natural place to begin looking is in areas that are susceptible to frost. Cocks and Ehrman (1987) examined frost tolerance for seventy-nine indigenous and introduced species of Mediterranean pasture legumes. They found considerable ecotypic differentiation in most of the species studied. In general, accessions collected in cold areas were frost tolerant while those from mild environments were not. However, they did find tolerant accessions in mild environments and susceptible accessions in cold environments.

To facilitate the selection process in breeding for cold tolerance it is desirable to find physiological traits that are highly correlated because frost events are not predictable and it would be more reliable to select for morphological traits. The association of cold tolerance with morphological traits has been well documented for perennial forage legumes. For example, a negative correlation between top growth (Shih *et al.*, 1967), plant height (Smith, 1961) and cold tolerance was found with *Medicago sativa*. Few studies though, have looked at adaptive traits of annual pasture or forage legumes. Bounejmate (1991) hypothesised that the greater specific leaf weight reported for genotypes of *M. truncatula* and *M. aculeata* collected at >1000 m above sea level, represented an adaptation to frequent frost. In another study, Ratinam *et al.* (1994) found a consistent relationship between genotypes of *Lalthyrus ochrus* with a low shoot - root ratio and cold tolerance. They also found there was a significant positive correlation between sugar concentrations in the root and frost tolerance. However, given the expense and difficulty of measuring sugar contents this is not a good selection criterion. It is clear that this is an area that requires further investigation.

18.10.3 Edaphic Factors

Legumes in general respond favourably to phosphate application. For example, legume presence in natural pastures in northeast Syria increased from 17 to 46% and total dry matter increased from 200 kg/ha to 1000 kg/ha (Osman *et al.*, 1991; 1994) after fertilisation with phosphate. Indeed the rapid increase in pasture production and therefor wheat yields in Australia during the 1950s was a result of widespread phosphorus application (Puckridge and Carter, 1980). Although there is considerable intra and inter - species variation in responses to phosphate, breeding for phosphate response is a low priority because in agricultural eco-systems phosphate can be easily applied to rectify soil deficiencies.

Soil pH has a major effect on the distribution of annual pasture legumes. In an agricultural context in which there is considerable variation in soil pH it becomes important to find adapted genotypes. This is particularly the case in Australia. In general medics and *Vicia* species favour alkaline soils while clover favours acid soils. However, considerable variation in the ability to nodulate under acid stress, both between and within species has been reported for medics (e.g. Howieson and Ewing, 1989; Loi *et al.*, 1993) which makes it possible to develop genotypes with wider adaptation.

18.10.4 Rhizobium - Host Relations

The *Rhizobium* species and strain that a legume species associates with will influence its performance and adaptation. For example, the native rhizobia capable of symbiosis with potential pasture legume crops in the Turkish highlands were surveyed. Estimates were made of the numbers and N fixing efficiency of native *R. meliloti* with a range of annual *Medicago* species. While the numbers of bacteria were generally adequate to permit efficient nodulation, the N fixing efficiency of three of the four host species with the indigenous rhizobia was often low in comparison to commercial innoculum (Materon *et al.*, 1995). This indicates that scope exists for improvement by manipulating the host- Rhizobia relations. In fact, this was the approach used to increase the area under medic pastures in Australia. The naturalized strains of *R. meliloti* did not survive when they were introduced to the acid soils in the southeastern wheatbelt of western Australia. However, isolates were found in acid soils of the Mediterranean basin, which were utilized to extend the adaptation of medics by hundreds of thousands of hectares.

Genetic modification of Rhizobial strains has also been used to enhance the Host - Rhizobia relations and improve production in Alfalfa. Seven recombinant strains of *R. meliloti* that contained an extra copy of nifa and/or dct genes inserted into the chromosome of two wild-type strains were used in the study to examine the effects of modifying or increasing the expression of these genes on alfalfa biomass. Two of recombinant strains, increased alfalfa biomass by 12.9% compared with the wild type, and 17.9% over the uninoculated controls, at a site where soil nitrogen and organic matter content was lowest (Hadegorn, 1994).

18.10.5 Disease and Pest Resistance

Resistance to the major pests and diseases is a necessary feature if a cultivar is to deliver a stable performance. In the WANA region powdery mildew (caused by *Erysiphe pisi)*, botrytis blight (caused by *Botrytis cinerea*), ascochyta blight (caused by *Ascochyta pisi*), and downy mildew (caused by *Peronospora viciae*) pose a significant threat to the success of *Vicia* and *Lathyrus* forage species. A number of sources of resistance for each of these diseases have been found in screening trials carried out at ICARDA (Abd El-Moneim, 1999).

Root knot nematode (caused by *Meloidogyne artiellia*) is another limiting factor in the WANA region to the introduction of forage legumes. Several selections of *Vicia dasycarpa* and *V. narbonensis* were found to be highly resistant in screening trials also carried out at ICARDA (Robertson *et al.*, 1996).

Another economically important parasite in the WANA region is *Orobanche crenata* a parasitic weed of legume species. Once again screening, of genetic resource collections was used to look for exploitable variation in resistance. It emerged that *L. ochrus* is largely free of the parasite while accessions of *L. sativus, V. narbonensis* and *V. dasycarpa* showed varying levels of resistance ranging from parasite free to highly susceptible lines.

Although initially Australian pastures suffered primarily from lucerne flea (*Sminthurus viridus*) and red-legged earth mite (*Halotydeus destructor*), from 1943 onwards a rapid influx of diseases and pests from overseas coincided with increased air travel (Cocks *et al.*, 1980). Some of the economically important pests of pasture legumes include bean yellow mosaic virus, the root diseases *Rhizoctonia solani*, *Fusarium aveaceum,* clover scorch caused by *Kabatella caulivora,* clover stunt and clover red leaf viruses, the weevil (*Sitona humeralis*), the spotted alfalfa aphid (*Therioaphis trifolii),* and blue green aphid, (*Acyrthosiphon kondoi*). Cocks *et al.* (1980) reflect that prior to the arrival of a particular disease many of the selection criteria used to develop improved pasture genotypes may in fact have unintentionally left the genotype more susceptible to the newly introduced diseases. They go on to cite the example of subterranean clovers susceptible to *Kabatella caulivora.* Earliness, length and thickness of petioles and size of leaves were used as selection criteria to develop the cultivars, however, the same attributes are also positively correlated to susceptibility.

Despite the challenges, screenings of massive numbers of accessions were and still are being carried out to look for sources of resistance to include in breeding programs. For example, 22 medic and clover species were screened for Sitona weevil tolerance. A total of 7045 accessions were used in one trial to find 62 resistant genotypes (Cocks *et al.*, 1980). To date there are resistant varieties for most of the economically important pasture legume pests.

18.10.6 Reproductive Ecology

Pasture and forages differ significantly in that pastures usually rely on regeneration from a soil seed back for renewal while forages are sown each season. Thus if a pasture is to be persistent an well-adapted reproductive ecology is critically important. While the time to maturity is critical to both pastures and forage adaptation, what happens to the seed beyond maturity will also influence the success of pasture legumes. This subject has received an enormous amount of attention, particularly in Australian literature, and so no attempt here will be made to review it. However the important aspects of the seed ecology pertinent to legume pasture persistence will be outlined below.

18.10.7 Seed Dormancy

Seed dormancy and how it is reversed protects the plant population in a number of ways. Firstly, dormancy over summer protects the seed bank from germinating in summer rains that do not provide adequate moisture to sustain seedling growth. Long term dormancy that lasts a number of years ensures population survival in unproductive seasons (Ritland, 1983). This has obvious connotations for ley-farming because during the cropping phase most of the legume seeds that germinate are killed with herbicides or do not perform well under the crop canopy. Taylor and Downs (1995) calculated that at least 40% seed dormancy is necessary for a species to maintain its seed bank in a ley-farming system. When in the season the seeds lose their dormancy is also important, particularly in an environment where initial winter rains do not deliver adequate moisture, to sustain seedling growth to prevent false break (Smith *et al.*, 1996).

In Mediterranean legumes, hardseededness or seed coat impermeability to water, is the main dormancy mechanism (Quinlivan, 1971). Seed coat softening appears to occur in response to diurnal temperature fluctuations of a given magnitude only after a certain amount of thermal pre-conditioning at high temperatures (Taylor, 1981; Taylor and Ewing, 1992; Taylor, 1993). There is an enormous amount of variation both between and within legumes species in the initial hardseededness, the rate of softening, and the temperature regime that is needed to initiate softening. Table 18.10 shows the initial and final hardseededness and the time it takes for half the seeds produced in one year to become soft under field conditions. It demonstrates and is representative of the enormous variation that exists both between and within species for seed softening parameters.

Plant breeders in Australia have exploited this variation in dormancy to produce genotypes that have dormancy characteristics adapted to specific environment conditions and farming systems. For example, in dry environments with irregular rainfall patterns at the beginning of winter and for long cereal phases, a particularly hardseeded genotype that loses its dormancy late in the season would ensure a more persistent pasture than a soft seeded genotype. In this scenario the adapted genotype would still have an adequate seed bank to regenerate from after the cereal phase and its seed losses due to false breaks would be minimized in the pasture phase.

Table 18.10. Calculated half lives and initial and final soft seed levels for selected Mediterranean legumes. Source (Cocks *et al.*, 1996; Street, 1999).

Strain	Half life (days)	Initial soft seed %	Final soft seed %
Subclover cv *Nungarin*	46	10	47
M. polymorpha cv Serena	126	5	37
T. glomeratum accn. 30	136	1	54
T. glomeratum accn 42	104	3	95
V. sativa subsp. *amphicarpa (1)*	141	0.4	92
V. sativa subsp. *amphicarpa (2)*	42	74	38

18.10.8 Seed Number vs. Seed Size

For most plant species, seed size is an important fitness character that, under resource limiting conditions, is preserved at the expense of seed number (Harper and Gajic, 1961). However, seed size varies considerably between species and even between ecotypes within species. Given this, the balance between seed size and seed number for a pasture species will be an important fitness attribute. It is generally held that in productive environments that contain more competitive plant interactions, larger seed sizes are favoured while in harsh, disturbed or dry environments smaller seed sizes are favoured (Baker, 1972; Silvertown, 1981). For example, individual seed weights of many Mediterranean annual legumes decrease with increasing aridity and temperature (Piano *et al.*, 1996; Ehrman and Cocks, 1996). Despite this, evidence exists that show a certain amount of substitutability among reproductive traits. For example, early flowering would be necessary if long term seed dormancy is absent. Norman *et al.* (1998) demonstrated this when they grew pairs of species collected at the same sites in the east and west Mediterranean. Species from the same site differed in their adaptation by trade-offs between seed size, seed dormancy, germination rate and flowering time.

Once again these reproductive characteristics must be considered when seeking to find persistent pasture genotypes adapted to specific environments.

18.10.9 Interaction Between Seed Biology and Grazing

Both flowers and seeds of pasture plants are vulnerable to grazing animals. Clearly if animals are allowed to heavily graze pastures during flowering then there will be a significant reduction in seed yield which in turn will influence persistence. For most species very little can be done in terms of genetic improvement to limit flowers lost to grazing. Damage control in this case is more a management issue. Several species though exhibit amphicarpy in which they flower and produce seeds on both above and below ground stems. In this case the underground flowers and seeds are protected from grazing. One such species with considerable potential as a pasture species is underground vetch (*Vicia sativa* subsp. *amphicarpa*). In a trial where barley was rotated with underground vetch in northeast Syria, by the end of the second rotation cycle sufficient seeds remained in the soil to allow regeneration the following year (Street, 1999). Its draw back though is it does not produce a lot of biomass, or for that matter, a great quantity of underground seed (5-200 per plant). At ICARDA breeding is underway to cross underground vetch with *Vicia sativa* to increase biomass while still maintaining an underground seed yield.

Seeds are also prone to animal grazing. (Cocks, 1988), showed that sheep grazing medic pods have the capacity to reduce the seed bank from 500kg/ha to 10 kg/ha over summer. Although this to is a grazing management issue there is more scope for genetic solutions to limit seed loss. Thomson *et al.* (1990) showed that seeds larger than 2 mg are unlikely to survive in large numbers, while up to 60% of seeds smaller than 1 mg will survive. Although the majority of Mediterranean legumes have seed sizes larger than 2 mg there is also a significant number of species that have small enough seeds to escape intact. *M. minima* (1.4 mg), *Trifolium stellatum* (1.4 mg), *Trifolium scabrum* (0.9 mg), *Trigonella monspeliaca* (0.8 mg) and *Trifolium tomentosum* (0.6 mg) are example of legumes with very small seeds.

Other species escape grazing by putting seeds underground. The example of amphicarpic plants has been outlined above. Subclover is another group of species that escape grazing by putting their seed pods underground after flowering. This is possibly one reason why subclover has been such a persistent and successful commercial pasture species in Australia.

18.11 Genetic Resources

Most of the discussion above emphasized the importance of screening intra-specific variation for morpho-agronomic attributes that can be exploited in breeding programs. This process relies on collections of germplasm in which sufficient variation exists to exploit. The germplasm is usually assembled through plant collection missions that target a variety of environments in an attempt to capture ecotypic variation for the target species. The material is then characterized and the seed stored in such a way to maximize viability. Another important aspect of genetic resource collections that enhance their utility is the documentation that describes the material. Of particular importance here is information on the environment from where the material originated. Table 18.11 details the pasture and forage legume collection held in the ICARDA gene bank.

The germplasm held at ICARDA, a large portion of which originates from the WANA region, is freely available to the scientific community upon request. Table 18.12 illustrates the distribution of pasture and forage germplasm of the ICARDA collection in the past five years. More than 9000 samples of pasture and forage species germplasm have been distributed per year. Nearly half of the samples distributed are of *Vicia* species (4400 per year), mostly *V. sativa* and *V. narbonensis*. More than 1300 samples of *Lathyrus* species have been distributed per year, mostly *L. sativus* and 2200 accessions of *Medicago* have been distributed per year, mainly *M. aculeata*, *M. laciniata*, *M. littoralis*, *M. minima*, *M. orbicularis*, *M. polymorpha*, *M. rigidula* and *M. truncatula*. Most of the other distributions are of *Trifolium* species, over 800 samples per year.

Selections made within accessions of the germplasm collections and lines developed from crosses with germplasm accessions have been distributed by the forage-breeding program in the form of ICARDA International Nurseries (Table 18.13). In the past five years an average of 212 sets of these nurseries have been distributed per year, with sixteen test entries per trial.

18.12 Conclusions and Prospects

Any discussion of the utilization of pasture and forage legumes from the Mediterranean region must be within the context of the farming systems they are used in. The significance of these pasture and forage legumes is considerably more than just from their direct utilization for animal grazing and feeding. The important beneficial role of pasture and forage legumes in increasing the production of the system as a whole has been well documented in Australia and the WANA region. Cereal cropping systems using a legume rotation have been shown to have substantial increases in yield (up to 65%) and protein content (up to 3% higher protein). These beneficial effects are a consequence of several factors including an increase in soil fertility from nitrogen fixation, improvement of soil structure, control of pests and diseases by providing a break crop, and by providing weed control. Animal production is also greatly increased with a legume component in pastures.

Many pasture and forage cultivars used in Australia are either direct introductions from the Mediterranean region are material selected from within this germplasm. A number of varieties of forages have been released in the WANA region that are selected from germplasm from the region or selected from within this germplasm. The largest collection of pasture and forage germplasm from the Mediterranean is at ICARDA, and over the past five years more than 9000 samples of this germplasm has been distributed each year.

The increasing concern about long term sustainability of farming systems is leading to a renewed interest in the role of pasture and forage legumes in maintaining high productivity. Most of the important pasture and forage legume species useful in rotations in temperate agricultural systems originate from the Mediterranean region. Germplasm from this region has proved useful in developing legume rotation crops and the large number of samples distributed by ICARDA in the past five years shows the strong interest in this germplasm.

Table 18.11. Numbers of accessions and species of potential forage and pasture legumes held at ICARDA.

Genus	**Number of species**	**Number of accessions**
Astragalus	23	913
Hedysarum	6	166
Lathyrus	43	3045
Lotus	9	138
Medicago	65	8412
Pisum	5	4981
Trifolium	80	4358
Trigonella	25	646
Vicia	57	5573
Total	313	28232

Table 18.12. Distribution of pasture and forage germplasm by ICARDA from 1995-1999 inclusive.

Genus/species	1995	1996	1997	1998	1999	Mean
Vicia						
V. ervilia	287	433	196	47	24	197.4
V. hybrida	292	66	131	134	32	131
V. narbonensis	315	217	646	283	68	305.8
V. sativa	2153	4289	2547	2147	1818	2590.8
Others *Vicia*	1191	963	1759	1303	689	1181
Total	4238	5968	5279	3914	2631	4406
Lathyrus						
L. cicera	20	256	67	36	62	88.2
L. ochrus	15	235	52	38	49	77.8
L. sativus	209	1528	280	1032	524	714.6
Others *Lathyrus*	18	1335	483	100	371	461.4
Total	262	3354	882	1206	1006	1342
Medicago						
M. aculeata	36	116	369	194	16	146.2
M. laciniata	196	27	1	223	46	98.6
M. littoralis	302	93	16	254	103	153.6
M. minima	52	47	0	465	132	139.2
M. orbicularis	65	871	0	438	13	277.4
M. polymorpha	150	421	3	1072	103	349.8
M. rigidula	31	393	14	664	97	239.8
M. truncatula	101	385	199	354	144	236.6
Others *Medicago*	66	702	290	1438	556	610.4
Total	999	3055	892	5102	1210	2251.6
Other forages						
Trifolium	1261	1115	791	441	512	824.0
Astragalus	143	0	83	104	223	110.6
Scorpiurus	122	1	11	51	101	57.2
Trigonella	151	10	13	183	146	100.6
Others	300	47	146	362	251	221.2
Total	1977	1173	1044	1141	1233	1313.6
Forage Total	**7476**	**13550**	**8097**	**11363**	**6080**	**9313.2**

Table 18.13. ICARDA International Nursery distribution of forage species (16 entries per trial).

Crop	Season					
	1995/96	1996/97	1997/98	1998/99	1999/00	Mean
V. sativa	51	52	45	42	46	47.2
V. narbonensis	42	39	38	32	37	37.6
V. ervilia	35	32	30	34	26	31.4
V. villosa subsp. *dasycarpa*	36	35	35			21.2
L. sativus	41	47	37	36	45	41.2
L. cicera	31	28	26	-	-	17
L. ochrus	29	29	24	-	-	16.4
Total	**265**	**262**	**235**	**144**	**154**	**212**

ACKNOWLEDGEMENTS

The authors gratefully acknowledge the invaluable contributions of Dr. Clive Francis and Dr. Sarita Bennett (Centre for Legumes in Mediterranean Agriculture) for information concerning Australian pasture and forage varieties developed from Mediterranean germplasm, of Mr. Jan Konopka (Germplasm Documentation Officer, ICARDA) for data concerning the ICARDA pasture and forage germplasm collection and its distribution, and of Dr. Bruno Ocampo (International Trials Scientist, ICARDA) for information about distribution of ICARDA Forage International Nurseries and release of cultivars based on ICARDA supplied germplasm.

REFERENCES

Abd El Moneim, A.M., 1993. Selection for Non-Shattering Common Vetch *Vicia sativa* L. *Plant Breeding*, 110: 168-171.

Abd El Moneim, A.M. and Cocks, P.S., 1986. Adaptation of *Medicago rigidula* to a cereal-pasture rotation in north-west Syria. *Journal of Agricultural Science*, 107: 179-186.

Abd El-Moneim, A.M., 1999. *Forage Legume Improvement*. Aleppo, ICARDA.

Ackerman, E.A., 1941. The Koepen classification of climates in North America. *Geographical Review*, 31: 105-111.

Arndt, T.M., Dalrymple, D.G. and Ruttan, V.W., 1977. Resource allocation and productivity in national and international agricultural research. pp. 1-617. University of Minnesota Press, Minneapolis.

Baker, H.G., 1972. Seed weight in relation to environmental conditions in California. *Ecology*, 53(6): 997-1010.

Bounejmate, M., 1991. *Soil and climatic factors affecting the distribution of annual Medicago species in Morocco*. PhD Thesis. Plant Sciences. University of Western Australia, Perth.

Christian, C.S. and Shaw, N.H., 1952. *Australian Journal of Agricultural Research*, 28: 795.

Christiansen, S. *et al.*, 1994a. *On-Farm Trials comparing forage legume/barley rotation to fallow / barley rotations and continuous barley in northwest Syria.* Aleppo, ICARDA.

Christiansen, S. *et al.*, 1994b. *Medic pasture establishment using pods sown into barley in the year prior to pasture phase.* Aleppo, ICARDA.

Cocks, P.S., 1988. Seed production and seed survival under grazing of annual medics (*Medicago* spp.) in north Syria. *Journal of Agricultural Science, Cambridge*, 110: 455-463.

Cocks, P. S. 1992a. Plant attributes leading to persistence in grazed annual medics (Medicago spp.) growing in rotation with wheat. *Australian Journal of Agricultural Resources*, 43(7): 1559-1570.

Cocks, P.S., 1992b. Evolution in sown populations of subterranean clover (*Trifolium subterraneum*) in southern Australia. *Australian Journal of Agricultural Resources*, 43(7): 1583-1595.

Cocks, P. S. and Ehrman, T.A.M., 1987. The Geographic Origin of Frost Tolerance in Syrian Pasture legumes. *Journal of Applied Ecology*, 24: 673-684.

Cocks, P.S. and Gintzburger, G., 1993. Long-term sustainability of livestock-producing farming systems in contrasting regions with Mediterranean-type climates. In: *XVII International Grassland Congress*. Massey University, SIR publishing.

Cocks, P.S., Mathison, M.J. and Crawford, E.J., 1980. From wild plants to pasture cultivars: annual medics and subterranean clover in south Australia. In: *Advances in Legume Science* (eds. Summerfield, R.J. and Bunting, A.H.). pp. 569-596. Kew, Royal Botanic Gardens.

Cody, M.L. and Mooney, H.A., 1978. Convergence versus nonconvergence in Mediterranean climate ecosystems. *Annual Review of Ecological Systems*, 9: 265-321.

Cooper, P.J.M. and Bailey, E., 1990. *Livestock in Mediterranean farming systems, a traditional buffer against uncertainty: now a threat to the agricultural resource base*. Risk in agriculture, Washington.

Cooper, P.J.M., Gregory, P.J., Tully, D. and Harris, H.C., 1987. Improving water use efficiency of annual crops in the rainfed farming systems of West Asia and North Africa. *Experimental Agriculture*, 23: 113-158.

Craufurd, P.Q., Austin, R.B., Acevedo, E. and Hall, M.A., 1991. Carbon isotope discrimination and grain-yield in barley. *Field Crops Research*, 27(4): 301-313.

CSIRO, 1977. *Rural Research*. CSIRO, Canberra.

Di Castri, F., 1981. Mediterranean Type Shrublands of the World. In: *Ecosystems of the world, Mediterranean Type Shrublands* (eds. Di Castri, F. and Mooney, H.A.). New York, Elsevier Scientific Publishing Company.

Donald, C.M., 1963. *Australian Journal of Science*, 25: 386-395.

Donald, C.M., 1970. *Australian Grasslands*. Australian National University, Canberra.

Donald, C.M., 1981. Agriculture in the Australian economy. Sydney University Press, Sydney.

Ehrman, T.A.M. and Cocks, P.S., 1996. Reproductive patterns in annual legume species on an aridity gradient. *Vegetatio*, 122: 47-59.

Francis, C.M. and Millington, A.J., 1965. Wether bioassay of annual pasture legumes. IV. The oestrogenic activity of annual medic pastures. *Australian Journal of Agricultural Science*, 16: 565.

Gartell, A., 1998. *Why wheat grain protein levels vary*. Agriculture Western Australia, Perth.

Gladstones, J.S., 1966. Naturalized Subterranean Clover (*Trifolium subterraneum* L.) in Western Australia: The Strains, Their Distributions, Characteristics, and Possible Origins. *Australian Journal of Botany*, 14: 329-54.

Hadegorn, C. 1994. *Rhizobium meliloti field tests*. http://www.acceessexcellence.org/AB/BA/Rhizobium_Meliloti_Tests.html.

Harper, J. and Gajic, D., 1961. Experimental studies of the mortality and plasticity of a weed. *Weed Research*, 1: 91-104.

Howieson, J.G. and Ewing, M.A., 1989. Annual species of *Medicago* differ greatly in their ability to nodulate on acid soils. *Australian Journal of Agricultural Research*, 40: 842-850.

Jones, M.J., 1990. The role of forage legumes in rotation with cereals in Mediterranean areas. In: *The role of legumes in farming systems of Mediterranean areas* (eds. Osman, A.E., Ibrahim, M.H. and Jones, M.A.). pp. 184-204. ICARDA, Aleppo.

Keating, J.D.H., Chapanian, N. and Saxena, M.C., 1988. Effect of improved management of legumes in a legume - cereal rotation on field estimates of crop nitrogen uptake and symbiotic nitrogen fixation. *Journal of Agricultural Science, Cambridge*, 110: 651-659.

Loi, A., Howieson, J.G., Cocks, P.S. and Caredda, S., 1993. The adaptation of *Medicago polymorpha* to a range of edaphic and environmental conditions effect of temperature on growth and acidity stress on nodulation and nod gene induction. *Australian Journal of Experimental Agriculture*, 33: 25-30.

Loi, A., Cocks, P.S., Howieson, J.G. and Carr, S.J., 1997. Morphological characterization of Mediterranean populations of *Biserrula pelecinus* L. *Plant Breeding*, 116: 171-176.

Macindoe, S.L., 1975. History of wheat production in Australia. In: *Australian field crops. Volume 1: wheat and other temperate cereals*. (eds. Lazenby, A. and Matheson, E.M.). Sydney, Angus and Robertson: 99-121.

Materon, L.A., Keating, J.D.H., Beck, D.P., Yurtsever, N., Karug, K. and Altuntas, S., 1995. The role of rhizobial biodiversity in legume crop productivity in the west Asian highlands, III. *Rhizobium meliloti*. *Experimental Agriculture*, 31: 493-99.

McPhee, M.J., Bell, A.K., Meaker, G.P., Griffith, G.R. and Graham, P., 1998. PRO Plus: a whole-farm fodder budgeting decision support system. In: *Beef - the path forward, NSW Agriculture Beef Products Conference*, Armidale, Australia, November 1998. pp. 122-125.

Nordblom, T.L., 1987. *Farming practices in southern Idleb province, Syria: 1985 survey results*. ICARDA, Aleppo.

Norman, H.C., Cocks, P.S. and Smith, F.P., 1998. Reproductive strategies in Mediterranean annual clovers: germination and hardseededness. *Australian Journal of Agricultural Research*, 47: 141-159.

Osman, A. E., Pagnotta, M., Russi, L., Cocks, P.S. and Falcinelli, M., 1990. The role of legumes in improving marginal lands. In: *The role of legumes in the farming systems of the Mediterranean areas* (eds. Osman, A.E., Ibrahim, M.H. and Jones, M.A.). pp. 205-216. ICARDA, Aleppo.

Osman, A.E., Cocks, P.S., Russi, L. and Pagnotta, M., 1991. Response of Mediterranean grassland to phosphate and stocking rates biomass production and botanical composition. *Journal of Agricultural Science*, 116: 37-46.

Osman, A. E., Cocks, P.S. and Bahhady, F., 1994. Response of Mediterranean grassland to phosphate and stocking rates: Livestock production. *Journal of Agricultural Science*, 123(3): 319-326.

Papastylianou, I., 1987. Effect of proceeding legume or cereal on barley grain and nitrogen yield.

Journal of Agricultural Science, 108: 623-626.
Peoples, M.B., Gault, R.R., Angus, J.F., Bowman, A.M. and McCallum, M., 1999. *Comparisons of the efficiency of nitrogen fixation in pastures.* http://life.csu.edu.au/agronomy/papers/73/73.html.
Piano, E., Pecetti, L. and Carroni, A.M., 1996. Climatic adaptation in subterranean clover populations. *Euphytica*, 92(1-2): 39-44.
Puckridge, D.W. and Carter, E.D., 1980. Dryland agriculture in southern Australia and its relationship to other regions - A perspective. *International Congress on Dryland Farming*, Adelaide, Department of Agriculture, South Australia.
Puckridge, D.W. and French, R.J., 1983. The Annual Legume Pasture in Cereal-Ley Farming Systems of Southern Australia: A Review. *Agriculture, Ecosystems and Environment*, 9: 229-267.
Quinlivan, B.J., 1971. Seed coat impermeability in legumes. *Journal of the Australian Institute of Agricultural Science*, 37: 283-295.
Rashed, E., 1986. *The effects of fertilization and crop rotation on rainfed barley development, growth yields in a semi-arid Mediterranean environment.* McGill University Press, Montreal.
Rathke, B. and Lacey, E.P., 1985. Phenological patterns of terrestrial plants. *Annual Review of Ecology and Systematics*, 16: 179-214.
Ratinam, M., El-Moneim, A. and Saxena, M.C., 1994. Variations in sugar content and dry matter distributions in roots and their associations with frost tolerance in certain forage legumes species. *Journal of Agronomy and Crop Science*, 173: 345 - 353.
Ritland, K., 1983. The joint evolution of seed dormancy and flowering time in annual plants living in variable environments. *Theoretical Population Biology*, 24: 213-43.
Robertson, L.D., Singh, K.B., Erskine, W. and Abd El Moneim, A.M., 1996. Useful genetic diversity in germplasm collections of food and forage legumes from West Asia and North Africa. *Genetic Resources and Crop Evolution*, 43: 447-460.
Rowland, I.C. and Halse, N.J., 1980. The role of legumes in Australian dryland agriculture. *International Congress on Dryland Farming, Adelaide*. Department of Agriculture, South Australia.
Russell, J.S., 1960. Soil fertility changes in the long term experimental plots at Kybybolite, South Australia. I. Changes in pH, total nitrogen, organic carbon and bulk density. Australian *Journal of Agricultural Science*, 11: 902-926.
Saxena, M.C., Abd El Moneim, A. and Ratinam, M., 1993. Vetches (*Vicia* spp.) and chicklings (*Lathyrus* spp.) in farming systems in West Asia and North Africa and Improvement of these crops. In: *Potential for* Vicia *and* Lathyrus *as a New Grain and Fodder Legumes for Southern Australia* (eds. Garline, J.R. and Perry, M.W.). pp. 2-9. CLIMA, Perth.
Shih, S. C., Jung, G.A. and Shelton, D.C., 1967. Effect of temperature and photoperiod on metabolic changes in alfalfa in relation to cold hardiness. *Crop Science*, 7: 384-389.
Silvertown, J.W., 1981. Seed size lifespan and germination date as coadapted features of plant life history. *American Naturalist*, 118: 860-864.
Smith, D., 1961. Association of fall growth habit and winter survival in alfalfa. *Canadian Journal of Plant Science*, 41: 244 - 251.
Smith, F.P., Cocks, P.S. and Ewing, M.A., 1995. Variation in the morphology and flowering time of cluster clover (Trifolium glomeratum L.) and its relationship to distribution in southern Australia. *Australian Journal of Agricultural Research*, 46: 1027-1038.
Smith, F.P.,Cocks, P.S. and Ewing, M.A., 1996. Short-term patterns of seed softening in *Trifolium subterraneum*, *T. glomeratum*, and *Medicago polymorpha*. *Australian Journal of Agricultural Research*, 47: 775-785.
Street, K.A., 1999. *Adaptation of subterranean vetch (*Vicia sativa *spp.* amphicarpa *batt.) to Mediterranean environments: the importance of genetic variation and amphicarpy*. PhD Thesis. Faculty of Agriculture. Perth, University of Western Australia.
Taylor, G.B., 1981. Effect of constant temperature treatments followed by fluctuating temperatures on the softening of hard seeds of *Trifolium subterraneum* L. *Australian*

Journal of Plant Physiology, 8: 547-558.

Taylor, G.B., 1993. Effect of some characteristics of diurnal temperature fluctuations on the softening of hard seeds of Medicago polymorpha. In: *XVII International Grassland Congress*. Massey University, SIR Publisher.

Taylor, G.B. and Downes, P.A., 1995. *MEDSIST: a model for predicting the effect of hard seededness on the persistence of annual medics in ley farming systems in a Mediterranean-type environments*. CLIMA, Perth.

Taylor, G.B. and Ewing, M.A., 1992. Long-term patterns of seed softening in some annual pasture legumes in a low rainfall environment. *Australian Journal of Experimental Agriculture*, 32: 331-337.

Taylor, G. B., Maller, R.A. and Rossiter, R.C., 1991. A model describing the influence of hard seededness on the persistence of an annual forage legume, in a ley farming system, in a Mediterranean-type environment. *Agriculture, Ecosystems and Environment*, 37: 275-301.

Thomson, E.F. and Oglah, M., 1988. *On-farm experiments with forage legumes - six years results form Breda. Aleppo, Pasture forage and livestock program*. ICARDA, Aleppo.

Thomson, E.F., Rihawi, S., Cocks, P.S., Osman, A.E. and Russi, L., 1990. Recovery and germination rates of seeds of Mediterranean medics and clovers offered to sheep at a single meal or continuously. *Journal of Agricultural Science, Cambridge*, 114: 295-299.

Tully, D., 1984. *Land use and farmer strategies in al-Bab: feasibility of forage legumes in place of fallow*. ICARDA, Aleppo.

Watson, E.R., Lapins, P., Arnold, G.W., Barren, R.J.W. and Anderson, G.W., 1977. Effect on Cereal Crop and Sheep Production of Two Rotations in a Ley Farming System in the South-west of Western Australia. *Australian Journal of Experimental Agriculture and Animal Husbandry*, 17: 1011-1019.

Webber, G.D., Cocks, P.S. and Jefferies, B.C., 1976. Farming system in South Australia. Ministry of Agriculture and fisheries, Adelaide, South Australia.

Williams, C.D. and David, D.J., 1976. *Australian Journal of Agricultural Science*, 14: 85-93.

Zeghida, A.S., Hamrit, S. and Korichi, M.L., 1994. Experimental results of integration of crops and sheep husbandry in Algeria. *Al Awamia*, 89:109-118.

Zeide, B., 1978. Reproductive behaviour of plants in time. *American Naturalist*, 112: 636-639.

Section 5

Afterword

Chapter 19

Future Conservation and Utilisation Priorities

Nigel Maxted and Sarita Jane Bennett

19.1 INTRODUCTION

With the ever increasing human population, poverty alleviation is becoming an increasing priority throughout the world. The human population reached 6 billion towards the end of 1999, seventy eight percent of whom live in developing countries with associated poor economic development and widespread social depravation (McNeeley *et al.*, 1995). Although the Mediterranean region does not have the fastest growing population, it is only increasing at 1.3% per year, it does have one of the highest levels of population density at 111 per sq. km. (Cincotta and Engelman, 2000). It is the highest population in any biodiversity hotspot. Even more telling Cincotta and Engelman (2000) note that only 5% of the original Mediterranean vegetation remains extant; the highest level of taxonomic and genetic erosion of *any* world biodiversity hotspot.

As Hawkes *et al.* (2000) discuss, there has often been a conflict between unrestrained development and conservation, the so-called conservation / development paradox. Which simply stated means that as a direct result of countries developing their agriculture and industry, genetic diversity is either threatened or lost. A simple solution would be to control population growth, adopt more sustainable development practices and ensure the provision of more equitable division of the benefits that arise from the exploitation of natural resources. These are essentially political rather than biological solutions to the paradox, but one which the region will need to face in the future. It would clearly be naive of us, however, to imply that solutions to these kinds of issues are ever simple, easily resolvable or short term. However, as biologists we can help to provide the people of the Mediterranean region with the higher quality and more sustainable standard of living they deserve. The twin goals of conservation of biodiversity and economic development need not be mutually exclusive; the adoption of wise development practices linked to holistic poverty alleviation measures could enable these goals to be won. Leguminous crops and diversity could hold a key.

The crops and species of the plant family Leguminosae are one of the world's most important and essential components of human food production, whether directly for human consumption or indirectly via animal production. They contain relatively high levels of protein for human and animal feed and as a nitrogen-fixing break crop, assist diverse farming systems. For the temperate world the highest concentration of species and crop ancestors, as well as being where the crops originally developed, is found in the Mediterranean region. As has been established in earlier chapters the genetic diversity of these valuable legume species is being rapidly and uncontrollably eroded. This diversity is essential for the well being of temperate agriculture (and via agriculture the peoples of the temperate world who depend on them for their sustenance) and so has immense value to humankind. If the aspirations of the people of the

Mediterranean region, as well as other temperate regions of the world, for a better quality of life are to be met, we must more efficiently conserve and effectively exploit legume diversity now.

Throughout the chapters of this text the various contributors have made numerous general and specific suggestions concerning the improvement of the conservation and use of legume diversity, these will now be reiterated and summarised to reinforce the need for action.

19.2 FUTURE CONSERVATION PROSPECTS AND PRIORITIES

A fundamental point that needs to be established first is the need to work together to facilitate better legume conservation and use. Both conservation and use are international and interdisciplinary: meaning that plant genetic resources span the entire region, and so can only efficiently be managed in a collaborative manner, and that conservation and use today involves conservationists, breeders, ethnobotanists, social scientists, geneticists, extension experts, etc. from different institutions representing the formal and informal sectors.

Networks promoting the conservation and use of legume species in the Mediterranean already exist. These have largely been led by the CGIAR centres in the region, ICARDA, IPGRI (European Cooperative programme, for conservation and utilization of Plant Genetic Resources [EC/PGR] and West Asian North African Genetic Resources Network [WANANET]) and ICRISAT, along with ACSAD, EuroMAB and external organisations active in the region, such as USDA and CLIMA; each has worked with national institutions to conserve and sustain legume genetic diversity. Although these networks are well established many still work in a largely *ad hoc* manner and there is clearly room for improving collaboration at the national, regional and international levels.

- *Legume Networks*. There is a need to ensure that the grain legume and forage legume networks of EC/PGR and WANANET collaborate fully and integrate their activities to meet mutual conservation, characterisation and exploitation goals.
- *Regional Taxonomic Coverage*. At the regional level the efforts to conserve and utilise Mediterranean legume diversity is led by ICARDA and ICRISAT within the CGIAR system. ICARDA having prime responsibility for lentils, faba bean and forage legume species, while ICRISAT has joint responsibility with ICARDA for chickpea. Neither has responsibility for peas. The position over the primary agency for pea conservation and use needs to be addressed.
- *National programme strengthening*. It is somewhat ironic that several of the largest *ex situ* collections of food and forage legumes of the Mediterranean region are found outside the region and that many countries in the region have resource restricted national PGR programmes. There is a need for institution building in the national PGR programmes within the region.

Having made these general points about the need to improve collaboration, we can address more specific issues that would each be enhanced by greater collaboration.

19.2.1 Conservation Planning

As ever, resources for conservation are always limited, therefore we are forced to make an efficient selection of target taxa and areas. The more efficient these choices, the more likely it is that the material once conserved will be utilised and that should always be the ultimate goal. To identify where most effectively to spend our conservation dollars, we need to be able to answer three questions: the relative value of legume species for utilisation, the relative threat their gene pool is faced with *in situ* and their current conservation status.

- *Relative value of legume species* - All other things being equal the most valuable legume genetic resources will be those most widely cultivated or producing the largest income for their growers, therefore we need to conserve the germplasm of peas, chickpeas, lentils and faba beans, followed by minor legume species and then the under-utilised species. This analysis may be viewed as somewhat simplistic initially, given the well known but devastating examples of late blight (*Phytophthora infestans*) attack on potatoes or southern corn blight (*Helminthosporium maydis*) on maize. The conservationist must therefore balance the needs to conserve the primary, secondary and tertiary gene pools, but due to their immediate utilisation potential the resources are still likely to be focused in descending order on grain, forage, minor and under-utilised legume species.
- *Relative threat to legume gene pools* - It is very difficult to quantify precisely taxonomic or genetic erosion (loss of taxonomic or genetic diversity). There is no effective way of assessing taxonomic erosion of legumes in the Mediterranean region. The 1997 IUCN Red List of Threatened Plants (Walters and Gillett, 1998) indicates that 322 species from 15 legume genera are threatened in the Mediterranean basin. However, as is emphasised earlier, these figures merely reflect the relative levels of floristic study in the countries of the Mediterranean (Turkey has 198 threatened taxa while Syria has none) and can not be taken as a true representation of the threat throughout the region. However, the picture maybe better when viewed at an individual country level. Spain and Palestine, for examples, have been studied throughly in floristic terms and the figures are thought to be accurate. In terms of genetic erosion, there have been no general attempts to quantify genetic erosion for legume diversity in the Mediterranean region, though a few specific studies do exist, such as: Ferguson (1998) for the lentils (*Lens* species) and van de Wouw (2000) for common vetch (*Vicia sativa* L.). It would seem logical, however, that the highest levels of genetic erosion have occurred in the legume crop species and their close wild relatives: the crops because of the replacement of diverse land race material with modern uniform cultivars and their close relatives because of genetic pollution from the crops themselves.
  - *Assessment of taxonomic erosion*. There is a need for a systematic review of taxonomic erosion and IUCN Red List categories for legume taxa throughout the Mediterranean region. ILDIS working in conjunction with ICARDA would seem ideally placed to undertake this project.
  - *Assessment of genetic erosion*. There is a need for a systematic review of genetic erosion for, at least the grain and major forage legume taxa throughout the Mediterranean region.

- *Current conservation status of legume species.* There is a need to review just how comprehensive current conservation activities are for Mediterranean legume diversity in terms of target taxa and geographical areas. The previous chapters have attempted this review in an *ad hoc* manner for *ex situ* conserved legume germplasm and have highlighted which taxa and which geographical locations need to be collected or recollected. As *in situ* conservation of legumes in the region is so scant, it can be firmly stated that genetic reserves and on farm legume conservation projects are required for all taxa in each country of the region. The level of duplicate material both within and between collections should also be considered when assessing conservation priorities. Some preliminary work on the identification of duplicate material in European *Pisum* collections has indicated that it could be as high as 60-80% (Ambrose *et al.*, 1997).
  - *Gap analysis.* There is a need for a systematic gap analysis in terms of gaps in *ex situ* conserved taxa and in geographical areas. As far as herbaceous legumes are concerned, there are gaps in nearly all perennial legume species, even in those species that have been more heavily collected, such as *Medicago sativa*, *Trifolium pratense* and *T. repens*. However, the likely geographical gaps include:
    - *Turkey.* Although there have been more accessions collected from Turkey than in any other country in the region, the high levels of genetic diversity found in Turkey for all legume genera justifies even further collecting activities, particularly in north-east and south-east provinces which remain relatively uncollected. All temperate legume genera are a priority.
    - *Caucasian Republics (Georgia, Armenia and Azerbaijan).* Due to the years of isolation from the international PGR movement, Caucasian material remains under-represented in regional and international *ex situ* collections. Pre-glasnost the national PGR programmes were managed by the N.I. Vavilov Institute in St. Petersburg, post-independence these republics have yet to establish active PGR programmes themselves. All temperate legume genera are a priority.
    - *Central Asian Republics (Kazakstan, Uzbekistan, Kirghizia, Tadzhikistan and Turkmenistan).* Like the Caucasian republics the Central Asian republics PGR programmes were managed by the N.I. Vavilov Institute in St. Petersburg. Post-independence these republics have fledgling PGR programmes that are under-resourced. Priority target taxa here are *Pisum*, *Cicer*, *Lens* and *Vicia*, which are all, except *Cicer*, at the eastern edge of their distribution.
    - *Albania.* During the socialist era there was a PGR programme in Albania but this has collapsed post-socialism. Now there is rapid genetic erosion and no resources for conservation.
    - *Palestine.* The regional isolation of Israel from its neighbours will hopefully cease soon with the fruition of the peace movement and the legume PGR of Palestine should be incorporated into regional and international *ex situ* collections.
  - *Rhizobia collections.* The importance of collecting and using rhizobia is only just being fully appreciated. Many legume species have specific obligate relationships with rhizobia and will not grow healthily or achieve full potential yields without

the correct rhizobia species. In the past, it is undoubtedly true that legume species were impossible to regenerate or were falsely rejected for agricultural exploitation simply because as *ex situ* collections they were not associated with the correct rhizobia species. This is particularly true when the collections are being assessed outside of the Mediterranean region.

- *Rhizobia collection*. There is a need to ensure greater emphasis is given to collecting legume species with their associated *Rhizobium* species.
- *Rhizobia use*. As the relationship between legume and *Rhizobium* species is obligate there is a need to ensure the correct rhizobial strain is used in agricultural trials and exploitation of legume species.
- *Rhizobia research*. There is a continuing need for research into the relationship between legume species and their associated rhizobia as a means of enhancing future legume conservation and use.

– *Accession duplication*. There is a need for a systematic assessment of the level of legume accession duplication between and within legume *ex situ* collections. Over- or unrecognised duplication between and within collections can give a false impression of the level of actual conservation.

- *ICARDA*. It would be most efficient to start with the ICARDA collection because it is the largest in the region and for safety reasons duplicates many national collections.

– *Accession database*. There is a need to establish an accession level database of legume holdings either in or collected from the Mediterranean region for grain legume and forage legume species.

- *EC/PGR and WANANET*. This might be achieved by bridging the gap between the databases of the existing EC/PGR and WANANET networks, either by constructing one overall database or ensuring compatibility between the two databases.

It should be remembered that when setting conservation priorities that there is unlikely to be congruence between National, Regional and International agencies. Each have their own objectives, mode of operation and budgetary restraints. There is no necessary efficiency in enforcing conformity; in fact it might be thought that the setting of diverse conservation priorities will better ensure the conservation and use of legumes as a whole.

19..2.2 Taxonomic, Ecogeographic and Genetic Surveys

Having established priority taxa and regions for active conservation it is important to have the appropriate knowledge to aid the practical implementation of strategies. This information is obtained from taxonomic, ecogeographic and genetic studies of the legume taxa targeted.

- *Taxonomic characterisation* - It is perhaps surprising that there remains so much taxonomic confusion at the inter and intra genetic levels in a group of crop species which are generally studied in more detail compared to less economically important taxa.

- *Inter-generic taxonomic resolution.* There is a need for inter-generic taxonomic resolution of the relationships between:
  - *Biserrula pelecinus* L. and *Astragalus* L.
  - *Lupinaster* Fabr., *Chrysaspis* Desv., *Trifolium* L. and *Amoria* C.Presl. There has been much debate over whether or not *Trifolium* L. should be one diverse entity or split into four distinct genera. The latter option was formalised by Roskov (1990a and b) but has not been widely accepted.
  - *Medicago* L., *Trigonella* L. and *Melilotus* L.
  - Oroboid species of *Vicia* L. and *Lathyrus L.* Although not of economic importance a resolution of the generic placement of the oroboid species that bridge the two genera would be of academic interest.
  - *Pisum* L. and *Vavilovia* A Fed. As a closely related perennial form of pea the generic distinction needs to be tested.
- *Intra-generic taxonomic resolution.* There is a need for intra-generic taxonomic study of the:
  - *Lupinus* L. South American lupins are very poorly understood although a revision has just been started (C. Hughes, pers. comm.) A clarification of the specific validity of some old world lupins, such as *L anatolicus*.
  - *Lathyrus* L. *Lathyrus* as a whole is not well studied, there have been no comprehensive taxonomic studies, even for section *Lathyrus* which contains the bulk of the cultivated species.
  - *Vicia* subgenus *Vicilla* (Schur) Rouy. Although the majority of the cultivated species are in subgenus *Vicia*, *Vicilla* does contain some under-utilised species whose use could be expanded and therefore taxonomic clarification is justified.
  - *Vicia sativa* L. Although resources have been lavished on trying to find the wild progenitor of *Vicia faba* L. and resolving its relationship with the *V. narbonensis* complex, there seems little hope of significant progress unless further novel species are found. A much better investment would be a clarification of the relationships within the *Vicia sativa* complex, a group that has received much less attention and is worthy of increase use. This complex is currently being revised by M. van de Wouw (University of Birmingham, UK).
  - *Hedysarum* L., *Lotus* L., *Tetragonolobus* L., *Trigonella* L. *Pisum* L. and *Astragalus* L. No comprehensive taxonomic treatment of these genera exists. *Astragalus* with its approximate 2000 species distributed throughout the North Temperate world, extending to Patagonia, North India and Tropical African mountain ranges (Polhill, 1981) may prove a particularly challenging but rewarding project. The most appropriate approach maybe to look at the infra-generic structure first, then when this is understood more clearly, allow individuals to focus on individual subgenera, sections or series. ILDIS may be able to play a role in coordinating this revision. Small *et al.* (1984) suggested that of the *Lotus corniculatus* L. group. in Turkey, three of the species of the group should be combined as one species, or at least as sub-species. However, there has

been no further work to confirm this from other regions. A revision of *Pisum*, a more immediately manageable entity, has just been started (N. Maxted and M. Ambrose, pers. comm.).

Whether at the inter or intra generic level there is a real need for user-friendly identification aids. Generally the identification aids available for legumes are of poor quality and often only of use to trained botanists because of the highly technical and stylised identification formats presently employed. The inability of many conservationists to use contemporary identification aids has led to confusion of specimen identity and has undoubtedly resulted in the application of poor conservation strategies and the poor utilisation of much valuable material. Genetic resources not accurately identified, either remains uncollected (because the different species are not recognised when encountered) or is collected, but remains unidentified and is incorporated into genebanks as "legume species", or if collected, but misidentified, leads to spurious results when utilised. The subsequent growing out and re-identification of germplasm collections is costly and time consuming, but accessions that are either wrongly identified or are not identified at all are effectively useless.

– *User-friendly identification aids*. There is a need to produce user-friendly legume identification aids and ensure legume conservationists are trained in their use.

- *Ecogeographic characterisation* - Allied to the resolution of taxonomic problems is the gathering and analysis of ecogeographic information, so that for all of the taxa highlighted for taxonomic study there is a need for parallel ecogeographic investigation. Along with these there is an urgent need for ecogeographic studies of:
  – *Cicer* L. Perennial *Cicer* are still under collected and it seems likely that further species will be discovered in Central and West Asia.
  – *Lens* L. Now that the taxonomic picture of the lentils has been clarified (Ferguson et al., 2000) there is a need to clarify the ecogeographical distribution of the two recently recognised species, *L. lamottei* Czefr. and *L. culinaris* subsp. *tomentosus* (Ladizinsky) Ferguson, Maxted, van Slageren & Robertson, as well as the distribution of all wild *Lens* taxa in North Africa.
  – *Medicago* L. Although some initial studies (Broughton, 1997) have been undertaken the ecogeography of *Medicago* need further clarification and then made more widely available.

- *Genetic characterisation* - In order to understand the ecogeography of legume species and facilitate their conservation it is important to understand the pattern of genetic variation within a species, distinct pockets of genetic variation are often associated with ecotypic differentiation of populations. Ferguson *et al.* (1998) have shown for lentils, genetic diversity does not necessarily correlate with ecogeographic distribution pattern and that patterns of genetic diversity are likely to be unique for each gene pool. Therefore there is a need to clarify individual patterns of genetic variation for all legume taxa. Genetic characterisation can also be used to help resolve some of the taxonomic problems outlined above. Routine techniques for genetic characterisation are only now becoming available, so it is not surprising that the patterns of genetic variation for individual legume taxa are available for so few taxa.

– *Genetic characterisation.* There is a need for genetic characterisation of all legume gene pools, with priority being given to those of most economic importance (notably peas, chickpeas and faba beans, lentils having already been studied). This is important for a number of reasons but primarily it helps establish the pattern of genetic diversity and so identify 'patches' of unique germplasm, where to locate individual sites as part of an *in situ* network or to define population limits for *ex situ* sampling.

19.2.3 Field Exploration

If little or no ecogeographic data are available, the conservationist may be forced to undertake survey missions to collate these data before an effective conservation strategy can be formulated. Field exploration is also a necessary precursor to both *ex situ* and *in situ* conservation. The ecogeographic survey will provide data on the broad distribution of a taxon but is unlikely to provide precise collecting localities, hence field exploration is needed to search the broad localities predicted, as containing the desired target taxon for actual populations.

- *Survey missions.* There is an urgent need for a survey mission for forage legume species in Central Asia, particularly focusing on *Cicer*, and to Albania for all legume species.

19.2.4 *Ex Situ* Conservation

It is obviously true that no *ex situ* collection will ever accurately reflect the genetic diversity found in nature, but this proposition is particularly true for forage legume species where there are so many species utilised. Even though large numbers of accessions have been collected for *ex situ* conservation or breeding, these are often dominated by the major grain species and many potentially valuable species have been completely neglected. In addition, many of the older accessions have insufficient passport data (such as latitude and longitude of the collecting site), which limits the usefulness of the conserved material. As such, and has been noted above, there are *ex situ* conservation gaps for nearly all legume species, even the major grain species.

Setting priorities may be even further complicated as currently accessions held may not have been collected in the most appropriate manner. For example, in a mountainous area with diverse habitats, genetic diversity is likely to be high, and therefore the distance between populations and numbers of sites visited will be much closer together and higher respectively, than in a less variable environment. Ideally when defining future collection priorities we would be aware of the collection strategy used by each collector for each accession, but this is rarely the case. Having made these general points the current *ex situ* priorities are:

- Ex situ *grain legume priorities.*
  – *Faba bean and lentils in China and Indian Sub-continent.* Major regions of cultivation of both faba bean and lentils remain under-collected from this area, especially in the remote areas of China.

- *Faba bean in WANA*. Within this region the major gaps are in the collections from Algeria, Egypt and Yemen.
- *Wild relatives of lentil*. There is a general need to increase collections of wild species, especially of *L. culinaris* subsp. *orientalis*, to facilitate breeding. Priority countries include Algeria, Libya and Sudan in north Africa, as well as the new central and west Asian republics of the former Soviet Union. In addition, for *L. culinaris* subsp. *orientalis*, collection priorities lie in two broad regions, the first being southern Syria, north and west Jordan and the West Bank of the Palestinian Authority, and secondly from south Turkey across to south-east Turkey. For *L. culinaris* subsp. *odemensis* additional priorities lie in the Levant and parts of north Africa. The region surrounding the coastal junction of Turkey and Syria is a priority for *L. ervoides*, and, although of a lower priority, the Ukraine may yield valuable *L. nigricans* germplasm.
- *Lentils in WANA*. There is a major gap in collections from Algeria and to a lesser extend for Tunisia.
- *Wild relatives of chickpea*. There is a general need to increase the collections of wild species, particularly the progenitor species, *C. reticulatum*, because they possess genes for resistance to almost all stresses for which evaluations have been made. Of the 34 perennial *Cicer* species known, live seed is available only for nine species.
- *Geographic gaps for chickpea*. Collection gaps are from Russia, Caucasus and Central Asia, Turkey, Afghanistan, Iraq, and Ethiopia.
- *Peas*. Possibly because peas have not been a priority for the CG Centres, they remain under-collected and there is a urgent need for a systematic *ex situ* collecting programme throughout their range, focusing on collecting both cultivated forms as well as the wild relative. Perhaps because of its geographical isolation and the lack of existing *ex situ* conserved material the highest priority should be focused on the secondary centre of diversity in Ethiopia.
- *Decline of minor legume species* - There are reports of decreased production of several minor legume species, *Lathyrus sativus*, *L. cicera* and *V. narbonensis* in areas where they have been traditionally cultivated. The status of the remaining landraces of these species from throughout the Mediterranean region needs to be assessed and matched against conserved germplasm to clearly establish conservation gaps.

- Ex situ *forage legume priorities*.
  - *All forage legume genera*. It is recognised that this is such a diverse group of species (739 forage legume species native to the Mediterranean, see Chapter 15) that most deserve *ex situ* priority. Specifically, however, no accessions have been conserved *ex situ* for *Cytisopsis*, *Lotononis*, *Tripodion* or *Vavilovia*, even though each has utilisation potential.
  - *Perennial species*. In general the annual species have been more extensively collected than the perennials, possibly because the later species are often found growing in more inaccessible habitats. Therefore perennial species of *Trifolium*, *Cicer*, *Vicia* and *Lathyrus* species are under-represented in all collections

- *Regional gaps*. For the forage legume genera, there are two regions, north-west Africa and the north-east Mediterranean, that have been relatively well sampled. However, two further regions, - European west Mediterranean (especially France, northern Italy and the former Yugoslavian republics) and Iran, Iraq and to the north and east have distinct germplasm but remain relatively poorly sampled. However, it must be restated that simple counts of the numbers of species and numbers of accessions collected may in certain cases be misleading as patterns of genetic diversity are not necessarily spread throughout the distributional range of a taxon.
- *American taxa*. Several genera have a secondary centre of diversity in the Americas. Both *Lupinus* and *Trifolium* have significant numbers of species in North and South America, while, both *Vicia* (sects. *Americanae* and *Australes*), *Lathyrus* (sect. *Notolathyrus*) have sections endemic to South America. Each of these are under-collected, particularly the South American species, and so requires collection and *ex situ* conservation.
- *Rare species*. There are many rare species that are only known from the type collection (e.g. *Trifolium euxinum*) or species with very narrow distributions (e.g. *T. pichisermollii*, *Lathyrus belinensis*, *L. mulkak* and *Vicia kalakhensis*). For the majority of these species these is no *ex situ* conserved material. Although these species may have no known current use, to make a professional assessment of their potential they must be represented in *ex situ* collections.

• Ex situ *collection procedures*.
- *Efficient Population Sampling*. The better the quality (genetically representative, viable and of sufficient size) of the samples transferred *ex situ*, the more likely they are to be utilised. Collectors are referred to Brown and Marshall (1995) and Marshall and Brown (1999) for details of the most appropriate strategy to employ when collecting legume species.
- *Collect Associated Materials*. Seed collecting is not simply restricted to collecting seed; an efficient collector will also collect herbarium voucher specimens, detailed passport data and possibly vegetative plants. Collectors are referred to Guarino *et al.* (1995) and Hawkes *et al.* (2000) for a detailed discussion of the sampling and collection of these materials.
- *Collect Associated Knowledge*. As well as collecting physical materials the collector should allow time to talk to local farmers and collate ethno-botanical and socio-economic data associated with the germplasm they are sampling. This is particularly important when collecting cultivated material, as collecting indigenous knowledge will help potential users understand the patterns of diversity, social characteristics of farming system and the selection criteria employed by the farmers, and so assist further exploitation. Collectors are referred to Guarino and Friis-Hansen (1995) and Martin (1995) for methods of indigenous knowledge collation.
- *Rhizobia*. Modern studies of rhizobial diversity, particularly those using molecular techniques, have highlighted the fact that only a minute percentage of the available rhizobial genetic diversity of the Mediterranean basin has been captured in gene

banks *ex situ*. Whenever legumes are collected their associated rhizobia should be collected.

- *Quality Collection of Passport Data*. The more complete the passport data collected, the more useful the germplasm accession will prove to users and therefore the more likely they are to be used. Even the country of origin is not known for 21% of the entire ICARDA collections and for many other accessions there is very little information available. In the future accessions without appropriate passport data should not be incorporated into gene banks unless they have some special characteristics, e.g. endemic species or known resistances.

19.2.5 *In Situ* Conservation Strategies

The point has been made in earlier chapters that the task of conserving *ex situ* an adequate representation of the gene pool of all 739 forage legume species native to the Mediterranean is clearly problematic. In practise *ex situ* conservation is always likely to focus primarily on an economically important subset of species, as has occurred *de facto* in the past. Therefore for the majority of legume species in the Mediterranean *in situ* conservation is not only the most desirable alternative, it offers the only realistic short and medium term option. Post UNCED and the ratification of the CBD there is a growing interest in the application of *in situ* conservation techniques for plant genetic resources, but as has been pointed-out on several occasions (Hawkes, 1991; Maxted *et al.*, 1997a; and in Chapter 16) the science of genetic *in situ* conservation is still poorly understood. There is so little current activity in terms of *in situ* conservation of legume diversity that virtually anything will help, however, the following are considered priorities:

- *Improve the science of* in situ *conservation*. There is a need to improve the scientific location, designation, management and monitoring of genetic reserves and on farm conservation projects. There are several projects currently focusing in this area (Jarvis and Hodgkin, 1998; PLEC, 1999), but none are addressing the specific issues associated with forage conservation.
- In situ *case studies*. Just as the science of *in situ* conservation need improving, so there is a need for case studies based on experience of the establishment and maintenance of *in situ* reserves for legume species. These may then be used as models for the establishment of genetic reserves and on farm management programmes elsewhere.
- *Improved access to in situ conserved material*. A criticism often levelled at *in situ* conservation is that the germplasm user has less easy access to germplasm conserved *in-situ*. This need not necessarily be the case. Maxted *et al.*, (1997b) argue that the onus is on reserve managers, just as it is on gene bank managers, to promote utilisation of the material in their care. Just as gene banks and botanical gardens publish catalogues of their collections, so the reserve manager should regularly publish a catalogue and description of characterisation data for the germplasm held in the reserve to assist potential users and promote use.
- *More systematic funding*. The internationally funded plant conservation projects tend to be focussed on a small number of countries and not necessarily those with the highest

levels of biodiversity or legume diversity, there is a need for more spatially systematic funding of *in situ* conservation projects throughout the region.

- *Designation and implementation of novel genetic reserves.* Maxted (1995) recommended the establishment of four genetic reserves for *Vicia* diversity in Syria: Ain Dinar, Al Hasakah; Kessab town, Kessab; Qal'at Al Hosn, Homs and Mimas, Djebel Druze, as well as in the Olimpos Beydaglari National Park, Belin, Turkey. Although there have been no other specific recommendations for the establishment of legume genera genetic reserves, there is clearly a need for them. These recommendations are likely to follow from detailed ecogeographic studies and the completion of these studies, as recommended above, is underscored to help promote *in situ* conservation.
- *On farm projects.* On farm conservation is the most recently developed conservation technique and certainly could be used for legume species. It would appear that on farm conservation would be particularly important for grain legume species where the land races are more likely to be threatened by replacement with high yielding varieties. The on farm conservation of forage species is a lower priority because of the difficulty of conserving the diversity and of number of species involved, plus the fact that these species are less subject to genetic erosion.
- *Protected area inventories.* - Wasteland, field margins, primary forest, national parks, other protected areas contain much legume diversity (Maxted *et al.*, 1997b) that could be genetically conserved and used. But before the wealth of legume genetic resources currently found in these locations can be used, we must first know what is present. We need to encourage the managers of these sites to produce an inventory of the wild legume species they contain, particularly for national protected areas that have conservation as an objective and justification for continued funding.

19.2.6 Conserved Product Enhancement

Once the germplasm has been actively conserved there remain various activities that can engender prolonged preservation and facilitate better utilisation.

- *Improved regeneration of difficult material.* Periodically any germplasm conserved in a gene bank will require extraction from the gene bank and regeneration to replenish population numbers and improve viability (FAO, 1998). It is also important to maintain the genetic integrity and purity of accessions. Experience at ICARDA has show that wild perennial *Cicer*, *Vicia* and *Lathyrus* species are much more difficult to propagate than annual species and there remains little information available on the appropriate method of regeneration. Many gene banks are entirely composed of collected material and there has been no necessity for regeneration yet, this will change with time. Therefore there is a requirement to develop regeneration protocols appropriate for all legume species. However, before recommendations can be made there is a need to establish the breeding system of the species. If the species is out crossing then regeneration its-self can lead to unnecessary loss of genetic diversity, but even this basic information is unknown for many legume species.

- *Efficient gene bank practise*. As with any *ex situ* conserved germplasm, there is a need for efficient:
  - *Collected Material Processing*. The materials collected require appropriate and effective storage following established procedures. Samples that are mis-managed are likely to be of little or no exploitation use.
  - *Safety duplication*. There is a need to ensure that the conserved accessions are duplicated between at least two gene banks as a safety precaution against accidental loss of collections. For legumes, the material could be duplicated at ICARDA, in the region legume gene bank, and in the national gene bank of the country of origin. However, over-duplication of single accessions in numerous gene banks should be avoided. Ambrose *et al.* (1997) suspect that the level of duplication across all collections in Europe could be as high as 60-80% and this is obviously wasteful of resources and may, if unrecognised, give a false impression of the efficiency of conservation.
  - *Lack of passport data*. As mentioned above, conserved legume utilisation is being inhibited by lack of passport data.
  - *Collecting Reports and Other Publications*. Following the expedition the commissioning agency will require an expedition report, but the existence of novel diversity can be signalled to potential users by publishing collecting reports. The gene bank manager must be pro-active and advertise the presence of the new accessions in the collection. Web sites should be designed to make the material conserved more readily available.
  - *Developing Core Collections*. With the increasing size of many collections and the limited funds available for characterisation and evaluation, it may be necessary to develop a core collection of representative samples, particularly for the larger collections of grain legumes.
  - *Quality Service to Users* - The quality of the service provided to germplasm users by germplasm curators will affect their potential for utilisation, therefore if we wish greater use of legume germplasm we need to ensure the gene bank service is efficient.
  - *Adequate collection funding*. The health and safety of all grain and forage legumes is dependent on adequate resources. Unfortunately the maintenance of *ex situ* collections is not generally perceived as fashionable, it lacks innovation and can not be described as research. Gene banks provide a service to the user community, but this service is invaluable. It is imperative that adequate resources are provided by international, regional and national agencies to ensure the viability of these collections. This material once lost would be lost forever, as many of the original populations are now extinct, and all will have a different genetic composition to that originally sampled.
- *Pre-characterisation*. The user community is often swamped by potentially exploitable germplasm and any pre-characterisation data that is made available to them will aid their choice of germplasm to characterise and evaluate fully. If the germplasm manager can provide information on taxonomic distinction, ecogeography of the collection site, agro-ecology of the population sampled, ethnobiology of legume land races or initial results of field screening will promote use and should be encouraged.

19.3 FUTURE UTILISATION PROSPECTS AND PRIORITIES

There is a clear need to improve the link between those conserving and using legume diversity. A gene bank or genetic reserve that conserves germplasm with no thought of making that germplasm available to the user community is doomed to financial insecurity. Simmonds (1962) was the first to point out that mismanaged germplasm collections were largely worthless and were effectively no more than poorly displayed museum exhibits. How can we expect the public, who fund the major legume collections to continue their support and provide sustainable conservation unless we ensure a better link between conservation and use?

Given (1994) has estimated that 65% of conserved germplasm lacks basic passport data, 80% lacks characterisation and 95% lack evaluation data - which leaves only approximately 1% which are correctly catalogued and available for use. Maxted *et al.* (2000) proposed that the best ways of ensuring the long-term security of conserved germplasm is to establish a seamless gradation from conservation into utilisation. In other words the two communities (plant conservationists and plant users) should collaborate fully, integrate their activities and be aware of challenges facing each other.

19.3.1 Characterization and Evaluation

The most valuable research to encourage use of conserved germplasm is to study the pattern of genetic diversity, to characterise and evaluate the accessions by recording and analysing agro-morphological and genetic characteristics.

- *Better characterisation and evaluation.* Lack of characterisation and evaluation is often cited as the major constraint for germplasm exploitation (FAO, 1998). The obvious answer is improve characterisation and evaluation, either by the germplasm curator undertaking simple characterisation as a guide to potential users or the user community themselves being more open about the material they are prepared to screen and incorporate into pre-breeding programmes. Even though lack of trained personnel, facilities and finance will always be limiting, characterisation and evaluation of conserved germplasm deserve greater priority. If this recommendation goes unheeded the reason for conserving the diversity of material itself becomes more questionable.
- *Better characterisation and evaluation techniques.* Through a programme of technology transfer and institutional building in the region, the techniques used to characterise and evaluate germplasm holding should be upgraded to ensure that traditional agro-morphological, as well as more advanced molecular characterisation techniques, are applied.
- *Publications of characterisation and evaluation results.* Any characterisation and evaluation results should be published to ensure the potential users are aware of the diversity of material available and attention should be drawn to any interesting or extreme forms. This can be achieved via the traditional route of catalogues or journal papers but increasing information can be posted on the world wide web.
- *Promote cooperative characterisation.* The efficiency of characterisation and evaluation can be enhanced by regional and national agencies working together with multi-

disciplinary groups from the formal and informal sectors. The existing EC/PGR and WANANET networks already collaborate in screening grain and forage legume material, but not all national programmes are involved. The point is made above that some of the largest Mediterranean legume collections are found outside the region, these collections are being actively characterised and evaluated and this information, even though potential financially sensitive, should be repatriated to the region.

- *Trait specific screening.* There are specific requirements for desirable traits for use in legume breeding programmes and there is a need for this form of trait specific screening, some of which include:
  - *Disease resistance.* There are many problem diseases for legume crops, e.g. Ascocyta blight and Anthracnose, but some progress is being made in locating resistant accessions (Mayer *et al.*, 1997; Frencel, 1998; Sweetingham *et al.*, 1995).
  - Orobanche *resistance. Orobanche crenata* is a major cause of yield loss in all grain legumes and there is a requirement for resistant material.
  - *Broadening the genetic base.* Although resistance to various pests and diseases is available there is a need to broaden the base of that resistence so it is less likely to collapse, for example it is thought that chocolate spot and rust resistance in faba bean may be obtained by further collection of landraces from the Andean region.
  - *Higher biomass production.* This is a particular problem for lentils and an immediate priority for screening.
  - *Environmental adaptability.* Many cultivars are restricted by their susceptibility to environmental conditions, there is a need to produce more generalised varieties. For example, there is a need to screen for *Lupinus albus* var. *albus* germplasm adapted to alkaline soils, which will enable the expansion of the albus industry in Europe and the Mediterranean or cold-tolerant types that will benefit the production of winter lupins in northern Europe.

As molecular screening techniques become cheaper and are applied more routinely, there will be an increasing evaluation of the grain legume germplasm collections for biochemical and molecular markers to assist in the study of the genetic diversity of these crops; and to plan future research in areas such as further *ex situ* collections, *in situ* conservation, core collections and breeding. The work with wild species is also likely to assume greater importance, as gene transfer techniques are improved and this may itself affect the choice of priority materials for characterisation and evaluation.

19.3.2 Plant Breeding

Active breeding has been largely focused on the grain legume species. Historically forage legume cultivars have been largely selected rather than bred. However, the general breeding or selection goals remain the same as for any crop and these include: yield, larger biomass, disease and pest resistence, avoidance of environmental stress and productivity in novel environments. Specifically, however some of the breeding goals for grain and forage legumes are:

- *Yield.* As ever yield is always a priority. For example, there is a considerable disparity between the potential yield (3-5 t ha^{-1}) and the current world average yield of 0.7 t ha^{-1} for commercial chickpeas (Robertson *et al.*, 1997). There is a significant need for research to obtain stability of yield in different environments for all grain legume species.
- *Biomass production.* As mentioned above for lentil breeders the major goal is for increased biomass.
- *Improved harvest index. L. angustifolius* for example, has a poor harvest index that is limiting its cultivation. There is a need for further research into lupin development and the possible genetic restriction of branching to achieve better crop yields (Perry *et al.*, 1998).
- *Disease and pest resistence.* As with any crop there is a continual search for resistance to the range of crop pests and diseases, increasingly however with legume crops the need is to search for broader-based genetic resistance that will provided longer-term sustainable production. *Orobanche crenata* seriously limits much grain legume production and searching for resistance should be a priority.
- *Low toxin forms.* Several legume crops have problems with toxicity and this limits their uptake. A sever problem for human consumption is encountered with *Lathyrus sativus*, but recently very low or toxin-free *L. sativus* varieties have been found (Malek *et al.*, 1996). These low toxin forms need to be made more widely available so that the crop can reach its true potential.
- *Particpatory plant breeding.* In recent years the development of breeding programmes that incorporate multi-disciplinary groups, including farming communities has proven very beneficial.
- *Crop diversification.* Although there is already a wide range of cultivated forage legume species, there is the potential for even wider taxonomic use of previously undomesticated species. Increasingly important novel species include, Large podded vetch (*V. sativa* subsp. *macrocarpa*), Serradella (*Ornithopus* sp.) and Biserrula (*Biserrula pelecinus*), the latter two are deep rooted and so can withstand some water stress (Revell, 1992; Howieson *et al.*, 1998). There are also a number of perennial legumes that are showing potential for low rainfall areas (Douglas and Foote, 1994). As many of these potentially useful wild species have only recently been collected and evaluated it is appropriate to undertake a more systematic review of forage and fodder potential of the entire range of legume species, both for new species for cultivation but also for genes or gene complexes that might prove of use to agriculture. As the total number of accessions available for many of these is either low or non-existent fresh collecting may be a prerequisite.
- *Alternative uses.* There is scope for looking at non-traditional uses of existing cultivated species. For example, wild populations within the *M. sativa* complex and other annual medics are of great agronomic interest, not only as a forages, but potentially for industrial uses as diverse as fibre crops for power generation and to help rehabilitation of degraded ecosystems. Also within the Australian forage breeding programme there is a search for perennial and deep-rooted annual legume species that will help to halt the rising water tables and salinity in the agricultural areas. Potential candidates species, such as, *Trifolium michelianum*, *T. resupinatum*, *T. fragiferum* and *Hedysarum coronarium* (Latta *et al.*, 1999; Rumbaugh *et al.*, 1993; Martiniello & Ciola) exist but more research is urgently needed to characterise the range of conditions that they can tolerate.

- *Organic nitrogen.* In recent years there has been increasing concern over the financial cost, the drain on finite fossil-fuel energy and the environmental cost of leaching nitrates in water supplied of intensive use of nitrogen fertilisers (Frame *et al.*, 1998). Wood (1996) investigated the comparison between the fossil-energy expenditure in a grass/N-fertiliser system and a grass/clover system and found that when both received the equivalent of 200 kg N ha^{-1}, showed that the latter expended only 5% the fossil energy of the former. Simple economics indicates that intensive use of nitrogen fertilisers is unsustainable. Certainly in the developed world, where use of forage legumes in rotation or ley mixes has declined rapidly in the past forty years, there is a need to reassess the role of using natural organic nitrogen supplied by legume species in farming systems. There is a need to test different farming model as well as screen different species and accession to compare organic nitrogen production.
- *Basic research.* The improvement of certain species requires basic research but the results could have fundamental importance. For example, an understanding of the genetic control of β-*N*-Oxalyl-L-α,β-Diaminopropionic Acid (ODAP) content in *Lathyrus* species, could greatly facilitate the breeding of low- toxin or toxin free cultivars or an investigation of the ecological relationships between rhizobia and legumes from genera containing both annual and perennial species could enhance crop yield.
- *Ornamental potential.* Several legume species are cultivated for ornamental purposes, particularly in *Lathyrus*; sweet pea (*L. odoratus* L.), everlasting pea (*L. latifolius* L.) and *L. sylvestris* L. There are however several other species, particularly in *Lathyrus* section *Lathyrus*, that have potential for development as new horticultural species (Davis, 1970). Due to the showy nature and scent of the flowers of the genus, there is a need to screen (and if the germplasm is not available, collect) diverse wild species for their ornamental potential.

19.3.3 Biotechnology

As for any crop the adoption of improved breeding techniques will enhance the production of novel legume cultivars. The recent use of molecular data as a tool in marker assisted selection has proved very beneficial. The transfer of exotic genes for tolerance or resistance to biotic and abiotic stresses into legume crops is becoming increasingly routine, for example the transfer of insect and pathogen resistance from *C. echinospermum* to cultivated chickpea (Porto-Pugtia *et al.* 1993) has proven successful.

- *Improved breeding techniques.* There is a need to ensure the latest breeding techniques are applied where appropriate for breeding legume crops. The development of linkage maps for the main grain legume species is likely to be beneficial in making it easier to search for desirable traits. The use molecular data obtained by the use of RAPD or AFLP techniques have been used, for example, by Clua *et al.* (1997) to assist in the transfer of the genes for cold tolerance to birdsfoot trefoil (*Lotus tenuis*).

19.3.4 Adoption and Utilisation

Having ensured a 'true' reflection of legume diversity is conserved using complementary *in situ* and *ex situ* techniques and that this conserved diversity is being used by the range of potential user communities, we still need to address some strategic issues that could aid legume conservation and use, both within and outside the Mediterranean region.

- *Implementation of communal objectives.* Both WANANET and ECP/GR have detailed set of objectives that if implemented could enhance legume conservation and use.
  - *Identifying constraints.* Both WANANET and ECP/GR legume networks have as a priority the identification of common constraints to effective conservation and utilisation of PGR in their region.
  - *Legume collaboration.* There is a need to ensure that the grain legume and forage legume professionals working both within and outside the region collaborate more fully and integrate their activities to meet mutually agreed collecting, conservation, documentation, germplasm exchange, breeding, research and training strategies that benefit all.
- *Institutional strengthening.* The CGIAR centres, USDA and CLIMA have a strong commitment to conserving and exploiting legume diversity in the Mediterranean region, these activities should be linked to technology transfer and institutional strengthening of the National PGR Research Systems within the region. The Mediterranean region contains both developed and developing countries and the former should be providing greater developmental assistance to the latter.
- *Linking conservation and use in a seamless manner.* Unfortunately the people who conserve and utilise germplasm are still seen as belonging to two distinct professions, often located in two distinct locations; however, utilisation can be improved by bringing conservationists and germplasm users physically as well as professionally together.
- *Sustainable development.* The conservation and use of legume germplasm in the Mediterranean should work within development plans for the region as a whole, such as the Blue Plan (Dowdeswell, 1995), which incorporates economic development, social harmonisation with conservation of natural resources.
- *Staff training.* All of the above actions to improve conservation and use of legume germplasm are only feasible if appropriately trained personnel staff gene banks. Too often genetic resource training is *ad hoc* and in house, this is not an efficient or effective mode of training, there is a need for a more coherent approach to expanding and improving the education and training of PGR staff in the region.
- *Promote public awareness.* Lastly, but definitely not least, there is a continuing need to promote public awareness of our activities to conserve and exploit legume diversity. Hawkes *et al.* (2000) estimate a cost of US $ 5.3 billion for collecting and conserving the world's germplasm currently held in gene banks, a cost which is ultimately met by the public. Therefore, it is not unreasonable for the public to expect us to provide them with an explanation of what we are doing with their money. Public awareness on the importance of biodiversity should constitute a major action at all levels from individuals, communities, national, regional and international levels. This can be achieved through

activities, such as specialized courses at school, frequent talks with local communities at any gathering of peoples (mosque, fairs, etc.), training, use of different media supports etc. The tools to be used should provide an easy understanding of biodiversity importance to a given audience. Local non-governmental organizations can help a lot in this respect.

19.4 CONCLUSIONS

The recommendations detailed above are extensive and many of them are already enshrined in general terms in the Global Plan of Action for conservation of plant genetic resources for food and agriculture (FAO, 1996), the various actions to conserve biodiversity proposed by McNeely and Guruswamy (1998) and the more specifically plant genetic resource related actions urged by Hawkes *et al.*, (2000). In this chapter we have attempted to summarise some of the specific legume conservation and use recommendations made in the preceding chapters, however we are of course fully responsible if we have misrepresented the views of our colleagues. Our sole aim in editing this book has been to help improve legume conservation and use, both within the Mediterranean region and elsewhere where temperate legume species are cultivated.

We believe that only if the recommendations outlined above are implemented can we be reasonably certain that the world's temperate legume genetic resources will be preserved for the benefit of future generations. Though it needs to be stressed that the conservation of legume diversity should not ignore the needs of profitable and evolving farming systems now. This should influence the choices made by breeders and, as such, involve the full participation of local communities. To increase regional and national efficiency and help implement the recommendations there is a growing coordination role for the regional agencies with an interest in legume diversity, notably ICARDA, IPGRI, ICRISAT and ACSAD.

REFERENCES

Ambrose, M.J., Weeden, N.F. and Simon, C.J., 1997. *Pisum* spp. conserved resources, priorities for collection and future prospects. *International Food Legume Research Conference III*, Adelaide, Australia, September, 1997.

Broughton, D.A., 1997. Ecogeographic survey of *Medicago* species. PhD Thesis. School of Biological Sciences, University of Birmingham, Birmingham, UK.

Brown, A.D.H. and Marshall, D.R., 1995. A basic sampling strategy: theory and practice. In: *Collecting plant genetic diversity: technical guidelines* (eds. Guarino, L., Ramanatha Rao, V. and Reid, R.). pp. 75-92. CAB International, Wallingford.

Cincotta, R.P. and Engelman, R., 2000. *Nature's place.* Population Action International, Washington DC.

Clua, A.A., Gimenez, D.O. and Fernandez, L.V., 1997. Increase in forage yield in Narrowleaf birdsfoot trefoil (*Lotus tenuis* Waldst and Kit ex Willd) in a permanent pasture with foliar applied gibberellic acid (GA-3), and phosphorus. *Plant Growth Regulation*, 21(3): 223-228.

Davis, P.H., 1970. *Lathyrus* L. In: *Flora of Turkey and East Aegean Islands, Vol. 3* (ed. P.H. Davis). pp. 328-369. Edinburgh University Press, Edinburgh.

Douglas, G.B. and Foote, A.G., 1994. Establishment of perennial species useful for soil conservation and as forages. *New Zealand Journal of Agricultural Research*, 37: 1-9.

Dowdeswell, E. 1995. The cooperative ethic of the biodiversity convention poses a great challenge for the vital habitats and cultures of the Mediterranean mosaic. *Diversity*, 11(1 & 2): 9-13.

Food and Agriculture Organisation of the United Nations, 1996. Global plan of action for the conservation and sustainable utilisation of plant genetic resources for food and agriculture. pp. 1-63. FAO, Rome.

Food and Agriculture Organisation of the United Nations, 1998. *The State of the World's Plant Genetic Resources for Food and Agriculture*. FAO, Rome.

Ferguson, M.E., 1997. *Studies of genetic variation within the genus* Lens. PhD Thesis. School of Biological Sciences, University of Birmingham, Birmingham, UK.

Ferguson, M., Ford-Lloyd, B.V., Robertson, L.D., Maxted, N. and Newbury, H.J., 1998. Mapping the geographical distribution of genetic variation in the genus *Lens* for the enhanced conservation of plant genetic diversity. Molecular Ecology, 7:1743-1755.

Ferguson M.E., N. Maxted, van Slageren, M. and L. D. Robertson, (2000). A re-assessment of the taxonomy of *Lens* Mill. (Leguminosae, Papilionoideae, Vicieae). *Botanical Journal of the Linnean Society*. 133: 41-59.

Frame, J., Chalton, J.F.L. and Laidlaw, A.S., 1998. Temperate forage legumes. pp. 1-328. CAB International, Wallingford.

Frencel, I.M., 1998. Report on first detection of anthracnose (*Colletotrichum gloeosporioides*) on Lupins in Poland. *Plant Disease*, 82(3): 350.

Given, D., 1995. *Principles and practice of plant conservation*. Chapman and Hall, London.

Guarino, L., Ramanatha Rao, V. and Reid, R., 1995. *Collecting plant genetic diversity: technical guidelines*. CAB International, Wallingford.

Guarino, L. and Friis-Hansen, E., 1995. Collecting plant genetic resources and documenting associated indigenous knowledge in the field: a participatory approach. In: *Collecting plant genetic diversity: technical guidelines* (eds. Guarino, L., Ramanatha Rao, V. and Reid, R.). pp. 345-365. CAB International, Wallingford.

Hawkes, J.G., 1991. International workshop on dynamic *in situ* conservation of wild relatives of major cultivated plants: summary of final discussion and recommendations. *Israel Journal of Botany*, 40, 529-536.

Hawkes, J.G., Maxted, N. and Ford-Lloyd, B.V., 2000. *The ex situ conservation of plant genetic resources*. Kluwer, Dordrecht.

Howieson, J., Loi, A. and Paterson, J., 1998. *Casbah - a new and promising pasture legume for low to medium rainfall environments*. Farmnote No. 28/98. Agriculture Western Australia, Perth.

Jarvis, D. and Hodgkin, T., 1998. Strengthening the scientific basis of in situ conservation of agricultural biodiversity on-farm. Options for data collection and analysis. Proceedings of a workshop to develop tools and procedures for in situ conservation on-farm, 25-29 August 1997. IPGRI, Rome, Italy.

Latta, R., Casson, T., Carr, S. and Paterson, J., 1999. *Establishing Balansa and Persian clovers on waterlogged, mildly saline soils*. Farmnote No.26/99. Agriculture Western Australia, Perth.

Malek, M.A., Sarwar, C.D.M., Sarker, A. and Hassan, M.S., 1996. Status of grasspea research and future strategy in Bangladesh. In: Lathyrus *Genetic Resources in Asia* (eds. Arora, R.K., Mathur, P.N., Riley, K.W. and Adham, Y.). pp. 7-12. IPGRI, New Delhi.

Martin, G., 1995. *Ethnobotany: a methods manual*. Chapman and Hall, London.

Marshall, D.R. and Brown, A.H.D., 1999. Sampling wild legume populations. In: *Genetic Resources of Mediterranean Pasture and Forage Legumes* (eds. Bennett, S.J. and Cocks, P.S.). pp.78-89. Kluwer Academic Publishers, Dordrecht.

Martiniello, P. and Ciola, A., 1994. The effect of agronomic factors on seed and forage production in perennial legumes sainfoin (*Onobrychis viciifolia* Scop.) and French honeysuckle (*Hedysarum coronarium* L.). *Grass & Forage Science*, 49(2): 121-129.

Maxted, N., 1995. *An ecogeographic study of Vicia subgenus Vicia.* Systematic and Ecogeographic Studies in Crop Genepools 8. pp. 1-184. IBPGR, Rome.

Maxted, N., Erskine, W., Robertson, L.D. and Asthana, A.N., 2000. Are Our Germplasm Collections Museum Items? Conference paper given at the International Food Legume Research Conference III, Adelaide, Australia, September, 1997.

Maxted, N., Ford-Lloyd, B.V. and Hawkes, J.G., 1997a. *Plant genetic conservation: the* in situ *approach*. pp. 1-451. Chapman & Hall, London.

Maxted, N., Hawkes, J.G., Ford-Lloyd, B.V. and Williams, J.T., 1997b. A practical model for *in situ* genetic conservation. In: *Plant genetic conservation: the* in situ *approach* (eds. Maxted, N., Ford-Lloyd, B.V. and Hawkes, J.G.). pp. 545-592. Chapman & Hall, London.

Mayer, M.S., Tullu Abebe, Simon, C.J., Kumar, J., Kaiser, W.J., Kraft, J.M. and Muehlbauer, F.J., 1997. Development of a DNA marker for Fusarium wilt resistance in chickpea. *Crop Science*, 37(5): 1625-1629.

McNeely, J.A. and Guruswamy L.D., 1998. Conclusions: how to save the biodiversity of planet earth. In: *Protection of Global Biodiversity: Converging Strategies* (eds. Lakshman, D.G. and McNeely, J.A.). pp. 376-391. Duke University Press, Durham.

Perry, M.W., Dracup, M., Nelson, P., Jarvis, R., Rowland, I. and French, R.J., 1998. Agronomy and farming systems. In: *Lupins as Crop Plants: Biology, Production and Utilization* (eds. Gladstones, J.S., Atkins, C. and Hamblin, J.). pp. 291-338, CAB International, Oxon, UK.

People, Land Management and Environmental Change, 1999. Special issue on methodology. *PLEC News and Views*, 13: 1-34.

Polhill, R.M., 1981. Galegeae - Astragalinae (Adans.) Benth. In: *Advances in Legume Systematics* (eds. Polhill, R.M. and Raven, P.M.). pp. 357-363. Royal Botanic Gardens, Kew.

Porta-Puglia, Singh, K.B. and Infantino, A., 1993. Strategies for multiple stress resistance breeding in cool season food legumes. In: *Breeding for stress tolerance in cool season food legumes* (eds. Singh, K.B. and Saxena, M.C.). ICARDA, Wiley Sayce Co. Pub.

Revell, C., 1992. New yellow serradella varieties for low rainfall pastures. *Journal of Agriculture Western Australia*, 3: 121-127.

Robertson, L.D., Singh, K.B. and Ocampo, B.,1997. Morphological variation in wild annual *Cicer* species in comparison to the cultigen. *Euphytica*, 95: 309-319.

Roskov, Y.R., 1990a. *Revision of the genus* Trifolium *L. in the flora of the U.S.S.R.* Karmarov Botanical Institute, Leningrad.

Roskov, Y.R., 1990b. The new species and nomenclatural combinations in the genera *Lupinaster*, *Chrysaspis*, *Trifolium* and *Amoria* (Fabaceae). *Botaniceskij Zhurnal, (Leningrad)*, 75(5): 715-720.

Simmonds, N.W., 1962. Variability in crop plants, its use and conservation. *Biological Reviews*, 37: 422-465.

Small, E., Grant, W.F. and Crompton, C.W., 1984. A taxonomic study of the *Lotus corniculatus* complex in Turkey. *Canadian Journal of Botany*, 62: 1044-1053.

Sweetingham, M.W., Cowling, W.A., Buirchell, B.J., Brown, A.G.P. and Shivas, R.G., 1995. Anthracnose of lupins in Western Australia. *Australasian Plant Pathology*, 24(4): 271-272.

Rumbaugh, M.D., Pendery, B.M. and James, D.W., 1993. Variation in the salinity tolerance of strawberry clover (*Trifolium fragiferum* L.). *Plant and Soil*, 153(2): 265-271.

van de Wouw, M. 2000. *Studies of genetic variation within* Vicia *section* Vicia *L.* PhD Thesis. School of Biological Sciences, University of Birmingham, Birmingham, UK.

Walters, K.S. and Gillett, H.J., 1998. *1997 IUCN Red List of Threatened Plants*. Compiled by the World Conservation Monitoring Centre. pp. 1-862. IUCN - The World Conservation Union, Gland, Switzerland and Cambridge, UK.

Wood, M., 1996. Nitrogen fixation: how much and at what cost? In: legumes in sustainable farming systems, Occasional Symposium Np. 30 (ed. Younie, D.). pp. 26-53. British Grassland Society, Reading.

Current Plant Science and Biotechnology in Agriculture

1. H.J. Evans, P.J. Bottomley and W.E. Newton (eds.): *Nitrogen Fixation Research Progress.* Proceedings of the 6th International Symposium on Nitrogen Fixation (Corvallis, Oregon, 1985). 1985 ISBN 90-247-3255-7
2. R.H. Zimmerman, R.J. Griesbach, F.A. Hammerschlag and R.H. Lawson (eds.): *Tissue Culture as a Plant Production System for Horticultural Crops.* Proceedings of a Conference (Beltsville, Maryland, 1985). 1986 ISBN 90-247-3378-2
3. D.P.S. Verma and N. Brisson (eds.): *Molecular Genetics of Plant-microbe Interactions.* Proceedings of the 3rd International Symposium on this subject (Montréal, Québec, 1986). 1987 ISBN 90-247-3426-6
4. E.L. Civerolo, A. Collmer, R.E. Davis and A.G. Gillaspie (eds.): *Plant Pathogenic Bacteria.* Proceedings of the 6th International Conference on this subject (College Park, Maryland, 1985). 1987 ISBN 90-247-3476-2
5. R.J. Summerfield (ed.): *World Crops: Cool Season Food Legumes.* A Global Perspective of the Problems and Prospects for Crop Improvement in Pea, Lentil, Faba Bean and Chickpea. Proceedings of the International Food Legume Research Conference (Spokane, Washington, 1986). 1988 ISBN 90-247-3641-2
6. P. Gepts (ed.): *Genetic Resources of* Phaseolus *Beans.* Their Maintenance, Domestication, Evolution, and Utilization. 1988 ISBN 90-247-3685-4
7. K.J. Puite, J.J.M. Dons, H.J. Huizing, A.J. Kool, M. Koorneef and F.A. Krens (eds.): *Progress in Plant Protoplast Research.* Proceedings of the 7th International Protoplast Symposium (Wageningen, The Netherlands, 1987). 1988 ISBN 90-247-3688-9
8. R.S. Sangwan and B.S. Sangwan-Norreel (eds.): *The Impact of Biotechnology in Agriculture.* Proceedings of the International Conference The Meeting Point between Fundamental and Applied in vitro Culture Research (Amiens, France, 1989). 1990. ISBN 0-7923-0741-0
9. H.J.J. Nijkamp, L.H.W. van der Plas and J. van Aartrijk (eds.): *Progress in Plant Cellular and Molecular Biology.* Proceedings of the 8th International Congress on Plant Tissue and Cell Culture (Amsterdam, The Netherlands, 1990). 1990 ISBN 0-7923-0873-5
10. H. Hennecke and D.P.S. Verma (eds.): *Advances in Molecular Genetics of Plant–Microbe Interactions.* Volume 1. 1991 ISBN 0-7923-1082-9
11. J. Harding, F. Singh and J.N.M. Mol (eds.): *Genetics and Breeding of Ornamental Species.* 1991 ISBN 0-7923-1094-2
12. J. Prakash and R.L.M. Pierik (eds.): *Horticulture – New Technologies and Applications.* Proceedings of the International Seminar on New Frontiers in Horticulture (Bangalore, India, 1990). 1991 ISBN 0-7923-1279-1
13. C.M. Karssen, L.C. van Loon and D. Vreugdenhil (eds.): *Progress in Plant Growth Regulation.* Proceedings of the 14th International Conference on Plant Growth Substances (Amsterdam, The Netherlands, 1991). 1992 ISBN 0-7923-1617-7
14. E.W. Nester and D.P.S. Verma (eds.): *Advances in Molecular Genetics of Plant–Microbe Interactions.* Volume 2. 1993 ISBN 0-7923-2045-X

Current Plant Science and Biotechnology in Agriculture

32. S.M. Jain, D.S. Brar and B.S. Ahloowlia (eds.): *Somaclonal Variation and Induced Mutations in Crop Improvement.* 1998 ISBN 0-7923-4862-1
33. S.J. Bennett and P.S. Cocks (eds.): *Genetic Resources of Mediterranean Pasture and Forage Legumes.* 1999 ISBN 0-7923-5522-9
34. R. Knight (ed.): *Linking Research and Marketing Opportunities for Pulses in the 21st Century.* Proceedings of the Third International Food Legumes Research Conference. 1999 ISBN 0-7923-5565-2
35. C. Oropeza, J.L. Verdeil, G.R. Ashburner, R. Cardeña and J.M. Santamaria (eds.): *Current Advances in Coconut Biotechnology.* 1999 ISBN 0-7923-5823-6
36. A. Altman, M. Ziv and S. Izhar (eds.): *Plant Biotechnology and In Vitro Biology in the 21st Century.* 1999 ISBN 0-7923-5826-0
37. E.C. Lefroy, R.J. Hobbs, M.H. O'Conner and J.S. Pate (eds.): *Agriculture as a Mimic of Natural Ecosystems.* 1999 ISBN 0-7923-5965-8
38. F.O. Pedrosa, M. Hungria, M.G. Yates and W.E. Newton (eds.): *Nitrogen Fixation: From Molecules to Crop Productivity.* Proceedings of the 12th International Congress on Nitrogen Fixation. 2000 ISBN 0-7923-6233-0
39. N. Maxted and S.J. Bennett (eds.): *Plant Genetic Resources of Legumes in the Mediterranean.* 2000 ISBN 0-7923-6707-3

KLUWER ACADEMIC PUBLISHERS – DORDRECHT / BOSTON / LONDON